Sustainable Agriculture Reviews

Volume 5

Series Editor

Eric Lichtfouse

For other titles published in this series, go to
www.springer.com/series/8380

Other books by Dr. Eric Lichtfouse

Sustainable Agriculture
Volume 1, 2009

Organic Farming, Pest Control and Remediation of Soil Pollutants
Sustainable Agriculture Reviews. Volume 1, 2009

Climate Change, Intercropping, Pest Control, and Beneficial Microorganisms
Sustainable Agriculture Reviews. Volume 2, 2010

Sociology, Organic Farming, Climate Change and Soil Science
Sustainable Agriculture Reviews. Volume 3, 2010

Genetic Engineering, Biofertilisation, Soil Quality and Organic Farming
Sustainable Agriculture Reviews. Volume 4, 2010

Environmental Chemistry
Volume 1, 2005

Rédiger pour être publié! Conseils pratiques pour les scientifiques
2009

Forthcoming

Sustainable Agriculture
Volume 2

Environmental Chemistry
Volume 2

Eric Lichtfouse
Editor

Biodiversity, Biofuels, Agroforestry and Conservation Agriculture

Springer

Editor
Dr. Eric Lichtfouse
INRA-CMSE-PME
17 rue Sully
21000 Dijon
France
eric.lichtfouse@dijon.inra.fr

ISBN 978-94-007-3366-4 ISBN 978-90-481-9513-8 (eBook)
DOI 10.1007/978-90-481-9513-8
Springer Dordrecht Heidelberg London New York

Softcover reprint of the hardcover 1st edition 2011

Printed on acid-free paper

Cover illustration: Old world swallowtail (*Papilio machaon*) on flowers.
Copyright: Louis Vidal, INRA 2010

Springer is part of Springer Science+Business Media (www.springer.com)

Contents

Agroecology as a Transdisciplinary Science for a Sustainable Agriculture

Fabio Caporali

Abstract Today's agriculture has to face the new challenge that derives from a new evolutionary era of civilisation that has been called "Anthropocene". Human dominion on Earth with exploitation of natural resources and environmental pollution is not socially acceptable anymore, since it leads to self-destruction in a confined "spaceship" like the planet Earth. A different cultural attitude to guide human behaviour is required in order to set up the base of a sustainable development for man and the whole biosphere. Agriculture scientists can play a role in providing a step towards agriculture sustainability. However, agriculture scientists need to be able to educate both themselves and the civic society with a new systems paradigm that focuses more on relations than on single components of agriculture reality, as a disciplinary approach usually does. Transdisciplinarity in agriculture theory and practice is required in order to face the new challenge of sustainability. Agroecology is a transdisciplinary area of enquiry that has both a scientific and a philosophical base for promoting sustainability in agriculture.

Here I review the most important areas of interface that qualify agroecology methodology and contents. Agroecological achievements are presented according to their chronology in order to account for the developmental process that agroecology has undergone. Concerning methodological achievements, four pillars of agroecological epistemology have been identified: (1) the agroecosystem concept; (2) the agroecosystem hierarchy; (3) the farm system as a decision making unity; and (4) the representation of agriculture as a human activity system. These four epistemological tools are models of agricultural organisation that allow us to understand, project and manage it as a process. They constitute the theoretical base of agroecology derived from the systems approach, which is at the core of ecology. With the aid of these four tools of enquiry, an agroecosystem monitoring process worldwide started since the early 1970s and it is still running. Information on the processes of energy

F. Caporali (✉)
Department of Crop Production, University of Tuscia, Via S Camillo de Lellis, 01100 Viterbo, Italy
e-mail: caporali@unitus.it

E. Lichtfouse (ed.), *Biodiversity, Biofuels, Agroforestry and Conservation Agriculture*,
Sustainable Agriculture Reviews 5, DOI 10.1007/978-90-481-9513-8_1,

transfer (energetics), productivity, nutrient cycling and biodiversity dynamics at different levels of agroecosystem hierarchy is growing. This constitutes the first knowledge body package of agroecology, a science that links structure and functioning of agroecosystems. This data collection has allowed scholars to raise judgements about the resource use efficiency, the environmental impact and the sustainability of agroecosystems at different hierarchical level of organisation. Since the 1990s, agroecology research and applications have focused more and more on the issue of sustainability. Attention to the problem of agriculture sustainability has promoted a spontaneous dialogue between scholars of ecology, agronomy, economics and sociology. Matching ecology with agronomy has produced more awareness on the benefits of increasing biodiversity at field, farm and landscape levels. Increasing within-field biodiversity with policultural patterns, such as crop rotations, cover cropping and intercropping, and increasing between-field biodiversity with field-margin management, hedgerow maintenance or introduction, and agroforestry applications, are practical solutions to the problem of enhancing biophysical sustainability of agroecosystems. Recent research on the role of field size for evaluating the trade-off between machinery efficiency and loss of biodiversity-friendly habitats in arable landscapes shows that there is no need for bigger field size beyond an evaluated threshold of 1–2 ha above which machinery efficiency increases very little (Rodriguez and Wiegand 2009). Matching ecology with economics and sociology has instead revealed that contrasting paradigms are still at work. One of the most outstanding example of paradoxical contrast between economic and ecological outcomes is the CAFO (Confined Animal Feeding Operations) system for meat production. The CAFO system is economically regarded as the most advanced intensive feedlot system for livestock production, although it contributes large greenhouse gas emissions. If the use of CAFOs is expanded, meat production in the future will still be a large producer of greenhouse gases, accounting for up to 6.3% of current greenhouse gas emissions in 2030 (Fiala 2008). An ecological conversion of economics is demanded whether the end of sustainability has to be pursued. Organic farming worldwide is the most important example of agriculture regulated by law with the expressed end of integrating bio-physical and socio-economic requirements of sustainability. The latest report by IFOAM (International Federation of Organic Agriculture Movements) (2007) mentions increasing annual rates for organic farming and shows that 32.2 millions of hectare were certified as "organically grown" in 2007 with more than 1.2 millions of farmers involved in the world. The global debate about agriculture sustainability has enormously enlarged the cultural landscape for mutual criticism between different disciplinary, traditionally separate areas. The area of agroecology enquiry is now really operating as "glue" at a transdisciplinary level, bridging the gap between different disciplines and between theory and practice of agriculture. Measuring agriculture sustainability through indicators of both biophysical and socio-economic performances is now a common praxis of international, national and local institutions. New curricula in Agroecology at academic level are performed in order to give an institutional base for education towards agriculture sustainability. A final outlook section provides some examples of agroecological approaches and applications for making a crowed planet more sustainable.

Keywords Agroecology • Systems paradigm • Transdisciplinarity • Integration • Sustainable agriculture

1 Introduction

Science is one of the most meaningful achievements of human culture. Its main goal is to expand human knowledge in order to meet human requirements. Human beings are bound to know more in order to live better. Pursuing knowledge is an ontological property of being human. To do science is a cultural need as to breathe, to drink, and to eat are physiological needs for human survival and welfare. Knowledge develops as a relationship between a conscious observer and his/her context of life or environment. Better knowledge means better possibilities of adaptation, survival and success in the context of life. Thanks to science and its application, i.e. technology, the human population has spread all around the whole planet and has increasingly changed the natural into the artificial, nature being eroded by culture. A human dominion (Scully 2002) on the Earth is already established and a new evolutionary era, called Anthropocene (Steffen et al. 2007), is now developing. Dominion is not a balanced state, since it is source of exploitation between man and man, i.e. rich, poor, male, and female, and between man and nature. Exploitation is unfair, destabilizing and unsustainable. In the long run, it leads to self-destruction in a confined "space-ship" like the planet Earth (Boulding 1971).

To counteract the culture of dominion and exploitation, an alternative way of human development is required, and therefore, a different cultural attitude to guide human behaviour. A new culture based on a more realistic appreciation of the "physicality" of our context of life is required: a culture of limits, balances, synergies, cooperation, justice, ethics and aesthetic, in a word an *ecological culture*, for setting the base of a sustainable development for man and the whole biosphere. The construction of a science of sustainability is both an emergent cultural step and a challenge in the development of our human evolutionary society (Kates et al. 2001).

There is the need for a new social contract for science, which should help society move toward a more sustainable biosphere that means a biosphere which is more ecologically sound, more economically feasible, and more socially just (Lubchenco 1998). In this contract, scientists should comply with both epistemological and moral issues. The commitment to moral issues should include: (a) addressing the most urgent needs of society in proportion to their importance; (b) communicating knowledge and understanding widely in order to inform individual and institutional decision; (c) exercising good judgement, wisdom and humility. As far as epistemological issues are concerned, the commitment should be for: (a) promoting innovative transdisciplinary mechanisms in order to facilitate the investigations for problems that span multiple spatial and temporal scales; (b) encouraging interagency and international cooperation on societal problems; (c) constructing more effective bridges between policy, management, and science, as well as between the public and the private sectors (Lubchenco 1998; Norgaard and Baer 2005). The time is mature for the scientific community to take adequate responsibility in order to contribute significantly to creating a culture and a science of sustainability (Fischer et al. 2007).

In this paper, agroecology is reviewed as a science of sustainability that moves from agriculture, i.e. from the ancient, cultural innovation that has put under human control the trophic link between man and his "mother" Earth (Caporali 2000).

2 The Systems Paradigm as an Epistemology at the Core of Both Ecology and Agroecology

Agroecology is the science of ecology applied to agriculture (Altieri 1987, 1995; Caporali 1991; Gliessman 1990, 2007; Wezel and Soldat 2009; Wezel et al. 2009). Like ecology, agroecology uses a systems approach as its basic methodology of enquiry and the concept of ecosystem, in the form of agroecosystem, as its basic epistemological tool for representing and explaining the agricultural reality (Caporali 2004, 2008). Agroecology became a well established field of enquiry at international level when its object of study, defined by the concept of agroecosystem, was clearly identified, agreed upon and adopted as the basic paradigm of this new integrative science (Loucks 1977). The relevance of this achievement is documented by the publication of the newborn scientific journal "Agro-Ecosystems" in 1974, which was supported by the International Association for Ecology. General information about aims and scope stated that the new journal would deal with "fundamental studies on ecological interactions within and between agricultural and managed forest systems" and also with "the impact of agriculture ecosystems on the other parts of the environment". Moreover, it was emphasized that the journal would not accept "papers directly concerned with the specialist areas of soil, crop and livestock husbandry". Those specifications made clear that the journal was established as a forum for collecting studies inspired by a systems paradigm.

Systems thinking is not a discipline but an epistemology, i.e. a methodological way of looking at reality for both understanding and managing it. In systems thinking, both philosophical and scientific roots intermingle into a bundle of transdisciplinary knowledge. Recently, the philosophical foundations of the systems paradigm were reviewed in order to show how influencing they had been in promoting a dialectical confrontation among the ecologists that ended up with the creation of the new world "ecosystem" by Tansley (1935), (Caporali 2008). Ecosystem refers to a holistic and integrative concept of life organisation, i.e. a patterning of nature, where inorganic and organic components are hardly separable, although discernible, life being an interactive and interdependent process, even at the level of a single individual.

The meaning of the term ecosystem is so wide that includes (Golley 1993; Caporali 2004):

1. Both a scientific and epistemological model for representing the reality we live in
2. An element in a hierarchy of physical systems from the universe to the atom

3. The basic system and the unity of study of ecology
4. A universal unity of study, which is applicable to any dimension of spatiotemporal scales

An epistemological model is an intellectual construct that unveils the correspondence between reality and its human representation. The ecosystem concept is an epistemological model that connects separate things or components of reality in a coherent framework of interconnectedness that explains:

1. The order of nature
2. The organisation patterns behind that order
3. The possibility of changing that order through human intervention
4. The possibility of evaluating the consequences of human intervention within the ecosystem of study and between that ecosystem and its surrounding environment

Systems thinking differs from conventional analytical thinking because it deals more with the overall organisation and performance of the whole system than with the role of its individual components. The main focus is more on relationships among components and processes than on single factor cause–effect relationships. Indeed, according to the input/output model of the ecosystem representation, the most important processes to be investigated concern energetics, nutrient cycling and biodiversity dynamics, in order to explain performances in terms of overall productivity, efficiency of resource use, environmental impact and sustainability of the whole system (Caporali 2004, 2008). Since its origin in 1935, the term ecosystem – with its rich content behind it – has been increasingly adopted. Thanks to its heuristic value, it is now the main intellectual tool for expressing judgements about reality organisation and management (Caporali 2008).

The ecosystem model as a standard framework for representing the organisation of reality has rapidly become a scientific tool for comparisons. An outstanding example of an early application is documented by two papers of Ovington et al. (1963) and Ovington and Lawrance (1967), that compared structure and functioning of four ecosystems, three natural and one agricultural, in a protected area of Central Minnesota, USA. These papers presented data of plant biomass, productivity, chlorophyll and energy content, according to a chorological integrative view of ecosystems that pays attention to the structure and functioning of the whole plant community. Aboveground and belowground distribution and composition of plant materials are shown in order to both discover the basic principles of actual and potential productivity and collect information for increasing the organic productivity of the earth. These papers are to be regarded as pioneering examples of enquiry between ecology and agroecology.

The concept of ecosystem includes man and his activities. Dating back to the origin of the ecosystem concept formulation, it was clear the description of the man–nature-relationship in Tansley's words: "Regarded as an exceptionally powerful biotic factor which increasingly upsets the equilibrium of pre-existing ecosystems and eventually destroys them at the same time forming new ones of a very different nature, human activity finds its proper place in ecology… Ecology must be applied to conditions brought about by human activity" (Tansley 1935). Within the functional model of the interpretation of reality that the concept of ecosystem represents, all the creative potentiality inherent in nature, including the evolutionary process, which results in its self-consciousness, that is man, can be identified.

A development strategy taking place in nature as a self-organising, self-supporting and self-evolving entity can be inferred by the functional model of ecosystem (Levin 2005; Keller 2005). Such a development strategy, culminating in man's genesis as a self-referring and responsible entity, can be qualified as an eco-development or sustainable development strategy based on the use of solar energy flux, recycling of matter and on the promotion of biodiversity. Therefore, the environmental scenario as a whole qualifies itself as a total and collective good to be recognised and protected. It is a total good because it cannot be separated into functionally autonomous parts and a collective good because it is available for all its components. Nowadays, owing to the demographic and technological pressure man exerts on the planet, mankind is recognised as a real "geological force" able to basically modify life substrata (air, water, soil, biodiversity) in accordance with what was originally predicted by the great prophet–scientist Vernadskij (1999). The capacity to affect the same fundamental features of the life scenario structure and functionality raises, for the first time in man's history, the theme of ecological responsibility (Jonas 1979), i.e. the awareness of the action of human interference in the ecosystems and the need for its control. This circumstance leads to the widening of man's ethical sphere that is usually limited to personal and social relations. Relationships with the biophysical environment should develop a third dimension: the *eco-ethics*. For behaviour implications to be necessarily considered, they should emerge from the sense of belonging to nature and the awareness of affecting the processes of ecosystem evolution, remarkably both on the local and planetary scale. Ecosystems ecology seems to be an ethic science rather than a neutral one (Caporali 2000).

One role increasingly played by man in the ecosystem dynamics is that of "decomposer", by virtue of his technological activity mainly based on combustion processes of organic-derived substances, which is called "techno-respiration" (Boyden and Dovers 1992). The industrialization era could also be named the era of "hunger for carbon" (Caporali 2000). It was based not only on the use of fossil fuels like petroleum, carbon and methane – that represent organic carbon stocked in geological times – but also on humus oxidation promoted by ploughing of soils put under cultivation and finally the oxidation of fresh necromass during deforestation for the burning of deforestation residues.

The human dimension of ecosystem studies underlies the necessity of building intellectual bridges between the life, earth, engineering and social sciences because most aspects of the structure and functioning of Earth's ecosystems cannot be understood without accounting for the strong, often dominant influence of humanity (Vitousek et al. 1997). Therefore, the integration of the social sciences into long-term ecological research is an urgent priority (Redman et al. 2004). It has been suggested that it is not tenable to study ecological and social systems in mutual isolation because humans are an integral part of all ecosystems and all human activity

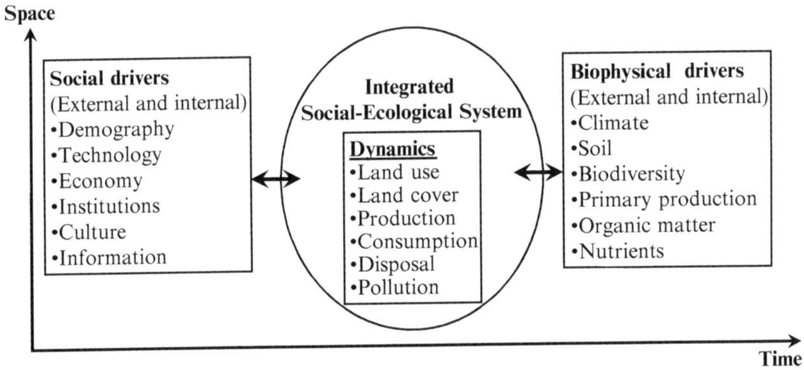

Fig. 1 A model of an integrated social-ecological system (SES) as a co-evolutionary system resulting from the interactions between human and non-human components and processes. Biophysical and social drivers regularly interact originating a perpetually dynamic, complex system with continuous adaptation (Modified from Redman et al. 2004)

has potential for affecting the local and the global environment. In this framework, what is usually divided into natural and human systems should be considered a single, complex social–ecological system (SES) (Redman et al. 2004) or a coupled human and natural system (CHANS) (Liu et al. 2007). In Fig. 1, a conceptual representation of an integrated SES is given, according to an input/output methodology which provides the process drivers in terms of external/internal constraints and feedbacks concerning its biophysical and socio-economic conditions, and the SES resulting dynamics.

3 Agroecology Establishment

The systems paradigm applied to agriculture defines a new transdisciplinary science that is called Agroecology (Caporali 2008). In a systems view, reality appears as a continuing development process characterized by a series of nested levels of organisation of increasing complexity and autonomy culminating in coevolving communities of living beings and ecosystems. If agriculture is the object of enquiry, a hierarchy of organisation levels due to human design and management can be identified with corresponding levels of agroecosystem patterning. In practice, it is possible to construct agroecosystem models representing real agroecosystems at different spatiotemporal scales in order to study them or design and manage their behaviour according to pre-established goals. In theory, a hierarchy of agroecosystems spans from the crop rhizosphere micro cosmos to the international networks of institutions that are connected in the food chain. Therefore, there is no limit to the extension of agroecological studies in this range of enquiry.

Many of the first articles in the already mentioned journal "Agro-Ecosystems" were devoted to starting discussion and promoting consensus about epistemological questions related to the meaning, modelling and use of the concept of agroecosystem (Harper 1974; Spedding and Brockington 1975; MacKinnon 1975; Loucks 1977; Frissel 1977).

In the paper of Spedding and Brockington (1975), the meaning of "ecosystem" is discussed within these two assumptions: (a) a system cannot be studied without some kind of model; (b) a model cannot be constructed unless and until its purpose is defined. In their words, "no ecosystem can be visualised except as a model: it is therefore necessary to establish the *purpose* of any study of an ecosystem, in order to have criteria for judging what is and is not essential to the content of the model." A crucial question is, for example, the establishment of system boundaries. Agricultural systems, with clearly defined purposes that are reflected in their organisation structure, are more easily modelled than other ecosystems.

In this frame of reference, MacKinnon's paper (1975) is innovative in many respects and grounds many points of interest for the future development of agroecological enquiry with providing:

1. An approach for analysing farm performance based upon concepts and principles from ecology, energetics and cybernetics
2. A representation model of a farm production system in an ecological context, i.e. a driving force producing environmental impact
3. A representation model of a farm unity as a energy-material transformer with internal interacting components, i.e. biological system, technical system and management system
4. Insights concerning the farmer's role as both designer and manager
5. Insights concerning the role of the socio-economic context on the farmer's decisions
6. Distinction between different sources of energy, i.e. native solar energy and imported, auxiliary, fossil fuel derived energy
7. Criteria for selection of optimal farm systems based on efficient use of energy and material, as expressed in the following point
8. Performance measures as ratios of energy flows, e.g. energy output as food divided by fossil fuel energy input, and magnitudes of energy transfer rates, e.g. energy flows as food or fertilizer per unit area

With Frissel's paper (1977), a big framework of cycling of mineral nutrients in agricultural ecosystems of the world was presented, as a result of a symposium of the Royal Netherlands Land Development Society, co-sponsored by the International Association for Ecology and Elsevier Scientific Publishing Company. An understanding of agricultural ecosystems was regarded as essential to the future work in land development. Therefore, a decision was made in order to have the subject reviewed by a large interdisciplinary group of international scientists. Before collecting data on nutrient cycles, it was necessary to define the scale, or level of organisation of the systems of interest. Such a level can vary from a field, to a crop, to a whole farm, to a regional or a national level of organisation. The solution chosen was to build upon a relatively simple unit, represented by a single farm, although it was recognised that, because of the transport of nutrients by water, a defined catchment area could also be regarded as a logical unit of study. To understand and quantify nutrient cycling of any element, it was necessary to design a conceptual model to represent the main transfer and compartments. Three main compartments or pools were adopted, plant, animal or livestock and soil, the latter being divided into three sub-pools, available, unavailable or soil minerals, and residues or soil organic matter. One of the main results of this enquiry was the description and classification of agroecosystems worldwide according to type of farming and the yield, and the evaluation of their potential for producing output of nutrients in the environment. For example, the output of nitrogen tended to increase in the most modern intensive horticultural and arable farming systems.

With studies on agroecosystem meaning, energetics, nutrient cycling and cybernetics, the phase of agroecology establishment started. In an extensive review paper, Loucks (1977) argued that agricultural ecosystems are more complex than other resource systems, because there are many man-manipulated processes going on, mostly modifying input and output, but also affecting rate relationships within the system. Human intervention is largely the result of economic and market processes that ultimately control the dominant characteristics of the systems, as he explains in the following words: "the economic system governing the intensity of input and exports, and the economic viability (survival) of the farm operator determining the inputs are essential, integral components of the agricultural ecosystem. Thus, students of agricultural ecosystems must recognize that we are only beginning the long process of moulding diverse viewpoints together as a coherent field of inquiry. For convenience, the generic term for the object of study has become *agro-ecosystem.* This paper reviews the recent development of an interfacing among the sciences involved and the emergence of an art and science of agro-ecosystem analysis".

In the phase of agroecology establishment, some important methodological questions were challenged. Among them, (a) a definition of universal components of agroecosystems, i.e. compartments and processes; (b) the use of agroecosystem modelling according to an input/output analysis; (c) the hierarchic approach for representing the continuum and the openness of agroecosystems; (d) the study of *culturally* dominated agricultural systems as ecosystems, amenable to intensive, holistic analysis (Loucks 1977).

3.1 Methodological Achievements

As a consequence of the agroecology establishment based on a systems view, four important epistemological achievements were reached:

1. The agroecosystem concept as an input/output model, representing both the basic epistemological tool and the basic object of study in agroecology
2. The representation of agriculture as a hierarchy of systems
3. The representation of the farm system as a decision-making unity
4. The representation of agriculture as a human activity system

Each of them is a meaningful component in the framework of systems thinking applied to agriculture and all together constitute the knowledge body which defines agroecology as a transdisciplinary science. These four pillars are briefly summarised in the following sections.

3.1.1 The Agroecosystem Concept

In analogy with the ecosystem concept, the agroecosystem is both a *real* ecosystem modified and used for agricultural purposes as well as a *model* that represents it. In this sense, agroecology exhibits a method that reflects contents, whereby ontology and epistemology coincides (Caporali 2008). The general model for representing structure and functioning of an agroecosystem is based on an input/output patterning (Fig. 2), where biophysical and socio-economic components are fully integrated in a process of continuing transformation of resources that happens at any spatial-temporal scale (Caporali 2007). While information and capital as both input and output are typical human-made constructs, matter, energy and biodiversity are primary natural resources, although more or less manipulated and changed by man. Crop components develop the function of primary productivity, that livestock components transform into secondary productivity; both are used up by the farmer who collects and sells out the produce. The semantics of the agroecosystem representation is that of a *process underway* limited by natural components, such as

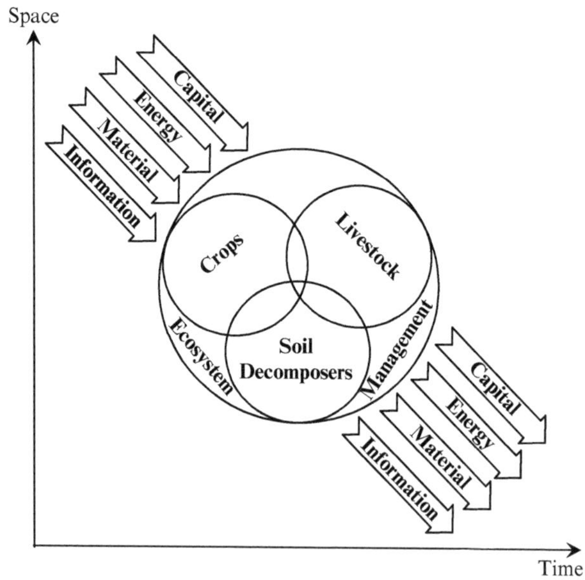

Fig. 2 The agroecosystem model as an input/output representation of mixed agriculture. Socio-economic and biophysical components are fully integrated in a process of continuous production and consumption that happens at any spatial-temporal scale

climate, soil and biodiversity, but deeply constrained by human information and intervention. Human organisation is as much a driving force in determining structure and functioning of agroecosystems as nature is.

3.1.2 Agroecosystem Hierarchy

"The agroecosystem concept is like a lens for focusing on rural reality at different levels of resolution" (Caporali 2007). It opens up the possibility of flexible enquiry at different scales. Hierarchy in agriculture is to be meant as a spatial–temporal continuum or an open, interconnected sequence of nested agroecosystems, which are isolated only for a necessity of study or management (Caporali 2008). The functional characteristics behind this hierarchical representation of agriculture is the *openness* of each level, being each level at the same time the context of the next lower level and a component of the next upper level. Therefore, integration between different levels is an ontological necessity of both each level and the whole system. A formalisation of a hierarchical classification of farm systems took place in the 1980s, in analogy with the ecological systems approach developed in ecology by Odum (1983). "By analogy with ecology, agriculture can be described as a hierarchy of systems, ranging from the cell at the lowest level, through the plant or animal organs, the whole plant or animal, the crop or herd, the field or pasture, and the farm, to complex ecosystems such as the village and watershed, culminating in the agricultural sector at the highest level" (Fresco and Westphal 1988). Scaling up hierarchical levels, new properties emerge that are the result of more communication and control between levels. Emergence of new properties is an effect of more integration and is useful for conferring more coherence among components within a hierarchical level and/or more correspondence between different levels. With the hierarchical approach or the process of flexible enquiry at different levels of agroecosystem organisation, the four fundamental elements of the systems paradigm, namely *hierarchy*, *emergence*, *communication* and *control* (Checkland 1981), are unveiled and confirmed.

3.1.3 The Farm System as a Decision-Making Unity

In the agroecosystem hierarchy, the farm is the management unit with a biophysical base, easily identifiable because of its boundaries, and which represents the meeting point between human interests and the natural environment (Caporali et al. 1989). The agroecosystem approach defines a farm as a unity of study and reveals both its internal organisation in terms of components and their interactions, and its exchange relations, i.e. input and output, with the environmental context. Indeed, "the farm system is a decision-making unit comprising the farm household, cropping and livestock systems, that transform land, capital (external input) and labour (including genetic resources and knowledge) into useful products than can be consumed or sold" (Fresco and Westphal 1988). As a consequence, the activity of each farmer results from a decision-making process which weighs the complex of both local and global information obtained by the biophysical and the socio-economic environment (Fig. 3). Biophysical factors are relatively stable in comparison with socio-economic factors that are subject to large fluctuations and shifts even in a short term. Thus, traditional farming systems have been recently both challenged and changed by a wider context of international information and constraints.

The identification of a farm as an agroecosystem permits a clear perception of the influence of human action on both its internal structure and functioning, and its external environmental impact. In the case of conceiving of a farm as a resource management unity, modern agroecology could trace some of its unexplored roots into the documented memory of early academic studies in Italy. In 1844, before the set up of the Italian state, an academic 3-year course in agriculture and animal husbandry was established at the University of Pisa, then belonging to the Great-Duchy of Tuscany. The focus of that curriculum was designing and implementing a farm as a "system" made up by different components including fields and crops, fodder and livestock, human labour, equipment, and buildings to be organised in accordance with socio-economic and environmental goals (Cuppari 1862). The pioneering work of Cuppari was later re-assessed and expanded by Alfonso Draghetti (1948), with a seminal book entitled "Principles of physiology of a farm" (in Italian).

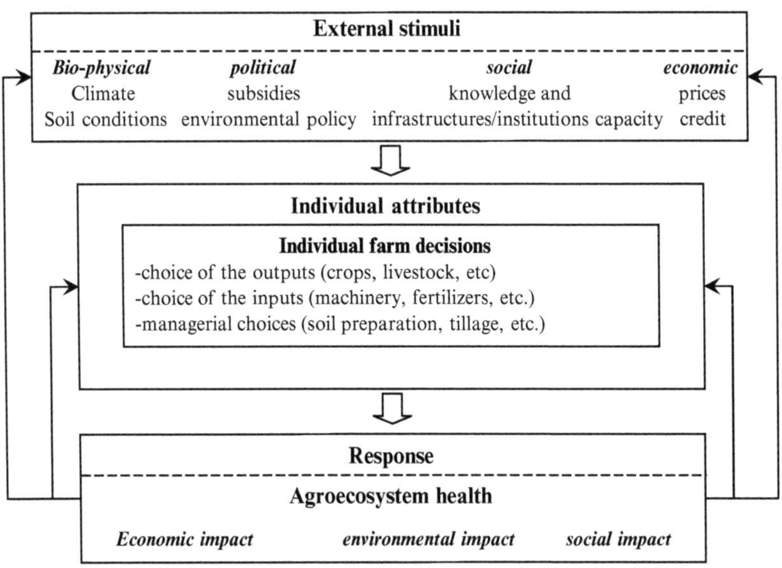

Fig. 3 The decision-making process at the farm level. The individual knowledge that the farmer integrates as a synthesis of external and internal stimuli results in decision-making at organizational and farm management level. Decisions made as farming design and practices produce continuous variations in the environment components that subsequently act as input to remodel the individual knowledge. The individual decision-making process is therefore cyclic, dynamic, and continuously subjected to remodelling by new elements of knowledge and therefore hardly predictable (Modified from Smit et al. 1996)

Draghetti expresses the vision of the farm as a "body", i.e. a whole composed of many parts harmonically organised in order to achieve integrity and autonomy of functioning. This theoretical tradition has been recognised as an important agroecological base for both enquiry in agricultural study and inspiration for sustainable farming by Caporali et al. (1989) and Caporali and Onnis (1992).

3.1.4 Agriculture as a Human Activity System

Systems thinking, enlightening the hierarchical organisation of reality, also helps perceive the interactive relationships between a farm and the institutions constituting its socio-economic and cultural context, i.e. political and administrative institutions, research institutions, industries throughout the field of agricultural activity, credit and marketing institutions, and finally the relationship with the individual consumer who depends on agricultural products trophically (Fig. 4). All this network of institutional involvement defines a contextual framework for agriculture which can be defined as a *human activity system* derived from the integration of natural and anthropogenic systems, both abstract and physical (Checkland 1981; Caporali 2000). This activity system is the outcome of operational connections

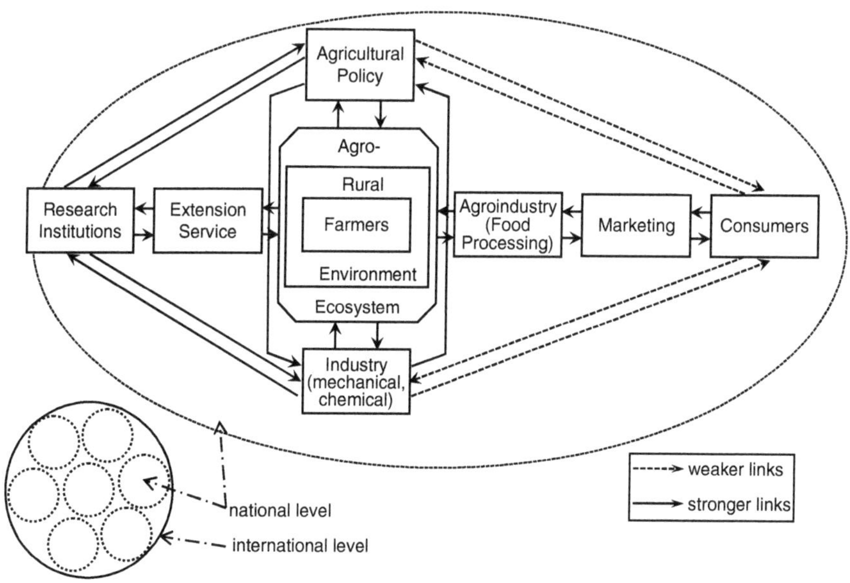

Fig. 4 Agriculture as a human activity system. The system agriculture pervades society as a whole and, following the market expansion through the process of globalization, it spreads, nowadays, from the local to the international level. Farmers, who are at the core of the system, organize their activity to interact with the other components of the agricultural system. These interactions manifest themselves firstly through information exchanges and secondly in the form of energy-matter exchanges between the agroecosystem and its context (Modified from Caporali 2004)

among the agriculture stakeholders that elaborate and distribute both information and energy-matter flows. All the important nodal points, represented by the boxes in Fig. 4, converge in a central one consisting of the farmers operating on the original ecosystems and transforming them into agroecosystems. The increasing globalisation of communications, economics and politics, has already extended the informative context to the entire planet. In addition to local conditioning of human activity by nature through climate and soils, a wider type of conditioning – a cultural conditioning by man – is occurring (Caporali 2000).

These four pillars constitute the theoretical base of agroecology and justify the statement that the systems paradigm is an epistemology at the core of both ecology and agroecology. They have paved the way for an enquiry attitude that regards agricultural reality as a *process* to be studied with an input/output methodology at different hierarchical levels of organisation.

4 Agroecosystem Monitoring Worldwide Between the 1970s and the 1980s

With the instrument of knowledge constituted by the agroecosystem concept and its just mentioned epistemological derivatives, a new phase of agroecology enquiry has been started, that of monitoring agroecosystems worldwide. Main elements of monitoring have been energetics, nutrient cycling, biodiversity, and related biophysical and socio-economic effects, at different spatial-temporal scales. Collection of data in this field of agroecosystem enquiry has allowed agroecologists to construct step by step pieces of knowledge on agricultural organisation. Most of this detailed information has been stored in specialised scientific journals, such as *Agro-Ecosystems*, which has become the current *Agriculture, Ecosystems and Environment*; *Agricultural Systems* and *Biological Agriculture and Horticulture*, but also in global science journals, such as *Science* and *BioScience*. Chronologically, this first phase of intensive agroecosystem monitoring has lasted for about 2 decades, from the 1970s to the 1980s, and its development has produced important insights of knowledge in order to construct the concept of sustainability in agriculture.

4.1 Energetics

In modern industrialised societies, energy is the physical driving force of human activity systems, including agriculture. Starting early in the 1970s, enquiries based on an energy accounting in agriculture were developed in order to assess sources and amount of energy involved with associated benefits and risks. At that time, concerns raised by the large use of fossil-fuel derived energy inputs in agriculture were already shared in society due to the seminal book on environmental risks of pesticides use in agriculture by Rachel Carson (1962), who suggested that "against the interest of food-production we have to balance other interests, like human health, watershed protection and recreation". Eco-energetics analysis is a good systems approach for tracing the pathway of energy transfer and efficiency along the food chain in agriculture, allowing scholars a better understanding of the interactions between the biophysical and socio-economic components of the agroecosystem hierarchy (Deleage et al. 1979). These studies have been carried out from local level to national levels.

A first big picture on the relationship between energy consumption and food production was presented for the USA conditions by Pimentel et al. (1973). With the data on input and output for maize as a model, they were able to make an examination of energy needs for a world food supply that depends on modern energy intensive agriculture. Alternatives in crop production technology were also considered which might reduce energy inputs in food production. Maize was chosen because it typifies the energy input in U.S crop production, being intermediate in energy inputs between the extremes of high energy-demand fruit production and low energy-demand small grain production. In 1970, about 2.9 million kilocalories was used by farmers to grow an acre of maize, which represents a small portion of the energy input when compared with the solar energy input. During the growing season, about 2,043 million kilocalories reaches a 1-acre maize field; about 1.26% of this is converted into maize biomass and about 0.4% in maize grain to be harvested. The 1.26% represents about 26.6 million kilocalories. Hence, when solar energy input is included, man's 2.9 million kilocalories fossil fuel input represents about 11% of the total energy input in maize production. The important point is that the supply of solar energy is unlimited in time, native, free and safe, whereas fossil fuel energy is finite, imported, costly and polluting.

In a mechanised crop, such as maize, all management practices are supported by fossil fuel derived energy, whether direct, i.e. energy costs for machinery functioning or indirect, i.e. energy costs for machinery construction. This analytical approach showed that machinery and nitrogen fertilizers contributed the largest shares of the total energy requirement. An increment of share in time by nitrogen fertilizers was paralleled by an increment of maize yield, although with a decreasing efficiency rate of the energy input. For selected years between 1945 and 1970, two primary conclusions are drawn from Pimentel's paper: (1) U.S. maize agriculture was using large absolute amounts of fossil energy; and (2) the yield crop energy per unit of input fossil energy was declining with time, i.e. the marginal product of fossil energy was decreasing as use of this energy was increasing. Declining conversion efficiencies are considered by many to be clear warnings to seek new directions for techno-economic change (Smil et al. 1983). These trends in energy input and maize yields in U.S. represent the general crop trend of the so-called "green revolution" style in agriculture, where crop production has been gained through large inputs of fossil-fuel derived energy.

The classic output/input ratio of energy can be used to show efficiency of a system or a sub-system, at local or national levels. In a detailed case of the French national system, with reference to human consumption and export, the efficiency was 2.0 for the sub-system of plant production, but fell to 0.19 for the whole of the rearing sub-system when only the meat, milk and egg output was considered. For the product of human food, the overall efficiency of the French agricultural system was 0.69, whereby one unit of produced energy, i.e. kcal or MGJ, required 1.45 kcal or MGJ as input of imported fossil-fuel derived energy (Deleage et al. 1979). For other industrialised national systems, this input was of the same order of magnitude (Stanhill 1974, 1979).

Things can change when considering comparisons between industrialised agroecosystems and other agricultural systems in developing countries. For example, Egyptian agriculture was shown to be very productive and efficient in its exploitation of solar and fossil fuel energy in a comparison with industrialised national systems (Stanhill 1979). The energetic efficiency, i.e. human diet output/non-solar energy input, for Egypt, California and Israel was 1.83, 0.90 and 0.41, respectively. In terms of food production per unit land area, the Egyptian agroecosystem was also the most productive of the seven other national systems for which data were available. This high potential for both productivity and energy efficiency of the Egyptian agroecosystem was attributed to the very intensive land use, sustained by abundant supplies of water and labour, fertile alluvial soils and a climate suitable for year-round cropping (Stanhill 1979).

4.2 Nutrient Cycling

Nutrient cycling involves the uptake, utilisation, release and re-utilisation of a given nutrient by various processes in an ecosystem. The efficient cycling of nutrients within a farming system is a prerequisite for long term-maintenance of both soil fertility and crop productivity, while it is a necessary condition for reducing adverse environmental impacts (Tomlinson 1970; Sharpley et al. 1987). Nutrient cycling is a process occurring step by step, being involved nutrient mobility between exchange pools of nutrients that can be air, water, soil and biomass of micro-organisms, plants and animals. Man heavily interferes in nutrient cycling altering the structure of ecosystems and the rate of natural pathways of nutrient transfer and transformation. Micro-organisms are usually the ubiquitary and leading driving force for fast nutrient dynamics in agroecosystems. To adopt a systems view in nutrient cycling means to be able to connect nutrient dynamics in a complex framework of multi-spatial and temporal scales, like that reported in Fig. 5a (Vlek et al. 1981).

With this approach in mind, it is easier to understand that, in agroecosystem design and management, a preference should be given to the establishment of synergetic effects among organic and inorganic agroecosystem components. This is a strategy for capturing nutrients and strengthening nutrient cycling into the soil-crop sub-system for a self-sustaining productivity. Figure 5b provides a good example of a framework for integration of both components and processes in a Mediterranean environment.

The objective of the farmer is to maximize crop production or yield at the farm level through manipulation of the factors that influence the availability of soil moisture and plant nutrients in the soil, for facilitating crop uptake while maintaining the

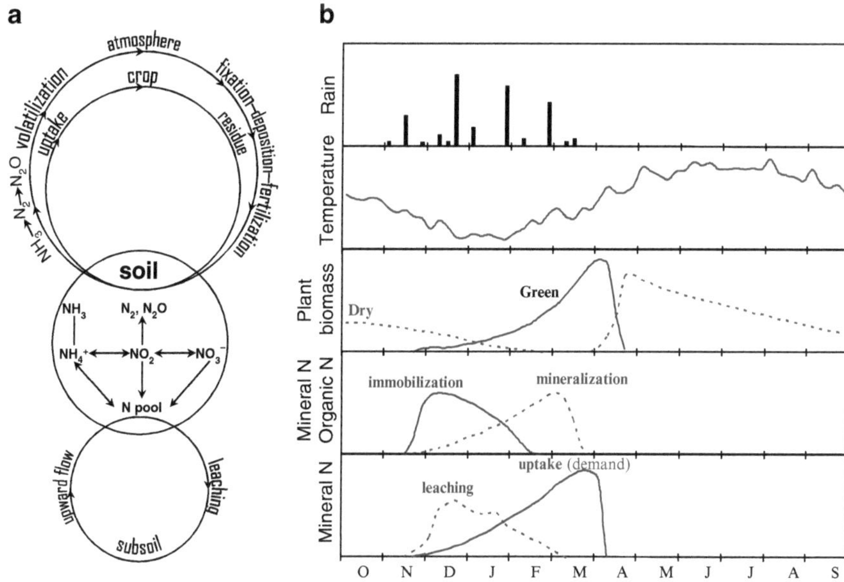

Fig. 5 Nitrogen cycles in agricultural ecosystems. (**a**) A general diagram representing the various cyclic N-processes operating in a soil–plant system, where accession, transformation, and loss of nitrogen take place. (**b**) depiction of typical seasonal patterns of climate, plant growth, and nitrogen flows in the arid-Mediterranean climate of the Northern Negev (Israel) (Modified from Vlek et al. 1981)

necessary conditions for soil fertility renewal. Today, important environmental effects, such as the greenhouse effect, water eutrophication and water nitrate pollution, are connected with nutrient cycling disturbances and unbalances. Therefore, there is the need for agricultural practices to be more conservative in the use of natural resources and less likely to spread chemicals into the fields.

Compared to natural soil ecosystems, agroecosystems have open cycles of mineral transfers due to both the frequent cultivation disturbances and the period of bare fallow between the seed bad preparation and the crop emergence. Comparisons between forested and agricultural watersheds showed that a permanent vegetation cover is an important condition for maintaining nutrient and avoiding losses through draining water or into the atmosphere (Armitage 1974; Johnson et al. 1976; Miller et al. 1979; Caporali et al. 1981). Monthly N concentration during the year has a typically seasonal trend in rivers draining agricultural land, with maximum concentrations during the period of bare soil or with limited uptake by crops for constraining temperature, and when rainfall is the heaviest, like in autumn and winter months. This pattern can be explained in terms of interplay between the climate and the factors that control the release of nitrate from soil reserves, i.e. mineralization and the absence or inactivity of crop roots (Caporali et al. 1981).

As to nitrate pollution of groundwater with regards to agricultural activities, three potential reasons can be considered: (a) the conversion of large areas with permanent grassland or forests to arable land use; (b) the intensification of crop production through frequent tillage and increasing nitrogen applications; (c) the intensification of livestock husbandry with increasing animal density and production of liquid manure per area cultivated land (Cameron and Wild 1984; Steenvoorden et al. 1986; Strebel et al. 1989). In the humid regions of Europe, it was shown that nitrate leaching takes place mainly during the colder and rainy seasons, with higher concentrations for sandy soils with arable crops, intensively managed grazed grassland and field cropping of vegetables (Strebel et al. 1989). To reduce the nitrate load of groundwater, it was suggested to perform agricultural practices in order to: (a) minimize the residual nitrate content in the root zone at the harvest time; (b) preserve the nitrate during the main leaching period in the form of biologically fixed plant N within the N cycle (Strebel et al. 1989).

The increasing size and concentration of agricultural livestock production units has made disposal of manure a major problem. In the 1970s, U.S. farmers had to deal with 1.5 billion metric tons of animal waste annually and ammonia-N losses from animal wastes were reported to range from 10% to 99% when surface-applied to soil (Hoff et al. 1981). The absorption of atmospheric NH_3–N volatilized from a nearby feedlot by surface was estimated as high – 73 kg/ha per year – as to promote eutrophication (Hutchinson 1969; Hoff et al. 1981).

Husbandry methods and farm systems in industrialised countries with a very simplified structure and higher level of inputs are bound to alter nutrient cycling and cause adverse environmental effects (Wagstaff 1987).

4.3 Biodiversity

Energy transfer and nutrient cycling happen through biodiversity development and make an ecosystem self-sustainable, although subject to continuing adaptation and change. The creative power of nature is expressed by biodiversity at different hierarchical levels: genes, individuals, species and ecosystems (MEA 2005). The whole planet earth is a mosaic of ecosystems or biomes, each of them being the result of adaptation, i.e. co-evolution between its biophysical and organic components. Life is the outcome of the integrative power of nature at both micro-and macro-scale. Ecosystems perform ecological services for a sustainable life. In short, they maintain clean air, pure water, fertile soil and a balance of creatures, including man (Westman 1977; Ehrlich and Mooney 1983). The development of an ecosystem, or ecological succession, unveils the strategy of nature, which is to maximize solar energy capture, i.e. gross primary productivity, in order to maintain a complex trophic web of consumers and decomposers, that make the whole ecosystem more bio-diverse, resilient and self-sustainable (Odum 1969). With agriculture, man has learnt to channel the use of solar energy through crops to his own exclusive benefit, net primary productivity being the main goal of agroecosystem management. Continuing increasing of human population on the earth has demanded more land for cultivation, with the result of dramatic erosion of natural ecosystems in every continent (Vitousek et al. 1997).

Natural biodiversity and natural ecological services have declined, while agroecosystems have thrived and yield boosted with the "green revolution" technology. On the other hand, agroecosystem management under the push of increasing mechanisation and chemical assault, has been threatening both environmental and human health. The challenge for the future is to find a balance for biodiversity management to be effective at both large and small scales. To influence biodiversity management at the large scale of sovra-national, national and regional levels, there is the necessity to develop policy measures that can have positive effects at local level, where farmers decide and operate in the fields. It is strategic to set up a top-down process of informational constraining in order to obtain a bottom-up construction of a sustainable landscape. Improvements of biodiversity at field, farm and landscape levels have potential to make agroecosystems more sustainable (Noss 1983; Gliessman 1990; Altieri 1999).

In the 1980s, the process of establishment of agroecology as "the science and art of agroecosystem design and management" (MacRae et al. 1989) was booming. Seminal papers defined the properties of agroecosystems, the ways of agroecosystem assessment and how the agroecosystem principles could be applied worldwide (Conway 1987; Marten 1988; Ewel 1986). An ecological paradigm has been recognised as a unifying concept in theory and practice of agroecosystem since the 1980s (Lowrance et al. 1984). Agroecosystems differ from natural system in that: (a) they are under human control; (b) human management has reduced their species diversity; (c) crops and livestock are artificially selected; (d) they are partly powered by fossil-fuel derived energy; (e) most produce is harvested and exported.

Agroecosystem functions regarded in relation to human requirements define the agroecosystem properties as productivity, stability, sustainability and equitability. Altogether, they outline a framework of agroecosystem science, or agroecology, where the human factor is the main focus of enquiry since agroecosystems are human-designed processes. Productivity is a measurable final step of a production process, which involves natural, semi-natural and man-made resources. Stability, or constancy/consistency of productivity, reflects the resilience of the production process over time against fluctuations in the surrounding environment of both biophysical and socioeconomic variables. Equitability, or the fair share of agricultural productivity among the human beneficiaries, is to be meant at different hierarchical scales, i.e. the farm, the village, the nation, the world. Sustainability, or capacity of the whole system to last, is a processual property concerning the overall organisation of the system and the efficiency of resource transformation according to an input/output scheme."Agroecologists study these characteristics both ecologically and socio-culturally…. Concern for the whole and for the study of relationships as they exist within their environment, are features that distinguish ecology and agroecology from most other scientific disciplines" (MacRae et al. 1989).

Agroecosystem monitoring between the 1970s and the 1980s on energetics, productivity, nutrient cycling and biodiversity dynamics has provided the first knowledge body package of agroecology as a science linking structure and functioning of agroecosystems worldwide.

5 Agroecology Development: the Challenge of Sustainability from the 1990s Onwards

Science is a very successful human activity because it is a globally open institutional system for the creation of knowledge and its technological applications. Technology modifies society and the environment, offering new input for science development. Science nurtures society, but it is also nurtured by society. Inevitably, science and society are mutually influencing and co-evolving in the progressing spiral of civilisation. New leading cultural concepts are produced in this process of continuing adaptation. Sustainability is a leading concept for the twenty-first century that UN organisation adopted in the Agenda 21 (UN 1992) for promoting sustainable development worldwide. The concept of sustainability is a transdisciplinary one, which emerges whenever different fields of enquiry meet. Sustainability is like traffic-lights at each cross-roads; however, the green or the red signals are not automatically imposed, but they must be agreed upon by a shared institutional consensus.

Agriculture, as any other human activity system, must face the challenge of sustainability and try to adapt to it, according to a process of informed evolution. Starting from the 1980s, agroecology is frequently assumed to be strongly linked to sustainable agriculture. As argued by Anderson (1991) in a book review article concerning the books of agroecology by Gliessman (1990), Carroll et al. (1990) and Tivy (1990), "agroecology is supposed to be the science that will provide the knowledge required to achieve and quantify agricultural sustainability". An article (Altieri et al. 1983) and a book by Altieri (1987) already anticipated this assumption, emphasizing the role of agroecology as "the scientific basis of alternative agriculture". The search for sustainable agroecosystems has become a constant goal in the agroecology agenda for the last 2 decades and therefore, agroecology has been defined as the science of sustainable agriculture (Stinner and House 1989; Altieri et al. 1983; Altieri 1995, Gliessman 2007; Caporali 2008). Some authors also refers to agroecology as the science for the sustainability of the whole food system (Francis et al. 2003; Poincelot 2003; Gliessman 2007).

The history of agriculture teaches us how the nature of its impact on natural ecosystems has changed dramatically during the past half-century in industrialised countries (Auclair 1976; Odum 1984; Caporali 2000). Relatively inexpensive fossil fuel-derived energy has boosted the process of increasing industrialisation of the whole food system (Steinhart and Steinhart 1974). In this process, most of profit has benefited more industry and trainers than farmers worldwide, with heavy distortions between rich and poor nations and between city and countryside (Rappaport 1971; Murdoch 1990). However, neither industry, nor traders and nor farmers have paid for social and environmental externalities (Clark 1985; Tilman et al. 2002; Pretty et al. 2000; Pretty 2008).

During the 1980s, agroecological studies had already produced the awareness of both the reasons and the effort behind the conventional agriculture development (MacRae et al. 1989). Therefore, the need was shared for a change towards a more sustainable agriculture (Papendick et al. 1986; Stinner and House 1989; Yunlong and Smit 1994; Pretty 1995). Sustainable agroecosystems were conceived of in terms of input/output processes delivering the following performances: (1) producing the necessary quantity of high quality food and fibre; (2) to be profitable to the grower; (3)conserving non-renewable resources; (4) to be harmonious with biological, physical and social environments (Geng et al. 1990). Also recognized was the difficulty of constructing sustainable agroecosystems in that not all their objectives are compatible, and trade-offs among the objectives are often necessary in a defined context (Geng et al. 1990). Agroecology implies a systems approach to farming, integrating technology, economy, and natural processes to develop productive systems. Fundamentally, sustainable farming systems are knowledge-based systems of farming (Ikerd 1993).

5.1 Matching Ecology with Agriculture

Due to its transdisciplinary foundation, agroecology stems from dialectic, i.e. the confrontation between different fields of knowledge, and the theory and the practice of agriculture.

The epistemology of the ecosystem concept reveals that sustainability is a systems property, because the turnover of the organisms is functional to the maintenance of the whole system where they live in (Caporali 2008). On a human scale, the planet earth or the living planet is self-sustainable through self-organisation. Biosphere is self-maintaining through continuing creativity and biodiversity renewal at different hierarchical levels, i.e. cells, organisms, species and ecosystems. Sustainability to and for human beings means possibility of maintaining the natural creative process that, at its climax, has produced man itself. Sustainability finds its proper roots in the ontology of planetary life. Only man, through a free choice, can undermine the process of creative life, thereby he must accept the weight of his both power and responsibility.

A recent contribution to the debate on sustainability argues that "sustainability must be conceptualised as a hierarchy of considerations, with the biophysical limits of the earth setting the ultimate boundaries within which social and economic goals must be achieved" (Fischer et al. 2007). Two sets of action are suggested as strategies for progress: (a) integration across academic disciplines; (b) integration of academic insights with societal action.

In 1989, the scientific journal *Ecology* devoted a 13-page Special Feature to "Ecology, Agroecosystems and Sustainable Agriculture" (Coleman 1989). Different reasons were given for justifying "the necessary marriage between ecology and agriculture", " if and just how a schism was created between the two disciplines of ecology and agronomy", if ecology and the agricultural sciences make up a false dichotomy, and "a perspective on agroecosystem science" (Jackson and Piper 1989; Coleman 1989; Paul and Robertson 1989; Elliot and Cole 1989). Compelling reasons reported for such a melding of disciplines and outlooks were: (1) problems of non-point pollution, such as soil erosion, ground water contamination, etc.; (2) concerns about global environmental issues, with gaseous emissions from agricultural fields; (3) the need to have a long-term, *sustainable* (lower-input) resource base for production of food and fibre. As to the relationships between ecology and agronomy, it was mentioned that "both disciplines have historically focused on narrow, and at times competing goals at the practical level: agronomy mainly on crop production and ecology mainly on environmental protection.... The emergence of sustainable agriculture now makes the acceptance of mutual goals a necessity" (Paul and Robertson 1989).As a common conclusion, all the contributors agreed on the necessity of improving the dialogue between the different fields of enquiry in agriculture, because " goals of ecologists and agricultural scientists are converging within agro-ecosystems science" (Elliot and Cole 1989). The action framework for sustainability in agriculture finds its appropriate theoretical base in the science of agroecology.

Since agriculture is a human activity system of a larger civil society, its goals are multifunctional and changing in accordance with evolving human conditions and requirements. These goals range from social functions, e.g. production of goods, revenue, employment, culture, etc. to environmental functions, e.g. preservation of natural resources and landscape, and reduction of environmental impact of imported resources, like machinery, materials and manipulated organisms (Vereijken 1992; Caporali 2004). Agroecology as a transdisciplinary science for sustainability should make a contribution to finding criteria and solutions for balancing goals in order to make agriculture sustainable.

As a general methodology for the search for sustainable agroecosystems, it was suggested that agroecosystems should be perceived and managed hierarchically (Lowrance et al. 1987; Stinner and House 1989). "Viewing croplands and pasturelands (and also plantation forestlands) as dependent ecosystems that are a functional part of larger regional and global ecosystems (i.e., a hierarchical approach) is the first step in bringing together the disciplines necessary to accomplish long-term goals. The so-called world food problem cannot be mitigated by efforts of any one discipline, such as agronomy, working alone. Nor does ecology as a discipline offer any immediate or direct solutions, but the holistic and system-level approaches that underlie ecological theory can make a contribution to the integration of disciplines" (Odum 1984). Indeed, agroecology can be considered as "an integrative discipline that includes elements from agronomy, ecology, sociology and economics" (Dalgaard et al. 2003).

5.2 Matching Ecology with Agronomy

At the turn of the last century, many contributions to the scientific journal *Advances in Agronomy* recognised the utility of an ecological approach to both agricultural science and practice to solving the problem of sustainability at different hierarchic levels of organisation, i.e. from the field to the globalised networks of food systems (Magdof et al. 1997; Haygarth and Jarvis 1999; Karlen et al. 1994; den Biggelaar et al. 2001; Lal 2001).

Ecology meets agronomy at the farm level by offering the basic scientific principles for pursuing a sustainable management of both the whole farm system and its subsystems, i.e. cropping and livestock systems. The benefit of constructing sustainable farming system will be consequently acknowledged at the higher landscape level. Agronomy is often taught without explicitly discussing its ecological foundation, forgetting that agronomy is an applied sub-discipline of ecology (Hart 1986). Ecology offers some basic principle in order to design and manage agroecosystems at field and farm levels:

1. Ecological principles relating to *adaptation* of fields, crops and animals to site-specific characteristics, such as climate and soils, in order to maximize productivity with the use of natural resources
2. Ecological principles relating to *population interactions*, such as complementary use of resources, symbiosis, and reduction of competitive ability
3. Ecological principles relating to *biodiversity management* at both the field and the farm levels in order to reduce infestations of weeds, pests and diseases
4. Ecological principles relating to on-farm *integration of components*, i.e. crops, livestock and other vegetation components, in order to maximize internal nutrient re-cycling, while reducing environmental impacts

Most of these principles are fundamentally based on the assumption that an increase of biodiversity at field and farm levels will correspond to better use of natural resources and better performances in terms of agroecosystem productivity, stability and sustainability (Swift and Anderson 1994; Lichtfouse et al. 2009).

5.2.1 Agroecosystem Site-Specificity

Ecology and agronomy have an overlapping area of enquiry in the basic event of *primary productivity* or plant productivity. It is curious to observe that one of the first and best detailed accounts of primary productivity made in accordance with ecological principles, i.e. where productivity is meant as accumulation of biomass or energy by plants, was made on a maize crop (Transeau 1926). Maize was selected because "it probably represents the most efficient annual of temperate regions". An energy budget of a hypothetical acre of maize was made with some conclusions that are of interest even today. "Plants are very inefficient gatherers of energy…the suggestion that our liquid fuels, petroleum and gasoline, may some day be replaced by alcohol made from plant is quite unreasonable…the solution of our future fuel energy supplies lies rather in the discovery of the physics and chemistry of photosynthesis". These considerations recall present controversial debate on the use of crops as bio-fuels for solving the increasing demands of energy by the global human society (Lawson et al. 1984; Pimentel and Patzek 2005; Milder et al. 2008; Thompson 2008).

Crop productivity of an agroecosystem such as a farm depends on the interactions between all natural and anthropic factors involved in its management: the climate, the soil and plants, the allocation of the crops in space and in time, and the

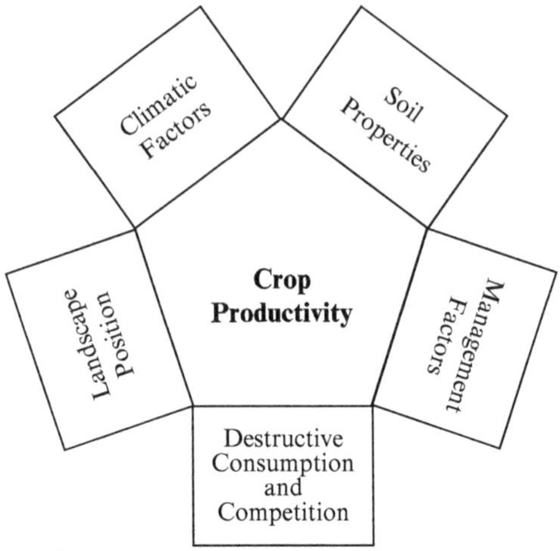

Fig. 6 Factors affecting crop productivity. Natural factors, such as climate, soil and landscape position hierarchically constrain both management factors and biological agents responsible for adverse effects on crop production (Modified from den Biggelaar et al. 2001)

interaction between the crops and the animal breeding. Moreover, the biomass accumulated on the fields must be defended by natural enemies, predators and parasites (Fig. 6). A farm is a piece of land to be organised as an agroecosystem, i.e. an ecosystem constituted of fields (field = *ager*, in Latin). Within a farm, fields are sub-systems or agroecosystems grown to crops, i.e. the elementary units of cultivation. Within a farm organisation, fields are mechanically constructed by fixing their boundaries in a way that defines their number, seize, and integration into a network of waterways, such as ditches and channels, rural roads and paths, or vegetation ecotones, such as hedges or woodlots. Efficient mobility of water, animals, human beings and machines must be assured for a good farm management, functioning and performance.

Because terrestrial productivity is heavily constrained by physical ecosystem components, such as climate and soil, a sound guideline for constructing sustainable agroecosystems is to adapt both their land-unit structures, i.e. fields, and their productive components, i.e. crops and livestock, to the site-specificity determined by the climate-soil interactions. Indeed, agriculture as a historical human experience worldwide has much to teach in terms of sustainability, because farmers generations have been operating on the land for centuries in the search for sustainable agricultural ecosystems on the base of a trial–error experience (Harper 1974). Traditional agroecosystem still running in each continent are to be considered an important source of information for constructing the knowledge useful for designing and implementing sustainable agroecosystems for the future.

Soil and water are the natural resources on which agriculture is based and soil and water conservation is the first measure to be adopted in a perspective of sustainable agriculture. Soils are finite resources created and degraded through natural and human-induced processes (Larson et al. 1983; den Biggelaar et al. 2001).

Fig. 7 Terraces planted to vineyards on steep slopes at the "Cinque Terre National Park", La Spezia, Italy. This steep coast on the Tyrrhenian see has been historically protected by man-made stone-terraces cultivated by hand. This landscape represents both an agrarian and a cultural heritage

Scientists estimate that 2.5 cm of new topsoil is formed every 100–1,000 years (Pimentel et al. 1976). Soil conservation measures govern the hydrological cycle at the soil level for the purpose of both the control of water flow and the achievement of an adequate water balance within cultivated fields. The web of ditches and channels has the task to regulate water circulation in agroecosystems and, as a consequence, marks the fields, determining their number and size. This man-made infrastructure contributes a mosaic-shaped structure to the whole agroecosystem landscape, where the fields represent the elementary units of cultivation.

In hilly and mountainous areas man-made infrastructures also concern the layout of slopes, since the fundamental principle to follow for water control in sloping areas is to allow for maximum rainfall infiltration and minimum runoff to prevent hydro-geological erosion. In these conditions the slope of the plots can be modified through terracing with stones or by embankments, if spontaneous grass-growing is encouraged on terracing walls. Terracing of sloping areas has been traditionally adopted in all the regions of the world in order to keep soil and water in situ and to properly utilize them for agricultural purposes (Lal 2001). Terraced slopes represent a historical heritage of the agricultural civilization as well as a precious testimony of the co-evolution between man and nature where the aim is an environmentally friendly development. The beauty of the terraced landscape is widely recognized and conveys a positive image of the agricultural impact on the environment to the observer. It is a sign that conveys information about civilization and socio-ecological values put into practice on the territory. Aesthetic and functional values are combined and synergistically classify the landscape as a highly valuable scenario, thereby making it both a tourist and educational destination (Fig. 7). Agricultural and aesthetic functions can be widely strengthened by the synergy between beauty and functionality as happens in many countries, including Italy (Caporali 2000; Hampicke 2006).

The size of cultivated fields has undergone a drastic evolution especially since the second half of the nineteenth century with the advent of the acute phase of agricultural industrialization, according to the paradoxical principle of adapting the territory to machinery instead of machinery to the territory. The classical layouts of hills and plains, the pride of the academic agricultural culture constructed over at least two centuries in Europe, have been progressively erased from the territory. The negative consequences on the water cycle, eutrophication and pollution have not been slow to manifest. The socio-economic spiral of the paradigm of industrialized agriculture has overwhelmed the socio-ecological defences carried out on the territory during the previous historical stage of traditional agriculture, both in capitalist and communist countries. The fields have started to become wider and most of the infrastructure boundaries have been destroyed. The consequences of such an evolution have been detailed in a territory where it was possible to historically verify the effect of the enlargement of cultivated fields, which resulted from the setting up of large farm cooperatives under the communist regime of collective rural areas (Van der Ploeg and Schweigert 2001). During the communist regime, which was established in East Germany after World War II, small and medium sized private farms were joined together and converted into larger collective farms. In the state of Saxony, for example, prior to collectivization in 1955 the average farm size was 7.3 ha, whereas after collectivization and before the reunification of Germany in 1989, the average farm size had risen to 565 ha. At the same time in West Germany the average farm surface area was 7.5 ha in 1955 and 18.2 ha in 1989. The size of fields had increased parallel to farm size. In collective farms, prior to 1974 the average surface area of fields was already 10–40 ha, one field being much larger than an entire previous farm; between 1974 and 1978 it had increased to 40–50 ha and after 1978 to between 90 and 100 ha. The accumulated power of all the tractors working in Saxony in the same period was 329 MW in 1960 and 1,543 MW in 1989. The combination between the wider size of fields and their intensive cultivation through big machines determined a significant soil degradation resulting in the increase in runoff in the Elba basin and in subsequent floods. Furthermore, the loss of the physical structure of the soil caused both wind and hydro geological erosion as well as the leaching of nutrients and pesticides which accentuated eutrophication and pollution phenomena in surface water and groundwater respectively. According to the authors' estimates, damage caused by the environmental impact, the so-called externalities, amounted to 10,000 million former German marks per annum, as a whole.

Recent research on the role of field size for evaluating the trade-off between machinery efficiency and loss of biodiversity-friendly habitats in arable landscapes shows that there is no need for bigger field size beyond an evaluated threshold of 1–2 ha above which machinery efficiency increases very little (Rodriguez and Wiegand 2009).

Water is a precious resource which must be retained as far as possible within the agroecosystems to encourage crop productivity. Water that leaves the agroecosystems through the network of surface water as well as the soil profile that supplies underground waters should leave the farm boundaries with the minimum amount of nutrients and pollutants.

In a situation of increasing incidence of the green house effect, like the present, climatic changes are also recorded in Italy (Colacino 2001; Maracchi and Cresci 2001), which imply greater attention to the aspect of soil control measures in the agricultural land. Among climatic anomalies of interest in the last decades, the increase in the extreme autumn rainfall events has been emphasized. That means the increase in frequency of intense rainfall exceeding 60 mm per hour, which can be considered as the risk thresholds for small floods. Such rainfall results in the so-called flash-floods (sudden local floods due to heavy rain) often with catastrophic effects as well as increase in run off and decrease in groundwater. Due to all these events, the authors estimate that damages to private and public heritage amount to an average of about 8,000 billion former lira per annum. The intensification of convective events, which is regarded as the tropicalisation of the Mediterranean, provides an explanation for this situation and it is related to the greater amount of energy available due to the general increase in the temperature of the earth and the warming of sea temperature (Maracchi and Cresci 2001). Farm management orientated towards increased biodiversity within and between the cultivated field can help in keeping soil and water on the spot for increasing productivity and reducing adverse environmental impact. More traditional forms of agriculture can offer greater potential for adapting to global climate change than current intensive systems do (Lin et al. 2008).

5.2.2 Within-Field Biodiversity Management

Crop productivity is the result of crop biomass accumulation in the field over time, and therefore is clearly a biological function strictly dependent on the farmer's capacity to organize a cropping system able to intercept solar radiation as long as possible year round. A crop field is to be viewed as a truly solar power plant, where plant leaves are to be considered proper collectors of solar energy that increase in number and size thanks to photosynthesis (Caporali 2004). All this energy contained in the crop biomass serves as potential energy for all the other components of the agroecosystem, animals, man and soil decomposers. A greater exploitation of solar energy is possible only by increasing duration and extension of crop coverage in the field during the year. In agronomic term, annual crop productivity depends largely on the Leaf Area Duration (LAD) (Loomis and Connor 1992). A greater use of solar energy is to be regarded as a major improvement of biological efficiency in agriculture, solar radiation being the most important driving force for allowing CO_2, water and nutrients to be converted into organic compounds by crops. This function of collecting solar energy for transforming inorganic compounds into organic ones is today referred to as carbon sequestration, the process being an important step for mitigating the greenhouse effect, to which conventional agriculture is actively contributing (Marinari and Caporali 2008).

Books and papers concerning the importance of biological efficiency in agriculture and the agronomic way to improve it started to appear in the 1980s (Spedding et al. 1981; Mead and Riley 1981; Francis 1986, 1989).The current mono-cropping and monoculture systems, both herbaceous and tree crops, typical of the specialised agro-ecosystems of conventional agriculture, do not meet the need of greater exploitation of solar energy. Therefore, these agroecosystems are extensive in relation to the use of direct solar energy and other native resources, and intensive in relation to the use of auxiliary energy derived from fossil fuels. Sole crop systems result from the need of specialization for industrialized agriculture, i.e. for homogenizing and simplifying technical itineraries according to work-time and methods of big machinery. From the perspective of natural resources use, a sole crop system is bound to use solar energy only in the period when the crop canopy has developed, which is strictly dependent on its phenology. In the temperate zone, for instance, a sole crop develops its canopy either in the autumn–winter time or in the spring–summer time, and its seasonality prevents it from using solar energy all year round. Moreover, most auxiliary energy replaces direct solar energy for the implementation of agricultural practices, such as crop nutrition and protection. Catabolic processes, i.e. combustion and predominance of heterotrophic respiration processes, are enhanced rather than the anabolic process of photosynthesis. Energy dissipation and the increased entropy inherent in this technological style give rise to negative environmental impact manifestations revealed in today's specialized agroecosystems, such as loss of soil fertility, erosion, soil and water pollution, etc. (Papendick et al. 1986).

The strategy of ecosystem development shows that the net result of a complex biological community is maximum gross primary productivity, symbiosis, nutrient conservation, stability, a decrease in entropy, and an increase in information (Odum 1969). Even if agricultural systems aim at maximizing net primary productivity, a certain level of biological complexity should be maintained for conserving self-sustaining capability at both field and farm levels. In order to get a more constant soil cover, it is necessary to have recourse to design and management that favour within-field biodiversity increment, or the implementation of multiple cropping systems with improved biological efficiency (Francis 1986, 1989). Cropping systems based on the use of more that one species in temporal and/or spatial dimensions are necessary to intercept solar radiation with more continuity. Cover cropping, intercropping and crop rotation are the most usual typology of multiple cropping systems to be mentioned in order to get benefit from more integral use of native resources due to the ecological principle of facilitation, i.e. positive species interactions (Bruno et al. 2003).

However, criticism has been raised about the way agriculture has evolved worldwide by relying most on sequences of annual grain crops, such as annual cereals and pulses (Glover 2005). As a matter of fact, more than two thirds of global cropland features annual grain crops typically grown in monoculture, which provide roughly 70% of humanity's food energy needs (Glover 2005). A more or less prolonged fallow period in between annual crops constitutes bare areas of failed photosynthesis with added loss of soil and nutrients. A shift towards perennial grain production would allow for a more integral functioning of field agroecosystems on a temporal scale, with more intensive use of natural resources and less pollution effects. Environmental benefits due to perennial cropping include: (a) reduced tillage that increases soil carbon sequestration and reduces fossil fuel and accompanying emissions; (b) reduced soil erosion and nutrient leaching (McCarl and Schneider 2001). More research is needed for achieving high-yielding perennial grain crops and perennial grain breeders are integrating ecological principles and traditional plant breeding methods in their effort to develop perennial grain crops, such as wheat, rice, sorghum and sunflower (DeHaan et al. 2005; Crews 2005). Significant commercial release of perennial grain crops can occur within the next 25 years (Glover 2005).

Cover Cropping

The biotic components of agroecosystems can be classified in relation to the role they play for the target of the farmer and may be regarded as productive, beneficial or destructive components (Swift and Anderson 1994). The productive biota is constituted by the crop plants and livestock producing some useful product for consumption, use or sale. This component is deliberately chosen by the farmer and it is the main determinant of the biodiversity and complexity of the system. The resource biota is constituted by the organisms that contribute positively to the productivity of the system but do not generate a product for harvesting. Examples are just cover crops grown for multi-functional purposes and the fauna and flora of the decomposer sub-system in the soil. The destructive biota is constituted by weeds, animal pests and microbial pathogens. Management is aimed at reducing the entity of this components.

Cover crops are to be intended as components of the resource biota and means to make the field system more complex in its vegetation composition in a way that natural resources can be better exploited, productivity enhanced, soil protected and environmental impact reduced. Cover crops also have the potential to suppress weeds, control pests, and create new sources of income for farmers (Haynes 1980; Clark 2007). In practice, they can be introduced in both herbaceous or tree cropping systems with patterns that vary according to the season of growth, e.g. winter or summer cover crops, the modality of sowing, e.g. inter-sowing or clean-sowing, and the modality of final treatment, e.g. dead mulching or green manure. Their use was highly recommended until the establishment of conventional agriculture as it has always been in traditional farming systems (MacKee 1947; Dale and Brown 1955). The renewed interest in cover crops is largely justified by their poly-functional role for agroecosystem sustainability.Important institutions, like the Soil and Water Conservation Society in USA, have supported the sharing of information about cover crop research and practice over the last decade with several conferences and numerous publications on the topic (Anderson-Wilk 2008; Delgado et al. 2007).

Legumes are the most profitable candidates to be adopted as cover crops, due to their capacity to fix atmospheric nitrogen and therefore enrich the soil with both C and N (LaRue and Patterson 1981). Different legumes are naturally widespread in all geographical regions of the world and constitute a powerful biological resource for cover cropping. Winter annual legumes native to the Mediterranean environment, like self-reseeding species of *Trifolium* and *Medicago*, have the potential as cover crops to change the presently specialised cropping systems into more structurally complicated and sustainable ones (Lanini et al. 1989; Altieri 1991; Caporali and Campiglia 2001). Subclover (*Trifolium subterraneum* L.) has been profitably used for improving the performance of both specialised, autumn-fruiting tree crops, such as hazel groves and vineyards, and cash crop sequences in Mediterranean rainfed conditions (Campiglia et al. 1991; Caporali and Campiglia 2001). Basically, many different traits of these legumes can be conveniently exploited when used as cover crops, as they are able: (a) to grow during the cool season; (b) to die in the early summer; (c) to regenerate at fall rains and therefore to provide cover even for several years; (d) to be shade-tolerant; (e) to provide weed control though good growth coverage; (f) to provide significant quantity of fixed N while conserving soil and water resources; (g) to consent the use of minimum or no-till practices; (h) to increase profits through higher yield and/or lower production costs.

In Mediterranean countries, the conventional soil management system in orchards is clean cultivation (Chisci 1980; Tropeano 1983). This is often accompanied by the elimination of weeds from a strip of orchard floor along the rows with the use of herbicides. The regime of repeated cultivations that tree crops undergo for both weed control and mechanical harvesting from the floor, such as in the case of hazel groves, inevitably results in soil structure degradation and more hazard of hydro-geological erosion. In addition, the increasing use of both pesticides and fertilizers is bound to result in their increasing diffusion throughout the environment, especially via water. These management criteria have undoubtedly led to increased yield, but the sustainability of the system has substantially decreased.

According to the view of regarding a crop field as a truly solar power plant, research has shown that the introduction of an herbaceous layer of sub-clover with a growth pattern complementary to that of the tree crop component makes the resulting cover cropped system more photosynthetically efficient on an annual basis (Fig. 8) (Campiglia et al. 1991). In addition to the effect of N fixation by the legume, a complementary use of native resource is accomplished, due to an early use of soil water and nutrients by the clover from fall to spring, when the tree component active growth does not occur. Since only a partial niche overlapping between the two species occur during late spring, when rainfall is still relatively abundant, damages by drought would probably be not so critical to the yield of the tree component. Appropriate interventions of defoliation, such as cutting or grazing, could be provided in order to lower water competition by sub-clover. The sub-clover dead mulch during summer would reduce evaporation and conserve more residual water for the tree crop. The sub-clover reseeding capacity would finally assure its re-establishment as a cover crop in the following years without any tillage, providing both energy savings and biomass accumulation useful for grazing or whatsoever forage utilization.

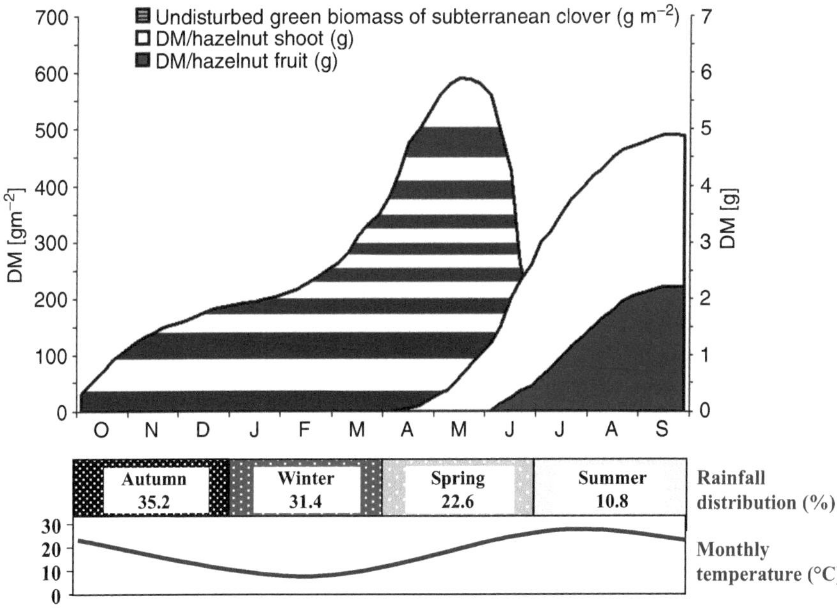

Fig. 8 Biomass accumulation as result of complementary use of natural resources in a cover cropped system. *Top*: average growth trends of subterranean clover under stored to hazelnut in a hilly area of Central Italy. *Bottom*: mean annual weather conditions (mean annual rainfall, 1,000 mm; mean annual temperature, 14.5°C) (Modified from Campiglia et al. 1991)

Associated with legume cover crops use, green manuring is the final step of the process for biologically delivering a N supply to the succeeding crop in a herbaceous crop sequence. This traditional practice for providing crops with N from organic sources was replaced by the use of cheap synthetic fertilizers during the green revolution era (Parson 1984). In grain crop sequences, it is desirable to encourage legume use instead of the traditional fallow phase, since efficiency and utilization of N is generally greater following a legume green-manure crop than following fallow (Badaruddin and Meyer 1990). A legume cover crop can supply most of the N required for maximum maize yield (Decker et al. 1994; Clark et al. 1997a, b). Because their high nitrogen content and low C:N ratio, legume cover crops can decompose rapidly in the soil, supplying large amount of mineral nitrogen, and increase maize yield similarly to nitrogen fertilizer rates ranging from 66 to 200 kg/ha (Kuo et al. 1996, 1997; Sainju and Singh 1997; Caporali et al. 2004).

Cover crops can be also grown as catch crops which occupy the land briefly between two main crops and serve to scavenge and recycle nutrients while protecting the soil. Such a role is important as it prevents nutrient leaching, especially nitrogen, avoiding N losses and the consequent pollution of underground water. In such systems, nitrogen is made available through the mineralization of organic material after green manuring (Atallah and Lopez-Real 1991; Magdof et al. 1997; Thorup-Kristensen et al. 2003; Delgado et al. 2007).

Intercropping

Intercropping is the practice of growing more than one crop in a field at the same time, which is still very common in traditional agriculture of every continent (Horwith 1985; Francis 1986). The main ecological basis of intercropping is provided by the assumption that growing populations of different species on the same ground can be advantageous if they manifest complementarity in the use of natural resources, such as light, water, nutrients, etc. (Willey 1979a, b; Hart 1986; Trenbath 1986; Vandermeer 1988). This condition can be purposely constructed in the intercropping system by managing: (a) differences in the phenologies of intercrops that allow for continuing plant cover; (b) differences in the morphology of intercrops as to the diversity of aboveground canopy development or different root systems that explore different habitat layers; (c) differences in the physiology of intercrops, with at least one N fixing species.

Four main types of intercropping can be identified (Ofori and Stern 1987):

(1) Mixed intercropping: growing components crops simultaneously with no distinct spatial arrangement; (2) row intercropping: growing components crops simultaneously in rows for mechanized agriculture; (3) strip intercropping: growing components crop simultaneously in different strips to permit the independent cultivation of each crop; (4) relay intercropping: growing component crops in relay, so that growth cycles partially overlap.

The case of realizing an intercrop advantage in comparison with the relative monoculture can be easily measured by appropriate indexes, such as RYT (relative yield total) and LER (land equivalent ratio) (Mead and Riley 1981). The relative yield of an intercrop is the ratio of its intercrop yield to its yield in monoculture. The sum of the yields of both species is the relative yield total. When RYT is more than 1.0, there is overyielding by the intercropping system. LER is a term equivalent to RYT in that it is a measure of the amount of land that would need to be planted in monocultures to give a yield comparable to a unit area of land planted as an intercrop. When LER is more than 1.0, overyielding occurs.

"If interspecific competition were less than intraspecific competition for all resources between two species, than it would be clearly advantageous to grow them as intercrops" (Horwith 1985). Beneficial effects for overyielding in intercropping can occur belowground with positive changes to biologically based nutrient sources (Magdof et al. 1997). Legume components may indeed contribute to enhancing biological belowground diversity and abundance because they form mutualistic associations with vesicular arbuscolar mycorrhizae (VAM), bringing about better nutrient uptake, particularly of phosphorus, by host plants (Hayman and Mosse 1972; Harrier and Watson 2003). The presence of a legume crop can actually enhance nutrient availability for a companion crops not only for N, but also for P and other nutrients (Chiariello et al. 1982; Magdof et al. 1997). Recent studies concerning plant diversity, soil microbial communities and ecosystem function confirm that soil microbial communities and the key ecosystem processes that they mediate are significantly altered by plant richness. Microbial community biomass, respiration, and fungal abundance significantly increase with greater plant biodiversity, as do mineralization rates, with beneficial effects on plant productivity (Zak et al. 2003).

In a more diversified field agroecosystem, such as that of an intercropping system in comparison with the relative mono-cultural field systems, the destructive biota (sensu Swift and Anderson 1994) is also affected. More biological interactions can occur favouring a biological control of infestations by weeds, pests and diseases (Altieri et al. 1983; Risch et al. 1983; Matteson et al. 1984; Altieri and Liebman 1986; Liebman and Dyck 1993; Theunissen 1997; Wolfe 2000). Two main mechanisms of pest deterrence have been proposed: the "resource concentration hypothesis" and the "enemy hypothesis" (Risch 1981; William 1981). In the first case, the presence of non host plants interferes with the search by a given pest for its host crop; in the second one, the diverse crop environment provides shelter and food sources for insect predators and parasites.

Intercropping has lost application in conventional agriculture because it is not easy to be completely mechanised in a conventional management scheme that provides specific treatment for each crop from the sowing to the harvest. However, today's expectations recognising monoculture as expensive economically, energetically, and environmentally, question the wisdom of depleting resources while ignoring alternatives like intercropping (Horwith 1985). A renewal of interest in intercropping is currently running especially in organic farming research and applications (Bergkvist 2003; Bellostas et al. 2003; Bath et al. 2006; Germeier 2006).

Crop Rotation and Reduced Tillage

Crop rotation is a cyclical system of crops organised in a temporal sequence on the same field. Each cycle of crops involves not only crop succession in between fallow periods, but also tillage operations, such as seed-bed preparation, fertilizer incorporation into the soil and mechanical weeding. In a framework of cropping system sustainability, each cycle of rotation should be organised as a whole, i.e. in a way to use the maximum of natural resources and the minimum of external resources of fossil fuel-derived energy. According to this principle, the period of bare fallow should be reduced in order to get more crop coverage throughout the year, more crop photosynthesis, more organic matter in the soil, more soil protection against erosion and leaching, and less infestations by weeds, pest and diseases. Crop rotation is unanimously considered one of the most important technological breakthroughs in the history of agriculture development (Karlen et al. 1994; Robson et al. 2002). Crop rotation represents a constitutional element of agroecosystem sustainability because is based on a balance between crops with complementary characteristics and allows for an integration among different farm components, i.e. crops and animals. Crop rotation meets the requirements of creating a system of organized crop diversity in order to serve multiple objectives like: self-sustained crop productivity; maintenance of soil fertility, biological and agronomic control of crop adversities (Burdon and Shattock 1980; Batra 1982; Francis 1989; Liebman and Dyck 1993). Crop rotation needs to be site-specific, because it has to comply with the environmental constraints of climate and soil in order to save energy and money, and reduce adverse environmental impact.

Rotation, owing to the fertility balance induced by the alternation of different crops in the soil, promotes productivity mainly because it is based on a synergy between crops and between crops and soil. In short, rotation occurs between soil-enhancing and soil-depleting crops with respect to the state of soil fertility. Poly-annual meadows of leguminous fodder crops fall into the first category whereas cereal crops (wheat, barely, oats, maize) fall into the second. The fundamental pillar of rotation is made up of the soil-enhancing crops, which belong to the family of leguminous crops. Through the cultivation of leguminous crops, large amounts of atmospheric carbon and nitrogen are fixed in the tissues of leguminous plants and can be used along the grazing chain through the foraging of the animals and along the detritus chain through the decomposers' action on crop residues which remain in the soil (Baldock et al. 1981; Spiertz and Sibma 1986; Bruulsema and Christie 1987; Hesterman et al. 1987). The amounts of fixed nitrogen vary according to the type of leguminous plant and above all, according to the cultivation cycle. Grain leguminous crops which have a life span of less than 1 year, fix the least amounts (around 100–150 kg/ha), which are concentrated above all in the grain and are taken away with harvesting. Poly-annual fodder crops, such as alfalfa and poly-annual clovers fix the highest amounts, which can reach 500 kg/ha in alfalfa meadows that are 4 or 5 years old. Thanks to the mineralization effected by micro-organisms, successive crops can benefit from the release of inorganic nitrogen for their nutrition. However, the soil disturbance through frequent and deep tillage operations that occur with a sequence of annual crops is cause of rapid decrease of organic matter in the soil, which in the long term undermines soil fertility and cropping system sustainability (Tiessen et al. 1982). Instead, since most of the biomass of cultivated poly-annual fodder crops is left in the field, a high potential of nitrogen supply is placed at the disposal of successive rotating crops (Caporali and Onnis 1992; Caporali 2000).

The emphasis on crop rotation is well documented in the great Italic tradition of agriculture of the past, in particular, in ancient Latin literature beginning with Authors like Varro, Cato, Virgil and Columella. The first work of great scientific and technical value on the innovative application of crop rotation with the introduction of leguminous fodder crops dates back to the Italian Renaissance. It is dealt with in the work of Camillo Tarello, "Ricordo di Agricoltura" (Memory of Agriculture) presented to the Senate of the Republic of Venice in 1565, in which he listed all the agronomic and economic advantages gained with the simple introduction of red clover (*Trifolium pratense*) into a sequence of cereal crops. Being aware of the importance of the proposal for the new "bio-technology" applied on the field, Tarello negotiated with the Senate of the Republic of Venice for a long time to receive compensation (a type of patent royalty) from farmers who were interested in applying this practice on their farms.

Academic science of the nineteenth century and the first half of the twentieth century has greatly contributed to the establishment of efficient rotation schemes for each region, firstly through Academies of Agriculture and then through

Faculties of Agriculture at universities (Cuppari 1862). Cuppari had already expressed a systemic view of the farm in 1862, then he identified 'a biological need for crop rotation which 'must be related to the rural management of the farm it belongs to, because the sequence of crops summarizes the necessity of integration in a whole'. Actually, crop rotation was related to animal breeding, in that it provided food for livestock, from which it drew in turn labour and manure. Later in Italy, Draghetti (1948) also identified the farm as a 'super organism' where the various parts must interact properly to guarantee a balanced whole. Crop rotation represents the pillar of this 'organism', which must find the reasons for its own sustainability within its own structure.

There are leguminous crops suitable for any type of soil and climate. In the Mediterranean environment, it is particularly the soil type that limits the fundamental choice of leguminous crops to be put under rotation. In the presence of deep soils with a balanced amount of sand, silt and clay, alfalfa will be chosen for the rotation, since it is the most productive fodder crop in a temperate climate (Spiertz and Sibma 1986; Caporali 2004). In the rain-fed situation of central Italy, for example, a common rotation for a good fertility balance is: sunflower–wheat–alfalfa (3 years)–wheat. In this rotation which occurs over a span of 6 years, 3 years are grown to annual crops, like wheat and sunflower, and the remaining 3 years to the poly-annual alfalfa. In the 3 years of alfalfa cultivation, the soil is sod and therefore soil micro-organisms provide for humification rather than for mineralization thus rebuilding higher levels of organic matter and consequently of soil fertility for the benefit of the annual crops that follow after soil tillage and the mineralization of the accumulated organic matter.

Instead, if the soil is more clayey and of marine origin, as in the case of many hilly areas of central and southern Italy, the less aerobic conditions of the soil call for leguminous crops more suitable for this particular substratum, like sweetvetch (*Hedysarum coronarium* L.) which serves as the pioneer plant for these soils since it is able to modify their structure due to the features of its own root system. In this situation the typical rotation can be as follows: wheat–sweetvetch (1–2 years).

It is not necessary to sow sweetvetch directly into the soil subject to this rotation since its hard seeds can remain in the soil for many years and therefore germinate when it is required to, e.g. on wheat stubble fields at the first autumn rains. In this case a sweetvetch meadow appears early in autumn and it is available for grazing of sheep even for 2 years. Pliocene clay areas in Italy lend themselves to a mixed agriculture regime, i.e. based on the wheat and sweetvetch rotation where wheat can also be replaced by horse-bean used for fodder in order to prepare feedstuffs containing proteins for livestock.

On the contrary, if the soil tends to be sandy and of volcanic origin, as is the case in the Latium area in particular in the provinces of Rome and Viterbo, its acidity (pH 5–6) limits the choice of leguminous crops to adapted species such as crimson clover (*Trifolium incarnatum*). Therefore, the typical rotation in this area can be as follows: wheat – crimson clover (1 year).In this case the clover is sown in autumn and is continuously grazed upon by sheep until the end of its growth cycle (June–July), then it is ploughed in before the autumn wheat sowing. In soils more suitably located in plains, where irrigation is practiced, crimson clover can be rotated with a summer vegetable crop, such as tomato.

In the extreme conditions of Mediterranean environments, where rainfall is particularly scarce and does not amount to more than 400–450 mm/annum, the most suitable leguminous crops to be rotated with cereals in mixed farms are the so-called self-reseeding legumes, i.e. annual species belonging to the *Trifolium* and *Medicago* genera (Caporali and Campiglia 2001). The introduction of these species on the part of the colonizers in the Australian continent where the south-western part is characterized by a Mediterranean climate, gave rise to one of the most widespread systems based on cereal growing and animal husbandry at the continental level. This system is called *ley farming* that is based on a cycle of 1-2-3 years of cereals rotating with 1-2-3 years of self-reseeding leguminous crops (Puckridge and French 1983).

More intensive systems of crop diversification than *ley farming* can be created with the use of self-reseeding legumes in crop rotations suitable for conventional farming. This is the case of contemporary sowing of an intercropping system of wheat and subclover in a bi-annual cash crop rotation wheat–sunflower, which is very usual in rain-fed conditions of central Italy where it is sustained by N fertilizers and herbicides in conventional farming (Caporali and Campiglia 2001). After harvesting wheat, subclover self-reseeds and generates an autumn canopy that serves as cover crop and can therefore be used as green manure for the subsequent sunflower crop. Subclover has multiple functions in this crop rotation: intercrop or living mulch in wheat, cover crop after harvesting wheat and green manure before sowing sunflower. In more stressed climatic conditions, the sunflower crop can be replaced by crops which are more resistant to drought, e.g. sorghum, or, if irrigation is available, by vegetables, e.g. tomato. This alternative crop rotation reflects the following major agroecological requirements: (a) increase in the use of renewable natural resources and decrease in the use of fossil fuel-derived resources for reduced tillage and treatments; (b) more intensive use of legumes in crop rotation in order to store more solar energy, conserve soil moisture, and fix atmospheric nitrogen; (c) more intensive soil coverage by the cropping system in order to assure a permanent plant canopy during the year and prevent soil erosion; (d) potential for weed suppression (Caporali and Campiglia 2001).

5.2.3 Between-Field Biodiversity Management

The achievement of appropriate agroecosystem structure for a multifunctional role of both production and protection is also related to the degree of complexity occurring between cultivated fields. In this framework, the management of the margins of cultivated fields is receiving increasing attention (Marshall 2002; Llausàs et al. 2009; Musters et al. 2009). The dogma that diversity creates stability is deeply rooted in ecological theory and it is frequently quoted in connection with pest problems and the simplifying influences of agriculture (van Emden and Williams 1974). The dogma is based on the concept that a food web of interactions between trophic levels acts to resist change in the abundance of individual species more effectively than simple food chains. Therefore, the diversification of crop ecosystems is regarded as a means of controlling pests. This diversity can be sought at two levels, either by diversifying the habitat surrounding the crop or/and by diversifying the crop habitat itself (Dempster and Coaker 1972).

Whoever appreciates agroecosystems based on the principles of agroecology sees in the appropriate distribution of hedges a major element to enhance farm biodiversity and environmental performances. Unfortunately, hedgerow removal has been associated with farming intensification in conventional arable systems. In an attempt to replace their function, new habitat can be created, such as strip margins of herbaceous cover and "island" habitats (Thomas et al. 1992).

Hedges basically play three fundamental roles in the agroecosystem hierarchy, i.e. at the field, farm and landscape levels (Caporali 1991):

- Mechanical barrier, which produces micro-climatic effects and modifies land slopes
- Biological filter, which intercepts the flow of water and nutrients
- Biological reservoir, which strengthens the realization and maintenance of trophic web and biological balances

The mechanical barrier effect concerns climatic stabilization which is mainly due to a greater interception of wind and to the subsequent decreased evaporation. The wind-breaker effect positively affects the climatic condition of cultivated fields giving rise to an increase of crop yield. This effect is particularly important in regions with a Mediterranean climate which are characterized by a dry summer period. The range of influence of a hedge extends leeward around 30 times its height. Therefore, an adequate mosaic of hedges on the territory may induce important micro-climatic modifications. Furthermore, the hedge basal part may act as a barrier for surface run off and give rise to deposit of muddy materials. In sloping areas, the deposit of materials may modify slopes in the long run and encourage a natural process of terracing if hedges are arranged perpendicularly to the lines of maximum slope. Such an arrangement of hedges is of great benefit for the stabilization of the hydrological cycle and for soil protection against erosion in mountainous and hilly areas. In areas with snowy precipitations, hedges serve to protect snow masses against solar radiation, thereby facilitating a slower snowmelt and consequently a higher and more continuous infiltration of water into the sub-soil and a more efficient refill of ground water. The mechanical barrier effect also concerns the interception of suspending materials such as dusts and aerosol. This effect may prove to be particularly useful for the protection against air-borne contaminations from the surrounding environment, such as drift resulting from pesticide-based treatments or the deposit of combustion fumes resulting for example from the traffic on congested roads on the border of the farm (Caporali 1991).

The biological filter effect of a hedge is performed by the action of plant roots that extend below the soil surface and effectively intercepts water that flows in the sub-soil towards drainage ditches. This action takes place in any kind of climate, from temperate (Lowrance 1998) to desert climates (Schade et al. 2002). There are two types of mechanisms involved in the process. Firstly, the hedge canopy catches water flow and absorbs part of the nutrients useful for the development of plant structures and considerably reduces the load of nutrients flowing into ditches, in particular in terms of N and P. Secondly, the slowing down of the flow into the subsoil near the hedge fosters the denitrification process through micro-organisms, hence part of water nitrates are converted into molecular nitrogen which returns to the atmosphere. As a whole, the hedge serves as a filtering and purifying structure for water and it is suitable for playing a protective role against eutrophication. This action becomes more capillary in the territory as the network of hedges continues to extend.

Finally, if properly distributed on the territory to form a network of biological corridors, hedges play the role of a biological reservoir in the rural environment. They can make up for the functions carried out by more natural areas, such as woods and grasslands, to house many forms of plant and animal life which find in the hedges the environment suited to feed, reproduce and protect themselves against predators. In rural arable areas, hedges can be regarded as ecotones, i.e. border areas rich in biodiversity among more homogeneous areas with a lower biological diversity (Risser 1995). The plant community constituting hedges reflects the community of adjacent habitats due to the deep interactions with the type of landscape, farm structure and crop practices of cultivated fields (Le Coeur et al. 2002). Today, the number and quality of the biological interactions amongst hedges and cultivated fields are considered more positive as a whole than negative for the performance of the agricultural activity (Marshall and Moonen 2002). Even though some biological components may migrate from hedges to the adjacent cultivated fields, such as weeds or plant-eating insects or plant pathogens, hedges can be of great importance to constantly maintain biological agents that are useful for the control of crop pests and predators in the rural environment. They guarantee in this way the long-term sustainability of biological balances in the agroecosystem that, instead, are impaired by the use of pesticides in conventional agriculture.

As for the biological control of plant-eating insects, the presence of plant structures situated on the margins of a field, such as hedges and strips of permanent herbaceous plant cover, are of great strategic importance. They supply the indispensable habitat for the over wintering of the community of arthropods, such as carabids, staphylinids, etc., which prey upon plant-eating insects harmful to agricultural crops (Sotherton 1985; Wallin 1985; Andersen 1997; Wissinger 1997; Altieri 1999). With reference to the spatial distribution of predatory arthropods in hedges, a recent survey (Maudsley et al. 2002) has shown that spiders mainly populate the shrubby portion of a hedge as well as the layers of basal plant cover. Carabids populate both the leaf litter and the soil layer ranging from 0 to 10 cm deep; staphylinids are exclusively allocated to the soil layer ranging from 0 to 10 cm deep. As for the time of colonization of the cultivated fields by predatory arthropods, it has been highlighted (Alomar et al. 2002) that, for the success of biological control, it is important that predators colonize the cultivated fields as soon as possible and settle when the density of the plant-eating insect population is still low. The type, the abundance and the richness of plant cover that surround the cultivated fields are positively related to the abundance, diversity and early occurrence of plant-eating insects and predators. All the functions of hedges can be further strengthened if their vegetation is associated with a 2–3 m wide herbaceous strip located at the margin of the cultivated field (Marshall and Moonen 2002).

Recent research on the development of biodiversity in field margins has shown that newly established grass margins are less species-rich than field boundaries or road verges with a long history, although the plant species richness of the field margins taken out of production increased in the 4 years following their initial establishment (Musters et al. 2009). On the other hand, traditional hedgerow landscapes and their ecological services are at risk in every kind of rural environment and their maintenance needs appropriate environmental programmes supporting both biodiversity and agrarian activity (Llausàs et al. 2009).

5.2.4 Farm Biodiversity Management Through Crop and Livestock Integration

If the field is represented as the agroecosystem hierarchical level where the process of primary productivity takes place, the farm is the next upper level where decisions are taken on the organisation of the whole input/output process that actually governs structure and function of both the whole farm and its components. Therefore, integration among farm components is a major aspect of successful organisation with a view to achieving sustainability goals. Coherence and complementary action amongst the parts is a prerequisite for an efficient functioning of the whole system. The search for and the implementation of mutualistic relationships among farm components is a key-factor for achieving farm sustainability.

The main criterion to adopt for pursuing sustainability is "designing farming system to mimic native ecosystems" (Huyck and Francis 1995). As the ecosystem model is a representation of both structure and functioning underlying nature, the derived agroecosystem model, such as that reported in figure 2 shows what are the components to be integrated at the farm system level. Integration of different components that interact functionally in order to promote a self-sustained productivity is the key operational principle for a sustainable agriculture. The crop/livestock combination is therefore one logical application of the principles of ecology to agriculture, since essentially all ecosystemic processes operate on the basis of functional integration amongst producers, consumers, and decomposers, the former two being represented in mixed agroecosystems by crops and livestock as productive biota, sensu Swift and Anderson (1994). This is a core principle of agroecology, especially under the narrow definition of combining agriculture with ecology (Von Fragstein and Francis 2008).

Therefore, crops and livestock are the fundamental components of a mixed farm and ensure the integration between the chains of grazing and detritus, which allows the farm to operate as a self-sustaining agroecosystem. Draghetti (1948) was one of the first scholars in Italy to clearly define the model of circulation of matter-energy in a farm and to carry out experimental trials at farm level in order to quantify the benefits of integration of crop and animal husbandry for output increase and soil fertility maintenance (Caporali 2008). Forage crops, especially perennial legumes like Lucerne and Trifolium spp., constitute the key cropping system components that meet animal nutrition requirements while achieving many advantages in a crop rotation, such as biological nitrogen fixation, tillage frequency reduction, building up of soil organic matter and biological weed control (Caporali and Onnis 1992). The presence of forage legumes is regarded as a meaningful indicator of sustainability at farming system level (Caporali et al. 2003).

The farmland under cultivation constantly loses organic matter due to the effect of mineralization as a consequence of tillage operations. Since organic matter determines the state of soil fertility, it is necessary to replenish it through sources of materials that must result chiefly from the farm's internal resources (Francis and King 1988; MacRae et al. 1990). Traditionally, the main source of organic matter for restoring the conditions of soil fertility has always been related to the presence of livestock and the spreading of farm manure. Since manure is a bulky material derived by straw as a bedding litter for livestock, spreading is usually preferred on crops that generally open the rotation and therefore undergo deep ploughing which is suitable for the interment of manure. These crops are traditionally said to renew soil fertility through the combination of ploughing and the supply of manure. Unfortunately, the farm specialization process which has followed the industrialization of agriculture and promoted the separation of animal breeding from plant breeding, does not facilitate the large-scale adoption of this traditional model of farm resources integration. As a consequence, there is an excessive availability either in terms of straw or slurry in farms specialized in crop or animal husbandry, respectively. Today both these materials, which should be devoted to the preparation of manure, are either wasted, as in the case of straw burned in the fields, or sold for other purposes. This is the case of straw sold for paper mills, or supplied separately to the soil causing other agronomic and environmental problems (Caporali 1991, 2004).

Manure is a source of humic substances for the soil and it is mainly responsible for the recovery of its fertility. Actually, it acts positively on the physical, chemical and biological properties of the soil. Humic substances generate aggregation of soil particles into aggregates or peds, the so-called granular structure. This granular structure is the basis of the soil biological habitability for plants, micro-organisms and animals as it determines a dynamic biophysical balance between soil components, which is indispensable to the maintenance of primary productivity over time.

Mixed farming is considered among the most sustainable type of agriculture also from an environmental standpoint, because of its reliance on grazing areas that have very low rates of soil erosion, the nutrient cycling through forage crops, ruminant animals and the soil, and the varied landscape it promotes. For this reasons, mixed farming pattern is preferred for sustainable agricultural systems in general (Wagstaff 1987; Oltjen and Beckett 1996).

5.2.5 Agroforestry

Agroforestry is a term coined in 1977 with the foundation of the International Council for Research in Agroforestry (ICRAF) (Steppler and Lundgren 1988).

It denotes both an age-old practice of having trees mixed in agricultural landscape and a new science at its pioneer stage of conceptual and methodological development (Carruthers 1990). "Agroforestry is a collective name for all land use systems and practices where woody perennials are deliberately grown on the same land management unit as agricultural crops and/or animals, either in spatial mixture or in temporal sequence. There must be significant ecological and economic interactions between the woody and non-woody components" (Steppler and Lundgren 1988). The revival of agroforestry meets growing demands of increasing population to compensate increasing rate of deforestation, soil degradation and loss of biodiversity, both in the tropics and temperate regions of the world (Batish et al. 2008).

Agroforestry requires an integrated approach for the study of the problems arising at the interface of agriculture and forestry. In terms of broad definitions, an agroforestry system is normally classified as silvicultural or silvopastoral when the agricultural component is a crop or a livestock activity, respectively (Thomas 1990). The main feature of agroforestry that distinguishes it from all agricultural systems is the deliberate introduction of trees into the landscape with multi-purpose functions (Steppler and Lundgren 1988). Multi-purpose trees may be grown to supply nitrogen to the soil through mulch or nodulation, to provide browse or shelter for livestock, or control of soil erosion, or fuel wood for the household. According to the spatial/temporal layout of the trees planted in the field, agroforestry systems can be classified as zoned, mixed or rotational (Wood 1990).

Zooned systems can provide trees planted in rows at equal spacing, with a crop grown between the rows. Alternatively, trees may be planted around the edges of fields where a crop is grown. One example of the first case is the tropical alley cropping system of the legume tree *Leucena leucocephala* and maize, where the tree component provides nitrogen to the maize crop through mulching (Wood 1990; Nair 1991). One example of the second case is the traditional tutoring system in Mediterranean countries between a woody perennial, e.g. elm-tree, and a vine plant, with single rows of the "married" plants aligned along the field borders (Caporali 2004). This traditional agroforestry system has almost completely disappeared due to the pressure of monocultural stand of modern viticulture, even in the most renowned traditional area of wine production, such as the Chianti area in Italy (Fig. 9).

Mixed systems of woody perennials intermixed with agricultural crops are widespread in the tropics, especially with the use of shade trees for crops such as tee and coffee, but also with the use of cover crops under rubber and oil palm (Wood 1990; Nair 1991). In humid conditions, home gardens with multi-layered trees combine high productivity with intensive recycling of organic matter and nutrients, while protecting the soil against erosion and leaching. Where soils are poor and summer drought is severe, traditional agroforestry systems such as "dahesas" in west and south west Spain and "montedos" in Portugal, are agrosilvopastoral systems tailored to environmental constraints which combine the management of oak-based woodland with extensive livestock grazing and extensive cereal cropping (Carruthers 1990).

Rotational systems with the use of trees restoring soil fertility during fallow periods of shifting agriculture are common in tropical areas (Nair 1991). Raising of agricultural crops during the early stages of establishment of tree plantations is

Fig. 9 An example of ongoing change in agroecosystem patterning. New vineyard plantations to monocultural stands, such as those shown in both *photo's foreground* and *background*, have almost completely replaced traditional agroforestry plantations in the Chianti area (Siena, Italy). However, remnant fields with vine grape "married" to elm are still present, such as those in the *middle* of the photo

usual in the tropical agroforestry system named "taungya" mainly to reduce weeding costs and provide some cash returns to a planting scheme (Wood 1990; Nair 1991). Growing crops in poplar stands in Northern Italy, such as maize or soya for 2–3 years after the tree planting, is one of the few examples mentioned of agrosilviculture in the EU, although the practice has declined in recent years (Carruthers 1990). Instead, raising poultry in small areas of woody plantations is an agroforestry practice gaining room especially in organic farming (Fig. 10), due to the benefit for animal welfare (Jones et al. 2007). Silvopastoral agroforestry is credited to have a future in Europe (Sibbald 2006). Recent investigations on the feasibility of an agroforestry system that combines the establishment of an extensive, diverse native woodland and traditional sheep husbandry have given encouraging results in terms of flock economics, local labour, woodland establishment and vegetation and bird impacts (Morgan-Davies et al. 2008).

Sustainable land use requires a combination of production with protection. Agroforestry has potential to contribute to these multi-purpose activities due to the key role played by multi-purpose trees (Young 1990; Pimentel et al. 1992). Agroforestry is a site-specific, relatively cheap form of rural development in many cases appropriate for a family farm-based agriculture and participatory approach

Fig. 10 *Socialising* between hens and human beings (the farmer's daughter) in the Cupidi organic farm, Viterbo, Italy. In the *background*, an agroforestry plantation to *Juglans regia* with poultry rising, which was established 5 years earlier within a farmers scheme of the Lazio Region in Italy (EU Regulation 1257/99)

(Morgan-Davies et al. 2008). Environmental service functions are provided by increasing biodiversity. They include biomass, soil, water and nutrient conservation for the health and beauty of the landscape, and fencing, shelter and shade for the benefit of plant, animals and humans. Moreover, domestication of new woody plants for providing indigenous fruits is possible. Current agroforestry tree domestication efforts emerged as a smallholder farmer-driven, market-led process in the early 1990s and have became an important international initiative for eradicating poverty and hunger, promoting social equity and environmental sustainability (Leakey et al. 2005).

5.3 *Matching Ecology with Economics and Sociology*

Today, it is recognised that ecological systems and socioeconomic systems are linked in their dynamics, and these linkages are key to coupling environmental protection and economic growth. Most fundamentally, social-economic systems (SESs) are regarded as complex adaptive systems, integrating phenomena across multiple scales of space, time and organizational complexity (Levin 1998, 2005;

Redman et al. 2004). In such systems, macroscopic phenomena emerge from, and in turn influence, the individual and collective dynamics of individual agents. In practice, we live in a global commons, and what we use for our own benefit is often at the expense of what is available to others, both now and in the future (Levin 2006). The challenge is to develop social norms and a cooperative behaviour in a more and more crowded environment where consumptive patterns must be balanced by a more just resource partitioning. The concept of "social capital" expresses the idea that social bonds and norms are important for determining behaviour and performances of people and communities (Pretty 2003). Four features are important components of social capital: relations of trust; reciprocity and exchanges; common rules, norms, and sanctions; and connectedness in networks and groups (Pretty 2003).

In 1989, the International Society for Ecological Economics (ISEE) was established as an "organisation dedicated to advancing understanding of the relationships among ecological, social, and economic systems for the mutual well-being of nature and people" (http://www.ecoeco.org/). To this end, ISEE publishes since then a research journal, *Ecological Economics*, which has a transdisciplinary character, being concerned with extending and integrating the study of ecology and economics. "This integration is necessary because conceptual and professional isolation have led to economic and environmental policies which are mutually destructive rather than reinforcing in the long term" (http:/www.ecoeco.org/publications_journals.php).

Historically, integration between ecology and economics has been fruitful because it has produced new metaphors and concepts, i.e. cultural emergences, such as natural capital, ecological services, ecological footprint, and sustainable development (Daily 1997; Wackernagel and Rees 1997). These concepts are meaningful for defining fields of inquiry that belong to an area of common interest, both material and spiritual, where problems emerge and solutions need to be found.

A major attempt to formulate an unconventional hierarchical interpretation of the relationship between ecology and economy is represented in Fig. 11 as put forward by Daly (1991). In this representation, "macroeconomy is regarded as an open subsystem of the ecosystem and is totally dependent upon it, both as a source

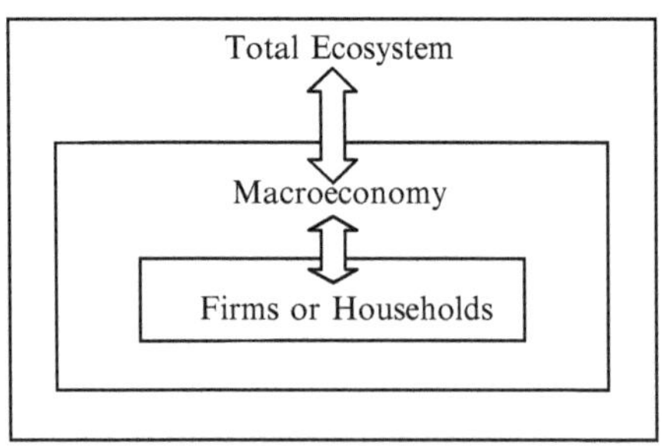

Fig. 11 Hierarchy of ecological and economic domains. Economy is a human activity system as a sub-system of the larger and global system represented by the "great economy of nature" (Daly 1991). That larger system, i.e. the "Earthsystem", is finite, nongrowing, and materially closed, although open to solar energy (Daly and Farley 2004)

for inputs of low-entropy matter-energy and a sink for outputs of high-entropy matter-energy. The physical exchanges crossing the boundary between the total ecological system and the economic subsystem constitute the subject matter of environmental macroeconomics" (Daly 1991). This is why ecological economics has been defined as "a new transdisciplinary field of study that addresses the relationship between ecosystems and economic systems in the broadest sense" (Costanza et al. 1991). Recognizing that human economy is a dependent subset of the biosphere is a stepping stone for addressing more truthful visions of sustainability (McMichael et al. 2003).

By transdisciplinarity it is meant that ecological economics goes beyond the conventional conceptions of scientific disciplines through: (a) trying to integrate and synthesize many different disciplinary perspectives; (b) focusing more directly on the problems, while ignoring arbitrary intellectual boundaries. Intellectual disciplinary tools are secondary to the goals of solving the critical problems of managing our use of the planet (Costanza et al. 1991). More recent contributions in the field of SESs (social-ecological systems) put emphasis on the implications of transdisciplinarity for sustainability research (Klein 2004; Hirsch Hadorn et al. 2006; Baumgartner et al. 2008; Faber 2008). "Transdisciplinarity means to reach out beyond science and to include aspects of practical contexts and values or normative judgements (sustainability, good-practice), as well as to feed back results into practical actions (politics, management)" (Baumgartner et al. 2008). In this perspective, transdisciplinary research is a bottom up process including participation and mutual learning for societal problem solving. Principles of integrating research and social change refers to another radical concept of science – in which theory and practice should be mutually beneficial – that is named *action research* (Hart and Bond 1995), i.e. an action-orientated research where action, research and education have to form an interlinked triangle. People's interpretation of reality and people's participation in the research process constitute the pillars for a successful action of both local knowledge building and assumption of responsibility for local management.

The hierarchical organisation of agriculture as a human activity system reveals that control for agroecosystem organisation depends on the interaction between an "internal controller", i.e. the farmer operating within the farm system, and more distant personal or institutional controllers, such as the market of input and output, governments, corporations, credit institutions, etc. Remote control has progressively expanded in a more and more globalised society. Unfortunately, these remote controllers cannot respond effectively to the positive and negative feedback that originate within cropping and farming systems themselves. Furthermore, the main goal of the remote controllers is obtaining the largest possible yield of a cash crop, not maintaining long-term productivity (Odum 1984). From this tension between local and distant controllers, with the latter being the most influential for determining local decisions, stem the reasons of environmental unsustainability of conventional agriculture.

In an early attempt to apply a hierarchical approach to sustainable agriculture, it was suggested that analysing agriculture as a hierarchical system is the appropriate way to incorporate different meaning of sustainability (Lowrance et al. 1987).

A hierarchical definition of sustainability was proposed where different disciplinary areas can meet at different levels of agriculture organisation. At the lower level of farm system organisation, agronomic sustainability meets with microeconomic sustainability and the farmer has to make trade-offs between them. Agronomic sustainability refers to the ability to maintain long-term productivity at field level, while microeconomic sustainability depends on the ability of the farmer to stay in business. At the upper levels, i.e. regional or national or international, ecological sustainability has to do with the maintenance of ecological services provided by SESs, including agriculture. Macroeconomic sustainability is regulated by factors such as international trade rules, national regime of land ownership, fiscal policies, interest rates, etc., which determine the viability of agricultural systems at national and international levels. In an ever more globalised socio-political-economic context, macroeconomic policy largely constraints microeconomic policy and decisions taken by the farmer, challenging both agronomic and ecological sustainability (Thompson 1985; Murdoch 1990).

In 1998, the scientific journal *Ecological Applications* devoted a 35-page Invited Features to "Ecology, the Social Sciences, and Environmental Policy". It was recognised that " new interdisciplinary connections will be required to conduct the needed research, to educate scientists and the public, and to ensure that the special expertise of ecological science is available to environmental decision-makers in all sectors of society" (Haeuber and Ringold 1998). While existing law generally does not speak in terms of ecosystems, ecosystems management or biodiversity, a legal perspective requires a better understanding of how the law can be employed to promote ecological and economic sustainability. However, "the challenge of integrating ecology into the law remains daunting" (Keiter 1998).

Sustainable development in agriculture means capability to harmonize goals for both the benefit of local rural society and the larger society, since no sustainable agriculture is possible without an effective convergence between the goals of a community of nations and those of a vital, local farmers' community (Buttimer 1998; Caporali 2008; Van Acker 2008; Wilson 2008). In an international scenario, i.e. in the whole planetary system of food production and delivery, agricultural development, poverty and environment merge in one big picture, that of a sustainable food system for the whole humanity (Hazell 1998; Van Acker 2008; Wilson 2008). "Progressive policy action must not only increase agricultural production, but also boost incomes and reduce poverty in rural areas where most of the poor live" (Rosegrant and Cline 2003). Since the ecosystem concept is a principle of organisation, it reveals in the case of agriculture how the performances of agroecosystems, such as sustainability, are strictly dependent on the principles of agroecosystem organisation, i.e. on the human factor that hierarchically determines agroecosystem performances at both local and global levels. Sustainable agriculture development requires a shift from an industrial to a multifunctional model of strong multifunctionality driven in each region and agricultural community by the existing governance structures (Van Acker 2008; Wilson 2008).

5.3.1 Factory Farms Versus Family Farms

In the modern, industrialised and globalised society, barriers to agriculture sustainability exist that are inherent to the principles of agriculture organisation themselves. These barriers need to be recognised and then amended. In any case, the epistemological question of "recognition" precedes the ethical question of "amendment". Agroecology, as a systemic or transdisciplinary science, can help in both these processes. Indeed, the whole human activity system of food production is not more currently definable with the narrower terms such as "agriculture" and "agri-food sector", but with the broader term "agribusiness". This term was coined in the United States to describe the vertical integration of certain companies to control the whole system of food production from seeds to fast-food outlets (Newby and Uttings 1984). In the already globalised scenario, powerful institutional private drivers, such as large corporations, are able to organise and control the food chain from the seed to the supermarket and from the local to the international levels. In this kind of organisation pattern, an increasing capital-intensive agribusiness is created, where farmers are made by agreements more and more dependent for the farm organisation on long-distance decision makers who buy the marketable farm output and sell the farm inputs. With this economic dependence, or "addiction" for economical survival, agriculture paradoxically becomes organised according to non-agricultural criteria, where production is only a step of a broader market-oriented business. The effects on a local level are of different kind, including reduced entrepreneurial autonomy of the farmer; farm specialisation and dependence on external inputs of seeds, machinery, agrochemicals and feedstuffs; enlargement of farm seize for complying with economies of scale; reduced number of smaller farmers for being excluded by market competition; reduced social and economic vitality at local level; contribution to the process of urbanisation of dismisses family farms. Effects on the global level are: concentration of economic power in the production of a basic human need like food; enlargement of disparity between rich and poor countries; technology control; dietary change; rise of global expenditure of energy for food transportation and increased pollution (Newby and Utting 1984; Murdoch 1990; Pretty et al. 2005; Fiala 2009). One of the most outstanding example of paradoxical contrast between economic and ecological outcomes is the CAFO (Confined Animal Feeding Operations) system for meat production. The CAFO system is economically regarded as the most advanced intensive feedlot system for livestock production, although it contributes large greenhouse gas emissions (Subak 1999; Steinfeld et al 2006). If the use of CAFOs is expanded, meat production in the future will still be a large producer of greenhouses gases, accounting for up to 6.3% of current greenhouse gas emissions in 2030 (Fiala 2008).

"As this large scale, capital-intensive and import-intensive agribusiness system is soaking up resources, the small-scale, labour intensive rural and urban industry - which would increase employment, economic well-being, economic linkages, and self-reliance – has largely been denied the resources and opportunity for growth" (Murdoch 1990).

Three main historical, widespread and convergent tendencies are recognised as driving forces of the agribusiness system patterning worldwide and barriers to agriculture sustainability: (1) "The Dual Economy" process at the international level due to the restructuring of the past colonial economy, where existing indigenous food-producing agriculture and local economy are replaced with exported-oriented agriculture and enclave; (2) the distortion between city and countryside, that has required the transfer of resources from the countryside to the city, to support the development of urban activities with cheap food policies whose purpose is to keep the wages of urban workers low; (3) the basic distortion in the distribution of land, where most part of agricultural land is owned by the least part of the landowners (Murdoch 1990; Kesavan and Swaminathan 2008). In most cases the dualism between rich and pour countries, city and countryside, and powerful landowners and powerless small farmers or landless people, or in other terms, the control of man upon man, and of man upon land not for care but for profit and exploitation, is the true cause of agriculture and rural unsustainability. Economic problems leading to ecological degradation and to social disintegration are widely witnessed (Hazell 1998; Kesavan and Swaminathan 2008; Gitau et al. 2009; Van Acker 2008).

"If the rich in distant places exploit ecosystem services and the local poor have to pay the costs, there are obvious equity implications – both internationally (South–North) and within countries (and indirectly inter-generational, as well)" (Munasinghe 2008).

"Today's conventional farmer ignores and overrides the natural ecosystem's heterogeneous characteristics with homogenized methods. He or she emulates the factory manager by standardizing procedures and technology to achieve results in mass quantity", instead, "the most endurable agriculture…is that which deviates least from the energy flows and nutrient cycles of the natural communities" (Bidwell 1986). Traditional family farms are instead credit with large consensus about their sustainability performance. They provide services such as better care of the soil, more energy and nutrient efficiency, social and economic support necessary to maintain vitality of nearby towns and cities, economic competition to avoid concentration of production on a few large farms which would practice monopoly pricing and raise food costs. As a conclusion, "society would be better off if publicly supported research and extension education were focused on small farms" (Tweeten 1983).

Traditional agroecosystems have been acknowledged as complex dynamic co-evolutionary processes generating adaptive plant–animal–human interactions (Harlan 1975; Oldfield and Alcorn 1987). The habitat complexity of traditional farms maintains high genetic variation due to temporal and spatial changes in selection pressure, inter-genotypic and interspecific competition, and interactions with pests and pathogens. Much of the world's biological diversity has been created by farmers and has been in the custody of farmers through millennia. Unlike modern farmers who do not use the harvest from their fields for seed, traditional farmers are both seed stock producers and crop breeders. Traditional agriculture generated variable, integrated and adapted populations as result of crop evolution, called *landraces*, that are regarded as an essential part of our crop genetic heritage (Harlan 1975; Oldfield and Alcorn 1987; Cleveland et al. 1994). With the advent of the "green revolution" process and the replacement of traditional agriculture with commercial agriculture, a rapid process of extinction or depletion of landraces, or "genetic erosion", has occurred, although genetic diversity is essential for evolution in nature as well as for improvement by plant breeding (Harlan 1975; Miller 1973; Damania 1994). Unfortunately, modern cultivars created for maximising production in ideal cultural conditions have often not met farmers' needs under the less optimal conditions of real farms (Janssens et al. 1990). In most cases, there is a conflict between optimization and maximization, while for the development of sustainable agroecosystems a balance for more resilience and long-term stability in the local conditions would be a better requirement than attempts to maximize outcomes of single processes and the glorification of this maximization (Janzen 1973). Modern trends for introducing genetically modified crops as cultivars for primary productivity maximization seem to follow the same logic of development by imposed homogenization of people and environment globally, while the strength of sustainability is on the contrary diversity improvement for local adaptation. The agribusiness globally extended contrasts with the principles of sustainability that move from the local to the global and not the way back (Caporali 2008). Breeding for low-input farming systems is being recently reconsidered in the framework of an unconventional evolutionary participatory breeding (EPB) that provides a partnership between plant breeders and local farmers, with the aim of developing high quality, genetically diverse, modern landraces (Murphy et al. 2005). "The involvement of farmers in the breeding process not only adds value to the conservation of local crop diversity but also helps maintain and enhance farmers' knowledge about how to select and manage local crop populations, and manage seed supply systems through social seed networks" (Sthapit et al. 2008).

Livestock are very important components of the livelihoods of at least 70% of the world's poorest people (Anderson 2003). The poor that live in unfavourable agricultural areas depends largely on the performances of their livestock keeping systems. For these systems to perform well, animal genetic resources (An GR) are needed that are flexible, resistant and diverse (Table 1).

Availability of low-priced external input for bettering the unfavourable environment has often led to trading-off "local" breed for "improved" breed, causing genetic erosion and loss in animal genetic diversity. "The relatively small gene pool of domestic AnGR (6,000–7,000 breeds of 40 species) is threatened by extinction, principally through crossbreeding and breed replacement" (Rege and Gibson 2003). Improved breed for productivity traits are often risky due to their lack of adaptive traits and subsequent poor reproductive rates and survivability, which

Table 1 Livestock biodiversity traits and functions for sustaining and improving the livelihoods of the poor (Anderson 2003)

Adaptive breed characteristics	Functions
Productivity, reproductivity rate, climatic tolerance, disease resistance	Regular cash income from sales of animals or their products
Size, power, docility, walking ability, water requirements	Input and services to crop production and other non-income functions
Appearance traits, such as hide and skin colour, horn size and shape, etc.	Social and cultural functions that provide status and identity

undermines the sustainability for the livestock system itself (Anderson 2003). There is an urgent call for the development and application of economic and policy tools to aid rational decision-making in the management of the global domestic AnGR (Rege and Gibson 2003; Mendelsohn 2003).

It is largely recognised that for accomplishing the dual objectives of sustainable agriculture, i.e. improving yield levels and food stability and preserving the quality and quantity of ecosystem services, appropriate incentives are needed. While in the past decades incentives have favoured increased agricultural production at the expense of ecosystem services, now many countries have instituted forms of "green payments", that is payments to farmers who adopt sustainable or environmentally benign farming practices (Tilman et al. 2002).

"Agriculturalists are the *de facto* managers of the most productive lands on Earth. Sustainable agriculture will require that society appropriately rewards ranchers, farmers and other agriculturalists for the production of both food and ecosystem services" (Tilman et al. 2002).

An economic investment in safeguarding the provision of ecological services by agriculturalists is now regarded as an important step to an "evergreen revolution" rooted in the principles of ecology, economics and social and gender equity (Kesavan and Swaminathan 2008; Gitau et al. 2009).

"What nations with small farms and resource-poor farmers need is the enhancement of productivity in perpetuity, without associated ecological or social harm" (Kesavan and Swaminathan 2008).

In the framework of agricultural sustainability, one of the most popular kind of subsided agriculture worldwide is *Organic Agriculture* (Lampkin 1990). It has been supported by an international body – IFOAM or International Federation of Organic Agriculture Movements – established in 1972 and regulated in Europe by law (EEC Reg. 2092/91; EC 834/2007), as in many other countries. Regulation by law means that (a) a series of technical implementation rules or standards detailing the disciplinary of production is identified; (b) credit is given to national inspection bodies for the control and certification of the productions that result in being labelled and are therefore recognizable as organic products by the buyers. The main agroecological criteria and technical rules that comply with the definition of the standards for organic farming are summarized in the following ten rules (Caporali 2004):

1. To create diversity within the farm
2. To integrate plant production with livestock
3. To adopt soil conservation measures and minimum tillage practices
4. To adopt crop rotations
5. To adopt intercropping and cover cropping
6. To use genotypes resistant to parasitic attacks
7. To treat the soil with manure and composted organic matter
8. To practice green manuring
9. To foster the biological control of weeds, phytofagous insects and phytopathogens
10. To plant and protect hedges

It is expected that organic farming promotes the re-establishment of environmental and social balances related to food quality, local economy and human and environmental health. The precondition essential for the implementation of organic farming is the maintenance of the rural population on the spot in a satisfactory economic and social condition that is favourable for farmers themselves and society as a whole. There are many advantages that society draws from the maintenance of an active agricultural population on the territory. They deal with:

(a) More favourable balances of population density
(b) A better distribution of animals on the territory
(c) The improvement of the agricultural landscape
(d) The improvement of the functionality of the agro-ecosystems
(e) The maintenance of the quality of natural resources as well as agricultural productions
(f) The enhancement of local economy through the direct employment and the strengthening of induced economical activities (handicraft, trade, tourism, etc.)

When the agricultural population resides on the rural territory and labour is not lacking, the maintenance and the aesthetic quality of the landscape is favoured. In such a situation both the human and territorial resources co-evolve in a symbiotic relationship with mutual benefit and increase in wealth and health. Policies that do not pursue the local maintenance of rural population are not far-reaching and undermine the foundation of social and environmental sustainability (Caporali 2004).

The establishment of organic farming in the local context guarantees consumers the origin of food that, in a system dominated by market globalisation, is hardly ascertainable today. A closer interaction between farmers and consumers can re-qualify both the quality of products and the quality of the production process. Indeed, organic farming can be a major factor for strengthening the protection of typical products and food traditions in a context of the turn to "the economy of qualities" (Fonte 2006). The link between typical products, i.e. produced locally, and their cultural identity is being widely advertised by movements such as Slow Food: "Quality rests in the small producers, into those realities that contribute to making quality and to forming a network. If agro-food products remain connected to the territory, they are not only agro-food products, but they become the cultural identity of our population" (Carlo Petrini, cited in Ferretti and Magaudda 2006).

The latest report by IFOAM (International Federation of Organic Agriculture Movements) mentions increasing annual rates for organic farming and shows that 32.2 millions of hectare were certified as "organically grown" in 2007 with more than 1.2 millions of farmers involved in the world (IFOAM 2007).

The global debate about agriculture sustainability has enormously enlarged the cultural landscape for mutual criticism between different disciplinary, traditionally separate areas. The area of agroecology enquiry is now really operating as "glue" at a transdisciplinary level, bridging the gap between different disciplines and between theory and practice of agriculture.

6 Agroecology Research and Teaching for Agriculture Sustainability

Sustainable agriculture research requires that scientists work with each other and with farmers, although there has been little successful institutionally-designed, multidisciplinary work to date (MacRae et al. 1989; Flora 1994). Agroecology can meet these new research expectations as an "integrative" science that looks for relationships in order both to promote more understanding of the agricultural reality and to improve agroecosystem management (Caporali 2007). Agroecology links theory with practice by integrating paradigms and techniques of ecology with agricultural practices (Edwards et al. 1993).

The field of agroecological enquiry is defined by its epistemological tool, i.e. the concept of agroecosystem. Indeed, the whole range of interest for research in agroecology stretches along the keyword sequence: ecosystem→ agroecosystem→ sustainable agriculture. Agroecological research can be carried out at any level of agroecosystem hierarchy, but the inherent goal of any kind of research should be the search for sustainability (Lowrance 1990). Since agroecosystem sustainability depends on human planning and management, a strong element of ethics emerges as a fundamental responsibility component of agroecological research. As a consequence, research in agroecology integrates also ethical principles both as efficient and final causes of its process of development (Caporali 2007).

A strategy of successful research in agroecology should be based on the following steps (Edwards et al. 1993):

1. Description of the target agroecosystem including its goals, boundaries, components
2. Functioning, interactions among components, and interactions across its boundaries
3. Detailed analysis of the agroecosystem to determine factors which limit or could contribute to attainment of productive and social goals
4. Design of interventions and identification of actions to overcome the constraints
5. On farm experimental evaluation of interventions
6. Review effectiveness of newly designed systems
7. Redesign as necessary

All this approach could be defined as participatory research, involving not only researchers but also farmers in the whole process, from planning to implementation and evaluation . All steps should be conducted on farms by an interdisciplinary team of agricultural, social and ecological scientists and with full participation of farmers. Understanding the farmer's goals is crucial, as the role of the proposed interventions is to help the farmer attain these goals.

6.1 Measuring Agriculture Sustainability Through Indicators

Political recognition of agriculture as a multipurpose human activity system requires more intellectual and financial investment in research for monitoring and measuring sustainability performances. This is necessary in order to appropriately inform decision-making processes. This necessity is well documented in international agreements, such as the Agenda 21 (UN 1992). In the chapter 8 of Agenda 21 "Integrating environment and development in decision-making", it is stated that: (a) prevailing systems for decision-making in many countries tend to separate economic, social and environmental factors at the policy, planning and management levels; (b) there is the necessity for a better integration among national and local government, industry, science, environmental groups and the public in the process of developing effective approaches to environment and development, (c) responsibility for bringing about changes lies with governments in partnership with the private sector and local authorities, and in collaboration with national, regional and international organisations; (d) the overall objective is to improve o restructure the decision-making process so that consideration of socio-economic and environmental issues is fully integrated and a broader range of public participation is assured. Moreover, for improving planning and management systems, it is recommended that "countries could develop systems for monitoring and evaluation of progress towards achieving sustainable development by adopting indicators that measure changes across economic, social and environmental dimensions". Indeed, such agriculture sustainability indicators (ASIs) are to be regarded as transdisciplinary tools for integrating knowledge into meaningful numerical forms that express relationships within and between agroecosystems components and processes (Smith et al. 2000).

As a follow up to the Agenda 21, the OECD Council approved in 1991, a "Recommendation on Environmental Indicators and Information" to further develop sets of reliable, readable, measurable and policy-relevant environmental indicators. The conceptual model inspiring the search for agri-environmental indicators (AEIs) was defined by OECD as Driving Force-State-Response (DSR) framework (OECD 1999).The DSR framework denotes much of its agroecological foundation. It can provide a flexible framework to improve understanding of the complexity of linkages and feedbacks between the causes and effects of the agriculture–environment relationship and the responses of the main stakeholders. Therefore, analysis of the linkages and feedbacks between driving forces, state and responses is a key element in shedding light on the dynamic functioning of agriculture as a human activity system.

The search for useful indicators is an evolving field of enquiry depending on societal pressures and political choices (Boody et al. 2005). Some environmental areas are gaining increasing importance, e.g. soil greenhouse gas sinks (Marinari and Caporali 2008). Due to the crucial linkages between policies, agricultural production and environmental quality, the interpretation of any one indicator may need to be complemented with other indicators and be seen within the overall context of the set or appropriate sub-set of indicators (OECD 2001).

Agriculture sustainability indicators (ASIs) have been largely promoted in research as a necessary instrument for monitoring agroecosystems' performances, facilitating judgements and suggesting solutions of improvement at different scales of enquiry (Caporali et al. 1989, 2003; Tellarini and Caporali 2000; OECD 1997,1999, 2001; van der Werf and Petit 2002; Bockstaller and Girardin 2003; Piorr 2003; Gerbenss-Leenes et al. 2003). Generally, ASIs have been developed on the base of the input/output model of agroecosystem representation (Edwards et al. 1993; Tellarini and Caporali 2000). ASIs can be subdivided into two large categories: (a) structural indicators and (b) functional indicators, the former describing the most relevant agroecosystem components and the latter the transformation process efficiency in agroecosystems (Tellarini and Caporali 2000). Indicators can be calculated in terms of energy, materials, information and monetary values. By relating each type or combination of output to each type or combination of input, it is possible to obtain a considerable amount of cross-information on the resource use within agroecosystem and between agroecosystems. In this way, it is possible to evaluate how the agroecosystem organisation influences the extent of dependency or autonomy of the whole agroecosystem and its parts on renewable or non-renewable inputs.

Research based on ASIs at the farm level is of great importance in informing those responsible for the decision-making processes, especially when groups of farms of contrasting management are involved. Organic farming systems are being considered as a long term benchmark for the evaluation of apparently environmentally benign agricultural production systems (OECD 1999). Therefore, the aim of some recent research includes the comparison of organic farming systems with conventional ones on the base of appropriate ASIs (Reganold et al. 2001; Mader et al. 2002; Caporali et al. 2003). Results of this kind of research can be easily shown graphically, with the help of a so-called sustainability polygon or web, which simultaneously displays scores for different indicators and avoids having to aggregate across different scales.

Several studies suggest that farmlands with higher landscape heterogeneity have higher biodiversity and more potential to build environmental sustainability (Olson et al. 1995; Ovenden et al. 1998; Stoate et al. 2001). Generally, a more heterogeneous landscape is one with a higher diversity of cover types and a higher complexity of their spatial patterning. In order to develop effective agri-environmental policies for biodiversity enhancement, it is important to determine which heterogeneity measures at landscape level are the most appropriate as the best indirect biodiversity indicators (Roschewitz et al. 2005; Caporali et al. 2009). With the development and use of ASIs at different hierarchical scales, agroecological research is getting more and more integrated in the structure of civil society, improving its role of scientific service for public utility.

6.2 Agroecology and Curriculum Development

In the context of higher educational systems, like University, a nested hierarchy of organisational levels can be detected from single scientific disciplines and their aggregates, e.g. degree courses, to the whole faculty or department. In the current academic organisation, agroecology as a science occupies different levels, from the discipline to the degree or the doctoral course levels (Caporali 2007; M.A. Altieri 2008, personal communication). This multilayered dimension denotes that its transdisciplinary method, i.e. the system paradigm, and its main goal, i.e. agriculture sustainability, are so strong and pervasive as to provide a broader and shared platform of content for teaching and research at the highest level of education (Caporali 2007; Lieblein and Francis 2007).

Agroecology is a systems science and its teaching/learning process at different hierarchical levels of higher education must reflect its systems methodology even in the curriculum structure. Agroecology as a systems science has potential for constructing more integrated academic curricula as well as promoting more integrated research. This potential for integration meets practical needs since a sustainable development in agriculture demands capacity of integration within and between levels of agroecosystem organisation from the field to the regional and global levels. The concept of integrated rural development has been recently created to revitalise rural environment and economy (Hampicke 2006).

At the discipline level, aspects of integration refer to the capacity to see even a single discipline as an elementary educational system, where the teaching/learning process occurs with reciprocal interactions among all the human components involved – teachers, students and other stakeholders in their context of action. A detailed description of the agroecological content can be found in Altieri and Francis (1992). A list of seminal publications dealing with the knowledge body of agroecology is provided by Francis et al. (2003).

At the curriculum level, agroecology has a role to play in society in order to provide agriculturalists with a culture of sustainability. In Caporali (2007), a model is provided representing the relationship between agroecology as a learning system at the curriculum level and agriculture as a real human activity system to be orientated toward sustainability. In this case, integration is a more complex issue in that has to do with harmonisation of relationships with both the other disciplines of the curriculum and the external context components. The transdisciplinary character of agroecology as a systems science emerges just in this potential for connection both of different disciplines by establishing internal coherence, and of theory and practice, by establishing external correspondence. Coherence means shared goals and methods of investigation; correspondence is the capacity to achieve goals through successful action (Röling 2003).

The relevance of a systems approach to the process of curriculum review in agriculture higher education was originally advocated and supported at Hawkesburry, one of Australia's oldest agriculture colleges. In this case, curriculum innovations would have had to challenge the mismatch between the competencies of graduates and the needs of the agricultural system (Bawden and Valentine 1985). According to this view, taking a systems approach to investigating problem situations provides a more useful paradigm for learning about agriculture than reductionist, discipline-based approaches (Bawden 1992; Bawden et al. 1984; Valentine 2005). The challenge was to provide programmes of learning appropriate to the complex issues of agriculture as a human activity system. Learning models appropriate to this approach were developed drawing heavily on the concept of systems thinking/systems practice relationship, experiential learning and problem solving (Checkland 1981; Kolb 1984; Maddison 1982; Bawden 1991). In this framework, transdisciplinarity implies full interaction between disciplines from a problem-based perspective.

Indeed, theory and practice are not separable in a systems view, but are mutually reinforcing in the processual construction of the knowledge–action cycle.

External methodological tools help introduce a broad concept of faculty and action-based learning. Integrating the expertise of farmers, business owners, government specialists, and non-profit- groups can enrich the educational process by offering different perspectives and ways of knowing (Francis et al. 2001). Case studies, interview and survey techniques, time-series measurements, and activity calendars can be taught and applied to answer questions about integration within the whole agro-ecosystem hierarchy, i.e., cropping systems – farming systems – regional systems – global systems. These approaches require several changes in attitude and organisation. Faculty members, administrators and others stakeholders must invest time and money to establish participatory research and learning opportunities (Stark 1995).

Tools are also needed in order to give more internal coherence to a curriculum. This means not only integration among the disciplines that belong to it, but also more integration between teachers and students, that are the basic components of a learning system such as a curriculum. Creating a truly integrated curriculum entails that the two groups become reciprocal members of a shared, mutually self-critical learning community. This can be achieved through: (a) creating a community that generates conversation, where knowledge is a process of continual negotiation and transformation; (b) creating a team-teaching context, since those involved find their intellectual life much enriched; (c)implementing intensive programmes or courses allocated in a short time, i.e. 2 or 3 weeks, since they can function as more flexible didactic tools for approaching different contextual experiences, provided that points (a) and (b) are met (Manley and Ware 1990; Francis and King 1994; Caporali 2007).

The outputs of a degree course can be described in terms of achievements to be pursued at a personal and an institutional level. More academic staff responsibility and more general societal benefits for public and private institutions are expected from improved networking and local sustainable development strategies. Advocates of new curricula in agroecology claim that the new epistemological, ontological and pedagogical tools based on a systems paradigm may allow university to successfully address the challenge of establishing new cultural basis for a sustainable development in agriculture and society (Francis et al. 2003; Lieblein and Francis 2007; Caporali 2007).

7 Outlook on Agroecology

According to alarming forecasts, agriculturally driven global environmental change can cause further habitat destruction involving 10 billions more hectares of natural ecosystems that would be converted to agriculture by 2050 (Tilman et al. 2001). This would cause unprecedented ecosystem simplification, loss of ecosystem services, species extinctions, eutrophication and pesticide pollution. Knowledge advances and regulatory, policy and technological changes are needed to control the environmental impact of agricultural expansion (Tilman et al. 2001). We really need an "Ecology for a Crowded Planet" (Palmer et al. 2004).

Agroecology is a systems science that uses the agroecosystem concept as an epistemological tool for investigating the relationships between agriculture and society. The main aim of agroecology is to produce knowledge for improving agroecosystem sustainability. Real agroecosystems are products of co-evolution between man and nature. New agroecological knowledge has proved that conventional agriculture in a conventional economy is no longer sustainable. The challenge for the future is to develop new patterns of human organisation for making the whole agri-food system sustainable. To be sustainable in a context of increasing human population and diminishing natural resources, agroecosystems must provide for human nutrition and other needs, while maintaining natural ecological services as soil fertility, biodiversity renewal, water and nutrient cycling, clean air, etc. Due to human choice, all the purposeful relationships between society and environment are regulated by economic rules. The cultural challenge for the future is to create a real area of hierarchical transdisciplinary knowledge for governance at every level of spatial-temporal scale, where economics is adapted to an ecological framework.

Agroecology can help create a purposeful "ecological knowledge system" as a driving force for sustainable development (Röling and Jiggins 1998). The dimensions of an ecological knowledge system are: (a) ecologically sound practices; (b) learning; (c) facilitation; (d) support institutions and networks; (e) conducive policy contexts. "The change to ecological sound farming is not only the outcome of technical intervention, but especially also a negotiated outcome based on accommodation among paradigms, coalitions, institutional interests and politics" (Röling and Jiggins 1998). Sustainable agricultural development is a true research topic to be pursued with a transdisciplinary method that is a combination of interdisciplinary and participatory approaches (Vandermeulen and Van Huylenbroeck 2008). By creating a transdisciplinary platform for action, boundaries of research disciplines are broken down into a systemic perspective that converge on problem solving, while a dialogue is created between stakeholders in the civic society. In this way, the research results can be more easily transformed into implementations, because people involved are included in the decision making process.

University, as a leader cultural institution in society, should take responsibility for adopting organisation patterns that comply with the end of networking for sustainability with stakeholders in the agri-food system (Perez and Sanchez 2003). Communication and extension are inclusive parts of a knowledge system of social learning to be put in action for constructing a culture of sustainability in agriculture and society (Warner 2006). For instance, agro-environmental partnerships have emerged in California as the primary strategy for extending alternative, agroecological knowledge in conventional agriculture. Partnerships are purposeful, multi-year relationships among growers, their organisations and scientists with the end to create and extend agroecological knowledge through negotiation of goals for research, education, production and field-scale demonstration (Warner 2006). "These partnerships are decidedly local in their attention to farming practices and environmental resource conservation, but they are global in their marketing ambitions" (Warner 2007).

However, there persists a gap of knowledge between ecosystem ecology and industrial agriculture that agroecology can contribute to bridging. The state-of-the-art is that our current Western culture of land commodification has produced farmers involved in industrial agriculture as business people. Their primary goal is to produce with a reasonable profit, whereas stewardship of resources is not usually their prime objective (Robertson 2000). As a consequence, many of the most serious environmental problems, such as pesticides pollution and biodiversity loss through habitat fragmentation, manifest themselves at the landscape scale (Baudry 1989). However, much of the research on the agricultural systems is done at the field level as many of the institutional programmes of agricultural policy operate at the farm scale. Agroecologists should look for more appropriate methods of communication between stakeholders and try to bridge the gap between scales and objectives of agroecosystem management (Christensen et al. 1996; Lautenschlager 1998; Norton 1998; Sutherland 2002; Rosegrant and Cline 2003; Otte et al. 2007; Merckx et al. 2009). For instance, experience of agroecosystem management initiatives in Australia, such as those of the "Landcare" programme and the "Total Catchment Management", served as an important forum for exchanging information, establishing demonstration findings and transferring research findings to farmers and the broader community, in order to achieve the awareness that rural ecosystems need to be managed both at the farm and landscape scale (Robertson 2000).

In Europe, new institutional initiatives for social networking and learning are included in the "Lifelong Learning Programme" (LLP), which is the overall framework of academic co-operation in the field of higher education and training. The "Leonardo" section of LLP provides funds for co-operation between universities and public and private institutions for developing initiatives of participatory research, education and training, with the main aim of developing knowledge, capacity and skills for trainers and trainees of human activity systems, including agriculture. Cooperation between small and medium enterprises (SMEs) and research centres can make a productive tissue more competitive and dynamic (Martin 2003). Effective positive experiences in the cooperation between rural small and medium enterprises and research institutions are documented in central Italy as pioneer examples of knowledge building and improved evolution of partnership in rural areas (Cannarella and Piccioni 2005).

An innovative proposal for the institution of an international curriculum in agroecology has been launched recently (Altieri 2007). "The idea is to train a critical mass of students from different parts of the world with the skills to deal with the intricacies of sustainable systems and to guide agriculture in various temperate and tropical regions through a path that sustains productivity while conserving natural resources, biodiversity and cultural traditions, in socially equitable and economically viable ways" (Altieri 2007). Beyond the agroecological basis for a sustainable agriculture production, processing and marketing, the new programme would also train students in participatory methods of research and development. Basic elements of the new student profile would be: (a) theoretical background in agroecosystem design and management for site-specific sustainable production; (b) grounding in quantitative methods to evaluate performance of agroecosystems with sustainability indicators; (c) relational skills for participating in multidisciplinary teams and in participatory processes; (d) capacity to appreciate traditional forms of agriculture and mobilize local human resources into scaling-up processes of agroecological initiatives; (e) capacity to understand and promote linkages in economic, social, cultural and political processes conducive to sustainable development locally and, hopefully, internationally. The programme would produce graduates well-equipped for careers in both local and international organisations of agricultural policy analysis, rural development, research, extension and consulting.

Agroecology is an open area of knowledge convergence and development of different cultural, scientific, philosophical, ethical, aesthetic, political and socio-economic interests. For such a nature, it is a cross-fertilisation or transdisciplinary meeting point where new knowledge "emerges" from dialogue. Agroecology is currently a democratic option both for connecting people in a naturally connected world and for developing a culture of sustainability.

References

Alomar O, Goula M, Albajes R (2002) Colonization of tomato fields by predatory mirid bugs (Hemiptera: Heteroptera) in northern Spain. Agric Ecosyst Environ 89:105–115

Altieri MA (1987) Agroecology, the scientific basis of alternative agriculture. Westview, Boulder, CO

Altieri MA (1991) How best can we use biodiversity in agroecosystems? Outlook Agric 20:15–23

Altieri MA (1995) Agroecology. In: Encyclopedia of environmental biology, vol. 1. Academic, New York, pp 31–36

Altieri MA (1999) The ecological role of biodiversity in agroecosystems. Agric Ecosyst Environ 74:19–31

Altieri MA (2007) How to teach agroecology: a proposal. Project, 47–55. Amersfoort, ILEIA

Altieri MA, Francis CA (1992) Incorporating agroecology into the conventional agricultural curriculum. Am J Altern Agric 7(1–2):89–93

Altieri MA, Letourneau DK, Davis JR (1983) Developing sustainable agroecosystems. BioScience 33:45–49

Altieri MA, Liebman M (1986) Insect, weed, and plant disease management in multiple cropping systems. In: Francis CA (ed) Multiple cropping systems. Macmillan, New York, pp 183–218

Andersen P (1997) Densities of overwintering carabids and staphylinids. (Col., Carabidae and Staphylinidae) in cereal and grass fields and their boundaries. J Appl Entomol 1212:77–80

Anderson M (1991) Book reviews. Am J Altern Agric 6(1):40–42

Anderson S (2003) Animal genetic resources and sustainable livelihoods. Ecol Econ 45:331–339

Anderson-Wilk M (2008) The gap between cover crop knowledge and practice. J Soil Water Conserv 63(4):96A

Armitage ER (1974) The runoff of fertilizers from agricultural land and their effects on the natural environment. In: Irvine DEG, Knights B (eds) Pollution and the use of chemical in agriculture. Butterworths, London, pp 43–60

Atallah T, Lopez-Real JM (1991) Potential of green manure species in recycling nitrogen, phosphorus and potassium. Biol Agric Hort 8:53–65

Auclair AN (1976) Ecological factors in the development of intensive management ecosystems in the Midwestern United States. Ecology 57:431–444

Badaruddin M, Meyer DW (1990) Green-manure legume effects on soil nitrogen, grain yield, and nitrogen nutrition of wheat. Crop Sci 30:819–825

Baldock JO, Higgs RL, Paulson WH, Jakobs JA, Shrader WD (1981) Legume and mineral N effects on crop yields in several crop sequences in the Upper Mississippi Valley. Agron J 73:885–890

Bath B, Malgeryd J, Richert Stintzing A, Akerhielm H (2006) Surface mulching with red clover in white cabbage production. Nitrogen uptake, ammonia losses and the residual fertility effect in ryegrass. Biol Agric Hort 23:287–304

Batish DR, Kohli RK, Jose S, Singh HP (eds) (2008) Ecological basis of agroforestry. CRC Press, Boca Raton, FL

Batra SWT (1982) Biological control in agroecosystems. Science 215:134–139

Baudry J (1989) Interactions between agricultural and ecological systems at the landscape scale. Agric Ecosyst Environ 27:119–130

Baumgartner S, Becker C, Frank K, Muller B, Quaas M (2008) Relating the philosophy and practice of ecological economics: the role of concepts, models, and case studies in inter- and transdisciplinary sustainability research. Ecol Econ 67:384–393

Bawden RJ (1991) Systems thinking and practice in agriculture. J Dairy Sci 74:2362–2373

Bawden RJ (1992) Systems approaches to agricultural development: the Hawkesbury experience. Agric Syst 40:153–176

Bawden RJ, Valentine I (1985) Learning to be a capable systems agriculturalist. Programmed Learn Educ Technol 21:273–287

Bawden RJ, Macadam RD, Packham RJ, Valentine I (1984) Systems thinking and practices in the education of agriculturalists. Agric Syst 13:205–225

Bellostas N, Nielsen-Hauggaard HA, Andersen MK, Jensen ES (2003) Early interference dynamics in intercrops of pea, barley and oilseed rape. Biol Agric Hort 21:337–348

Bergkvist G (2003) Influence of white cover traits on biomass and yield in winter wheat- or winter oilseed rape-clover intercrops. Biol Agric Hort 21:151–164

Bidwell OW (1986) Where do we stand on sustainable agriculture? J Soil Water Conserv 41(5):317–320

Bockstaller C, Girardin P (2003) How to validate environmental indicators. Agric Syst 76:639–653

Boody G, Vondracek B, Andow DA, Krinke M, Andow D, Westra J, Zimmerman J, Welle P (2005) Multifunctional agriculture in the United States. BioScience 55(1):27–38

Boulding KE (1971) The economics of the coming spaceship earth. In: Holdren JP, Ehrlich PR (eds) Global ecology. Readings toward a rational strategy for man. Harcourt Brace Jovanovich, New York, pp 180–187

Boyden S, Dovers S (1992) Natural-resources consumption and its environmental impacts in the western world. Impacts of increasing per capita consumption. Ambio 21(2):63–69

Bruno FJ, Stachowicz JJ, Bertness MD (2003) Inclusion of facilitation into ecological theory. Trends Ecol Evol 18(3):119–126

Bruulsema TW, Christie BR (1987) Nitrogen contribution to succeeding corn from alfalfa and red clover. Agron J 79:96–100

Burdon JJ, Shattock RC (1980) Disease in plant communities. Appl Biol 5:145–219. Academic, London

Buttimer A (1998) Close to home. Making sustainability work at the local level. Environment 40(3):13–40

Cameron KC, Wild A (1984) Potential acquifer pollution from nitrate leaching following the plowing of temporary grassland. J Environ Qual 13(2):274–278

Campiglia E, Caporali F, Paolini R, De Sanctis D, Anelli G (1991) Yield quality aspects of the hazelgrove (*Corylus avellana* L.) agroecosystem in central Italy. Agric Mediterr 121:1–7

Cannarella C, Piccioni V (2005) Knowledge building in rural areas: experiences from a research centre-rural SME scientific partnership in Central Italy. Int J Rural Manage 1(1):25–43

Caporali F (1991) Ecologia per l'Agricoltura. Utet-Libreria, Torino

Caporali F (2000) Ecosystems controlled by man. In: Frontiers of Life, vol. 4. Academic, New York, pp 519–533

Caporali F (2004) Agriculture and health. The challenge of organic farming. Editeam, Cento (FE), Italy
Caporali F (2007) Agroecology as a science of integration for sustainability in agriculture. Ital J Agron/Riv Agron 2(2):73–82
Caporali F (2008) Ecological agriculture: human and social context. In: Cini C, Musu I, Gullino ML (eds) Sustainable development and environmental management. experiences and case studies. Springer, Dordrecht, The Netherlands, pp 415–429
Caporali F, Campiglia E (2001) Increasing sustainability in Mediterranean cropping systems with self-reseeding annual legumes. In: Gliessman SR (ed) Agroecosystem sustainability. Developing practical strategies. CRC Press, New York, pp 15–27
Caporali F, Onnis A (1992) Validity of rotation as an effective agroecological principle for a sustainable agriculture. Agric EcosystEnviron 41:101–113
Caporali F, Nannipieri P, Pedrazzini F (1981) Nitrogen content of streams draining an agricultural and a forested watershed in central Italy. J Environ Qual 10(1):72–76
Caporali F, Nannipieri P, Paoletti MG, Onnis A, Tomei PE, Telarini V (1989) Concepts to sustain a change in farm performane evaluation. Agric Ecosyst Environ 27:579–595
Caporali F, Mancinelli R, Campiglia E (2003) Indicators of cropping system diversity in organic and conventional farms in Central Italy. Int J Agric Sustain 1:67–72
Caporali F, Campiglia E, Mancinelli R, Paolini R (2004) Maize performances as influenced by winter cover crop green manuring. Italian J Agron 8(1):37–45
Caporali F, Mancinelli R, Campiglia E, Di Felice V, Vazzana C, Lazzerini G, Benedetti A, Mocali S, Calabrese J (2009) Indicatori di Biodiversità per la Sostenibilità in Agricoltura. ISPRA, Roma, Italy
Carson R (1962) Silent spring. Penguin Books, England
Carroll CR, Vandermeer JH, Rosset PM (1990) Agroecology. McGraw-Hill, New York
Carruthers P (1990) The prospects for agroforestry: an EC perspective. Outlook Agric 19(3):147–153
Checkland PB (1981) Systems thinking, systems practice. Wiley, New York
Chiariello N, Hickman JC, Mooney HA (1982) Endomycorrhizal role for interspecific transfer of phosphorus in a community of annual plants. Science 217:941–943
Chisci G (1980) Phisical soil degradation due to hydrological phenomena in relation to change in agricultural systems in Italy. Ann Istituto Sperimentale Stud Difesa Suolo 11:271–283
Christensen NL, Bartska AM, Brown JH, Carpenter S, D'Antonio C, Francis R, Franklin JF, MacHaon JA, Noss RF, Parson DJ, Peterson CH, Turner MG, Woodmansee RG (1996) The report of the Ecological Society of America Committee on the scientific basis for ecosystem management. Ecol Appl 6(3):665–691
Clark A (2007) Managing cover crops profitability, 3rd edn, Handbook series book 3. Sustainable Agriculture Network, Beltsville, MD
Clark AJ, Decker AM, Meisinger JJ, McIntosh MS (1997a) Kill date of vetch, rye, and vetch-rye mixture: I. Cover crop and corn nitrogen. Agron J 89:427–434
Clark AJ, Decker AM, Meisinger JJ, McIntosh MS (1997b) Kill date of vetch, rye, and vetch-rye mixture: II. Soil moisture and corn yield. Agron J 89:427–434
Clark EH II (1985) The off-site costs of soil erosion. J Soil Water Conserv 40:19–22
Cleveland DA, Soleri D, Smith SE (1994) Do folk crop varieties have a role in sustainable agriculture? BioScience 44(1):740–750
Colacino M (2001) Andamento del clima in Italia negli ultimi cinquanta anni. Legno Cellulosa Carta 7(2):8–13
Coleman DC (1989) Ecology, agroecosystems, and sustainable agriculture. Ecology 70(6):1590
Conway GR (1987) The properties of agroecosystems. Agric Syst 24:95–117
Costanza R, Daly HE, Bartholomew JA (1991) In: Costanza R (ed) Ecological economics: the science and management of sustainability. Columbia University Press, New York, pp 1–20
Crews TE (2005) Perennial crops and endogenous nutrient supplies. Renew Agric Food Syst 20(1):25–37
Cuppari P (1862) Saggio di ordinamento dell'azienda rurale. Cellini, Firenze

Daily GC (1997) Nature's services: societal dependence on natural ecosystems. Island, Washington
Dale T, Brown GF (1955) Grass crops in conservation farming. Farmers'Bullettin N° 2080. USDA, Washington
Dalgaard T, Hutchings NJ, Porter JR (2003) Agroecology, scaling and interdisciplinarity. Agric Ecosyst Environ 100:39–51
Daly HE (1991) Elements of environmental macroeconomics. In: Costanza R (ed) Ecological economics: the science and management of sustainability. Columbia University Press, New York, pp 32–46
Daly HE, Farley J (2004) Ecological economics principles and applications. Island Press, Washington
Damania AB (1994) *In situ* conservation of biodiversity of wild progenitors of cereal crops in the Nera East. Biodivers Lett 2:59–60
DeHaan LR, Van Tassel DL, Cox TS (2005) Perennial grain crops: a synthesis of ecology and plant breeding. Renew Agric Food Syst 20(1):5–14
Deleage JP, Julien JM, SaugetNaudin N, Souchon C (1979) Eco-energetics analysis of an agricultural system: the French case in 1970. Agro-Ecosystems 5:345–365
Delgado JA, Dillon MA, Sparks RT, Essah SYC (2007) A decade of advances in cover crops. J Soil Water Conserv 62(5):111A–117A
Decker AM, Clark AJ, Meisinger JJ, Mulford FR, McIntosh MS (1994) Legume cover crop contributions to no-tillage corn production. Agron J 86:126–135
Dempster JP, Coaker TA (1972) Diversification of crop ecosystems as a means of controlling pests. In: Jones DP, Solomon ME (eds) Biology in pest and disease control. Wiley, New York, pp 107–114
den Biggelaar C, Lal R, Wiebe K, Breneman V (2001) Impact of soil erosion on crop yields in North America. Adv Agron 72:1–52
Draghetti A (1948) Principi di Fisiologia dell'Azienda Agraria, Istituto Edizioni Agricole, Bologna
Edwards CA, Grove TL, Harwood RR, Pierce Colfer CJ (1993) The role of agroecology and integrated farming systems in agricultural sustainability. Agric Ecosyst Environ 46:99–121
Ehrlich PR, Mooney HA (1983) Extinction, substitution, and ecosystem services. BioScience 33(4):248–254
Elliot ET, Cole CV (1989) A perspective on agroecosystem science. Ecology 70(6):1597–1602
Ewel JJ (1986) Designing agricultural ecosystems for the humid tropics. Ann Rev Ecol Syst 17:245–271
Faber M (2008) How to be an ecological economist. Ecol Econ 66:1–7
Ferretti MP, Magaudda P (2006) The slow pace of institutional change in the Italian food system. Appetite 47:161–169
Fiala N (2008) Meeting the demand: an estimation of potential future greenhouse gas emission from meat production. Ecol Econ 67:412–419
Fiala N (2009) The greenhouse hamburger. Sci Am, February 2009:62–65
Fischer J, Manning AD, Steffen W, Rose DB, Daniell K, Felton A, Garnett S, Gilna B, Heinsohn R, Lindenmayer DB, MacDonald B, Mills F, Newell B, Reid J, Robin L, Sherren K, Wade A (2007) Mind the sustainability gap. Trends Ecol Evol 22(12):621–624
Flora CB (1994) Science and sustainability: an overview. Am J Altern Agric 9(1–2):72–75
Fonte M (2006) Slow food presidia: what do small producers do with big retailers? Between the local and the global: confronting complexity in the contemporary agri-food sector. Res Rural Sociol Dev 12:203–240, Elsevier, Oxford
Francis CA (ed) (1986) Multiple cropping systems. Macmillan, New York
Francis CA (1989) Biological efficiencies in multiple cropping systems. Adv Agron 42:1–42
Francis CA, King JW (1988) Cropping systems based on farm-derived renewable resources. Agric Syst 27:67–75
Francis CA, King JW (1994) Will there be people in sustainable ecosystems? Designing an educational mosaic for the 22nd century. Am J Altern Agric 9(1–2):16–22

Francis CA, Lieblein G, Helenius J, Salomonsson L, Olse H, Porte J (2001) Challenges in designing ecological agriculture education: a Nordic perspective on change. Am J Altern Agric 16(2):89–95
Francis C, Lieblein G, Gliessman S, Breland TA, Creamer N, Harwood R, Salomonsson L, Helenius J, Rickerl D, Salvador R, Wiedenhoeft M, Simmons S, Allen P, Altieri M, Flora C, Poincelot R (2003) Agroecology: the ecology of food systems. J Sustain Agric 22(3):99–118
Fresco LO, Westphal E (1988) A hierarchical classification of farm systems. Exp Agric 24:399–419
Frissel MJ (1977) Cycling of mineral nutrients in agricultural ecosystems. Agro-Ecosystems 4:1–354
Geng S, Hess CB, Auburn J (1990) Sustainable agricultural systems: concepts and definitions. J Agron Crop Sci 165:73–85
Gerbenss-Leenes PW, Moll AJM, Schoot Uiterkamp AJM (2003) Design and development of a measuring method for environmental sustainability in food production systems. Ecol Econ 46:231–248
Germeier CU (2006) Competitive and soil fertility effects of forbs and legumes as companion plants or living milch in wide sowed organically grown cereals. Biol Agric Hort 23:325–350
Gitau T, Gitau MW, Waltner-Toews D (2009) Integrated assessment of health and sustainability of agroecosystems. CRC Press, Boca Raton, FL
Gliessman SR (ed) (1990) Agroecology: researching the ecological basis for sustainable agriculture, vol 78, Ecological studies. Springer Verlag, New York
Gliessman SR (2007) Agroecology. The ecology of sustainable food systems, 2nd edn. CRC Press, Boca Raton, FL
Glover (2005) The necessity and possibility of perennial grain crop production systems. Renew Agric Food Syst 20(1):1–4
Golley FB (1993) A history of the ecosystem concept in ecology. Yale University Press, New Haven, CT
Haeuber R, Ringold P (1998) Ecology, the social sciences, and environmental policy. Ecol Appl 8(2):330–331
Hampicke U (2006) Efficient conservation in Europe's agricultural countryside. Rationale, methods and policy reorientation. Outlook Agric 35:97–105
Harlan JR (1975) Our vanishing genetic resources. Science 188:618–621
Harper TL (1974) Agricultural ecosystems. Agro-Ecosystems 1:1–6
Harrier LA, Watson CA (2003) The role of arbuscolar mycorrhizal fungi in sustainable cropping systems. Adv Agron 79:186–225
Hart RD (1986) Ecological framework for multiple cropping research. In: Francis CA (ed) Multiple cropping systems. Macmillan, New York, pp 41–56
Hart E, Bond M (1995) Research for health and social care: a guide to practice. Open University Press, Buckingham
Haygarth PM, Jarvis SC (1999) Transfer of phosphorus from agricultural soils. Adv Agron 66:195–249
Hayman DS, Mosse B (1972) Plant growth responses to vesicular-arbuscolar mycorrhiza. III. Increased uptake of labile P from soil. New Phytol 71:41–47
Haynes RJ (1980) Influence of soil management practice on the orchard agro-ecosystem. Agro-Ecosystems 6:3–32
Hazell P (1998) Agricultural growth, poverty, and the environment: introduction. Agric Econ 19:ix–xii
Hesterman OB, Russelle MP, Sheaffer CC, Heichel GH (1987) Nitrogen utilization from fertilizer and legume residues in legume-corn rotations. Agron J 79:726–731
Hirsch Hadorn G, Bradley D, Pohl C, Rist S (2006) Implications of transdisciplinarity for sustainability research. Ecol Econ 60:119–128
Hoff JD, Nelson DW, Sutton AL (1981) Ammonia volatilization from liquid swine manure applied to cropland. J Environ Qual 10(1):90–95
Horwith B (1985) A role of intercropping in modern agriculture. BioScience 35(5):286–291

Hutchinson GL (1969) Nitrogen enrichment of surface water by absorption of ammonia loss from cattle feedlots. Science 166:514
Huyck L, Francis CA (1995) Designing a diversified farmscape. In: Exploring the role of diversity in sustainable agriculture. ASA, Madison, WI, pp 95–120
IFOAM (International Federation of Organic Agriculture Movements) (2007) The World of Organic Agriculture: Statistics and Emerging Trends 2007. IFOAM
Ikerd JE (1993) The need for a systems approach to sustainable agriculture. Agric Ecosyst Environ 46:147–160
Jackson W, Piper J (1989) The necessary marriage between ecology and agriculture. Ecology 70(6):1591–1593
Janssens MJJ, Neumann IF, Froidaux L (1990) Low-input ideotypes. In: Gliessman SR (ed) Agroecology. Ecol Study 78:130–145. Springer Verlag, Berlin
Janzen DH (1973) Tropical agroecosystems. Science 182:1212–1219
Jonas H (1979) Das Prinzip Verantwortung. Insel Verlag, Franfurt am Main
Jones T, Feber R, Hemery G, Cook P, James K, Lambert C, Dawkins M (2007) Welfare and environmental benefits of integrating commercially viable free-range broiler chicken into newly planted woodland: a UK case study. Agric Syst 94:177–188
Johnson AH, Bouldin EA, Goyette EA, Hedges AM (1976) Nitrate dynamics in Fall Creek, New York. J Environ Qual 5:386–391
Karlen DL, Varvel GE, Bullock DG, Cruse RM (1994) Crop rotations for the 21st century. Adv Agron 53:1–45
Kates RW, Clark WC, Corell R, Hall JM, Jaeger CC, Lowe I, McCarthy JJ, Schellnhuber HJ, Bolin B, Dickson NM, Faucheux S, Gallopin GC, Grubler A, Huntley B, Jager J, Jodha NS, Kasperson RE, Magobunje A, Matson P, Mooney H, MooreIII B, O'Riordan T, Svedin U (2001) Sustainability science. Science 292:641–642
Keiter RB (1998) Ecosystems and the law: toward an integrated approach. Ecol Appl 8(2):332–341
Keller EF (2005) Ecosystems, organisms, and machines. BioScience 55(12):1069–1074
Kesavan PC, Swaminathan MS (2008) Strategies and models for agricultural sustainability in developing Asian countries. Phil Trans R Soc B 363:877–891
Klein JT (2004) Prospects for transdisciplinarity. Futures 36:515–526
Kolb D (1984) Experiential learning: experience as the source of learning and development. Prentice Hall, Englewood Cliffs, NJ
Kuo S, Sainju UM, Jellum EJ (1996) Winter cover cropping influence on nitrogen mineralization, preside dress soil nitrate test, and corn yields. Biol Fertil Soils 22:310–317
Kuo S, Sainju UM, Jellum EJ (1997) Winter cover cropping influence on nitrogen in soil. Soil Sci Soc Am J 61:1392–1399
Lal R (2001) Managing world soils for food security and environmental quality. Adv Agron 74:155–192
Lampkin N (1990) Organic farming. Farming Press, Ipswich, UK
Lanini WT, Pittenger DR, Graves WL, Munoz F, Agamalian HS (1989) Subclovers as living mulches for managing weeds in vegetables. Calif Agric 43(6):25–27
LaRue TA, Patterson TG (1981) How much nitrogen do legume fix? Adv Agron 34:15–38
Larson WE, Pierce FJ, Dowdy RH (1983) The threat of soil erosion to long-term crop production. Science 219:458–465
Lautenschlager RA (1998) From rhetoric to reality: using specific environmental concerns to identify critical sustainability issues. Ecosystems 1:176–182
Lawson GJ, Callaghan TV, Scott R (1984) Renewable energy from plants: bypassing fossilization. Adv Ecol Res 14:57–114
Leakey RB, Tchoundjeu Z, Schreckenberg K, Shackleton S, Shackleton CM (2005) Agroforestry tree products (AFTPs): targeting poverty reduction and enhanced livelihoods. Int J Agric Sustain 3(1):1–23
Le Coeur DL, Baudry J, Burel F, Thenail C (2002) Why and how we should study field boundary biodiversity in an agrarian landscape context. Agric Ecosyst Environ 89:23–40

Levin SA (1998) Ecosystems and the biosphere as complex adaptive systems. Ecosystems 1:431–436
Levin SA (2005) Self-organization and the emergence of complexity in ecological systems. Bioscience 55(12):1075–1079
Levin SA (2006) Learning to live in a global commons: socioeconomic challenges for a sustainable environment. Ecol Res 21:328–333
Lichtfouse E, Navarrete M, Debaeke P, Souchere V, Alberola C, Menassieu J (2009) Agronomy for sustainable agriculture. A review. Agron Sustain Dev 29:1–6
Lieblein G, Francis CA (2007) Towards responsible action through agroecological education. Ital J Agron/Riv Agron 2(2):83–90
Liebman M, Dyck E (1993) Crop rotation and intercropping strategies for weed management. Ecol Appl 3(1):92–122
Lin BB, Perfecto I, Vandermeer J (2008) Synergies between agricultural intensification and climate change could create surprising vulnerabilities for crops. BioScience 58(9):847–854
Liu J, Dietz T, Carpenter SR, Folke CM, Redman CL, Schneider SH, Ostrom E, Pell AN, Lubchenco J, Taylor WW, Ouyang Z, Deadma P, Kratz T, Provencher W (2007) Coupled human and natural systems. Ambio 36((8):639–648
Llausàs A, Ribas A, Varga D, Vila J (2009) The evolution of agrarian practices and its effects on the structure of enclosure landscapes in the Alt Empordà (Catalonia, Spain), 1957–2001. Agric Ecosyst Environ 129:73–82
Loomis RS, Connor DJ (1992) Crop ecology. Productivity and management in agricultural systems. Cambridge University Press, Cambridge
Loucks OL (1977) Emergence of research on agro-ecosystems. Ann Rev Ecol Syst 8:173–192
Lowrance R (1990) Research approaches for ecological sustainability. J Soil Water Conserv 45:51–54
Lowrance R, Stinner BR, House GJ (eds) (1984) Agricultural ecosystems. Unifying concepts. Wiley, New York
Lowrance R, Hendrix PF, Odum EP (1987) A hierarchical approach to sustainable agriculture. Am J Altern Agric 1:169–173
Lowrance RR (1998) Riparian forest ecosystems as filters for non-point source pollution. In: Pace ML, Groffman PM (eds) Successes, limitations and frontiers in ecosystem science. Springer Verlag, New York, pp 113–141
Lubchenco J (1998) Entering the century of the environment: a new social contract for science. Science 279:491–497
MacKee R (1947) Summer crops for green manure and soil improvement. Farmers'Bulletin No. 1750, USDA, Washington
MacKinnon JC (1975) Design and management of farms as agricultural ecosystems. Agro-Ecosystems 2:277–291
MacRae RJ, Hill SB, Henning J, Mehuys GR (1989) Agricultural science and sustainable agriculture: a review of the existing scientific barriers to sustainable food production and potential solutions. Biol Agric Hort 6:173–219
MacRae RJ, Hill SB, Mehuys GR, Henning J (1990) Farm-scale agronomic and economic conversion from conventional to sustainable agriculture. Adv Agron 43:155–198
Maddison D (1982) Innovation, ideology and innocence. Soc Sci Med 16:623–628
Mader P, Fliessback A, Dubois D, Gunst L, Fried P, Niggli U (2002) Soil fertility and biodiversity in organic farming. Science 296:1694–1697
Magdof F, Lanyon L, Liebhardt B (1997) Nutrient cycling, transformations, and flows: implications for a more sustainable agriculture. Adv Agron 60:1–73
Manley JC, Ware N (1990) How do we know what we have done? Assessment and faculty development within a learning community. In: Clark ME, Wawrytko SA (eds) Rethinking the curriculum – toward an integrated, interdisciplinary college education. Green Wood Press, New York, pp 243–252
Maracchi G, Cresci A (2001) Impatto dei cambiamenti climatici negli agroecosistemi italiani. Legno Cellulosa Carta 7(2):2–7
Marinari S, Caporali F (eds) (2008) Soil carbon sequestration under organic farming in the Mediterranean environment. Transworld Research Network, Trivandrum, Kerala, India

Marshall EJP (2002) Introducing field margin ecology in Europe. Agric Ecosyst Environ 89:1–4
Marshall EJP, Moonen AC (2002) Field margins in northern Europe: their functions and interactions with agriculture. Agric Ecosyst Environ 89:5–21
Marten GG (1988) Productivity, stability, sustainability, equitability and autonomy as properties for agroecosystem assessment. Agric Syst 26:291–316
Martin S (2003) The evaluation of strategic research partnerships. Technol Anal Strateg Manage 15(2):159–176
Matteson PC, Altieri MA, Gagnè WC (1984) Modification of small farmer practices for better pest management. Ann Rev Entomol 29:383–402
Maudsley M, Becky s, Owen L (2002) Spatial distribution of predatory arthropods within an English hedgerow in early winter in relation to habitat variables. Agric Ecosyst Environ 89:77–99
McCarl BA, Schneider UA (2001) Climate Change – Greenhouse Gas Mitigation in U.S. Agriculture and Forestry. Science 294:2481–2482
McMichael AJ, Butler CD, Folke C (2003) New visions for addressing sustainability. Science 302:1919–1920
MEA (Millenium Ecosystem Assessment) (2005) Ecosystems and human well-being: biodiversity synthesis. World Resources Institute, Washington, DC
Mead R, Riley J (1981) A review of statistical ideas relevant to intercropping research. J R Stat Soc Ser A (general) 144(4):462–509
Mendelsohn R (2003) The challenge of conserving indigenous domesticated animals. Ecol Econ 45:501–510
Merckx T, Feber RE, Riordan P, Townsend MC, Bourn NAD, Parsons MS, Macdonald DW (2009) Optimizing the biodiversity gain from agri-environment schemes. Agric Ecosyst Environ 130:177–182
Milder JC, McNeely JA, Shames SA, Scherr SJ (2008) Biofuels and ecoagriculture: can bioenergy production enhance landscape-scale ecosystem conservation and rural livelihoods? Int J Agric Sustain 6(2):105–121
Miller J (1973) Genetic erosion: crop plants threatened by government neglect. Science 182:1231–1233
Miller HG, Copper JM, Miller JD, Pauline OJL (1979) Nutrient cycles in pine and their adaptation to poor soils. Can J For Res 9:19–26
Morgan-Davies C, Waterhouse A, Pollock ML, Holland JP (2008) Integrating hill sheep production and newly established native woodland: achieving sustainability through multiple land use in Scotland. Int J Agric Sustain 6(2):133–147
Munasinghe M (2008) Mainstreaming and implementing the Millennium Ecosystem Assessment results by integrating them into sustainable development strategy: applying the Action Impact Matrix methodology. In: Rangainathan J, Munasinghe M, Irwin F (eds) Policies for sustainable governance of global ecosystem services. Edward Elgar, Cheltenham, UK, pp 73–107
Murdoch W (1990) World hunger and population. In: Carrol CR, Vandermeer JH, Rosset PM (eds), Wayne M Getz (Ser ed). Agroecology. University of California, Berkeley, CA, pp 3–20
Murphy KL, Lammer D, Lyon S, Carter B, Jones SS (2005) Breeding for organic and low-input farming systems: an evolutionary-participatory breeding method for inbred cereal grains. Renew Agric Food Syst 20(1):48–55
Musters CJM, van Alebeek F, Geers RHEM, Korevaar H, Visser A, de Snoo GR (2009) Development of biodiversity in field margins recently taken out of production and adjacent ditch banks in arable areas. Agric Ecosyst Environ 129:131–139
Nair PKR (1991) State-of-the-art of agroforestry systems. In: Jarvis PG (ed) Agroforestry: principles and practice. Elsevier, Amsterdam, pp 5–29
Newby H, Utting P (1984) Agribusiness in the United Kingdom: social and political implications. In: The social consequences and challenges of new agricultural technologies. Rural Studies Series. Westview Press, Inc., UK, pp 265–289
Norgaard RB, Baer P (2005) Collectively seeing complex systems: the nature of the problem. BioScience 55(11):953–960

Norton BG (1998) Improving ecological communication: the role of ecologists in environmental policy formation. Ecol Appl 8(2):350–364
Noss RF (1983) A regional landscape approach to maintain diversity. BioScience 33(11):700–706
Odum EP (1969) The strategy of ecosystem development. Science 164:262–270
Odum HT (1983) Systems ecology. Wiley, New York
Odum EP (1984) Properties of Agroecosystems. In: Lowrance R, Stinner BR, House GJ (eds) Agricultural ecosystems. Unifying concepts. Wiley, New York, pp 5–11
OECD (Organisation for Economic Cooperation and Development) (1997) Environmental indicators for agriculture: concepts and frameworks, vol. 1. OECD, Paris
OECD (Organisation for Economic Cooperation and Development) (1999) Environmental indicators for agriculture: issues and design, vol. 2. OECD, Paris
OECD (Organisation for Economic Cooperation and Development) (2001) Environmental indicators for agriculture: methods and results, vol 3. OECD, Paris
Ofori F, Stern WR (1987) Cereal-legume intercropping systems. Adv Agron 41:41–90
Oldfield ML, Alcorn JB (1987) Conservation of traditional agroecosystems. BioScience 37(3):199–208
Olson R, Francis C, Kaffka S (1995) Exploring the role of diversity in sustainable agriculture. ASA, Madison, WI
Oltjen JW, Beckett JL (1996) The role of ruminant livestock in sustainable agricultural systems. J Anim Sci 74(6):1406–1409
Otte A, Simmering D, Wolters V (2007) Biodiversity at the landscape level: recent concepts and perspectives for multifunctional land use. Landscape Ecol 22:639–642
Ovenden GN, Swash ARH, Smallshire D (1998) Agri-environment schemes and their contribution to the conservation of biodiversity in England. J Appl Ecol 35:995–960
Ovington JD, Heitkamp D, Lawrance DB (1963) Plant biomass and productivity of prairie, savanna, oakwood, and maize field ecosystems in Central Minnesota. Ecology 44:52–63
Ovington JD, Lawrance DB (1967) Comparative chlorophyll and energy studies of prairie, savanna, oakwood and maize field ecosystems. Ecology 48:515–524
Palmer M, Bernhardt E, Chornesky E, Collins S, Dobson A, Duke C, Gold B, Jacbson R, Kingsland S, Kranz R, Mappin M, Marinez ML, Micheli F, Morse J, Pace M, Pascual M, Palumbi S, Reichman OJ, Simons A, Townsend A, Turner M (2004) Ecology for a crowded planet. Science 304:1251–1252
Papendick R, Elliot L, Dahlgren RB (1986) Environmental consequences of modern production agriculture: how can alternative agriculture address these issues and concerns? Am J Altern Agric 1:3–10
Parson JW (1984) Green manuring. Outlook Agric 13(1):20–23
Paul EA, Robertson GP (1989) Ecology and the agricultural sciences: a false dichotomy? Ecology 70(6):1594–1597
Perez MP, Sanchez AM (2003) The development of university spin-offs: early dynamics of technology transfer and networking. Technovation 23:823–831
Pimentel D, Patzek TW (2005) Ethanol production using corn, switchgrass, and wood: biodiesel production using soybean and sunflower. Nat Resour Res 14(1):65–76
Pimentel D, Hurd LE, Bellotti AC, Forster MJ, Oka IN, Sholes OD, Whitman RJ (1973) Food production and the energy crisis. Science 182:443–449
Pimentel D, Terhune EC, Dyson-Hudson R, Rocherau S, Samis R, Smith EA, Denman D, REifschneider D, Shepard M (1976) Land degradation: effects on food and energy resources. Science 276:149–155
Pimentel D, Stachow U, Takas DA, Brubaker HW, Dumas AR, Meaney JJ, O'Neil JAS, Onsi DE, Corzilius DB (1992) Conserving biological diversity in agricultural/forestry systems. BioScience 42(5):354–362
Piorr HP (2003) Environmental policy, agri-environmental indicators and landscape indicators. Agric Ecosyst Environ 98:17–33
Poincelot R (2003) Agroecology and agroecosystems: the ecology of food systems. J Sustain Agric 22:99–118

Pretty J (1995) Regenerating agriculture: policies and practice for sustainability and self-reliance. Earthscan/National Academic Press, London
Pretty J (2003) Social capital and the collective management of resources. Science 302:1912–1914
Pretty J (2008) Agricultural sustainability: concepts, principles and evidence. Phil Trans R Soc B 363:447–465
Pretty J, Brett C, Gee D, Hine R, Mason CF, Morison JI, raven H, Rayment M, van der Bijil G (2000) An assessment of the total external costs of UK agriculture. Agric Syst 65:113–136
Pretty J, Lang T, Ball A, Morison J (2005) Farm costs and food miles: an assessment of the full cost of the weekly food basket. Food Policy 30:1–20
Puckridge DW, French RJ (1983) The annual legume pasture in cereal-ley farming system of southern Australia: a review. Agric Ecosyst Environ 9:229–267
Rappaport R (1971) The flow of energy in agricultural society. Sci Am 9:116–132
Redman CL, Grove MJ, Kuby LH (2004) Integrating social sciences into the long-term ecological research (LTER) network: social dimensions of ecological change and ecological dimensions of social change. Ecosystems 7:161–171
Reganold JP, Glover JD, Andrews PK, Hinman HR (2001) Sustainability of three apple production systems. Nature 410:926–930
Rege JEO, Gibson JB (2003) Animal genetic resources and economic development: issues in relation to economic valuation. Ecol Econ 45:319–330
Risch SJ (1981) Insect herbivore abundance in tropical monocultures and polycultures: an experimental test of two hypotheses. Ecology 62:1325–1340
Risch SJ, Andow D, Altieri MA (1983) Agroecosystem diversities and pest control: data, tentative conclusions, and new research directions. Environ Entomol 12:625–629
Risser PG (1995) The status of the science examining ecotones. BioScience 45(5):318–325
Robertson AI (2000) The gaps between ecosystem ecology and industrial agriculture. Ecosystems 3:413–418
Robson MC, Fowler SM, Lampkin NH, Leifert C, Leitch M, Robinson D, Watson CA, Litterick AM (2002) The agronomic and economic potential of break crops for ley/arable rotations in temperate organic agriculture. Adv Agron 77:369–427
Rodriguez C, Wiegand K (2009) Evaluating the trade-off between machinery efficiency and loss of biodiversity-friendly habitats in arable landscapes: the role of field size. Agric Ecosyst Environ 129:361–366
Röling N (2003) From causes to reasons: the human dimension of agricultural sustainability. Int J Agric Sustain 1(1):73–88
Röling NG, Jiggins J (1998) The ecological knowledge system. In: Röling NG, Wagemakers MAE (eds) Facilitating sustainable agriculture. Cambridge University Press, Cambridge, UK
Roschewitz I, Thies C, Tscharntke T (2005) Are landscape complexity and farm specialisation related to land-use intensity of annual crop fields? Agric Ecosyst Environ 105:87–99
Rosegrant MW, Cline SA (2003) Global food security: challenges and policies. Science 302:1917–1919
Sainju UM, Singh BP (1997) Winter cover crops for sustainable agricultural systems: influence on soil properties, water quality, and crop yields. Hort Sci 35:21–28
Schade JD, Marti E, Welter JR, Fisher SG, Grimm NB (2002) Sources of nitrogen to the riparian zone of a desert stream: implications for riparian vegetation and nitrogen retention. Ecosystems 5:68–79
Scully M (2002) Dominion. The power of man, the suffering of animals and the call to mercy. S. Martin's Press, New York
Sharpley AN, Smith SJ, Naney JW (1987) Environmental impact of agricultural nitrogen and phosphorus use. J Agric Food Chem 35:812–817
Sibbald AR (2006) Silvopastoral agroforestry: a land use for the future. Scott For 60:4–7
Smil V, Nachman P, Long TV II (1983) Energy analysis and agriculture. An application to U.S. corn production. Westview Press, Boulder, CO
Smit B, McNabb D, Smithers J (1996) Agricultural adaptation to climatic variation. Clim Change 33(1):7–29

Smith OH, Petersen GW, Needelman BA (2000) Environmental indicators of agroecosystems. Adv Agron 69:75–97
Sotherton NW (1985) The distribution and abundance of predatory Coleoptera overwintering in field boundaries. Ann Appl Biol 106:17–21
Spedding CRW, Brockington NR (1975) The study of ecosystems. Agro-Ecosystems 2:165–172
Spedding CRW, Walshingham JM, Hoxey AM (1981) Biological efficiency in agriculture. Academic, London
Spiertz JHJ, Sibma L (1986) Dry matter production and nitrogen utilization in cropping systems with grass, lucerne and maize. 2. Nitrogen yield and utilization. Neth J Agric Sci 34:37–47
Stanhill G (1974) Energy and agriculture: a national case study. Agro-Ecosystems 1:205–217
Stanhill G (1979) A comparative study of the Egyptian agro-ecosystem. Agro-Ecosystems 5:213–230
Stark CR (1995) Adopting multidisciplinary approaches to sustainable agriculture research: potentials and pitfalls. Am J Altern Agric 10(4):180–183
Steenvoorden J, Fonk H, OOsterom HP (1986) Losses from intensive grassland systems by leaching and surface runoff. In: Gwan der Meer H, Ryder JC, Ennik GC (eds) Nitrogen fluxes in intensive grassland systems. Martinus Nijhof Publishers, Dordrecht, The Netherlands, pp 85–97
Steffen W, Crutzen PJ, McNeill JR (2007) The Anthropocene: are humans overwhelming the great forces of nature? Ambio 36(8):614–621
Steinfeld H, Gerber P, Wassenaar T, Castel V, RosalesM, De Haan C (2006) Livestock's long shadow: environmental issues and options. FAO, Rome, Italy. ISBN 978-92-5-105571-7
Steinhart JS, Steinhart CE (1974) Energy use in the U.S. food system. Science 184:307–316
Steppler HA, Lundgren BO (1988) Agroforestry: now and in the future. Outlook Agric 17(4):146–152
Sthapit B, Rana R, Eyzaguirre P, Jarvis D (2008) The value of plant genetic diversity to resource-poor farmers in Nepal and Vietnam. Int J Agric Sustain 6(2):148–166
Stinner BR, House GJ (1989) The search for sustainable agroecosystems. J Soil Water Conserv 44(2):111–116
Stoate C, Boatman ND, Borralho RJ, de Carvalho CR, Snoo GR, Eden P (2001) Ecological impacts of arable intensification in Europe. J Environ Manage 63:337–365
Strebel O, Duynisveld WHM, Bottcher J (1989) Nitrate pollution of groundwater in Western Europe. Agric Ecosyst Environ 26:189–214
Subak S (1999) Global environmental costs of beef production. Ecol Econ 30:79–91
Sutherland WJ (2002) Restoring a sustainable countryside. Trends Ecol Evol 17(3):148–150
Swift MJ, Anderson JM (1994) Biodiversity and ecosystem function in agricultural systems. In: Schulze ED, Mooney HA (eds) Biodiversity and ecosystem function. Springer-Verlag, Berlin, pp 15–41
Tansley AG (1935) The use and abuse of vegetational concepts and terms. Ecology 16(3):284–307
Tellarini V, Caporali F (2000) An input/output methodology to evaluate farms as sustainable agroecosystems. An application of indicators to farms in Central Italy. Agric Ecosyst Environ 77:111–123
Theunissen J (1997) Intercropping in field vegetables as an approach to sustainable horticulture. Outlook Agric 26(2):95–99
Thomas TH (1990) Agroforestry – does it pay? Outl Agric 19(3):161–170
Thomas MB, Wratten SD, Sotherton NW (1992) Creation of "island" habitats in farmland to manipulate populations of beneficial arthropods: predator densities and species composition. J Appl Ecol 29:524–531
Thompson PB (2008) The agricultural ethics of biofuels: a first look. J Agric Environ Ethics 21:183–189
Thompson RL (1985) The effect of monetary and fiscal policy on agriculture. In: Marton LB (ed) U.S. agriculture in a global economy.. USDA, Washington, pp 314–321
Thorup-Kristensen K, Magid J, Jensen LS (2003) Catch crops and green manures as biological tools in nitrogen management in temperate zones. Adv Agron 79:227–302

Tiessen H, Stewart JWB, Bettany JR (1982) Cultivation effect on the amounts and concentration of carbon, nitrogen and phosphorous in grasslands soils. Agron J 74:831–835
Tilman D, Fargione J, Wolff B, D'Antonio C, Dobson A, Howarth R, Schindler D, Schlesinger WH, Simberloff D, Swackhamer D (2001) Forecasting agriculturally driven global environmental change. Science 292:281–284
Tilman D, Cassman KG, Matson PA, Naylor R, Polasky S (2002) Agricultural sustainability and intensive production practices. Nature 418:671–676
Tivy J (1990) Agricultural ecology. Longman Scientific & Technical, Harlow, Essex, England
Tomlinson TE (1970) Trend in nitrate concentrations in English rivers in relation to fertilizer use. Water Treat Exam 19:277–289
Transeau EG (1926) The accumulation of energy by plants. Ohio J Sci 26(1):1–11
Trenbath BR (1986) Resource use by intercrops. In: Francis CA (ed) Multiple cropping systems. Macmillan Publishing Company, New York, pp 57–81
Tropeano D (1983) Soil erosion on vineyards in the Terziary Piemontese Basin (Northwestern Italy): studies on experimental areas. In: De Ploey J (ed) Rainfall simulation, runoff and soil erosion. Catena Suppl 4: 115–127
Tweeten L (1983) The economics of small farms. Science 219:1037–1041
UN (United Nations)(1992). Agenda 21. Sustainable Development
Valentine I (2005) An emerging model of a systems agriculturalist. Syst Res Behav Sci 22:109–118
Vandermeer JH (1988) The ecology of intercropping. Cambridge University Press, New York
Vandermeulen V, Van Huylenbroeck G (2008) Designing trans-disciplinary research to support policy formulation for sustainable agricultural development. Ecol Econ 67:352–361
Van Acker RC (2008) Sustainable agriculture development requires a shift from an industrial to a multifunctional model. Int J Agric Sustain 6(1):1–2
Van der Ploeg RR, Schweigert P (2001) Elbe river flood peaks and postwar agricultural land use in East Germany. Naturwissenschaften 88:522–525
Van der Werf HMG, Petit J (2002) Evaluation of the environmental impact of agriculture at the farm level: a comparison and analysis of 12 indicator-based methods. Agric Ecosyst Environ 93:131–145
Van Emden HF, Williams GF (1974) Insect stability and diversity in agro-ecosystems. Ann Rev Entomol 19:455–475
Vereijken P (1992) A methodic way to more sustainable farming systems. Neth J Agric Sci 40:209–233
Vernadskij VI (1999) La biosfera e la noosfera. Sellerio Editore, Palermo
Vitousek PM, Mooney HA, Lubechenco J, Melillo JM (1997) Human domination of earths's ecosystems. Science 277:494–499
Vlek PLG, Fillery IRP, Burford JR (1981) Accession, transformation and loss of nitrogen in soils of the arid region. Plant Soil 58:133–175
Von Fragstein P, Francis CA (2008) Integration of crop and animal husbandry. In: Caporali F, Lieblein G, von Fragstein P, Francis CA (eds) Teaching and research in agroecology and organic farming: challenges and perspectives. Università della Tuscia, Viterbo, Italy, pp 44–54
Wackernagel M, Rees WE (1997) Perceptual and structural barriers to investing in natural capital: economics from an ecological footprint perspective. Ecol Econ 20:3–24
Wagstaff H (1987) Husbandry methods and farm systems in industrialised countries which use lower levels of external inputs: a review. Agric Ecosyst Environ 19:1–27
Wallin H (1985) Spatial and temporal distribution of some abundant carabid beetles (Coleoptera: Carabidae) in cereal fields and adjacent habitats. Pedobiologia 28:19–34
Warner KD (2006) Extending agroecology: grower participation in partnerships is key to social learning. Renew Agric Food Syst 21(2):84–94
Warner KD (2007) The quality of sustainability: agroecological partnerships and the geographic branding of California winegrapes. J Rural Stud 23:142–155
Westman WE (1977) How much are nature's services worth? Science 197:960–964

Wezel A, Soldat V (2009) A quantitative and qualitative historical analysis of the scientific discipline of agroecology. Int J Agric Sustain 7(1):3–18
Wezel A, Bellon S, Dorè T, Francis C, Vallod D, David C (2009) Agroecology as a science, a movement and a practice. A review. Agron Sustain Dev 29(4): 503–515
William RD (1981) Complementary interactions between weeds, weed control practices, and pests in horticultural cropping systems. Hort Sci 16:508–513
Willey RW (1979a) Intercropping – its importance and research needs. Part 1. Competition and yield advantages. Field Crop Abstr 32:1–10
Willey RW (1979b) Intercropping – its importance and research needs. Part 2. Agronomy and research approaches. Field Crop Abstr 32:73–85
Wilson GA (2008) Global multifunctional agriculture: transitional convergence between North and South or zero-sum game? Int J Agric Sustain 6(1):3–21
Wissinger M (1997) Cycle colonization in predictability ephemeral habitats: a template for biological control in annual crop systems. Biol Control 10:4–15
Wolfe M (2000) Crop strength through diversity. Nature 406:681–682
Wood PJ (1990) The scope and potential of agroforestry. Outlook Agric 19(3):141–146
Young A (1990) Agroforestry, environment and sustainability. Outlook Agric 19(3):155–160
Yunlong C, Smit B (1994) Sustainability in agriculture: a general review. Agric Ecosyst Environ 49:299–307
Zak DR, Holmes WE, White DC, Peacock AD, Tilman D (2003) Plant diversity, microbial communities, and ecosystem function: are there any links? Ecology 84(8):2042–2050

Measuring Agricultural Sustainability

Dariush Hayati, Zahra Ranjbar, and Ezatollah Karami

Abstract Sustainability in agriculture is a complex concept and there is no common viewpoint among scholars about its dimensions. Nonetheless various parameters for measuring agricultural sustainability have been proposed. This manuscript reviews some aspects of agricultural sustainability measures by referring to measuring difficulties, components of sustainability measurement and their interaction. Criteria to select sustainability indicators are discussed. Agricultural sustainability scales at national level and farm level are reviewed. A large number of indicators have been developed but they do not cover all dimensions and levels. Therefore, indicators used for agricultural sustainability should be location specific. They should be constructed within the context of the contemporary socioeconomic and ecological situation. Some recommendations to select indicators in order to better measure agricultural sustainability are presented.

Keywords Agricultural sustainability • Measuring sustainability • Sustainability indicators • Sustainability components

1 Introduction

For any study on sustainable agriculture, the question arises as to how agricultural sustainability can be measured. Some argue that the concept of sustainability is a "social construct" (David 1989; Webster 1999) and is yet to be made operational (Webster 1997). The precise measurement of sustainability is impossible as it is site-specific and a dynamic concept (Ikerd 1993). To some extent, what is defined

D. Hayati (✉), Z. Ranjbar and E. Karami
Department of Agricultural Extension and Education, College of Agriculture,
Shiraz University, Shiraz, Iran
e-mail: hayati@shirazu.ac.ir

E. Lichtfouse (ed.), *Biodiversity, Biofuels, Agroforestry and Conservation Agriculture*,
Sustainable Agriculture Reviews 5, DOI 10.1007/978-90-481-9513-8_2,

as sustainable depends on the perspectives of the analysts (Webster 1999). Although precise measurement of sustainable agriculture is not possible, "when specific parameters or criteria are selected, it is possible to say whether certain trends are steady, going up or going down" (Pretty 1995).

Practices that erode soil, remove the habitats of insect predators, and cut instead of plant trees can be considered unsustainable compared to those that conserve these resources. According to Altieri (1995), farmers can improve the biological stability and resilience of the system by choosing more suitable crops, rotating them, growing a mixture of crops, and irrigating, mulching and manuring land. According to Lynam and Herdt (1989), sustainability can be measured by examining the changes in yields and total factor productivity. Beus and Dunlop (1994) considered agricultural practices such as the use of pesticides and inorganic fertilizers, and maintenance of diversity as measures of sustainability. For sustainable agriculture, a major requirement is sustainable management of land and water resources.

Reviewing the aspects of agricultural sustainability measures, by referring to measuring difficulties, components of sustainability measurement and criteria for indicators selection were the main objectives of this manuscript. It should be declared that the article has inevitably had to take a bias toward cropping because of the huge amount of literature on sustainability indicators in various disciplines.

2 General Issues

Considerable efforts have been made to identify appropriate indicators for agricultural sustainability. In the realm of practice, the most influential model of environmental reporting is the causality chain of Pressure-State-Response (PSR). Although its conceptual development can be traced back to the 1950s, the PSR model was pioneered by the Organisation for Economic Cooperation and Development (OECD) (OECD 1991). The PSR model and variants have been extensively used to organise a menu of indicators. Examples of applications include the State-of-Environment (SOE) reporting (Australia, Canada and New Zealand) and the set of sustainability indicators proposed by the United Nations Commission on Sustainable Development (CSD). The latter has been tested in selected developed and developing countries. This sets a new precedent of cross-nation sustainability indicator comparability which has been followed recently by other international initiatives such as the Environmental Sustainability Index and OECD Environmental Performance Review. In effect, indicators become a policy instrument to exert peer pressure among nations to perform better.

Recently, OECD has developed a common framework called "driving force state response" (DSR) to help in developing indicators. Driving force indicators refer to the factors that cause changes in farm management practices and inputs use. State indicators show the effect of agriculture on the environment such as soil, water, air, biodiversity, habitat and landscape. Response indicators refer to the actions that are taken in response to the changing state of environment. Using the DSR framework, OECD (1997) identified 39 indicators of issues such as farm financial resources, farm management, nutrient use, pesticide use, water use, soil quality, water quality, land conservation, greenhouse gases, biodiversity, landscape, wildlife habitats, and farm's contextual information, including socioeconomic background, land-use, and output. Similarly, the British Government suggested 34 indicators under 13 themes such as nutrient losses to fresh water, soil P levels, nutrient management practices, ammonia emissions, greenhouse gas emissions, pesticide use, water use, soil protection, and agricultural land resource, conservation value of agricultural land, environmental management systems, rural economy and energy (MAFF cited in Webster 1999).

Most of the indicators mentioned above are suitable to evaluate agricultural sustainability at aggregate level. They cannot, however, be used to assess sustainability at the farm level, although individual farmers take the major decision in land-use including mode of use and choice of technology (Webster 1999). Sands and Podmore (2000) used environmentally sustainability index (ESI) as an indicator of assessing agricultural sustainability and applied it to farms in the United States. ESI represents a group of 15 sustainability sub-indices including soil depth, soil organic carbon, bulk density and depth of ground water. Tellarini and Caporali (2000) used the monetary value and energy value to compare the sustainability of two farms, high-inputs and low-inputs in Italy. Gowda and Jayaramaiah (1998) used nine indicators, namely integrated nutrient management, land productivity, integrated water management, integrated pest management, input self-sufficiency, crop yield security, input productivity, information self-reliance and family food sufficiency, to evaluate the sustainability of rice production in India. Reijntjes et al. (1992) identified a set of criteria under ecological, economic and social aspects of agricultural sustainability. Ecological criteria comprise the use of nutrients and organic materials, water, energy, and environmental effects, while economic criteria include farmers' livelihood systems, competition, factor productivity, and relative value of external inputs. Food security, building indigenous knowledge, and contribution to employment generation are social criteria (Rasul and Thapa 2003). Various parameters for measuring agricultural sustainability have been proposed by scholars. Their emphasis and tendency has been classified in three groups of components (social, economic, and ecological) as part of a review of literature and the result has been presented in Table 1.

Theoretical discussions are attending the challenges of disciplinary and methodological heterogeneity. The quest to define sustainability through biophysical assessment has brought distributional issues to the fore, initiating preliminary interaction with the social sciences and humanities (see Hezri 2005; Miller 2005). Another important theoretical output is the availability of various methodologies in aggregating raw and incongruent sustainability variables through indices development.

The existing indicator systems in the realm of policy are becoming instrumental in mainstreaming sustainable development as a policy goal. Following persistent applications across time at various levels of government, the PSR model has pooled

Table 1 Classification of scholars' emphasis and their tendency toward three components of agricultural sustainability according to a review of literatures

Sources	Component	Parameters
Herzog and Gotsch (1998); Van Cauwenbergh et al. (2007)	**social**	• The education level of the household members
Herzog and Gotsch (1998)		• Housing facilities
Herzog and Gotsch (1998)		• Work study
Herzog and Gotsch (1998); Rasul and Thapa (2003); Van Cauwenbergh et al. (2007)		• Nutritional/health status of the family members
Ingels et al. (1997); Pannell and Glenn (2000); Horrigan et al. (2002); Rasul and Thapa (2003)		• Improved decision making
Karami (1995); Ingels et al. (1997); Rezaei-Moghaddam (1997); Norman et al. (1997); Lyson (1998); Van Cauwenbergh et al. (2007)		• Improved the quality of rural life
Ingels et al. (1997); Van Cauwenbergh et al. (2007)		• Working and living conditions
Becker (1997); Ingels et al. (1997); Van Cauwenbergh et al. (2007)		• Participation/social capital
Becker (1997); Rigby et al. (2001); Rasul and Thapa (2003); Rasul and Thapa (2004)		• Social equity
Hayati (1995); Nambiar et al. (2001); Rasul and Thapa (2003)	**Economic**	• Average of crop production
Becker (1997); Herzog and Gotsch (1998)		• Expenses for input
Herzog and Gotsch (1998); Van Cauwenbergh et al. (2007)		• Monetary income from outside the farm
Herzog and Gotsch (1998); Pannell and Glenn (2000); Nijkamp and Vreeker (2000); Van Cauwenbergh et al. (2007)		• Monetary income from the farm
Becker (1997); Herzog and Gotsch (1998); Nijkamp and Vreeker (2000); Van Cauwenbergh et al. (2007)		• Economic efficiency
Karami (1995); Herzog and Gotsch (1998); Lyson (1998); Smith and McDonald (1998); Comer et al. (1999); Pannell and Glenn (2000); Rigby et al. (2001); Koeijer et al. (2002); Rasul and Thapa (2003); Van Passel et al. (2006); Gafsi et al. (2006)		• Profitability
Herzog and Gotsch (1998)		• The salaries paid to farm workers
Herzog and Gotsch (1998); Rasul and Thapa (2003)		• Employment opportunities
Smith and McDonald (1998); Van Cauwenbergh et al. (2007)		• Market availability
Karami (1995); Nijkamp and Vreeker (2000); Van Cauwenbergh et al. (2007)		• Land ownership
Hayati (1995); Becker (1997); Ingels et al. (1997); Bouma and Droogers (1998); Pannell and Glenn (2000); Sands and Podmore (2000); Bosshard (2000); Nambiar et al. (2001); Horrigan et al. (2002); Rasul and Thapa (2003); Van Cauwenbergh et al. (2007)		• Soil management

(continued)

Table 1 (continued)

Sources	Component	Parameters
Hayati (1995); Ingels et al. (1997); Gafsi et al. (2006); Van Cauwenbergh et al. (2007)	**Ecological**	• Improve water resource management
Hayati (1995); Rezaei-Moghaddam (1997); Ingels et al. (1997); Norman et al. (1997); Pannell and Glenn (2000); Rasul and Thapa (2004)		• Usage of pesticides, herbicides and fungicides
Saltiel et al. (1994); Hayati (1995); Norman et al. (1997); Bosshard (2000)		• Usage of animal/organic manures
Senanayake (1991); Saltiel et al. (1994); Hayati (1995)		• Usage of green manures
Ingels et al. (1997); Herzog and Gotsch (1998)		• Physical inputs and efficient use of input
Herzog and Gotsch (1998); Rasul and Thapa (2003)		• Physical yield
Senanayake (1991); Saltiel et al. (1994); Ingels et al. (1997);Comer et al. (1999); Praneetvatakul et al. (2001); Nambiar et al. (2001); Horrigan et al. (2002); Rasul and Thapa (2003)		• Crop diversification
Saltiel et al. (1994); Rasul and Thapa (2003)		• Use of alternative crop
Saltiel et al. (1994)	**Ecological**	• Usage of fallow system
Saltiel et al. (1994); Hayati (1995); Comer et al. (1999); Horrigan et al. (2002); Rasul and Thapa (2003)		• Crop rotation
Nijkamp and Vreeker (2000); Rasul and Thapa (2003); Rasul and Thapa (2004)		• Cropping pattern
Smith and McDonald (1998); Van Cauwenbergh et al. (2007)		• Trend of change in climatic conditions
Hayati (1995); Rezaei-Moghaddam (1997);Ingels et al. (1997)		• Usage of chemical fertilizer
Hayati (1995); Ingels et al. (1997); Comer et al. (1999); Horrigan et al. (2002);		• Conservational tillage (no/minimum tillage)
Hayati (1995); Ingels et al. (1997); Rasul and Thapa (2003); Gafsi et al. (2006); Van Cauwenbergh et al. (2007)		• Control erosion
Senanayake (1991); Pannell and Glenn (2000)		• Microbial biomass with in the soil
Senanayake (1991);); Ingels et al. (1997); Norman et al. (1997); Nambiar et al. (2001); Van Cauwenbergh et al. (2007)		• Energy
Ingels et al. (1997); Norman et al. (1997); Comer et al. (1999); Horrigan et al. (2002); Rasul and Thapa (2003)		• Cover crop/Mulch
Pannell and Glenn (2000); Sands and Podmore (2000); Van Cauwenbergh et al. (2007)		• Depth of groundwater table
Pannell and Glenn (2000)		• Protein level of crops
Comer et al. (1999); Praneetvatakul et al. (2001); Horrigan et al. (2002); Rasul and Thapa (2003)		• Integrated pest management

an enormous amount of data previously inaccessible, a prelude for the much needed long-term trend monitoring that is important for governments to prioritize actions. The recent global interest in ecological monitoring not only contributes in improving information accessibility, but in generating more data for environmental policy-making (Hezri and Dovers 2006).

3 Measuring Difficulties

The multifaceted nature of sustainable agriculture, with three interdependent and interactive components (ecological, social, and economic) causes difficulty in monitoring. Therefore, a number of indicators are currently emerging the measurement of the different components. Norman et al. (1997) noted, at least three major challenges remain:

- The measures currently available generally fall short in terms of assessing the interactions and interdependencies among the three components and the trade-offs of pursuing one component at the expense of another.
- Many of the measures or indicators currently available are not particularly useful to farmers or are too time-consuming to measure in their day-to-day work, making it difficult for them and their families to monitor progress in terms of agricultural sustainability. This is particularly regrettable because many of the issues relating to sustainable agriculture are location or situation specific.
- Most indicators show progress or no progress towards specific components of sustainability, but they fall short in terms of helping to determine cause/effect relationships to help assess current problems and provide ideas on what needs to be done to ensure continued progress towards sustainability. An additional complication is that some strategies relating to sustainable agriculture require 5–10 years (e.g., a full crop rotation) of implementation before they result in visible or measurable signs of payoff.

Although a large number of indicators have been developed, they do not cover all dimensions and those levels noted in Table 2. Due to variation in biophysical and socioeconomic conditions, indicators used in one country are not necessarily applicable to other countries (Rasul and Thapa 2003). Therefore, indicators should be location specific, constructed within the context of contemporary socioeconomic situation (Dumanski and Pieri 1996).

Moreover, sustainable agriculture is a dynamic rather than static concept. What may contribute towards sustainability today may not work as the system changes, thus requiring a high level of observation and skills that can adapt to change. Consequently, sustainability is a direction/process and does not by itself result in a final fixed product, making it even more difficult to monitor and/or measure (Norman et al. 1997).

Table 2 Basic dimensions and conforming levels to assess agricultural sustainability

Dimensions	Levels
Normative	Ecological aspects
	Economic aspects
	Social aspects
Spatial	Local
	Regional
	National
Temporal	Long-term
	Short-term

von Wirén-Lehr 2001

4 Components of Sustainability Measurement

System theory has proven valid for sustainability assessment. First, it contributes to clarifying the conditions of sustainability. By definition, system theory forces one to define the boundaries of the system under consideration and the hierarchy of aggregation levels. In agricultural land use systems the most relevant subsystems (or levels) are the cropping system (plot level); farming system (farm level); watershed/village (local level); and landscape/district (regional level). Higher levels (national, supranational, and global) influence agriculture more indirectly by policy decisions or large-scale environmental changes (e.g., acid rain or global warming).

By identifying the system hierarchy, externalities between levels and tradeoffs among components can be traced and explicitly taken into consideration. For example, in an agro-ecological system analyzed at the farm level, the effects of national policies are externalities as long as they are outside the decision context of the farmer (Olembo 1994). Typical tradeoff among components within a farming system includes unproductive fallow lands in a rotation system for the sake of soil recovery for future use. In resource economics the aspect of externalities has gained great importance in that methodologies are being developed to convert such externalities into accountable quantities (Steger 1995), as well as the assignment of "opportunity costs" to tradeoff effects.

Similarly, the "tragedy of the commons" i.e., individual use of common resources can be analyzed adequately only by considering the higher system level to find proper policies for sustainable use e.g., the case of overgrazing in pastoral societies. Such conflicting interests among different groups – or hierarchical levels of the system – is a typical problem in sustainability strategies. Problem analysis is greatly facilitated by system theory to derive alternative scenarios of future development, depending on the policy chosen (Becker 1997).

Thus, agricultural sustainability not only is a difficult concept to define but also is difficult to implement and monitor/measure. This complexity is demonstrated in Table 3 which shows the expected interactions among the three components of sustainability and the five levels of influence. Although sustainability tends to be

Table 3 Interacting components of sustainability[a]

Levels influencing sustainability	Components of sustainability		
	Ecological	Economic	Social/institutional
International	Secondary	Secondary	Secondary
National	Secondary	Secondary	Primary
Community	Secondary	Primary	Primary
Farm	Primary	Primary	Primary
Field	Primary	Secondary	Secondary

[a]The 'primary' cells represent where the component of sustainability is mainly expressed, and the 'secondary' cells represent other factors that can influence sustainability (Norman et al. 1997)

locational or site specific (at the field, farm, and community levels), as Norman et al. (1997) noted, it is very much influenced by:

1. *What happens at the higher levels?* National policies have a great influence on ecological and economic sustainability at the field/farm levels. Other policies at that level related to social/institutional issues also can have major effects on the viability/welfare of communities and, hence, on quality of life. International markets and influences (particularly in smaller countries) are increasingly affecting what happens at the lower levels. Such influences tend to be relatively greater in countries that are poor (low income) and/or where agricultural production is influenced heavily by the export market. Thus, it is necessary to understand the interaction between these levels, because "each level finds its explanations of mechanism in the levels below, and its significance in the levels above" (Bartholomew 1964; Hall and Day 1977).
2. *Interactions among the sustainability components.* In the focus group discussions with Kansas farmers, some of them indicated that those who were in conventional agriculture were often on an economic treadmill e.g., having to raise enough money to service debts and hence had little time to consider ecological sustainability issues. They also had to make compromises concerning quality of life because of having to work very long hours. In fact, the prevailing attitude among the farmers was that all three components of sustainability (environmental, economic, and social) had to be pursued at the same time, if progress was to be achieved (Norman et al. 1997). A more extreme example of the potentially negative interactions among the components of sustainability occurs in many low income countries, where a close link has been established between poverty and ecological degradation. In parts of West Africa, for example, population pressures and low incomes are forcing farmers to cultivate land that is not suitable for agriculture. They are aware of the problems of doing this, but the short-run economic needs of survival are forcing them to sacrifice long-run ecological sustainability (Ibid). In such a situation, ensuring ecological sustainability without solving the problems of poverty and population pressure on the land is impossible (World Bank 1992).

According to three components of sustainability, Zhen and Routray (2003), proposed operational indicators for measuring agricultural sustainability. These indicators are summarized in Fig. 1:

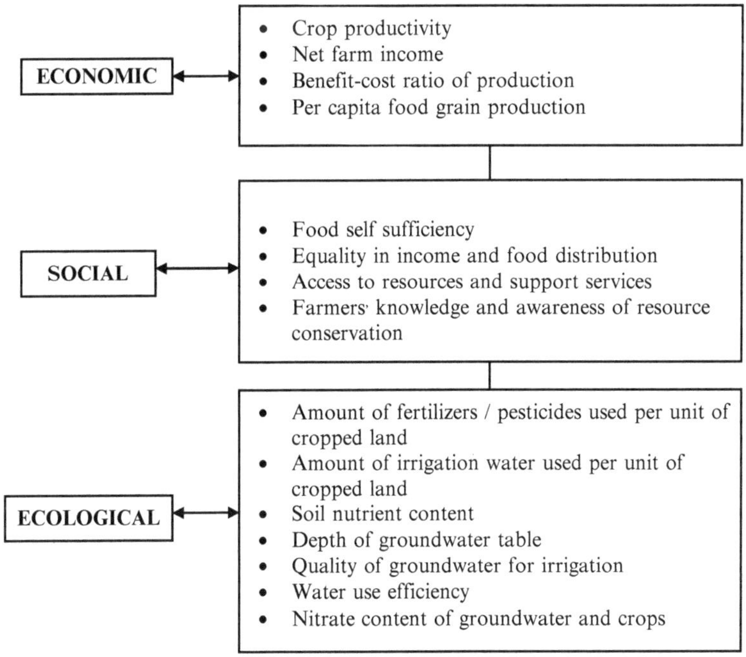

Fig. 1 Proposed agricultural indicators for measuring sustainability (Zhen and Routray 2003)

5 Criteria for Indicators Selection

Considering sustainable agriculture in the global context, preliminary indicators were developed for assessing agricultural sustainability. The preliminary indicators meet the following suitability criteria (Nambiar et al. 2001):

1. Social and policy relevance (economic viability, social structure, etc.)
2. Analytical soundness and measurability
3. Suitable for different scales (e.g. farm, district, country, etc.)
4. Encompass ecosystem processes and relate to process oriented modeling
5. Sensitive to variations in management and climate
6. Accessible to many users (e.g. acceptability)

Table 4, developed by Becker (1997), presents criteria for the selection and evaluation of sustainability indicators. The first demand on sustainability indicators is their scientific validity (BML 1995). Bernstein (1992) demanded that "the ideal trend indicator should be both ecologically realistic and meaningful and managerially useful." These two key properties should be complemented by the requirement that appropriate indicators be based on the sustainability paradigm (cf. RSU 1994). This last property explicitly introduces the normative element, guiding selection of the indicator according to the value system of the respective author, institution, or society (Becker 1997).

Table 4 Criteria for the selection and evaluation of sustainability indicators (Becker 1997)

Scientific quality	Ecosystem relevance	Data management	Sustainability paradigm
• Indicator really measures what it is supposed to detect • Indicator measures significant aspect • Problem specific • Distinguishes between causes and effects • Can be reproduced and repeated over time • Uncorrelated, independent • Unambiguous	• Changes as the system moves away from equilibrium • Distinguishes agro-ecosystems moving toward sustainability • Identifies key factors leading to sustainability • Warning of irreversible degradation processes • Proactive in forecasting future trends • Covers full cycle of the system through time • Corresponds to aggregation level • Highlights links to other system levels • Permits tradeoff detection and assessment between system components and levels • Can be related to other indicators	• Easy to measure • Easy to document • Easy to interpret • Cost effective • Data available • Comparable across locus and over time quantifiable • Representative • Transparent • Geographically relevant • Relevant to users • User friendly • Widely accepted	• What is to be sustained? • Resource and efficiency • Carrying capacity • Health protection • Target values • Time horizon • Social welfare • Equity • Participatory definition • Adequate rating of single aspects

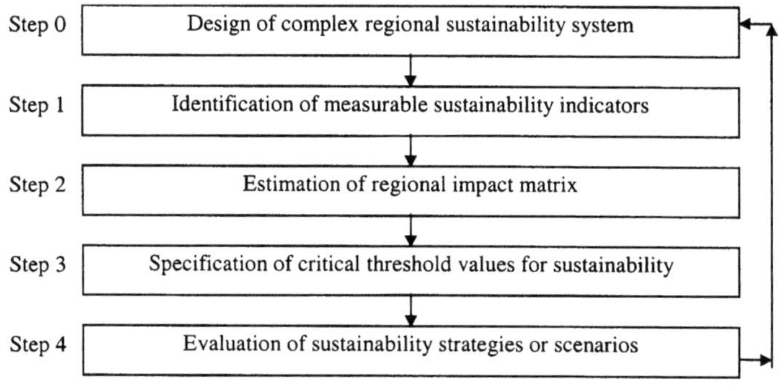

Fig. 2 Steps in a sustainability assessment procedure (Nijkamp and Vreeker 2000)

In the regional sustainability assessment Nijkamp and Vreeker (2000) presented the following steps (Fig. 2). Clearly, various feedback mechanisms and/or iterative steps may also be envisaged and included in this stepwise approach. It goes without saying that the above simplified and schematic general framework for a regional sustainability assessment study is fraught with various difficulties of both a theoretical/methodological and empirical/policy nature (Bithas et al. 1997).

6 Indicators of Agricultural Sustainability

Two basic approaches to sustainability assessment have been developed: First, the exact measurement of single factors and their combination into meaningful parameters. Second, indicators as an expression of complex situations, where an indicator is "a variable that compresses information concerning a relatively complex process, trend or state into a more readily understandable form" (Harrington et al. 1993).

The term sustainability indicator will be used here as a generic expression for quantitative or qualitative sustainability variables. According to WCED (1987) and Conway's (1983) definitions, which focuses on productivity trends, both quantitative and qualitative variables concentrate on the dynamic aspect of sustainability over time. Indicators to capture this aspect belong to the group of trend indicators, while state indicators reflect the condition of the respective ecosystem (Bernstein 1992). In developing environmental indicators for national and international policies it has become common practice to distinguish pressure, state, and response indicators (OECD 1991; Adriaanse 1993; Hammond et al. 1995; Pieri et al. 1995; Winograd 1995). An overview on current sustainability indicators is presented in Table 5.

Extensive set of indicators including biophysical, chemical, economic and social can be used to determine sustainability in a broader sense (Nambiar et al. 2001). These indicators are:

Table 5 Indicators and parameters for sustainability assessment (Becker 1997)

Economic indicators	Environment indicators
• Modified gross national product	• Yield trends
• Discount rates	• Coefficients for limited resources
– Depletion costs	– Depletion rates
– Pollution costs	– Pollution rates
• Total factor productivity	• Material and energy flows and balances
• Total social factor productivity	• Soil health
– Willingness to pay	• Modeling
– Contingent valuation method	– Empirical
• Hedonic price method	– Deterministic-analytical
• Travel cost approach	– Deterministic-numerical
	• Bio-indicators
Social indicators	Composite indicators
• Equity coefficients	• Unranked lists of indicators
• Disposable family income	• Scoring systems
• Social costs	• Integrated system properties
• Quantifiable parameters	
• Participation	
• Tenure rights	

6.1 Crop Yield

Long-term crop yield trends to provide information on the biological productive capacity of agricultural land and the ability of agriculture to sustain resource production capacity and manage production risks.

6.2 Agricultural Nutrient Balance

Excessive fertilizer use can contribute to problems of eutrophication, acidification, climate change and the toxic contamination of soil, water and air. Lack of fertilizer application may cause the degradation of soil fertility. The parameters of agriculture nutrient balance are gross nutrient balance (*B*) and input: output ratio (I/O). Gross nutrient balances of the total quantity of N, P and K, respectively, applied to agricultural land through chemical fertilizers and livestock manure, input in irrigation, rain and biological fixation minus the amount of N, P and K absorbed by agricultural plants, run-off, leaching and volatilization.

6.3 Soil Quality

Soil quality indicators include physical properties, e.g. soil texture, soil depth, bulk density, water holding capacity, water retention characteristics, water content, etc., chemical properties, e.g. total organic C and N, organic matter, pH, electrical conductivity, mineral N, extracted P, available K, etc., and biological properties, e.g. microbial biomass C and N, potentially mineralisable N, soil respiration, biomass C/total organic C ratio, respiration: biomass ratio, etc.

6.4 Agricultural Management Practices

Management and the type of fertilizers and irrigation systems will affect the efficiency of fertilizer, pesticide and water use. Agricultural management indicators here include efficiencies of fertilizer, pesticide, and irrigated water uses.

6.5 Agri-Environmental Quality

These agri-environmental indicators provide information on environmental impacts from the production process. Degrees of soil degradation and water pollution are included. The degree of soil degradation is measured by the effects of water and wind erosion, Stalinization, acidification, toxic contaminants, compaction, water logging and declining levels of soil organic matter. The quality of surface, ground and marine water is measured by concentrations in weight per liter of water of nitrogen, phosphorus, dissolved oxygen, toxic pesticide residues, ammonium and soil sediment.

6.6 Agricultural Biodiversity

Biodiversity of plants and livestock used for agricultural production is important to conserve the agro-ecosystem balance. However, the dependence on a limited number of varieties and breeds for agricultural production may increase their susceptibility to pests and diseases. Biodiversity measurement is reflected by the total number of varieties/breeds used for the production of major crops/livestock, and the number of animals and microorganisms in the production.

6.7 Economic and Social

Aspects and sustainable agriculture sustainability of agroecosystems is reflected not only in environmental factors but also in economic soundness and social considerations. These aspects are included as real net output (real value of agricultural production minus the real cost), and the change in the level of managerial skills of farmers and land managers in income and farming practice.

6.8 Agricultural Net Energy Balance

Agriculture not only uses energy such as sunlight and fossil fuels, but also is a source of energy supply through biomass production.

Principles and criteria derived from the function of the agro-ecosystem have been presented in Table 6. With respect to the “environmental pillar”, its function is connected with the management and conservation of natural resources and fluxes within and between these resources. Natural resources provided by ecosystems are water, air, soil, energy and biodiversity (habitat and biotic resources).

Regarding the “economic pillar”, its function in the agro-ecosystem is to provide prosperity to the farming community. In addition, each agro-ecosystem has several social functions, both at the level of farming community and at the level of society. The definition of these functions is based on present-day societal values and concerns. Farming activities should be carried out with respect of the quality of life of the farmer and his family. The agro-ecosystem needs to

Table 6 List of principles and criteria derived from the functions of the agro-ecosystem

Principles	Criteria
	Environmental pillar
	Air
Air quality is maintained or enhanced.	Supply (flow) of quality air function
Wind speed is adequately buffered.	Air flow buffering function
	Soil
Soil loss is minimized.	Supply (stock) of soil function
Soil chemical quality is maintained or increased.	Supply (stock) of quality soil function
Soil physical quality is maintained or increased.	
Soil mass flux (mudflows, landslides) are adequately buffered.	Soil flow buffering function
	Water
Adequate amount of surface water is supplied.	Supply (flow) of water function
Adequate amount of soil moisture is supplied.	
Adequate amount of groundwater is supplied.	
Surface water of adequate quality is supplied.	Supply (flow) of quality water function
Soil water of adequate quality is supplied.	
Groundwater of adequate quality is supplied.	
Flooding and runoff regulation of the agro-ecosystem is maintained or enhanced.	Water flow buffering function
	Energy
Adequate amount of energy is supplied.	Supply (flow) of energy function
Energy flow is adequately buffered.	Energy flow buffering function
	Biodiversity
Planned biodiversity is maintained or increased.	Supply (stock) of biotic resources function
Functional part of spontaneous biodiversity is maintained or increased.	
Heritage part of spontaneous biodiversity is maintained or increased.	
Diversity of habitats is maintained or increased	Supply (stock) of habitat function
Functional quality of habitats is maintained or increased.	Supply (stock) of quality habitat function
Flow of biotic resources is adequately buffered.	Biotic resource flow buffering function
	Economic pillar
	Viability
Farm income is ensured.	Economic function
Dependency on direct and indirect subsidies is minimized.	
Dependency on external finance is optimal.	
Agricultural activities are economically efficient.	
Agricultural activities are technically efficient.	
Market activities are optimal.	
Farmer's professional training is optimal.	
Inter-generational continuation of farming activity is ensured.	
Land tenure arrangements are optimal.	
Adaptability of the farm is sufficient.	

(continued)

Table 6 (continued)

Principles	Criteria
	Environmental pillar
	Social pillar
	Food security and safety
Production capacity is compatible with society's demand for food.	Production function
Quality of food and raw materials is increased.	
Diversity of food and raw materials is increased.	
Adequate amount of agricultural land is maintained.	
	Quality of life
Labour conditions are optimal.	Physical well-being of the farming community function
Health of the farming community is acceptable.	
Labour conditions are optimal.	Psychological well-being of the farming
Health of the farming community is acceptable.	
Internal family situation, including equality in the man–woman.	Community function
relation is acceptable.	
Family access to and use of social infrastructures and services is acceptable.	
Family access to and participation in local activities is acceptable.	
Family integration in the local and agricultural society is acceptable.	
Farmer's feeling of independence is satisfactory.	
	Social acceptability
Amenities are maintained or increased.	Well-being of the society function
Pollution levels are reduced.	
Production methods are acceptable.	
Quality and taste of food is increased.	
Equity is maintained or increased.	
Stakeholder involvement is maintained or increased.	
Educational and scientific value features are maintained or increased.	Cultural acceptability Information function
Cultural, spiritual and aesthetic heritage value features are maintained or increased.	

Van Cauwenbergh et al. 2007

be organized in such a way that social conditions are optimal for the people who work on farms. This refers to the physical well-being (labour conditions and health) and the psychological well-being (education, gender equality, access to infrastructure and activities, integration and participation in society both professionally and socially, feeling of independence) of the farm family and its workers.

7 Agricultural Sustainability Scales at National Level

Assessing and implementing sustainability in agriculture can be undertaken by using goal-oriented strategy approaches according to von Wirén-Lehr (2001). These approaches outlined in Fig. 3 include four fundamental steps, which are:

7.1 Goal Definition

Since goal definition represents the basis of strategies, it determines all subsequent steps as well as the whole methodological framework. Corresponding to

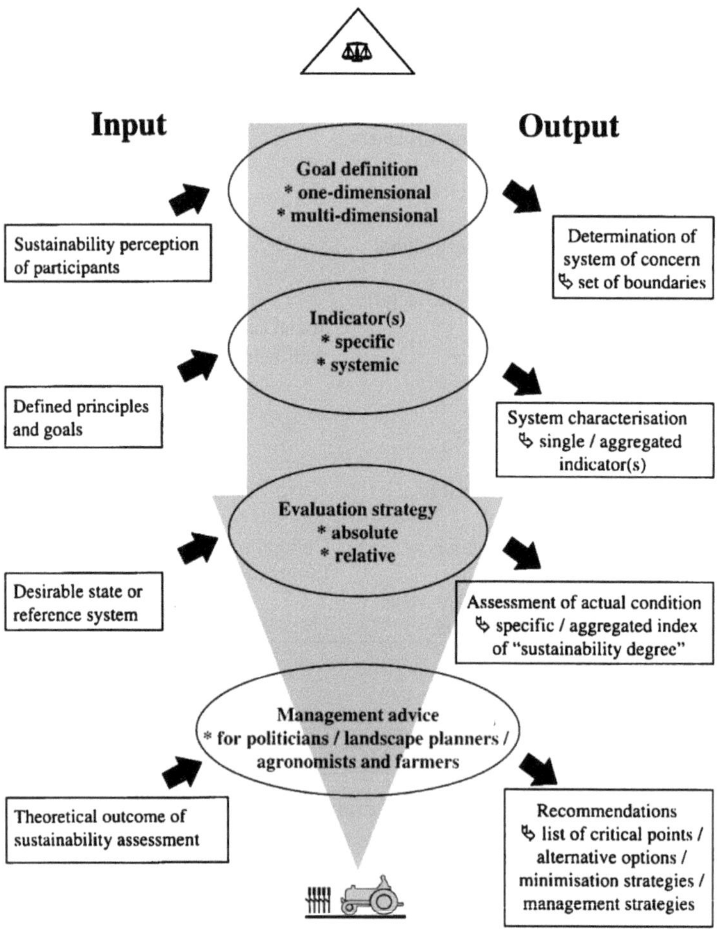

Fig. 3 Basic features of four-step strategies to assess and implement sustainability in agriculture. Frames present required data influx (*left frames*) and expected outcome (*right frames*) of feature derivation (von Wirén-Lehr 2001)

the general multidimensional sustainability paradigm, definitions of sustainable agriculture have to include ecological, economic and social aspects with respect to their diverse spatial and temporal scales (Allen et al. 1991; Herdt and Steiner 1995; Christen 1996). Even though this holistic approach integrates all principles of the theoretical term, its applicability is considerably reduced by the high complexity. Hence, a first step must be to condense the holistic sustainability perception, to restrain definitions on single selected principles and to define aims and systems of concern.

Depending on the priorities of participants and target groups, goal definitions may concentrate on one single (one-dimensional goal definition) or various selected dimensions (multidimensional goal definition). In the agricultural sector, the normative focus of sustainability perception is predominantly based on ecological and/or economic aspects (Crews et al. 1991; Dunlap et al. 1992; Neher 1992; Farshad and Zinck 1993). However, to ensure successful implementation of sustainable systems, management advice has to be strongly adapted to the requirements and abilities not only of target groups but of all groups concerned, for example, also political stakeholders or customers. They should be included in the conceptual work from the beginning. Consequently, concepts to assess and implement sustainability in agriculture have to enhance co-operation not only between different scientific sections but also between divergent socio-professional groups (Giampietro and Bukkens 1992; Flora 1995). Essential for this interdisciplinary work is a separate survey of normative options, e.g. setting of goals and objective parameters (e.g. agro-technical options) permitting every participant or user to verify the fundamental conditions of the work.

7.2 *Indicators*

All goal-oriented concepts deduce single indicators or indicator sets to 'translate' the defined principles. Indicators represent a powerful tool both to reduce the complexity of system description and to integrate complex system information (Giampietro 1997). Hence, indicators have to be deduced for different systems such as agricultural production systems or other ecosystems, e.g. forests or lakes and at diverse spatio-temporal scales. If the agricultural production system is considered as one compartment of a whole cultured landscape, indicator sets have to provide information not only on imbalances, e.g. releases and deficits of the agricultural production system itself, but also on the external deposition and off-site effects of emissions resulting from agricultural production, e.g. toxic effects in natural aquatic ecosystems due to pesticide residues. Two types of indicators can be distinguished according to their focus of characterization such as:

- Specific indicators, characterising single parts of the system of concern (Nieberg and Isermeyer 1994; Bockstaller et al. 1997)
- Systemic indicators, describing key functions and processes of systems as a whole (Beese 1996; Müller 1998; Xu et al. 1999)

7.3 Evaluation Strategies

Evaluation strategies enable the determination of the sustainability of systems under investigation. They are based on the previously characterised sustainability perception, goal definitions and selected indicators or indicator sets. The evaluation process represents one of the most delicate parts of the concept. First, evaluation ultimately depends on normative options concerning setting of goals, selection of systems of concern and deduction of threshold values or ranges of tolerance (Finnveden 1997). Second, the evaluation of systems based on sets of single indicators ultimately remains inadequate since systemic sustainability represents ‘more than the sum of the parts’.

Two strategies of sustainability evaluation may be distinguished – absolute and relative strategies.

- Absolute evaluation procedures exclusively investigate indicators and corresponding data derived from one single system. Hence, validation is based on a comparison with previously defined margins of tolerance or distinct threshold values for each selected indicator (Mitchell and McDonald 1995). These limits are determined either by estimation, e.g. resulting from expert interviews or referring to socio-political postulates for the reduction of emissions or by scientific deduction, e.g. elaboration of critical loads/levels based on eco-toxicological experiments. Therefore, absolute evaluation assesses distinct datasets e.g. the phosphorus content of the soil compared to the maximum tolerable content. This transparent presentation of results permits end-users to verify the assessment and – if necessary – to adapt the presented data to alternative threshold values.
- Relative evaluation procedures are established on a comparison of different systems among themselves or with selected reference systems. Due to this comparative assessment of systems, there is no need to define distinct margins of tolerance or threshold values. Frequently the results of a relative evaluation are presented as normative point scores.

7.4 Management Advice for Practical Application

The development of management advice for practical application represents the last step for adapting the theoretical outcome of sustainability assessments into implementation of agricultural practice. These recommendations support end-users either in planning new, sustainable production systems or to improve the sustainability of existing systems. The elaboration of management advice considerably varies with respect to the needs and knowledge of the target group, e.g. farmers, political stakeholders or landscape planners.

Table 7 Applied indicators in the agricultural policy scenario analysis (Lehtonen et al. 2005)

Applied indicator	Measured quality	Indicator reflecting	Strategic goal of indicator
Total number of animal units up to 2020	Animal units	The scales and long-term economic viability of aggregate animal production	To conclude the relative economic viability of animal production in different policy scenarios
Number of bovine animal units	Animal units	The scales and long-term economic viability of dairy and beef production	To conclude the relative economic viability of dairy and beef production in different policy scenarios
Number of pig animal units	Animal units	The scales and long-term economic viability of pig production	To conclude the relative economic viability of pig production in different policy scenarios
Number of poultry animal units	Animal units	The scales and long-term economic viability of poultry production	To conclude the relative economic viability of poultry production in different policy scenarios
Total cultivated area (excluding set-aside) up to 2020	Hectares	Incentives for active crop production	Changes in incentives for active crop production
Set-aside area	Hectares	Incentives for fulfilling cross compliance criteria and minimizing costs	Changes in incentives in fulfilling cross compliance criteria and minimizing costs in different policy scenarios
Unused area	Hectares	Share of abandoned agricultural land due to unprofitable production	Changes in the share of abandoned land due to unprofitable production in different policy scenarios
Grass area	Hectares	The scales of gross feed production; incentive for gross feed use and bovine animal production	Changes in scales and incentive for gross feed production in different policy scenarios
Grain area	Hectares	The scales and incentive for grain production	Changes in scales and incentive for grain production in different policy scenarios
Nitrogen balance on cultivated area[a]	Kilogram per hectare	Nitrogen leaching potential from cultivated land	Changes in nitrogen leaching potential in different policy scenarios
Phosphorous balance on cultivated area[a]	Kilogram per hectare	Phosphorous leaching potential from cultivated land	Changes in phosphorous leaching potential in different policy scenarios
Agricultural income	Money unit	The level of economic activities in agriculture	Changes in the level of economic activities in different policy scenarios

(continued)

Table 7 (continued)

Applied indicator	Measured quality	Indicator reflecting	Strategic goal of indicator
Profitability coefficient[b]		Profitability of agricultural production	Changes in profitability of agricultural production in different policy scenarios
Labour hours in agriculture	Million hours	Social sustainability of farmers, the working conditions of agricultural labour	Changes in the number of people employed in agriculture in different policy scenarios
Agricultural income per hour of labour	Money per hour	Economic and social welfare of farmers	Changes in the economic and social viability of agriculture in different policy scenarios

[a]The soil surface nitrogen and phosphorus balances are calculated as the difference between the total quantity of nitrogen or phosphorus inputs entering the soil and the quantity of nitrogen or phosphorus outputs leaving the soil annually, based on the nitrogen or phosphorus cycle

[b]The Profitability coefficient is a ratio obtained when the agricultural surplus is divided by the sum of the entrepreneur family's salary requirement and the interest requirement on capital invested

Further management advice is provided by lists of critical points indicating parts of systems which diverge from the desired state and consequently should be improved. However, lists of critical points which result from a separate evaluation of selected indicators represent case- and site-specific information with limited transferability to different agricultural systems. Since they do not provide any information on how to improve the indicated 'hot spots', their direct applicability in agricultural practice is considerably restricted. It obligates end-users, e.g. farmers and agronomists to interpret and weigh by themselves the presented set of results to develop a corresponding improvement strategy. 'One-solution strategies' resulting from lists of critical points (like strategies exclusively improving nutrient balances) are considered inappropriate to reflect the systemic aspect of sustainability. To enhance successful implementation, case- and site-specific advice should be provided indicating alternative management strategies to optimise the system under investigation.

The most elaborate assistance to the target group is supplied by the formulation of entire improved management strategies. Since the management of agricultural systems is strongly dependent on variable natural conditions, e.g. soil or climate but also on socio-political constraints, e.g. subventions of certain crops or statutory limitations of factor input, final design of these management strategies has to be performed in a case- and site-specific manner in co-operation with end-users (von Wirén-Lehr 2001).

A set of applied indicators for sustainability in different agricultural policy scenarios at the national level is presented by Lehtonen et al. (2005). Their purpose is to provide material for an interactive policy dialogue rather than assemble a comprehensive and conclusive assessment of sustainability of various agricultural policy alternatives (Table 7). They also present what kind of agricultural development each indicator is reflecting and the strategic goal of each specific indicator. It is important to realize that not only the numerical values of the calculated indicators but also their relative changes over time are important when evaluating the sustainability of alternative agricultural policies.

8 Agricultural Sustainability Scales at Farm Level

The indicators discussed here draw on Taylor et al. (1993). In their paper the index is constructed for a sample of 85 agricultural producers in Malaysia with points scored under the headings of (i) insect control, (ii) disease control, (iii) weed control, (iv) soil fertility maintenance and (v) soil erosion control. Gomez et al. (1996) also construct a farm level index of sustainability where six aspects of sustainability are monitored: (i) yield, (ii) profit, (iii) frequency of crop failure, (iv) soil depth, (v) organic C and (vi) permanent ground cover. The following indicators were then constructed for a sample of ten farms from the Guba region of the Philippines (Rigby and Caceres 2001):

- Improved farm-level social and economic sustainability
  - Enhances farmers' quality of life (US Farm Bill 1990)
  - Increases farmers' self-reliance (Pretty 1995)
  - Sustains the viability/profitability of the farm (Pretty 1995; US Farm Bill 1990; Ikerd 1993)
- Improved wider social and economic sustainability
  - Improves equity (Pretty 1995), 'socially supportive' (Ikerd 1993)
  - Meets society's needs for food and fiber (US Farm Bill 1990)
- Increased yields and reduced losses while
  - Minimising off-farm inputs (Hodge 1993; Pretty 1995; US Farm Bill 1990)
  - Minimising inputs from non-renewable sources (Hodge 1993; Ikerd 1993; Pretty 1995; US Farm Bill 1990)
  - Maximising use of (knowledge of) natural biological processes (Pretty 1995; US Farm Bill 1990)
  - Promoting local biodiversity/'environmental quality' (Hodge 1993; Pretty 1995; US Farm Bill 1990).

Senanayake (1991) proposed that agricultural systems have varying degrees of sustainability according to the level of external inputs required to maintain the system that the state of the biotic community within a system operates. His index was in the shape of an equation:

$$S = f\left(E_i, E_r, P_e, S_e, R_s, R_b\right)$$

S = Index of ecological sustainability
E_i= External input
E_r = Energy ratio
P_e = Power equivalent

S_e = Efficiency of solar flux use
R_s = Residence time of soil
R_b = Residence time of biotic

Each parameter has its own possible states ranging from two to three. For instance, the three possible states of E_i are listed as 0.1, 0.5 and 1.0. E_i is seen to be more sustainable at lower values.

The terms R_s and R_b are such that only two possible states exist, namely zero and one. In the zero state the farming category is unsustainable no matter what its other measures are. In the value state, the farming type is sustainable, but the degree of sustainability depends on the values of other parameters. In terms of agricultural sustainability:

$$S = R_s \times R_b / R_s \times R_b \left[f\left(v_e\right) - f\left(v_d\right) \right]$$

where
$v_e = f\,(S_e, P_r)$
$V_d = f\,(E_i, E_r, P_e)$

Thus, any farming system type that contributes to physical erosion or a high rate of soil biomass loss will yield a value of zero and can be termed non sustainable. A farming type that conserves these basic resources will demonstrate a positive value, and therefore be termed potentially sustainable.

Hayati and Karami (1996) suggested an operational index to measuring agricultural sustainability trend in farm level. The parameters measured in that method are those factors that intervene in the crop production process and could have positive effect in the process. The measurement is summarized in below equation:

$$S = f\left[\sum_{i=1}^{8} Xi, \sum_{j=1}^{3} Yj \right]$$

S = Trend of sustainability
X_1 = Average of crop production per hectare
X_2 = Execution of crop rotation
X_3 = Usage of organic manures
X_4 = Usage of green manures
X_5 = Usage of crop stubble
X_6 = Usage of conservational plough
X_7 = Trend of change in water resources (at the farm)
X_8 = Trend of change in soil resources (at the farm)
Y_1 = Amount of pesticides, herbicides, and fungicides consumption in the farm in one cultivational season
Y_2 = Amount of nitrate fertilizer consumption per 1 t of crop production
Y_3 = Amount of phosphate fertilizer consumption per 1 t of crop production

In fact, parameters of X_1 till X_8 could lead to more sustainability if they increase and parameters of Y_1 till Y_3 could lead to unsustainability if they increase. Thus the below equation is established:

$$S = \sum_{i=1}^{8} Xi - \sum_{j=1}^{3} Yj$$

In order to measure agricultural sustainability at the farm level, Saltiel et al. (1994) presented an index which is constituted of seven components. They are: cultivation of sustainable crops, conservational cultivation, crop rotation, diminishing of pesticides and herbicides usage, soil mulching, and use of organic fertilizers.

9 Conclusion

The main difficulty in measuring and monitoring agricultural sustainability is that it is a dynamic rather than static concept and needs high level of observation and skills that can adapt to change. Whereas most agricultural scholars believe that measuring sustainability at the farm level is the most precise method, policies at the higher levels (such as national) are increasingly affecting at the lower levels (such as farm). It is necessary to understand the interaction between all levels because each level finds its explanation of mechanism in the level below, and its significance in the levels above.

Moreover, the level of analysis chosen can be a significant influence on the diagnosis of sustainability. At the field level, particular soil management, grazing and cropping practices will be the most important determinants of sustainability. At the farm level, sustainable resource use practices need to support a sustainable farm business and family household. At the national level, there may be broader pressures on the use of agricultural land from non-farming sectors, and at the global level, climatic stability, international terms of trade and distribution of resources also become important determinants.

Although sustainability is a global concept and a farm is only a small subsystem that interacts in various ways with surrounding systems, indicators are needed to know whether a farm system is moving towards or away from sustainability. Indicators can also be used to educate farmers and other stakeholders about sustainable production. Furthermore, indicators provide farmers with a tool to measure their achievements toward sustainability. Further, indicators allow for comparisons between farms' performance in the economic, social and environmental aspects of their production. Indicators also inform policy makers about the current state and trends in farm performance or sector performance. Sustainability performance measures can be used as input for policy tools and stimulate better integration of decision-making. Finally, sustainability indices can encourage public participation in sustainability discussions.

While no measure of sustainability can be perfect, the sustainable value is a useful measure and describes the current sustainability performance. On the other hand, the 'sustainable efficiency' indicator can be used to compare and rank farms. Besides, in view of the fact that biophysical and socioeconomic conditions of countries are different to each other, those indicators which are developed and used in one country may not applicable to other countries.

Some recommendations to selecting indicators in order to more appropriate measuring of agricultural sustainability are:

- Necessity to adoption of a systemic approach
- Establishment and gathering appropriate data base and other necessary information in shape of time series in developing countries
- More emphasis on determining of sustainability trend instead of precision determining amount of sustainability, especially with respect to lack of accessing such data in developing countries
- Launch of professional institutes to monitoring and measuring sustainability of agricultural and industrial systems
- Develop those indicators which be feasible to implementing, meanwhile responsive and sensitive toward any stresses and manipulation on system

References

Adriaanse A (1993) Environmental pool performance indicators. A study on the development of indicators for environmental policy in the Netherlands. Uitgeverij, The Hague

Allen P, van Dusen D, Lundy J, Gliessman S (1991) Integrating social, environmental and economic issues in sustainable agriculture. Am J Altern Agric 6:34–39

Altieri M (1995) Agroecology: the science of sustainable agriculture. West View Press, Boulder, CO

Bartholomew GA (1964) The role of physiology and behavior in the maintenance of homeostatic in the desert environment. In: Hughes GM (ed) Homeostasis and feedback mechanism. A symposium of the society for experimental biology, vol 18, Cambridge University Press. Cambridge, UK, pp 7–29

Becker B (1997) Sustainability assessment: a review of values, concepts, and methodological approaches. Consultative Group on International Agricultural Research, The World Bank, Washington, DC, USA, p 70

Beese F (1996) Indikatoren für eine multifunktionelle waldnutzung. Forstw Cbl 115:65–79

Bernstein BB (1992) A framework for trend detection: coupling ecological and managerial perspectives. In: McKenzie DH, Hyatt DE, McDonald VJ (eds) Ecological indicators, 2 vols. In: Proceedings of the international symposium on ecological indicators, Ft. Lauderdale, Florida, 16–19 Oct 1990. Elsevier, London/New York, pp 1101–1114

Beus CE, Dunlop RE (1994) Agricultural paradigms and the practice of agriculture. Rural Sociol 59(4):620–635

Bithas K, Nijkamp P, Tassapoulos A (1997) Environmental impact assessment by experts in cases of factual uncertainty. Proj Apprais 12(2):70–77

BML (Bundesministerium fur Ernahrung Landwirtschaft und Forsten) (1995). Synoptic portrait of similarities of the contents of existing criteria and indicator catalogues for sustainable forest management. Background paper by the Federal Ministry of Food, Agriculture, and Forestry, Bonn, prepared with assistance from the Federal Research Center for Forestry and Forest Industry

Bockstaller C, Girardin P, van der Werf HMG (1997) Use of agro-ecological indicators for the evaluation of farming systems. Eur J Agron 7:261–270
Bosshard A (2000) A methodology and terminology of sustainability assessment and its perspectives for rural planning. Agric Ecosyst Environ 77:29–41
Bouma J, Droogers P (1998) A procedure to derive land quality indicators for sustainable agricultural production. Geoderma 85:103–110
Christen O (1996) Nachhaltige landwirtschaft (sustainable agriculture). Ber Landwirtschaft 74:1–21
Conway GR (1983) Agroecosystem analysis. Imperial College of Science and Technology, London
Comer S, Ekanem E, Muhammad S, Singh S, Tegegne F (1999) Sustainable and conventional farmers: a comparison of socio-economic characteristics, attitude, and beliefs. J Sustain Agric 15(1):29–45
Crews TE, Mohler CL, Power AG (1991) Energetic and ecosystem integrity: the defining principles of sustainable agriculture. Am J Alter Agric 6:146–149
David S (1989) Sustainable development: theoretical construct on attainable goal? Environ Conser 16:41–48
Dumanski J, Pieri C (1996) Application of the pressure-state-response framework for the land quality indicators (LQI) program. In: Land quality indicators and their use in sustainable agriculture and rural development, p 41. Proceedings of the workshop organized by the Land and Water Development Division FAO Agriculture Department, Agricultural Institute of Canada, Ottawa, 25–26 Jan 1996
Dunlap RE, Beus CE, Howell RE, Waud J (1992) What is sustainable agriculture? an empirical examination of faculty and farmer definitions. J Sustain Agric 3:5–39
Farshad A, Zinck JA (1993) Seeking agricultural sustainability. Agric Ecosyst Environ 47:1–12
Finnveden G (1997) Valuation methods within LCA – where are the values? Int J LCA 2:163–169
Flora CB (1995) Social capital and sustainability: agriculture and communities in the great plains and corn belt. Res Rural Sociol Dev: A Res Annu 6:227–246
Gafsi M, Legagneux B, Nguyen G, Robin P (2006) Towards sustainable farming systems: effectiveness and deficiency of the French procedure of sustainable agriculture. Agric Syst 90:226–242
Giampietro M, Bukkens SGF (1992) Sustainable development: scientific and ethical assessments. J Agric Environ Ethics 5:27–57
Giampietro M (1997) Socioeconomic pressure, demographic pressure, environmental loading and technological changes in agriculture. Agric Ecosyst Environ 65:201–229
Gomez AA, Kelly DE, Syers JK, Coughlan KJ (1996) Measuring sustainability of agricultural systems at the farm level. Methods Assess Soil Qual SSSA Special Publication 49:401–409
Gowda MJC, Jayaramaiah KM (1998) Comparative evaluation of rice production systems for their sustainability. Agric Ecosyst Environ 69:1–9
Hall CA, Day JW (1977) Ecosystem modeling in theory and practice: an introduction with case studies. Wiley, New York
Hammond A, Adriaanse A, Rodenburg E, Bryant D, Woodward R (1995) Environmental indicators: a systematic approach to measuring and reporting on environmental policy performance in the context of sustainable development. World Resources Institute, Washington, D.C
Harrington L, Jones IG, Wino M (1993) Indicators of sustainability. Report brad. Measurements and a consultancy team. Centro International de Agricultural Tropical (CIAT), Cali, Colombia, p 631
Hayati D (1995) Factors influencing technical knowledge, sustainable agricultural knowledge and sustainability of farming system among wheat producers in Fars province, Iran. M.Sc. thesis presented in College of Agriculture, Shiraz Univ., Iran
Hayati D, Karami E (1996) A proposed scale to measure sustainability at farm level in socio-economic studies. Paper presented at first agricultural economic conference of Iran, Zabol, Iran, 5–7 April
Herdt RW, Steiner RA (1995) Agricultural sustainability: concepts and conundrums. In: Barnett V, Steiner R (eds) Agricultural sustainability: economic, environmental and social considerations. Wiley, Chichester/New York/Brisbane/Toronto/Singapore, p 257

Herzog F, Gotsch N (1998) Assessing the sustainability of smallholder tree crop production in the tropics: a methodological outline. J Sustain Agric 11(4):13–37
Hezri AA (2005) Utilization of sustainability indicators and impact through policy learning in the Malaysian policy processes. J Environ Assess Policy Manage 7(4):575–595
Hezri AA, Dovers SR (2006) Sustainability indicators, policy and governance: issues for ecological economics. Ecol Econ 60:86–99
Hodge I (1993) Sustainability: putting principles into practice. An application to agricultural systems. Paper presented to `Rural Economy and Society Study Group', Royal Holloway College, December
Horrigan L, Robert SL, Walker P (2002) How sustainable agriculture can address the environment and human health harms of industrial agriculture. Environ Health Perspect 110(5):445–456
Ikerd J (1993) Two related but distinctly different concepts: organic farming and sustainable agriculture. Small Farm Today 10(1):30–31
Ingels C, Campbell D, George MR, Bradford E (1997) What is sustainable agriculture? www.sarep.ucdavis.edu/concept.htm. Last accessed: October 12, 2008
Karami E (1995) Agricultural extension: the question of sustainable development in Iran. J Sustain Agric 5(1/2):61–72
Koeijer TJD, Wossink GAA, Struik PC, Renkema JA (2002). Measuring agricultural sustainability in terms of effciency: the case of Dutch sugar beet growers. J Environ Manage 66:9–17. http://www.idealibrary.com
Lehtonen H, Aakkula J, Rikkonen P (2005) Alternative agricultural policy scenarios, sector modeling and indicators: a sustainability assessment. J Sustain Agric 26(4):63–93
Lynam JK, Herdt RW (1989) Sense and sustainability: sustainability as an objective in international agricultural research. Agric Econ 3:381–398
Lyson TA (1998) Environmental, economic and social aspects of sustainable agriculture in American Land Grant Universities. J Sustain Agric 12(2,3):119–129
Mitchell G, McDonald A (1995) PICABUE: a methodological framework for the development of indicators of sustainable development. Int J Sustain Dev World Ecol 2:104–123
Miller CA (2005) New civic epistemologies of quantification: making sense of indicators of local and global sustainability. Sci Technol Hum Values 30(3):403–432
Müller F (1998) Ableitung von integrativen indikatoren zur bewertung von Ökosystem-Zuständen für die umweltökonomischen Gesamtrechnungen. Metzler-Poeschel, Stuttgart
Nambiar KKM, Gupta AP, Fu Q, Li S (2001) Biophysical, chemical and socio-economic indicators for assessing agricultural sustainability in the Chinese coastal zone. Agric Ecosyst Environ 87:209–214
Neher D (1992) Ecological sustainability in agricultural systems: definitions and measurement. J Sustain Agric 2:51–61
Nieberg H, Isermeyer F (1994) Verwendung von Umweltindikatoren in der Agrarpolitik. Report No. COM/AGR/CA/ENV/ EPOC (94)96. OECD, Paris
Nijkamp P, Vreeker R (2000) Sustainability assessment of development scenarios: methodology and application to Thailand. Ecol Econ 33:7–27
Norman D, Janke R, Freyenberger S, Schurle B, Kok H (1997) Defining and implementing sustainable agriculture. Kansas Sustainable Agriculture Series, Paper #1. Kansas State University, Manhattan, KS
Organization for Economic Co-operation and Development (OECD) (1997) Environmental indicators for agriculture. OECD Publication, Paris
Organization for Economic Co-operation and Development (OECD) (1991) Environmental indicators: a preliminary set organization for economic cooperation and development. OECD Publication, Paris
Olembo R (1994) Can land use planning contribute to sustainability? In: Fresco LO, Troosnijder L, Bouma J, Van Keulen H (eds) The future of the land: mobilising and integrating knowledge for land use options. Wiley, Chichester, U.K., pp 369–376
Pannell DJ, Glenn NA (2000) Framework for the economic evaluation and selection of sustainability indicators in agriculture. Ecol Econ 33:135–149

Pieri C, Dumanski J, Hamblin A, Young A (1995). Land quality indicators. World Bank Discussion Paper 315. Washington, D.C.

Praneetvatakul S, Janekarnkij P, Potchanasin C, Prayoonwong K (2001) Assessing the sustainability of agriculture: a case of Mae Chaem Catchment, northern Thailand. Environ Int 27:103–109

Pretty JN (1995) Regenerating agriculture: policies and practice for sustainability and self-reliance. Earthscan, London

Rasul G, Thapa GB (2003) Sustainability analysis of ecological and conventional agricultural systems in Bangladesh. World Dev 31(10):1721–1741

Rasul G, Thapa GB (2004) Sustainability of ecological and conventional agricultural systems in Bangladesh: an assessment based on environmental, economic and social perspectives. Agric Syst 79:327–351

Reijntjes C, Bertus H, Water-Bayer A (1992) Farming for the future: an introduction to low external input and sustainable agriculture. Macmillan, London

Rezaei-Moghaddam K (1997) Agricultural extension, poverty, and sustainable agriculture in Behbahan county. Thesis presented for M.Sc. degree in agricultural extension, College of agriculture, Shiraz University, Shiraz, Iran

Rigby D, Caceres D (2001) Organic farming and the sustainability of agricultural systems. Agric Syst 68:21–40

Rigby D, Woodhouse P, Young T, Burton M (2001) Constructing a farm level indicator of sustainable agricultural practice. Ecol Econ 39:463–478

RSU (Der Rat von Sachverstandigen fur Umweltfragen) (1994) Umweltgutacbten 1994. Verlag Metzler-Poesche, Stuttgart

Saltiel J, Baunder JW, Palakovich S (1994) Adoption of sustainable agricultural practices: diffusion, farm structure and profitability. Rural Sociol 59(2):333–347

Sands GR, Podmore H (2000) A generalized environmental sustainability index for agricultural systems. Agric Ecosyst Environ 79:29–41

Senanayake R (1991) Sustainable agriculture: definitions and parameters for measurement. J Sustain Agric 4(1):7–28

Smith CS, McDonald GT (1998) Assessing the sustainability of agriculture at the planning stage. J Environ Manage 52:15–37

Steger U (1995) Nachhaltige und dauerhafte Entwicklung aus wirnchaftswissenschaftlicher Sicht. In: Fritz I, Huber J, Levi HW (eds) Nachhaltigkeit in naturwirsenschaJtlicber und sozialwissenschaftlicher perspektive. Hirzel, Stuttgart, pp 91–98

Taylor D, Mohamed Z, Shamsudin M, Mohayidin X, Chiew E (1993) Creating a farmer sustainability index: a Malaysian case study. Am J Altern Agric 8:175–184

Tellarini V, Caporali F (2000) An input/output methodology to evaluate farms as sustainable agroecosystems: an application of indicators to farms in central Italy. Agr Ecosys Environ 77:111–123 doi: 10.1016/S0167-8809(99)00097-3

U.S. Farm Bill. Food, agriculture, conservation, and trade act of (1990) Public low 101-624, title XVI, subtitle A, Section 1603. Government Printing Office, Washington DC

Van Cauwenbergh N, Biala K, Bielders C, Brouckaert V, Franchois L, Cidad VG, Hermy M, Mathijs E, Muys B, Reijnders J, Sauvenier X, Valckx J, Vanclooster M, der Veken BV, Wauters E, Peeters A (2007) SAFE – a hierarchical framework for assessing the sustainability of agricultural systems. Agric Ecosyst Environ 120:229–242

Van Passel S, Nevens F, Mathijs E, Huylenbroeck GV (2006) Measuring farm sustainability and explaining differences in sustainable efficiency. Ecol Econ 62(1):149–161

von Wirén-Lehr S (2001) Sustainability in agriculture: an evaluation of principal goal oriented concepts to close the gap between theory and practice. Agric Ecosyst Environ 84:115–129

World Commission on Environment and Development (WCED) (1987) Our common future The Brundtland Report. Oxford University Press, Oxford, UK

Webster JPG (1997) Assessing the economic consequences of sustainability in agriculture. Agric Ecosyst Environ 64:95–102

Webster P (1999) The challenge of sustainability at the farm level: presidential address. J Agric Econ 50(3):371–387

Winograd M (1995) Environmental indicators for Latin America and the Caribbean: towards land-use sustainability. GASE (Ecological Systems Analysis Group, Bariloche, Argentina), in collaboration with IICA (Instituto Interamerican de Cooperacibn para la Agricultura)-GTZ (Deutsche Gesellschaft fiir Technische Zusammenarbeit) Project, OAS (Organisation of American States), and WRI (World Resources Institute). Washington, D.C.
World Bank (1992) World development report 1992. Oxford University Press, New York
Xu FL, Joergensen SE, Tao S (1999) Ecological indicators for assessing freshwater ecosystem health. Ecol Model 116:77–106
Zhen L, Routray JK (2003) Operational indicators for measuring agricultural sustainability in developing countries. Environ Manage 32(1):34–46

Sustainable Bioenergy Production, Land and Nitrogen Use

Enrico Ceotto and Mario Di Candilo

Abstract In recent years the challenge of reducing the reliance on petroleum and natural gas with the energy produced by agricultural crops has received a renewed interest. However, many scientists have expressed serious reservations about the real benefit of a widespread diffusion of crops grown for energy feedstocks. While a diversification of energy portfolio is strongly needed, one of the greatest scientific challenge for the near future is to identify land use options that minimize negative impact on food prices and greenhouse gases emissions. The objective of this article is to discuss the following topics: (i) competition for land: bioenergy versus food; (ii) bioenergy crops and nitrogen cycling; (iii) plant traits to be targeted for improving land and nitrogen use efficiency; and (iv) the debated role of legumes. Because fertile land, suitable for food production, is a dwindling resource, the production of feedstocks for biofuels should be enhanced by exploiting favourable plant characteristics in marginal land areas. We point out that a rethinking of the concept of marginal land is necessary: not only areas poorly suited to grain crops production owing to low soil fertility, but also land unsuited to produce food owing to food safety reasons. Yet, whether a land area is marginal or not should be evaluated not only from the economic standpoint, but also from the ecological and environmental points of view. Moreover, grain crops residues should be exploited for bioenergy production providing that well devised height of cuttings assure the maintenance of soil organic matter. The main message of this review is that bioenergy should be seen as a complementary product of food and feed production, to be attained by optimized land and nitrogen use. Emphasis is given to the contribution that dedicated perennial lignocellulosic crops might provide in sustainable bioenergy production.

Keywords Ecological footprint • Fertile and marginal land • Greenhouse gas emissions • Plant characteristics • Crop residues • Legumes • Nox • N_2O • Miscanthus

---

E. Ceotto (✉) and M. Di Candilo
C.R.A. – CIN, Centro di ricerca per le colture industriali, Bologna, Italy
e-mail: enrico.ceotto@entecra.it

E. Lichtfouse (ed.), *Biodiversity, Biofuels, Agroforestry and Conservation Agriculture*,
Sustainable Agriculture Reviews 5, DOI 10.1007/978-90-481-9513-8_3,

1 Introduction

The amount of energy and carbon that enters ecosystems in form of plant biomass is referred to as net primary production (Chapin and Eviner 2003). Plant biomass is predominantly used for food and feed, but it can also be used as a source of energy. A fuel can be considered biofuel if it is derived from recently produced biomass (Granda et al. 2007). Conversely, petroleum and natural gas are fossil fuels because they were formed over million of years from the decay of land and marine organisms. Biomass has major advantages over other renewable sources of energy, because it does not suffer from intermittency of supply, unlike wind and photovoltaic systems, and in case of liquid fuels for transportation no other options are currently available (Rowe et al. 2009). Woody and lignocellulosic biomass can be directly used for generating heat and electricity by combustion. However, our modern industrialized societies require a considerable amount of liquid fuels suitable for transportation. Unfortunately, the conversion processes of plant biomass into ethanol or biodiesel require energy themselves (Nonhebel 2005). In fact, an early comparative study on the suitability of plant biomass to produce energy indicated that a short rotation forestry of poplar is the most efficient energy system compared to wheat and sugarbeet for ethanol and rapeseed for biodiesel (Lövenstein et al. 1992). Accordingly, Campbell et al. (2009) estimated that bioelectricity produces an average 81% more transportation kilometers and 108% more emissions offsets per unit area cropland than cellulosic ethanol.

The interest in exploiting biomass for energy production is not new nor novel. Throughout the history of mankind wood has been used as the primary source of energy for heating and cooking (Nonhebel 2005). One century ago, 20% of arable land in Europe and North-America was devoted to oats, the cereal that made up the largest part of diets for horses and mules that powered agricultural and urban transportation (Gressel 2008). In 1890 about 75% of the energy was still provided by biomass fuels, whilst coal provided the remaining part (Smil 1994). During the period 1890–1990, owing to the availability of petroleum at low price, much of the world energy production in the world was transformed from a biofuel to a fossil fuel economy (Galloway and Cowling 2002). During recent years growing concerns for energy security and global climate change have stimulated a renewed interest in bioenergy. In fact, many countries worldwide have actively pursued policy aimed at increasing bioenergy production. Koh and Ghazoul (2008) summarized the major "raisons d'être" behind the booming of biofuels as follows:

1. Achieving energy security: the political instability in petroleum- and gas-rich countries, combined with the rising energy prices, stimulated many countries to diversify their energy portfolio aiming to reduce their reliance on imported fuels. The threat of freezing in the dark is certainly the primary motivation for fostering bioenergy.
2. The general perception that displacement of fossil fuels by biofuels can contribute in curbing the rise of greenhouse gas emissions, although the current debate in this area is heated.
3. Promoting rural development: a widespread diffusion of biomass for energy production is regarded as a mean to increase employment and income of rural populations.

Even the rise of cereal prices, attributable to higher grain demand for biofuels, can be viewed as a benefit for rural populations (Koh and Ghazoul 2008).

The increasing diffusion of biofuels drew a firestorm of criticisms from many researchers. Some contend that the conversion of incoming solar radiation into plant biomass has an inherently low efficiency, hence the production of energy from this source has high land demand. Because fertile land area, suitable for grain crops cultivation, is a limited and vulnerable resource in our planet, the production of biofuels compete against the demands for food production and nature conservation (Cassman and Liska 2007; Searchinger et al. 2008; Stoeglehner and Narodoslawsky 2009). Others claim that intensive agricultural management, needed to sustain productivity of bioenergy crops, negates the energy and greenhouse gases benefits of replacing fossil fuels (Pimentel 1992; Tilman et al. 2006). In particular, some warned against the increased use of industrial nitrogen fertilizers, because of their energy-intensive manufacture and their subsequent release of nitrous oxide, a potent greenhouse gas (Crutzen et al. 2008). Given that the level of human-induced reactive nitrogen in the biosphere had already exceeded critical threshold, it is not advisable to increasing the global use of nitrogen fertilizer with the purpose of producing biofuels (Galloway 2005).

Recently, Tilman et al. (2009) pointed out that in a world searching solutions for energy and food security, our society cannot accept the undesirable effects of biofuels when they are done wrong, but also cannot afford to miss the benefits of biofuels when they are done right. In particular, these authors stressed that biofuels done right must be produced with little or no competition with food production. It is useful to distinguish first and second generation biofuels. The basis for this difference is the potential competition with food and the efficiency of carbon balance: first generation biofuels are obtained from food crops and allow limited, if any, reduction of greenhouse gas emission; second generation biofuels are produced by non-food crops or non-food parts of crops, and have at least a reduction of 50% of greenhouse gas emissions (Erisman et al. 2009). Notably dedicated perennial lignocellulosic crops may offer significant contribution to the environment, compared to arable crops (Rowe et al. 2009). The objective of this article is to discuss the following topics: (i) competition for land: bioenergy versus food; (ii) bioenergy crops and nitrogen cycling; (iii) plant traits to be targeted for improving land and nitrogen use efficiency; (iv) the debated role of legumes. Emphasis is given to the contribution that dedicated perennial lignocellulosic crops might provide to sustainable bioenergy production.

2 Competition for Land: Bioenergy Versus Food

2.1 Ecological Footprint of Energy Use

From an ecological perspective the use of fossil fuels was defined as "borrowing land from the past" because thc carbon stored in fossil reserves has been accumulated by photosynthesis occurred millions of years ago (Wackernagel and Rees 1993). In this view, a substitution of fossil energy by biofuels implies that the borrowed land from the past becomes a real portion of land used in the present, consuming actual bioproductive capacity (Stoeglehner and Narodoslawsky 2009). These authors used the concept of ecological footprint, which is the amount of land necessary to sustain a given human activity. In case of fossil fuels the footprint can be estimated in two ways: the first way is to determine the land area needed to produce the same amount of energy by agriculture resources; the second way is to estimate the area required to sequester the CO_2 emitted by burning fossil fuels (Stoeglehner and Narodoslawsky 2009).

2.2 Cropland, a Dwindling Resource

Spiertz (2009) pointed out that cropland is a finite and vulnerable resource at global scale. Currently, the average amount of arable land per capita is about 0.45 ha, although this area is unevenly distributed among regions: China, for instance, has only 7% of the global arable land despite to the fact it has to feed about 20% of the global population. But then the crucial questions is: (i) how much of this 0.45 ha per person for producing food, could be displaced for producing bioenergy? However, the rising demand of biofuels is not the only pressure on cultivated land. Between now and 2050 the global demand for food is expected to double (Koning et al. 2008). The main drivers of such demand are further growth in world population and the current trend in changing diet, notably in developing countries. Spiertz and Ewert (2009) indicated that the current trend in diet change is likely to result in doubling of meat consumption in developing countries in the next 3 decades, rising from the actual 40 to 80 kg per capita per year. Moreover, the growing production of non ruminants, e.g. pigs and chicken, compared to ruminants, e.g. cattle and sheep, is shifting the demand from forages to grain crops, leading to increased pressure on arable land. Moreover, additional claims to land for non-food purposes should be taken into account. Koning et al. (2008) estimated that human settlement (buildings, roads, parkings) requires about 50 ha per 1,000 people. They surmised that by 2050 human settlement would claim 3% of all potential farmland.

Nitrogen fertilizers have improved the productivity of agricultural systems, raising the efficiency of utilization of solar radiation in crop production from about 0.25% to 2%. Because without nitrogen supply much more land would be required to achieve the same production target, fertilizers can be seen as land sparing tools (Loomis and Connors 1992; IFA 2009). Despite to such an improved efficiency the amount of land required to provide a substantial contribution to the current energy consumption is enormous. The UN special rapporter on the right to food, warned against biofuels, who defined as a "crime against humanity" and called for a 5-year ban on the practice, suggesting that within that time technological advances would enable to use agricultural wastes, rather than crops themselves to produce energy (Ziegler 2007). While for Brazil to produce 10% of its entire fuel consumption would require just 3% of its agricultural land, for the US to meet that target would require 30% of its agricultural land, and for Europe this portion raise to a staggering

72% (Pearce 2006). Different, but still impressive calculation were provided by the International Energy Authority: a 10% substitution of petrol and diesel fuel is estimates to require 43% and 38% of the current cultivated area in the United States and Europe, respectively (IEA 2004). Since these areas cannot be displaced from food and fodder production, forests and natural grasslands would need to be cleared to allow the cultivation of bioenergy crops. This would imply a rapid oxidation of the carbon stores present in soils and biomass, with an overall CO_2 emissions that would out-weigh the emissions avoided by biofuels (Righelato and Spracklen 2007). In good agreement, two companion reports of Searchinger et al. (2008) and Fargione et al. (2008) disentangled, in convincing way, the unintended, perverse effect of fostering biofuels at global scale. Searchinger et al. (2008) pointed out that prior studies have failed to consider the carbon emissions that occur as farmers worldwide respond to higher prices and convert forest and semi-natural grassland to new cropland to replace the grain (or cropland) diverted to biofuels. The rising demand of biofuels boost the prices of cereals on global market and this lead to increased demand for food. In response to such demand, in tropical developing countries vast land area are converted from forest into arable land. Searchinger et al. (2008) introduced the concept of "*carbon uptake credit*", which is the carbon sequestration achieved by biomass feedstock production in a given land area. Fargione et al. (2008) supported this view, and introduced the intriguing concept of "*carbon debt*" of land conversion, defined as the amount of CO_2 released by burning or microbial decomposition of organic C stored in plant biomass and soil, during the first 50 years of land conversion from forest to arable agriculture. Searchinger et al. (2008) using a worldwide agricultural model to estimate emissions from land use change, estimated that corn-based ethanol would need up to 167 years to restore the carbon credit derived by converted forests. Therefore, effective use of land areas, either for food or carbon sequestration purposes, is an overwhelming criteria for evaluating land use strategies at local and global level.

2.3 *Wildlife Friendly or Land Sparing Farming?*

The search for wise tradeoffs between agricultural activities and the protection of the environment has been a long debated issue. According to Green et al. (2005) two conflicting land use approaches has been proposed: (i) wildlife-friendly farming, whereby agricultural practices are made as benign as possible to the environment, at the cost of productivity per unit area, with increased pressure to convert marginal land to agriculture; (ii) land-sparing farming, in which productivity per unit area is increased to potential levels and pressure to convert land to agriculture is consequently decreased, at the cost of higher risk of environmental pollution from smaller areas and threat to wildlife species on farmland. Nevertheless, this topic is inherently more complex than it may seems from bio-physical considerations. In fact, Ewers et al. (2009), using 20 years of statistical data, concluded, in convincing way, that in developed countries agricultural subsidies override any land-sparing pattern that might otherwise occur. Moreover, in developed countries, there was no evidence that higher staple crop yields were associated with decreases in per capita cropland area. Hence, the overwhelming nature of agricultural subsidies might play a crucial role in the diffusion of bioenergy crops. Sharing this view, Robertson et al. (2008) pointed out that a science-based policy is essential for guiding an environmentally sustainable approach to bioenergy. The critical challenge of agriculture is to produce enough food to meet the rising demand from population, and also to do so in environmentally sound manner (Cassman and Liska 2007). Therefore, the total land area required for food supply will ultimately determine the biofuel production capacity that can be achieved without causing food shortages and high food prices.

3 Bioenergy Crops and Nitrogen Cycling

Supplementing nitrogen to the crops is a key agronomic practice for improving productivity and economic income, in food as well as in bioenergy cropping systems. Erisman et al. (2009) indicated that the grain productivity of a well fertilized wheat crop is 4.5 times higher than one receiving no nitrogen supply. Moreover, the total net CO_2 assimilation of a crop growing under non-limiting conditions amply makes up the emissions involved in the production of the mineral fertilizers used to sustain its growth (Ceotto 2005). This implies that much more carbon is embedded into plant biomass and soil organic matter in cultivated land. Enhanced yields are particularly important to prevent deforestation, which is the most critical contribution of greenhouse gas emissions related to agriculture (IFA 2009). Hence, nitrogen fertilization can be regarded as an effective tool for improving the efficiency with which cropland is used. Unfortunately, the nitrogen applied to the crops as fertilizers and manure is inefficiently used (30–60%) in most cropping systems (Pierce and Rice 1988). As a consequence, the unused fraction can contaminate surface and ground water resources, or it can be lost in the atmosphere (Kitchen and Goulding 2001). Several ^{15}N recovery experiments have reported losses of nitrogen fertilizers in cereal production from 20% to 50%. These losses have been attributed to the combined effects of denitrification, volatilization and leaching (Raun and Johnson 1999).

Galloway et al. (2002) defined as "*reactive nitrogen*" all biologically active, photochemically reactive and radiatively active nitrogen compounds present in the biosphere and atmosphere of the Earth. Therefore, the term reactive nitrogen includes: (i) inorganic reduced forms of nitrogen, as NH_3 and NH_4^+; (ii) inorganic oxidized forms of nitrogen as NO_X, HNO_3, NO_3^-, nitrous oxide (N_2O); organic compounds as urea, amines and amino acids.

The environmental multifaced role of reactive nitrogen was summarized in a brilliant manner by Galloway et al. (2003) who introduced the concept of nitrogen cascade: "*The same atom of reactive nitrogen can cause multiple effects in the atmosphere, in terrestrial ecosystems, in freshwater and marine systems, and on human health. We call this sequence of effects the nitrogen cascade. As the cascade progresses, the origin of reactive nitrogen becomes unimportant. Reactive nitrogen does not cascade at the same rate through all environmental systems; some systems have the ability to accumulate reactive nitrogen, which leads to lag times in the continuation of the cascade. These lags slow the cascade and result in reactive nitrogen accumulation in certain reservoirs, which in turn can enhance the effects of Nr on that environment. The only way to eliminate reactive nitrogen accumulation and stop the cascade is to convert reactive nitrogen back to nonreactive* N_2". It is important to notice that even the legume-derived reactive nitrogen, once it has entered into the agro-ecosystems, has equally negative environmental impacts as the reactive nitrogen derived from synthetic fertilizers.

3.1 *Historical Outline of Nitrogen Use in Agriculture*

Even though nitrogen has been for millennia the most common yield-limiting factor in agriculture, the element nitrogen and its role were discovered at the end of the eighteenth century. Until that time, human societies relied mostly on the amount of reactive nitrogen entering the biosphere via biological nitrogen fixation and, to a lesser extent, lighting (Galloway and Cowling 2002). Yet, at the end of the nineteenth century several nitrate and guano natural deposit of Pacific Islands and South America were exploited to provide fertilizers to the crops. A fundamental breakthrough occurred in 1909, when Fritz Haber discovered that ammonia, a chemically reactive form of nitrogen, could be obtained by reacting atmospheric dinitrogen with hydrogen in the presence of iron at high pressures and temperatures. Subsequently, Carl Bosch developed the process at industrial scale in 1931 (Smil 2002). Nowadays this reaction is known as the Haber-Bosch process. A large scale diffusion of synthetic nitrogen fertilizers has begun after the Second World War, in 1950 (Smil 2001). Since that time a potentially unlimited source of nitrogen has been made available for improving crop production and effective land use. This ameliorated the standard of life of human populations. During the last decades global human population increased sharply together with the increase of reactive nitrogen. Galloway (2005) estimated that in 1990 the human produced reactive nitrogen already exceeded the one entering agro-ecosystems via biological nitrogen fixation. But now, with the advent of biofuels, a crucial question arises: (i) can we allow to increase the human interference on the biospheric nitrogen cycle with the scope of producing bioenergy?

Galloway et al. (2008) proposed a multitasking strategy for reducing nitrogen reactive creation worldwide: (i) the control of NOx emissions from fossil fuels combustion might results in a decrease in reactive nitrogen from 25 to 7 Tg N year^{-1}; (ii) increasing nitrogen uptake efficiency of the crops would decrease reactive nitrogen by about 15 Tg N year^{-1}; (iii) improved animal management would decrease reactive nitrogen by about 15 Tg N year^{-1}; (iv) proper treatment of at least half of the sewage of the 3.2 billion people living in cities would convert about 5 Tg N year^{-1} in form of inactive N_2. Overall, these intervention represent a potential decrease of about 53 Tg N year^{-1}, which is equivalent to about 28% of the reactive nitrogen created in the year 2005. With this reduction, we might be able to compensate the future increase of nitrogen reactive required for future increase in food, feed, fuel production and energy use (Galloway et al. 2008).

3.2 *Energy and Greenhouse Gas Emissions Implications of Nitrogen Synthetic Fertilizers*

While about one third of the fossil energy used in modern agriculture is consumed directly in the farm, e.g. in form of diesel fuel and electricity, the remaining two third are consumed indirectly, for producing elsewhere goods that are used as agronomic inputs (Helsel 1992). The major indirect use of fossil energy is devoted to synthetic nitrogen fertilizers, because of their energy-intensive manufacture. The energy inputs for producing phosphorous and potassium fertilizers is much less with respect to nitrogen (Helsel 1992; Pimentel 1992). In most of the energy analysis at farm level, the energy attributed to nitrogen fertilizers ranged from 30% to 50% of the overall fossil energy use (Loomis and Connors 1992). Most of the industry's energy requirement is used in manufacturing of ammonia, the compound from which all nitrogen fertilizers are derived. The most common fossil fuel feedstock for fertilizers production is natural gas (67%), coal provide a significant portion (27%), while fuel oil (3%) and naphtha (2%) are less used. (IFA 2009). Overall, the production of nitrogen fertilizers accounts for about 1.2% of the fossil energy used worldwide (Erisman et al. 2009). Indeed, this is only a small parts of the total energy used by modern societies, especially if one considers the utmost importance of agriculture in human standard of living. While more parsimonious use of fossil energy use could be achieved in agricultural systems, a question arise naturally: should our societies rather concentrate their energy saving efforts in transportation and heating? The question is more tough if considered from the standpoint of reactive nitrogen and greenhouse gas emissions.

The greenhouse gases emissions related to the use of industrial nitrogen fertilizers is a twofold problem. It is useful to distinguish between emissions precedent and subsequent to field application of fertilizers. Greenhouse gases emissions precedent to field application are the consequence of the fossil energy used for the industrial manufacture of nitrogen fertilizers. Transportation and field distribution also implies additional emissions. The emissions subsequent to field application of nitrogen fertilizers are due to the release of nitrous oxide (N_2O). Among the undesired by-product of the nitrogen applied in agriculture, the most worrisome is the nitrous oxide (N_2O), a compound originated from terrestrial ecosystems by the process of denitrification. In agricultural soils, the increase of available mineral nitrogen lead to enhanced nitrification and denitrification, hence nitrous oxide emission (Mosier 2001). Nitrogen oxide is a greenhouse gas with a 100-year average global warming potential (GWP) 296 times larger than that of CO_2. As a source for NOx, it also plays a major role in stratospheric ozone chemistry

(Crutzen et al. 2009). The effectiveness of nitrous oxide in trapping heat is mainly due to its atmospheric residence time of 114 years (FAO 2006).

Crutzen et al. (2008) pointed out that biofuels may have a negative greenhouse gas balance due to the fact that N fertilization implies inevitable nitrous oxide emissions, a potent greenhouse gas. These authors, on the basis of global present and pre-industrial atmospheric concentration, estimated a nitrous oxide emission from agricultural acitivites of 4.3–5.8 Tg N_2O-N year^{-1}. Consequently, they propose a nitrous oxide emission factor of 3–5% for the newly fixed nitrogen (synthetic fertilizer and biologically fixed N). Such value of emission factors is apparently much higher compared to the one proposed by IPCC (2006). Nevertheless, Ammann et al. (2007) pointed out that the discrepancy between the emission factor of 3–5% derived by Crutzen et al. (2008) and the IPCC emission factor of 1% (or about 1.7% including indirect emissions) for fertilizer application is merely due to a difference in methodology: Crutzen et al. (2008), following the conceptual pathway of Galloway et al. (2003) consider the comprehensive N_2O emission of the "newly-fixed nitrogen" (synthetic fertilizer and biologically fixed N) along their nitrogen cascade, whereas the IPCC methodology accounts the N_2O emissions of each individual field application of N-fertilizers. According to Mosier et al. (1998), only about one third of N_2O emissions (direct and indirect) are due to newly-fixed N fertilizer, another third is due to the application of recycled organic fertilizer (plant residues and manure) and the last third is due to specific waste management in animal production. Therefore, about two thirds of agricultural N_2O emissions are due to internal recycling of nitrogen in animal production or by using plant residues as fertilizer. This means that there is no indication of an important unconsidered N_2O emission in the IPCC methodology. Hence, the higher emission factor derived by Crutzen et al. (2008) is likely due to the fact that, on a global average, each newly fixed nitrogen molecule is used several times as fertilizer, and also flows through animal production systems, before it undergoes denitrification. Consequently, the challenge is to minimize the use of "*newly fixed nitrogen*" in bioenergy crops fertilization.

Finally, it is important to note that Crutzen et al. (2008) focused their analysis on relatively N-intensive biofuel crops, like maize and rapeseed, but they postulated that "*energy plants*", such as switch grass (*Panicum virgatum*) and miscanthus (*Miscanthus* × *giganteus* hybrid), might have a moderately positive effects on climate, owing to their lower nitrogen to dry matter ratio.

4 Plant Traits to be Targeted for Improving Land and Nitrogen Use Efficiency

Aiming to achieve an effective use of bioenergy, two diverse, but not conflicting strategies can be devised: (i) improving yield of dedicated energy crops on limited land area; (ii) exploiting the potential of dual purpose crops on arable land.

4.1 *Improving Yield, and Net Energy Gain, of Dedicated Energy Crops*

Goudriaan et al. (1991) and Nonhebel (1997) have highlighted the key traits of the ideal plant for producing feedstock for energy generation in form of electricity and heat:

1. A high daily growth rate that should be maintained over a long growing period. Crop photosynthesis is closely related to the amount of light intercepted by the green foliage, thus an early canopy closure in the spring and a delay of leaves senescence in the autumn will enhance the amount of light captured during the season. An optimized canopy architecture also play a substantial role in absorbing solar radiation.
2. Biomass production should be aboveground, because harvesting belowground products requires too much energy.
3. A low nitrogen concentration in harvestable biomass, because the manufacture of industrial nitrogen fertilizers requires a lot of energy, which reduces the net energy yield, and substantially increases greenhouse gases emissions; We here stress that Crutzen et al. (2009) recently proposed the nitrogen to dry matter ratio of the harvested plant material, r(N/dm), as a quantitative criterion which show the degree to which the reduction of global warming, attributable to the use of plant biomass in substitution of fossil fuels, is counteracted by the release of N_2O.
4. The crop should be perennial in order to reduce the energy expenditure for soil tillage and sowing, and lessen the needs for nitrogen fertilizers; in perennial crops, the belowground storage organ typically act alternatively as a source and sink of both nitrogen and carbon, and this facilitate nutrient recycling within the plant.
5. It should be possible to harvest biomass relatively dry, because: (a) moisture in the product increases transportation costs; (b) extra energy is needed for drying.
6. The crop should be not susceptible to pathogens, because the necessity to spraying against fungi and insects implies the use of energy, which lowers the net energy output, and spraying tall crops (e.g. poplars) causes technical problems; yet pesticides are even more energy intensive than nitrogen fertilizers.
7. It should be a strong competitor against weeds; also for this scope the crop should start to grow early in the spring season.
8. The crop should be drought resistant and cold tolerant; a good resilience after dry spells is particularly important.
9. The crop should have low water use, because irrigation implies substantial energy use and also place a demand on limited fresh water resources; an extended, deep rooting system is a key trait for exploiting water resources in deep soil profile, normally unavailable to most of the cultivated plants.

It is interesting to note that a unique combination of the most of the above-mentioned characteristics is present in the herbaceous perennials giant reed (*Arundo donax* L.) (Fig. 1) and miscanthus (*Miscanthus × giganteus* hybrid) (Fig. 2). For both species the only shortcoming lies in their high moisture content at harvest, normally about 50% (our data unpublished).

Fig. 1 Giant reed (*Arundo donax* L.). A unique combination of favourable characteristics is present in this species: it is perennial; drought tolerant and cold resistant; it requires little nitrogen; it is not susceptible to pest and diseases; it is highly competitive against weed. At the end of the growing season the height of this crop is about 4.5 m (Photo: Mario Di Candilo)

A slightly different, but still very interesting approach was provided by Ragauskas et al. (2006) who focused on woody species for biofuels and biorefineries. As these author pointed out, while annual grain crops benefit from centuries of domestication efforts, wild perennial species that could play a central role in providing a renewable source of feedstock for conversion to fuels and materials have not received such attention until now. Hence, these authors propose that an accelerated domestication should be targeted to the following plant traits:

1. Increased photosynthesis via optimized photoperiodic response and optimized crown/leaf architecture.
2. Enhanced biomass production per unit area by reducing perception of nearest neighbour by manipulating photo-morphogenic responses of phytocrome red/far-red light perception systems; this assumption is questionable, however, because in mutual shaded leaves the carbon dioxide assimilation may be overweighed by the maintenance respiration, with and overall decrease of the light-use efficiency (Goudriaan and van Laar 1994).
3. Floral sterility is a key trait. Ragauskas et al. (2006) underscored that the grain plants normally invest substantial energy in making and filling their reproductive structure. Hence, if flowering can be delayed or prevented, this energy can

Fig. 2 Miscanthus (*Miscanthus* × *giganteus* hybrid). It is ironic that the floral sterility is, in the meantime, a favourable trait for high productivity and a major disadvantage because the propagation of this species is unpractical and monetarily expensive. At the end of the growing season the height of this crop is about 3 m (Photo: Mario Di Candilo)

be effectively used for increasing the overall plant biomass. As Johnson et al. (2007) pointed out, the benefit of floral sterility is twofold: on the one hand carbon and nitrogen are not diverted to reproductive structures; on the other hand the absence of seeds ensures that the species does not exhibit invasive characteristics in contiguous field. A typical example of convenient floral sterility is given by the previously mentioned herbaceous perennials giant reed and miscanthus. It is ironic, however, that the floral sterility is also a major disadvantage because the vegetative propagation of these species is unpractical and monetarily expensive.

4. Pest and diseases resistance, coupled to drought and cold tolerance and resilience.
5. Greater carbon allocation to stem diameter versus height growth (Ragauskas et al. 2006); indeed, this assertion is debatable if one refers to herbaceous perennials. In Mediterranean environments giant reed, with stem height about 4.5 m, is more productive than both miscanthus and fiber sorghum (*Sorghum bicolor* L. Moench), both species with a stem height of about 3 m, even though the latter two species have the potential advantage of C4 photosynthetic pathway. Example data were reported by Di Candilo et al. (2008) as average yields of 7-years comparison in Northern Italy: 39.6 Mg dry matter ha^{-1} for giant reed, 25.2 and 25.7 Mg dry matter ha^{-1} for miscanthus and fiber sorghum, respectively. Therefore, it

seems likely that the superior height of giant reed, might contribute, at least partially, to its high productivity, by increasing the sink strength of the plant.

6. Less extensive root system to maximize aboveground biomass, and optimal nitrogen acquisition and use. This point, indeed, is contradictory because one might argue that a deep and extended root systems normally allows better nitrogen and water uptake, combined with higher carbon storage in belowground biomass.
7. Optimized content of readily processable cellulose, hemicelluloses and lignin; tailored biomass composition with value-added components. The intriguing idea is to modify the plant composition with the target of cost-effective depolimerization and subsequent fermentation. However, it remains to be seen whether the genetic variability of the species is wide enough to allow such an ambitious domestication.

Gressel (2008) pointed out that most of the "*wild*" species recently proposed as second generation biofuels should be domesticated transgenically in order to increase their productivity, to improve their properties as a feedstock for fuels, and to remove toxins and environmental pollutants from their tissues. While we concur that enhancing the product quality of biomass feedstock is strongly needed, we contend that domestication might entail some undesired trade-offs. In fact, according to MacNeish (1992) a domesticated plant is one that differs from its wild ancestors because it has been changed genetically through human selection, either consciously or unconsciously. A relevant trait of domesticated species it that they may lose their ability to survive and reproduce in the wild Mannion A.M. (1997). Several lignocellulosic plant species, recently proposed as bioenergy crop are undeniably "*wild plants*". But then, they normally possess the virtues of rusticity, stress resilience, pest tolerance, and high competitive ability against weeds. It is important to underscore that these virtues are particularly well suited for dealing with poor soil fertility, and low agronomic inputs.

As far as productivity is concerned, Loomis and Amthor (1999) pointed out that there is no evidence that important beneficial in photosynthesis have occurred during domestication of crop plants or through recent breeding. Koning et al. (2008) agree and underscored that improvements in productivity can be expected by modification of crop architecture, but not in light and water-use efficiency. While the improvement of photosinthesys is an obvious objectives of genetical engineering, there is no scientific evidence that this will be achieved in the foreseeable future (Cassman and Liska 2007).

Furthermore, Porter et al. (2008) pointed out that the food crop ideotype, developed during the food revolution in the 1960s to have short stems, small erected leaves, and high carbon and nitrogen harvest index, is poorly suited to provide feedstock for bioenergy. Hence, the crop ideotype for bioenergy production requires a totally different characteristics, because the plant is harvested for its carbon but not for its nitrogen content:

1. Biomass crops intended for fermentation should posses high concentration of low molecular carbohydrates.
2. Low fertilizer requirement and high nitrogen use efficiency are desirable to reduce energy inputs and greenhouse gases emissions, notably N_2O.

3. Biomass energy production requires management that generate ecosystems services such as soil carbon sequestration, pollination, pest suppression and biodiversity conservation. Rowe et al. (2009) reviewed the environmental impact of dedicated second-generation biomass crops. These authors concluded that perennial ligno-cellulosic crops have the potential provide a range of benefits for both ecosystem services and carbon storage compared with the use of land for arable crop production. Nevertheless, these benefits are questionable when bioenergy crops are used to replace permanent unmanaged grassland (Rowe et al. 2009).

The most important agricultural grain crops, notably maize (*Zea mays* L.), soybean (*Glicine Max* (L.) Merr.) and wheat (*Triticum aestivum* L.) are poorly suited to serve as dedicated energy crops because: (i) they are annual, requiring large agronomic input for nitrogen fertilization, tillage, seeding and pest control, and this implies high greenhouse gases emissions (Farrel et al. 2006); (ii) their growing cycle is too short, allowing limited radiation interception on annual basis (Mannion 1997); (iii) they normally requires high agronomic inputs, and inevitably nitrogen and pesticides losses in the environment (Randal and Goss 2001). Hence, crops with less N demand, such as grasses and woody coppice species, should be selected for their favourable energy balance and climate impacts (Crutzen et al. 2008).

4.2 Exploiting the Potential of Dual Purpose Crops: Food and Biomass Feedstock on the Same Land Area

It is unfortunate that dedicated energy crop are always in competition for land with the overwhelming objective of agriculture, which is to provide food, in good quantity and quality, to human populations. We here stress that cereals, and in particular wheat can be regarded as a potential dual purpose crop. In fact, the high nitrogen harvest index of cereals, i.e. the fraction of grain nitrogen over the total aboveground nitrogen, which is 0.6–0.8 for wheat (López-Bellido et al. 2008), can be conveniently exploited to produce food and biomass feedstock on the same land area at the same time. As Ceotto (2008) pointed out, since about one-half of the dry matter produced by grain crops has no direct human nutritional value, crop residues have the potential to provide a strategic source of biofuels. Owing to its low nitrogen (~0.5% of dry matter) and low moisture content (10–13% of total weight) wheat straw is particularly well suited to be burned to obtain energy in form of heat and electricity, associated with little reactive nitrogen emissions (Fig. 3). Maize stover is also well suited although it is normally harvested at higher moisture content. Indeed, the major advantage of straw over other crop residues lies on its low moisture content at harvest time. This diminish the energy costs for transportation and for water evaporation during combustion. The use of straw for energy generation certainly does not threaten global food security, if soil fertility is maintained. On the contrary, an additional income derived from crop residues would stimulate farmers to produce more cereals. Yet, it is important to point out here that using straw for energy implies that the fossil energy and greenhouse gas emissions

Fig. 3 Wheat can be regarded as a potential dual purpose crop. It allows to produce edible grain and feedstock for bioenergy on the same land at the same time. The major advantage of straw over other crop residues lies on its low moisture content at harvest time

related to nitrogen fertilization are attributable to grain production, where most of the nitrogen is allocated. Wheat is grown on 200 million hectares of farmland worldwide (Ortiz et al. 2008), thus there is large potential to exploit straw as a source of energy. Nonetheless, crop residues are a precious commodity, essential to maintenance of soil fertility and preventing erosion (Lal 2004). Thus, a crucial question arises: what is the fraction of crop residues that could be collected from the field without either depleting soil organic matter or increasing soil erosion? Lafond et al. (2009) working in Canada, quantified the crop residues removal with baling in spring wheat, and evaluated the effect of straw removal on soil organic carbon and wheat productivity. They reported that, depending on the harvesting system, only 26–40% of total aboveground crop residues other than grain are removed with baling. They also indicated that 50 years of straw removal did not influence spring wheat grain yield and grain protein concentration, and there was no measurable impact on the amount of soil organic carbon. They concluded that potential exists to harvest cereal residues with a baler for industrial purposes from medium- to heavy textured without adversely affecting soil quality and productivity, providing that proper soil and crop management practices are employed. Lal and Pimentel (2007) disagree, pointing out that bioenergy crops should be rather established on specifically identified lands, e.g. on agriculturally marginal/surplus lands, and degraded or drastically disturbed soils. Gressel (2008) suggested that a wise trade-off could be cutting the straw at higher height, removing only the 80% for biofuels and leaving the rest to the soil.

4.3 High Response to Nitrogen Fertilizers or High Productivity at Low Nitrogen Supply?

Improving nitrogen use efficiency (NUE) for cereal crops has been a constant effort for crop scientists (Raun and Johnson 1999). One alternative strategy might be appropriate for minimizing the undesired environmental effects of nitrogen fertilizers application: cultivating energy crops that have high productivity at low nitrogen supply. An example is provided here. Erisman et al. (2009) indicated that with a nitrogen supply of 192 kg ha^{-1} it is possible to produce 9.3 Mg of wheat grain ha^{-1}, whilst the unfertilized crop produces 2.07 Mg ha^{-1} Thus, an overall nitrogen use efficiency, of 37.7 kg grain per kg N applied can be achieved for wheat. Similar results are obtained from maize. Spiertz (2009) indicated that maize grain yield can be raised from 3 to 14 Mg ha^{-1} with an adequate N supply. Assuming a nitrogen supply of 300 kg ha^{-1} the nitrogen use efficiency is around 36.7 kg grain per kg N applied. Interestingly, Angelini et al. (2005), working on giant reed, reported an average production of 27 Mg biomass ha^{-1} (years 1–6 mean value)with an annual fertilization of 200 kg of nitrogen. Surprisingly, the productivity of the unfertilized treatment was 23 Mg biomass ha^{-1}. While the nitrogen use efficiency of giant reed is lower than cereals, about 20 kg aboveground biomass per kg N applied, the productivity of this species was terrific in the light of a 6-years period with no nitrogen supply. The likely explanation lies on the conservative use of nitrogen by rhizomatous perennial crops: nitrogen is alternatively accumulated and released by belowground crop organs, and the aboveground biomass is normally harvested during the winter, when is has very low nitrogen content. Therefore, perennial lignocellulosic crops have much to offer if the objective is to produce biomass feedstock for bioenergy with sparing use of nitrogen.

5 The Debated Role of Legumes

In literature there are contrasting views about the potential role of legumes in cropping systems (Crews and Peoples 2004). Some authors suggest that legumes, which are able to support biological nitrogen fixation, offer a more environmentally sound and sustainable source of N to cropping systems compared to synthetic nitrogen fertilizers (Drinkwater et al. 1998). Hence, using legume crops as a main route of N in agro-ecosystems would reduce the reliance of agriculture on fossil energy and related greenhouse emissions (Tilman et al. 2006). In contrast to this view, other researchers contend that: (i) legume-derived N has equally negative environmental impacts as the N derived from synthetic fertilizers; (ii) the human population now exceeds the carrying capacity of agricultural systems that depend on legumes for N inputs. According to MacAdam et al. (2003) the industrial cost of producing ammonia from N_2 is equivalent to 3.8 g glucose per g N. In addition, the cost to the plant for assimilation of the nitrate formed in the soil to a usable $-NH_2$ is about 4.2 g glucose per g N, and this makes the global cost of 8 g glucose per g N. Although the measured cost for legume fixation is about twice the industrial cost (16 g glucose per g N), these authors pointed out it is less wasteful of resources, especially fossil fuel. In contrast, Ryle et al. (1979) pointed out that the pressures to utilize crop plants that can fix nitrogen by themselves must be balanced against the equally important objective of achieving optimal utilization of solar energy. In this view, Sinclair and Cassman (1999) contend that the increasing food demand from the human population already exceeds the low carrying capacity of legume-dependent cropping systems.

Referring specifically to a biobased economy, Brehmer et al. (2008) pointed out that legumes are not suited to bioenergy production since they are quite inefficient in terms of solar energy conversion, thus in land utilization. Although nitrogen fixation is indisputably an economy in terms of agronomic input, accruing nitrogen in this way entails unacceptable losses in term of captured solar energy. In principle, we concur. Nevertheless, we contend there is a pitfall in their analysis, because they referred solely to food and fodder common crops. In fact, our data, referring specifically to dedicated energy crops, suggest different conclusions. A poliennal experiment was undertaken in 2002 in the Low Po Valley, Northern Italy, comparing short rotations coppice of three woody species, hybrid poplar (*Populus x canadensis*), willow (*Salix alba* L.) and the legume black locust (*Robinia pseudoacacia* L.)(Di Candilo et al. 2008). The compared species were harvested every 2 years. In Fig. 4 are shown the patterns of productivity for woody plant species throughout the experiment. Interestingly, the yield of black locust over weighted the ones of non-legumes for the second and third subsequent harvest. Both hybrid poplar and willow were fertilized with 120 kg nitrogen ha^{-1} at crop establishment and then every 2 years, after every harvest, whilst black locust has not received nitrogen fertilization. It is apparent that black locust is more tolerant to repeated cuttings compared to both hybrid poplar and willow, but also other favourable plant characteristics might have contributed. Overall, our data suggest that important plant traits of woody species, like rusticity, stress resilience and cutting tolerance could overtake the theoretical disadvantage of legumes.

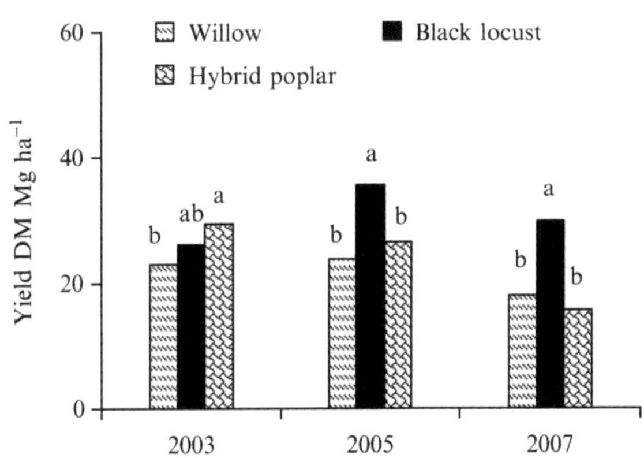

Fig. 4 Patterns of productivity for woody plant species throughout a multi-year experiment (redrawn from Di Candilo et al. 2008). Means sharing common letters within years are not significantly different at $p < 0.05$. REGWQ multiple range test. DM: dry matter

Recently, DeHaan et al. (2009), working in the Cedar Creek biodiversity experiment in Minnesota, reported that average yields of the biculture *Lupinus perennis*/C4 grass were similar to those of 16 species mixture (maximum biodiversity), and both were >200% greater than the average of monocultures. In this low-input grassland system for bioenergy production, one particular legume species, well adapted to the site, was critical to achieve maximum production. Hence, the role of legumes in improving productivity of multi-species bioenergy systems merit consideration and further investigation.

6 Conclusion

The amount of land that could be conveniently devoted to grow energy crops without detrimental effects of greenhouse gases emissions and food availability is limited, hence it is crucial to enhance productivity per unit land. Meanwhile, the challenge is to minimize the amount of synthetic nitrogen fertilizers used for growing biomass feedstock crops. There are several avenues for gaining efficiency in biomass feedstock production:

1. We believe that a proper classification and quantification of "land suitable for dedicated bioenergy production" is needed at regional level. Beside to marginal land, poorly suited for cultivation of grain crops owing to low fertility, it would be important to consider all the arable land that should not be used for food and fodder production owing to toxicity or sanitary reasons, e.g., polluted areas, vicinity of industrial plants, discharge or incineration plants of municipal wastes, strips surrounding high traffic highways. Moreover, the term "marginal land" normally refers to least economic value, but the trade-offs of converting pasture, wetlands or forests to biofuels cultivation should be carefully evaluated from the ecological and environmental points of view. The main message of this review is that bioenergy should be seen as a complementary product of food and feed production, to be attained with high land use efficiency and sparing use of nitrogen and water. Sustainable land use could be improved by integrating the complementary role of land areas suitable for food and fodder production and land areas suitable for bioenergy production.
2. On land areas available for bioenergy, only crops with the most favourable characteristics should be cultivated, and wild plant species with interesting traits should be considered for future cultivation. In particular, the cultivation of perennial bioenergy crops should be encouraged in order to achieve carbon sequestration and additional ecosystem services like pollination and wild species conservation.
3. The use of cereal crops residues for generating energy merits consideration. Using the same land area for producing grain for food and straw for bioenergy might be a valuable compromise. But then, well devised crop rotations and straw cutting height, are crucial to ensure that soil organic matter, and soil fertility, are not depleted over time. In order to accomplish these targets, a set of regulation

and subsidies is strongly needed for guiding a sustainable land use. This set of policy should be based on solid scientific knowledge, in order to minimize the unavoidable risk that ignorance and uncertainty will result in undesired global environmental consequences of land use.

References

Ammann C, Spirig C, Fischer C, Leifeld J, Neftel A (2007) Interactive comment on "N_2O release from agro-biofuel production negates global warming reduction by replacing fossil fuels" by P. J. Crutzen et al. Atmos Chem Phys Discuss 7:S4779–S4781. www.atmos-chem-phys-discuss.net/7/S4779/2007

Angelini GL, Ceccarini L, Bonari E (2005) Biomass yield and energy balance of giant reed (*Arundo donax* L.) cropped in central Italy as related to different management practices. Euro J Agron 22:375–89.doi:10.1016/j.eja.2004.05.004

Brehmer B, Struik PC, Sanders J (2008) Using an energetic and exergetic lifecycle analysis to assess the best applications of legumes within a biobased economy. Biomass Bioenerg 32:1175–1186

Campbell JE, Lobell DB, Field CB (2009) Greater transportation energy and GHG offset from bioelectricity than Ethanol. Science 324:1055–1057. doi:10.1126/Science.1168885

Cassman KG, Liska AJ (2007) Food and fuel for all: realistic or foolish? Biofuels Bioprod Bioref 1:18–23. doi:10.1002/bbb.3

Ceotto E (2005) The issues of energy and carbon cycle: new perspectives for assessing the environmental impact of animal waste utilization. Bioresource Technol 96:191–196. doi:10.1016/j.biortech.2004.05.007

Ceotto E (2008) Grasslands for bioenergy production. Agron Sustain Dev 28:47–55. doi:10.1051/agro:2007034

Chapin FS, Eviner VT (2003) Biogeochemistry of terrestrial primary production. In: Schlesinger WH (ed) Biogeochemistry, vol 8; Holland HD, Turekian KK (eds) Treatise on geochemistry. Elsevier-Pergamon, Oxford, pp 215–248

Crews TE, Peoples MB (2004) Legume versus fertilizer sources of nitrogen. Agr Ecosyst Environ 10:279–297

Crutzen PJ, Mosier AR, Smith KA, Winiwarter W (2008) N_2O release from agro-biofuel production negates global warming reduction by replacing fossil fuels. Atmos Chem Phys 8:389–395

Crutzen PJ, Mosier A, Smith K, Winiwarter W (2009) Atmospheric N_2O releases from biofuel production systems: a major factor against "CO_2 Emission Savings": a global view. In: Zerefos C, Contopoulos G, Skalkeas G (eds) Twenty years of ozone decline. Proceedings of the symposium for the 20th anniversary of the Montreal protocol, pp 67–70. doi:10.1007/978-90-481-2469-5

DeHaan LR, Weisberg S, Tilman D, Fornara D (2009) Agricultural and biofuel implications of a species diversity experiment with native perennial grassland plants. Agric Ecosyst Environ (in press). doi:10.1016/j.agee.2009.10.017

Di Candilo M, Ceotto E, Diozzi M (2008) Comparison of 7 ligno-cellulosic biomass feedstock species: 6-years results in the Low Po Valley. In: Rossi Pisa P (ed) 10th congress of the European society of agronomy, Bologna, multi-functional agriculture, agriculture as a resource for energy and environmental preservation. Ital J Agron/Rivista di Agronom 3(suppl. 3):481–482

Drinkwater LE, Wagoner P, Sarrantonio M (1998) Legume based cropping systems have reduced carbon and nitrogen losses. Nature 396:262–265

Erisman JW, van Grinsven H, Leip A, Mosier A, Bleeker A (2009). Nitrogen and biofuels; an overview of the current state of knowledge. Nutr Cycl Agroecosyst (in press). doi:10.1007/s10705-009-9285-4

Ewers RM, Scharlemann JPW, Balmford A, Green RE (2009) Do increases in agricultural yield spare land for nature? Global Change Biol. doi:10.1111/j.1365-2486.2009.01849.x
FAO (2006) Livestock long shadows, environmental issues and options. Food and agriculture organization of the United Nations, Rome. The livestock, environment and development (LEAD) Initiative website: http://www.virtualcenter.org. Chapter 3: Livestock role in climatic change and air pollution, pp 79–133
Fargione J, Hill J, Tilman D, Polasky S, Hawthorne P (2008) Land clearing and the biofuel carbon debt. Science 319:1235–1238. doi:10.1126/science.1152747
Farrell AE, Plevin RJ, Turner BT, Jones AD, O'Hare M, Kammen DM (2006) Ethanol can contribute to energy and environmental goals. Science 311:506–508. doi:10.1126/science.1121416
Galloway JN, Cowling EB (2002) Reactive nitrogen and the world: 200 years of change. Ambio 31:64–71
Galloway JN, Cowling EB, Seitinger SP, Socolow RH (2002) Reactive nitrogen: too much a good thing? Ambio 31:60–63
Galloway JN, Aber JD, Erisman JW, Seitzinger SP, Howarth RW, Cowling EB, Cosby BJ (2003) The nitrogen cascade. BioScience 53(4):341–356
Galloway JN (2005) The global nitrogen cycle. In: Schlesinger WH (ed) Biogeochemistry, vol 8; Holland HD, Turekian KK (eds) Treatise on geochemistry. Elsevier-Pergamon, Oxford, pp 557–583
Galloway JN, Townsend AR, Erisman JW, Bekunda M, Cai Z, Freney JR, Martinelli LA, Seitzinger SP, Sutton MA (2008) Transformation of the nitrogen cycle: recent trends, questions, and potential solutions. Science 320:889–892
Goudriaan J, Kropff MJ, Rabbinge R (1991) Mogelijkheden en beperkingen van biomassa als energiebron. Energie Spectrum 6:171–76
Goudriaan J, van Laar HH (1994) Modelling potential crop growth processes. Textbook with exercises. Kluwer, Dordrecht
Granda CB, Zhu L, Holtzapple MT (2007) Sustainable liquid biofuels and their environmental impact. Environ Prog 26:233–250. doi:10.1002/ep
Green RE, Cornell SJ, Scharlemann JPW, Balmford A (2005) Farming and the fate of wild nature. Science 308:550–555
Gressel J (2008) Transgenic are imperative for biofuel crops. Plant science 174:246–263. doi:10.1016/j.plantsci.2007.11.009
Helsel ZR (1992) Energy and alternatives for fertilizer and pesticide use. In: Fluck RC (ed) Energy in farm production, vol 6. In: Stout BA (ed) Energy in world agriculture. Elsevier Science, Amsterdam pp 177–202
IEA (2004) International energy authority: biofuels for transport: an international perspective, chap. 6. www.iea.org/textbase/nppdf/free/2004/ biofuels2004.pdf
IFA (2009) International fertilizer industry association: fertilizers, climate change and enhancing agricultural productivity sustainably, 1st edn. IFA, Paris
IPCC (2006) IPCC guidelines for national greenhouse gas inventories, prepared by the national greenhouse gas inventories programme. In: Eggleston HS, Buendia L, Miwa K, Ngara T, Tanabe K (eds) N_2O emissions from managed soils, and CO_2 emissions from lime and urea application, vol 4, chap 11. IGES, Hayama, Japan
Johnson JMF, Coleman MD, Gesch R, Jaradat A, Reicosky Don Mitchell R, Wilhelm WW (2007) Biomass-bioenergy crops in the United States: a changing paradigm. Am J Plant Sci Biotechnol 1(1):1–28
Kitchen NR, Goulding KWT (2001) On farm technologies and practices to improve nitrogen use efficiency. In: Follett RF, Hatfield JL (eds) Nitrogen and the environment: sources, problems and management. Elsevier Science B.V, Amsterdam, pp 335–369
Koh LP, Ghazoul J (2008) Biofuels, biodiversity, and people: understanding the conflicts and finding opportunities. Biol Conserv 141:2450–2460. doi:10.1016/j.biocon.2008.08.005
Koning NBJ, Van Ittersum MK, Becx GA, Van Boekel MAJS, Brandenburg WA, Van Den Broek JA, Goudriaan J, Van Hofwegen G, Jongeneel RA, Schlere JB, Smies M (2008) Long term global availability of food: continued abundance or new scarcity? NJAS – Wageningen J Life Sci 55:229–292

Lafond GP, Stumborg M, Lemke R, May WE, Holzapfel CB, Campbell CA (2009) Quantifying straw removal through baling and measuring the long-term impact on soil quality and wheat production. Agron J 101:529–537. doi:10.2134/agronj2008.0118x

Lal R (2004) Soil carbon sequestration impacts on global climate change and food security. Science 304:1623–1627

Lal R, Pimentel D (2007) Biofuels from crop residues. Soil and Tillage Res 93:237–238. doi:10.1016/j.still.2006.11.007

Loomis RS, Amthor JS (1999) Yield potential, plant assimilatory capacity, and metabolic efficiencies. Crop Sci 39:1584–1596

Loomis RS, Connor DJ (1992) Crop ecology: productivity and management in agricultural systems, chap 15. Cambridge University Press, Cambridge, pp 400–427

López-Bellido RJ, Castillo JE, López-Bellido L (2008) Comparative response of bread and durum wheat cultivars to nitrogen fertilizer in a rainfed Mediterranean environment: soil nitrate and N uptake and efficiency. Nutr Cycl Agroecosys 80:121–130

Lövenstein HM, Rabbinge R, van Keulen H (1992) World food production. Textbook 2: biophysical factors in agricultural production. Open University, Herleen

MacAdam JW, Nelson CJ (2003) Physiology of forage plants. In: Barnes RF, Jerry Nelson C, Collins M, Moore KJ (eds) Forages, 6th edn. Iowa State Press, Blackwell Publishing Company, pp 73–78

MacNeish RS (1992) The origins of agriculture and settled life. University of Oklahoma Press, Norman/London

Mannion AM (1997) Agriculture and environmental change. Temporal and spatial dimensions. Wiley, Chichester, England

Mosier AR (2001) Exchange of gaseous nitrogen compounds between terrestrial systems and the atmosphere. In: Follett RF, Hatfield JL (eds) Nitrogen and the environment: sources problems and management. Elsevier Science B.V, Amsterdam, pp 291–309

Mosier A, Kroeze C, Nevison C, Oenema O, Seitzinger S, van Cleemput O (1998) Closing the global N_2O budget: nitrous oxide emissions through the agricultural nitrogen cycle. Nutr Cycl Agroecosys 52:225–248. doi:10.1023/A:1009740530221

Nonhebel S. (1997) Harvesting the sun's energy using agro-ecosystems. Wageningen: DLO research institute for agrobiology and soil fertility. Quantitative approaches in systems analysis no.13

Nonhebel S (2005) Renewable energy and food supply: will there be enough land? Renew Sustain Energy Rev 9:191–201. doi:10.1016/j.rser.2004.02.003

Ortiz R, Sayre KD, Govaerts B, Gupta R, Subbarao GV, Ban T, Hodson D, Dixon JM, Ortiz-Monasterio JI, Reynolds M (2008) Climate change: Can wheat beat the heat? Agri Ecosyst Environ 126:46–58. doi:10.1016/j.agee.2008.01.019

Pearce F (2006) Fuels gold. Are biofuels really the greenhouse-busting answer to our energy woes? NewScientist 2570:36–41. www.newscientist.com

Pierce FJ, Rice CW (1988) Crop rotations and its impact on efficiency of water and nitrogen use. In: Hargrove (ed) Cropping strategies for efficient use of water and nitrogen. ASA Special Publ. 51. ASA, CSSA, and SSSA, Madison, WI, pp 21–42

Pimentel D (1992) Energy inputs in production agriculture. In: Fluck RC (ed) Energy in farm production, vol 6. In: Stout BA (ed) Energy in world agriculture. Elsevier Science, Amsterdam, NL, pp 13–29

Porter JR, Chirinda N, Felby C, Olesen JE (2008) Biofuels: putting current practice in perspective. Science 320:1421

Ragauskas AJ, Williams CK, Davison BH, Britovsek G, Cairney J, Eckert CA, Frederick WJJ, Hallet JP, Leak DJ, Liotta CL, Mielenz JR, Murphy R, Templer R (2006) The path forward for biofuels and biomaterials. Science 311:484–489

Randal GW, Goss MJ (2001) Nitrate losses to surface water through subsurface, tile drainage. In: Follett RF, Hatfiled JL (eds) Nitrogen in the environment: sources, problems, and management. Elsevier Science, Amsterdam, pp 95–122

Raun WR, Johnson GV (1999) Improving nitrogen use efficiency for cereal production. Agron J 91:357–363

Righelato R, Spracklen DV (2007) Carbon mitigation by biofuels or by saving and restoring forests? Science 317:902. doi:10.1126/science.1141361

Robertson GP, Dale VH, Doering OC, Hamburg SP, Melillo JM, Wander MM, Parton WJ, Adler PR, Barney JN, Cruse RM, Duke CS, Fearnside PM, Follett RF, Gibbs HK, Goldemberg J, Mladenoff DJ, Ojima D, Palmer MW, Sharpley A, Wallace L, Weathers KC, Wiens JA, Wilhelm WW (2008) Sustainable biofuels redux. Science 322:49–50. doi:10.1126/science.1161525

Rowe RL, Street NR, Taylor G (2009) Identifying potential environmental impacts of large-scale deployment of dedicated bioenergy crops in the UK. Renew Sustain Energy Rev 13:271–290. doi:10.1016/j.rser.2007.07.008

Ryle GJA, Powell CE, Gordon AJ (1979) The respiratory costs of nitrogen fixation in soybean, cowpea and white clover.II. Comparisons of the costs of nitrogen fixation and the utilization of combined nitrogen. J Exp Bot 30(114):145–153

Searchinger T, Heimlich R, Houghton RA, Dong F, Elobeid A, Fabiosa J, Tokgoz S, Hayes D, Yu TH (2008) Use of U.S. croplands for biofuels increases greenhouse gases through emissions from land-use change. Science 319:1238–1240. doi:10.1126/science.1151861

Sinclair TR, Cassman KG (1999) Green revolution still too green. Nature 398:556

Smil V (1994) Energy in world history. Westview Press, Boulder, CO

Smil V (2001) Enriching the earth. MIT Press, Cambridge, MA

Smil V (2002) Nitrogen and food production: proteins for human diets. Ambio 31:126–131

Spiertz JHJ (2009) Nitrogen, sustainable agriculture and food security. A review. Agron Sustain Dev. doi:10.1051/agro/2008064

Spiertz JHJ, Ewert F (2009) Crop production and resource use to meet the growing demand for food, feed and fuel: opportunities and constraints. NJAS Wageningen J Life Sci 56–64:281–300

Stoeglehner G, Narodoslawsky M (2009) How sustainable are biofuels? Answer and further questions arising from an ecological footprint perspective. Bioresource Technol 100:3825–3830. doi:10.1016/j.biortech.2009.01.059

Tilman D, Hill J, Lehman C (2006) Carbon-negative biofuels from low-input high-diversity grassland biomass. Science 314:1598–1600. doi:10.1126/science.1133306

Tilman D, Socolow R, Foley JA, Hill J, Larson E, Lynd L, Pacala S, Reilly J, Searchinger T, Somerville C, Williams R (2009) Beneficial biofuels – the food, energy, and environment Trilemma. Science 325:270–271. doi:10.1126/science.117790

Wackernagel M, Rees W (1993) How big is our ecological footprint – a handbook for estimating a community's carrying capacity. Vancouver

Ziegler J (2007) The impact of biofuels on the right to food. UN Report of the special Rapporteur on the right to food. http://www.righttofood.org/A/62/289

Biofuels, the Role of Biotechnology to Improve Their Sustainability and Profitability

Meenu Saraf and Astley Hastings

Abstract Energy supply and environmental change are the challenges facing humanity today. The need to develop new, carbon neutral forms of energy is now urgent. Bio-energy is a promising option that offers both energy sustainability and greenhouse gas emissions mitigation. In 2005 biomass provided 13.4% of current world energy needs, mainly as a heating and cooking fuel in rural communities, However, even with the increasing contribution from industrially produced biomass fired heat and power, the use of bio-diesel from oil crops and ethanol derived from maize and sugar-cane, the overall percentage contribution of bio-energy to world energy needs diminishes with time. Increasing bio-energy use to supplement the world's energy needs will require the growing of energy crops on a large scale, entailing changes to agricultural and forestry production techniques and the identification of new bio-energy feed-stocks. The use of ethanol from maize and sugar cane to replace petrol has more than tripled in 6 years and biodiesel from the esters of oil from crops such as palm, soya and oil seed rape now contributes 4% of Europe's diesel needs. However these first generation biofuels use valuable arable land and food crops in competition with human food needs and are not efficient energy producers per hectare of land nor do they effectively mitigate greenhouse gas emissions. Ligno-cellulosic ethanol, bio-butanol, bio-gas and biodiesel from gas to liquid bio-refineries using native grasses, *Miscanthus*, wood processing waste and short rotation coppice willow and non-food oil seed plants such as *Jatropha* enable non arable land and non food crops to be used for second generation biofuels. In addition, autotrophic algae have been shown to exceed productivity of many oil crops in using sunlight and carbon dioxide for oil accumulation.

M. Saraf (✉)
Department of Microbiology and Biotechnology, University School of Sciences,
Gujarat University, Ahmedabad 380009, Gujarat, India
e-mail: sarafmeenu@gmail.com

A. Hastings
Institute of Biological and Environmental Sciences, School of Biological Science,
University of Aberdeen, AB51 0LZ, Aberdeen, UK
e-mail: Astley.hastings@abdn.ac.uk

E. Lichtfouse (ed.), *Biodiversity, Biofuels, Agroforestry and Conservation Agriculture*,
Sustainable Agriculture Reviews 5, DOI 10.1007/978-90-481-9513-8_4,

Microbial oils can also be used as feedstock for biodiesel productions with advantages like short life cycle, less labour intensive, less seasonal, geographical and climatic variability and easy scale up. This paper reviews first and second generation bio-energy systems for producing biofuels and investigates the potential of biotechnology to improve the sustainability and profitability of existing and future biofuels bio-energy systems.

Keywords Biodiesel • Biotechnology • Microalgae • Lignocellulose • Biorefinery • Single cell oils • Bioethanol • Biobutanol • Biogas • Biohydrogen • Oil seeds

1 Introduction

Finding sufficient supplies of clean energy for the future is one of society's most daunting challenges and is intimately linked to global stability, economic prosperity and quality of life. Until recently, the need for biofuels remained a low priority as petroleum, gas and coal supply and demand were balanced. Nonetheless, global petroleum demand has increased steadily from 57 × 10^6 barrels day^{-1} in 1973 to 85 × 10^6 barrels day^{-1} in 2009 and will continue to increase in line with the world's economy. This is driven by the rapid industrialization of economies like Brazil, China and India. Recent rapid increases in the price of oil have been caused by the demand for energy being just balanced by the ability to produce it. Oil production capacity limitations have been offset by the increasing use and supply of natural gas and coal and this has avoided an energy crisis. However by 2025 the projected economic growth is anticipated to increase global demands for liquid fuels by ~50% (Ragauskas et al. 2006), so the pressure for more energy continues and even though the recent increase in price had an impact on the short term growth in demand and world economic growth, this projected future need is providing an impetus to develop alternative energy sources.

The commitment made by signatories of the Kyoto protocol (1998) and their future expansion in the up-coming Copenhagen meeting to reduce greenhouse gas emissions to mitigate global warming has specifically targeted combustion CO_2 from fossil fuels released into the atmosphere. This need to reduce emissions, coupled with concerns about future energy security due to rising prices of crude oil and continuing political instability in oil producing countries, has increased the impetus to develop sustainable alternative energy sources. As described by Hoffert et al. (2002) future reductions in the ecological footprint of energy generation will reside in a multifaceted approach that includes the use of hydrogen, wind, nuclear, solar power, fossil fuels from which carbon is sequestered and bio-energy.

Bio-energy, as a candidate for sustainable energy, presents many advantages as at first sight it is carbon neural, produced by a farming and forestry sector which exists in most countries and in terms of security is an indigenous source

(Goldemberg 2000; Davenport 2008). However it does compete for land use with food, fiber and forestry production. To put its potential into perspective in 2005 biomass provided 13.4% of current world energy needs, mainly as a heating and cooking fuel in rural communities. However, even with the increasing contribution from industrially produced biomass fired heat and power in industrialized countries with forestry resources like the USA, Canada and Sweden and the increasing use of first generation biofuels like bio-diesel from oil crops and ethanol derived from maize and sugar-cane, the overall percentage contribution of bio-energy to world energy needs diminishes with time. Even this moderate use of biofuels has caused a disruption in the supplies of food. This is manifested by the recent increase in grain prices due to the large scale use of maize as a feedstock for fuel ethanol in the USA. This caused riots in Mexico due to the increase in the price of tortillas a staple food. Increasing bio-energy use to supplement the world's energy needs will require the growing of bio-energy crops on a large scale, entailing changes to agricultural and forestry production techniques and the identification of new bio-energy feed-stocks. This has to be achieved in a way that does not disrupt or impact food production.

Using petroleum based fuels for transportation creates atmospheric pollution during combustion. Apart from emission of the greenhouse gas (GHG) CO_2, air contaminants like NOx, SOx, CO, particulate matter and volatile organic compounds are produced (Klass 1998). Their continued use is now widely recognized as unsustainable because of depleting supplies and the contribution of these fuels to accumulation of carbon dioxide in the environment. Renewable, carbon neutral, transport fuels are necessary for environmental and economic sustainability. Worldwide, about 27% of primary energy is used for transportation, which is the fastest growing sector (EIA 2006). Transportation fuels are thus promising targets for a reduction in GHG. The use of biofuels could reduce excessive dependence on fossil fuels and ensure security of supply, promoting environmental sustainability and meet the target of at least of 10% replacement by 2020 (Kyoto 1998) in the transport sector. Biofuel has received considerable attention recently, as it is made from non-toxic, biodegradable and renewable resources and provides environmental benefits, since in addition to reducing GHG emissions, its use can also decrease harmful emission of carbon monoxide, hydrocarbons and particulate matter and eliminates SOx emissions (Gouveia and Oliveira 2008).

However biofuels are not zero carbon fuels as energy is consumed and GHG's emitted during their production before they are actually burnt as fuels. Converting land from a natural eco-system to one for energy causes a carbon reduction in the soil and cultivation uses fertilizers, which emit N_2O, a GHG with 265 times more global warming potential than CO_2. Farm management such as tilling, sowing and harvesting and transporting the feedstock for processing and distributing the biofuel use energy and emit carbon to the atmosphere. Converting the feedstock to a useable transport fuel is usually energy intensive and hence emits CO_2 to the atmosphere. For example Patzek (2008) shows that, after performing a life cycle analysis (LCA) using all of these factors, the production of the first generation biofuel, bio-ethanol from maize, actually emitted more CO_2 and used more energy than the petroleum based gasoline it replaced, even when there was no land use change. Searchinger et al. (2008) demonstrated that when rainforest was replace by oil palm plantations, the emissions of carbon from the land use change exceeded the carbon saved by the use of the bio-diesel, which without this factor would be a strong carbon mitigator. To be sustainable and economical, second and future generations of bio-energy systems should produce biofuels with less energy consumed and greenhouse gasses emitted in their feedstock production and conversion to fuel and to increase the amount of energy produced per hectare of land. Metrics to judge the environmental and economic sustainability of biofuels should include the energy intensity, energy use efficiency and carbon intensity (Hastings et al. 2009a). Energy intensity (EI) is net energy produced per hectare of land (GJ ha^{-1}). The energy use efficiency (EUE) is the ratio of the net energy produced to the energy cost of producing it (J J^{-1}). The carbon intensity is the amount of green house gas emitted per unit of energy expressed in g CO_2 equivalent Carbon per GJ energy, considering the global warming potential of nitrous oxide to be 298 and methane to be 25 times that of carbon dioxide for a 100 year time horizon (IPCC 2007).

Advances in the realm of biotechnology could have a large effect on the environmental impact and economics of biofuels. So called "Green Biotechnology" will yield crops with better fatty acid profiles, more biomass for hectare, resistance to drought and other factors. "White Biotechnology" (Industrial biotechnology) should strive to produce organisms or enzymes that convert a wider range of substrates directly into variety of desired products. A comprehensive approach is needed for rapid development of alternative fuels, involving plant breeders, agronomists, bioprocess engineers biotechnologists and microbiologists. Research has intensified towards production of alternative fuels by thermo chemical means or by fermentation by microbes.

This review provides insight into the various sources of biofuels available along with ecological and environmental benefits. The possible contribution of new biofuels along with the research challenges, in particular the contribution of biotechnology is discussed in the following sections: use of agricultural crops for ethanol, biodiesel and bio-butanol, use of lignocellulosic biomass from agriwastes forestry and perennial grasses, and use of microorganisms and algae as source of biofuel.

2 Agricultural Crops as a Source of Biofuel

Traditional food crops are used as feed-stocks for bio-ethanol (corn, sugarcane, sugar-beet, etc), bio-diesel (soybean, oil palm, oil seed rape, etc.) and bio-butanol (sorghum, sugar-beet, etc.) production, used as biofuel. This use of agricultural crops is aggressively expanding and may represent 15–20% of biofuel demand, however, the limited availability of additional suitable land to grow traditional crops beyond their continuing primary uses as food and feed will limit its future potential. In addition the sustainability of this first generation of biofuel production in terms of its energy intensity, energy use efficiency and carbon intensity is questionable and requires a step change in technology to improve.

2.1 Bio-Ethanol

The barriers to the expansion of the use of bio-ethanol derived from food crops as a biofuel include the availability of suitable land. In addition its sustainability requires improvement as its economic use currently depends on government subsidies. The limiting factors are the energy cost of the production of anhydrous ethanol from the fermentation broth and the energy cost of producing the enzymes used in the fermentation process.

2.1.1 Corn Bio-Refinery

In the USA cereal based bio-refineries utilizing corn have been developed for the production of fuel ethanol and lactic acid or are under construction for production of polyhydroxyalkanotes (PHAs) and 1,3-propanedione (Bio-PDO) (Bevan and Franssen 2006). Corn's use for ethanol production has more than tripled in 6 years. Projections for 2010 indicate that 2.6 billion bushel will be required for ethanol production. Concerns have been raised over the availability and cost of corn grain for livestock feed globally. The ethanol industry is based on processing of corn grain (i.e. starch) through either dry grind or wet milling processes (Gibbons 2007). The US corn based fuel ethanol industry is experiencing unprecedented growth. Rosentrater (2007) has reviewed three areas where substantial potential lies in value added processing and utilization of bye products, including animal feeds, human foods and industrial product. This increase in value added products will lead to better economics and sustainability of ethanol processing.

2.1.2 Wheat Bio-Refinery

Wheat is the most widely cultivated cereal grain in EU and is regarded as a potential candidate for bio-refinery developments (Picture 1). The process being adapted begins with a simple dry milling stage leading to production of whole wheat flour, the starch content is then hydrolysed into glucose by commercial amylolytic preparation (α/β- amylase, glycoamylase, pullulanase). The resulting glucose solution is fermented into ethanol. (Koutinas et al. 2007). The economics and efficiency of a wheat based bio-refinery has been suggested by Webb et al. (2004), that optimizes the conversion of wheat components into arrange of products. It may be stressed that this is a generic concept proposed that could be modified in order to lead to the production of biofuels (fuel ethanol),

Picture 1 Wheat

platform chemicals (e.g. Succinic acid, xylitol), speciality chemicals (e.g. recombinant proteins) and/or biodegradable plastics (polyhydroxybutyrate).

2.1.3 Sugarcane Biorefinery

Hydrous ethanol, produced from sugarcane was first used on a large scale as a transportation fuel for spark ignition engines in Brazil in the 1970s in response to the first oil shock and a lack of petroleum resources in the country. Brazil has now one of the largest bio-ethanol industries and in spite of being self sufficient in petroleum, all its gasoline fuel contains 25% anhydrous ethanol overcoming the corrosion problems associated with hydrous ethanol use. Brazil is now the main suppler of bio-ethanol to Europe and also exports to the USA, however the expansion of this industry is at the cost of reducing the size of its forested area. Life cycle analyses of sugar cane ethanol produced in Brazil, show it to be the most sustainable of all bio-ethanol production due to the mature nature of the industry. All the heat and energy to produce the ethanol is derived from the burning of the bagasse, the lingo-cellulose part of the plant remaining after the sugar is extracted. Excess electricity is injected into the grid. Most other electricity used in Brazil is from low carbon hydro-generation and transportation uses 25% bio-ethanol. This results in a high energy use efficiency of 6% and a reduction of greenhouse gas emissions compared to gasoline of 30%. Most of the greenhouse gas emissions in its production come form land use change from the rain forest to sugarcane plantation (Goldemberg et al. 2008).

Picture 2 Sugarcane

The high moisture content biomass such as sugarcane, lends itself to a met/aqueous conversion process such as fermentation. Sugar in sugarcane is readily converted to ethanol by fermentation organisms (Bohlmann 2006) (Picture 2).

2.2 *Plant Oil as Source of Biofuels*

The first compression ignition engine was designed by Otto Diesel to run on vegetable oil and many current diesel engines can use it as a fuel in their unmodified form as long as the viscosity is reduced to that of diesel by additives or warming. Vegetable oils have a long history as lubricants, especially in steam engines, and are a renewable, biodegradable alternative to petroleum based oils. Compared to petroleum oils, vegetable oils show better performance in a number of areas, demonstrating higher lubricity, lower volatility a higher viscosity index and higher shear stability. A drawback in the use of vegetable oils has keen the presence of unsaturated fatty acids, which reduce the oils oxidative stability (Grushcow and Smith 2006). Before the invention of synthetic oils castor oil was used as a lubricant in high performance spark ignition engines such as those used in formula 1 racing cars.

Castor oil is one known to have excellent oxidative stability besides being a good source of ricinoleic acid (12-hydroxyoleic acid). A great degree of research has been directed to the cloning of the enzyme, referred to as castor hydroxylase for promoting its extensive use (Vande Loo et al. 1995). Heterologus expression of a fatty acid hydroxylase gene in developing seeds of *Arabidopsisthaliava* has been to cloned and seed oil has been produced that contains up to 20% hydroxyl fatty acids (Grushcow and Smith 2006).

2.2.1 Bio-Diesel from Oil Seeds

Currently the bio-diesel production using oil seeds as feedstocks is challenged by technologies for further processing of the oil seed cake and the glycerol by-product that is generated from the trans-esterification of vegetable oils. All oil seeds have been tried as a raw material but by far the most popular is soybean in the US and rapeseeds in Europe. In India, many companies are developing bio-energy systems using the new oil crop *Jatropha* to produce bio-diesel. Malaysia is aggressively developing palm oil as a bio-diesel feedstock and new palm oil conversion plants and plantations over 600,000 ha are being developed (Wyse 2005) (Picture 3). Most bio-diesel used in the UK is derived from palm oil.

Most types of vegetable oils can be used for the preparation of biodiesel like; peanut, soybean, palm, sunflower, etc. (Table 1). There are no technical restrictions on the use of any type and non-edible oils like *Jatropha* can be used. Palm oil has a yield of 4,000 kg ha^{-1} which is the highest, next comes coconut oil which is 2,250 kg ha^{-1}. Many different scientists have reported studies based on several other oils like olive oil (Sanchez and Vasudevan 2006), rich bran oil (Lai et al. 2005) and canola oil (Chang et al. 2005).

Jatropha, a member of the euphorbia family has long been used around the world as a lamp oil and soap and hedging plant. It thrives on eroded farmland or non-arable wasteland. These bushes can live up to 50 years, fruiting annually for more than 35 years and weathering droughts. The oil content of the seed

Picture 3 Jatropha Seeds

Table 1 Comparison between properties of biodiesel fuels produced from various vegetable oils (Zuhair 2007)

Vegetable oil used	Production yield (kg/ha^{-1})	Kinematic viscosity (mm^2 s^{-1})	Cetane number
Peanut	890	4.9 (37.8°C)	54
Soybean	375	4.5 (37.8°C)	45
Palm	4,000	5.7 (37.8°C)	62
Sunflower	655	4.6 (37.8°C)	49
Rapeseed	1,000	4.2 (40°C)	51–59.7
Used rapeseed	–	9.48 (30°C)	53
Used corn oil	–	6.23 (30°C)	63.9

Table 2 Choices of crops (Fairless 2007)

Biofuel crop	Liter of oil per hectare
Oil palm	2,400
Jatropha	1,300
Rapeseed	1,100
Sunflower	690
Soybean	400

ranges between 35–43%, seed cake can be used as fertilizer and seed husk into a high density brick that can be burnt for fuel (Fairless 2007). Approximately 175 species of *Jatropha* are available, but *Jatropha curcas* is the most widely used variety. This oil has at least cloud point of −2°C, preferred for use as fuel in temperate region. It is poisonous and to date this makes *Jatropha* the only source of non food vegetable oil for energy. Growth of *Jatropha curcas* in the heavy metal contaminated soil amended with industrial wastes and *Azotobacter* has been successfully studied by Kumar et al. (2008). Other researchers have supplemented with chitin and *Bacillus* spp. and observed stimulation in seedling growth of Jatropha curcas (Desai et al. 2007). It produces an excellent bio-diesel with a relatively high yield but its limitation is that it so far is very labour intensive as it requires manual harvesting as not all the fruits mature at the same time (Table 2).

2.2.2 Used Vegetable Oils

The use of waste cooking oil has also been promoted by several workers. This is important as it recycles waste oil. A fresh vegetable oil and its waste differ significantly in water (around 2,000 ppm) and free fatty acid (10–15%) contents. To use this resource as a bio-diesel feedstock the use of an acidic catalyst has been used (Zhang et al. 2003). On the other hand lipase enzyme is also capable of converting all free fatty acids contained in waste oils to esters. Other methods have been described like waste oil adsorbed on activated bleaching earth (ABE) and cellular biomass of oleaginous yeast and filamentous fungi.

2.3 Bio-Butanol

Bio-butanol, another fermentation product, has been described as a new 'fuel' but has been in almost continuous production since 1916 as a solvent and basic chemical. n-Butanol has many advantages and has emerged as a diesel and kerosene replacement (Schwarz and Gapes 2006). Substrates for bio-butanol production include hydrolysates of lignocellosics from agriculture waste material and waste potatoes (Gapes 2000). The agricultural residues and wastes that can be used for the production in acetone, butanol, ethanol (ABE) fermentations include; rice straw, wheat straw, corn fibre, other crop and grass residues etc. (Quershi and Blaschek 2005). Since these substrates contain five and six carbon sugars, feedstocks must be hydrolysed prior to the addition of *Clostridium acetobutylicum* and *Clostridium beijerinckii*. Butanol production from wheat straw hydrolysate using *Clostridium beijerinckii* P260 has shown an excellent capacity to convert biomass derived sugars to solvents producing 28gL-1 ABE (Quershi et al. 2007).

3 Ligno-Cellulose as a Source of Biofuel

Although the contribution of agro-energy crops and agricultural waste for biofuel production is being intensively developed commercially, the potential of the forest product industry to contribute to this effort is still in the pre-commercial trial phase for both cellulosic bio-ethanol production (Pimental and Patzek 2005) and biodiesel from biomass gasification and then liquid production (Blades et al. 2006) from the Fischer and Tropsch (1930) process.

3.1 Forestry Production By-Products and Waste

Wood is produced for both paper and wood production. In general only the round wood is used for both purposes with the bark and branches being discarded as waste which, although is normally left on the forest floor, can be utilized as a bio-energy feedstock. Wood processing at sawmills also produces up to 25% of the round wood as sawdust, shaving or off-cuts, this can be used as a fuel or a bio-energy feedstock. On an annual basis the US paper and pulp industry collects and processes approximately 108 million tons of wood. Wood extractives from pulping

Picture 4 Lignocellulosic waste

provide approximately 700 million liters of turpentine and tall oil annually that could be employed for biodiesel applications (Ragauskas et al. 2006). Wood is composed primarily of cellulose, hemi-cellulose, lignin and small amounts of extractives (Picture 4). The lignin and oils can be used as a bio-energy feedstock.

3.2 Dedicated Ligno-Cellulose Crops

Dedicated perennial herbaceous crops including pasture grasses, switch grass, elephant grass, *Miscanthus* and woody species including willow and poplar (Picture 5) can be grown as biofuel feed-stocks. They are characterized by fast growth, high biomass yields, low input costs and as they can be harvested for many years after establishment can be termed as robust and reliable. With second generation biofuel processes, they can provide large quantities of cost competitive biofuels. In addition, molecular technologies can be used to increase biomass yield, water use efficiency, ease of processing and chemical content. These crops have the ability to be grown on non-prime agricultural land, under utilized or abandoned farmland (NABC 2007) and so will not compete with arable land used for food production. Field scale trials and modeling of *Miscanthus* yields across Europe have show that when used as a bio-mass fuel it has an energy use efficiency of 6% and reduces greenhouse gas emissions by 80%, so it as a biofuel feed stock it has a low energy and carbon footprint (Hastings et al. 2009a). *Miscanthus* is a native of SE Asia and has not been historically cultivated on a large scale therefore it has not benefited from the breeding improvement that food crops like wheat, with a 5,000 year cultivating history, have had to improve yields and suitability for their end use. Molecular and genetic techniques are now being used to breed varieties that have higher yields and can grow in dryer and colder climates (Clifton-Brown et al. 2008).

Picture 5 Switch grass

3.3 Second Generation Bio-Diesel Production by Gasification

Gasification and combustion are readily applicable technologies for processing biomass of various kinds for production of biofuels. The potential feed-stocks include dedicated energy crops, farm waste like corn stover, forest product waste or other biodegradable waste such as that from an abattoir. Gasification is a thermo-chemical process that converts carbohydrates into hydrogen and carbon monoxide under oxygen starved conditions (Bricka 2007). Several perennial crops have been identified as potential gasification feedstocks like pasture grass, switch grass, *Miscanthus*, poplar and shrub willows. Once established, perennial grasses or woody crops can be harvested multiple times over the life of a planting with relatively low inputs on an annual basis. These plants also tend to accumulate carbon below ground over time and can provide valuable wild life habitat while diversifying the agricultural landscape (Volk et al. 2004). In this respect shrub willows can be planted on otherwise marginal agricultural soils that do not support high yields of corn or soybean due to poor drainage conditions, limited fertility or regular spring flooding. This is a fast growing plant that vigorously re-sprouts the spring after the stem biomass is harvested (Coppiced). This is then fed into the harvester, which delivers wood chips-uniformly 5 cm or less in size. Chips are trucked to the fuel yard of the power plant and are piled for storage and moderate drying before conversion by combustion and gasification (Smart et al. 2007). Lignocellulosic biomass is a mixture of phenolic lignin and carbohydrates- cellulose and hemicellulose. Lignocellulosic conversion processes are still expensive today, being competitive at crude oil prices between $50 and 100/bbl. Although Lignocellulosic can be a fairly cheap feedstock, cheaper than crude oil, its conversion requires large investments in the gasification and gas to liquids plant, which will benefit from cost reductions that will come with widespread implementation (Lange 2007).

3.4 *Second Generation Ligno-Cellulose Bio-Ethanol*

There are basically two techniques available for the conversion of wood cellulose and hemicelluloses into fermentable sugar solution. The first is an acid hydrolysis process which would relinquish monosaccharides for the production of ethanol via fermentation. Alternatively extracted wood polysaccharides could be enzymatically hydrolysed and fermented to ethanol. The enzymatic hydrolysis of pretreated cellulosic biomass has been commercialized for processing of wheat straw to bio-ethanol and is being actively pursued for other agricultural waste resources (Tolan 2003). In order to construct a strain that converts sugar mixture and resist/metabolize inhibitors in lignocellulosic dilute acid hydrolysate (softwood). The protoplast fusion between *Saccharomyces cerevisiae* and *Pachysol tannophilis* was performed. The result indicated that the hybrid strain was genetically stable, ethanol tolerant and promising prospect for industrial application (Yan et al. 2008).

The major problems in the use of plant biomass are the pentose sugars released from lignocellulosic biomass. The maize plant (without corn) contains about 24% (of dry mass) xylose and 2% arabinose (in comparison with 45% glucose) which cannot be neglected. Generally hardwoods and agricultural raw material contain a larger proportion of D-xylose and L-arabinose while softwood is less rich in pentoses. The state of metabolic engineering to yield pentose fermenting yeast strains is reviewed by Hahn Hagerdahl et al. (2007). Ethanol fermentation using cellulosic biomass to using thermophilic bacterium *Clostridium thermocellum* has been studied by Demain et al. (2005). However further work needs to be done as the strain is sensitive to higher ethanol concentration.

4 Algae as a Source of Biofuel

Microalgae are sunlight driven cell factories that convert carbon dioxide to potential biofuels, foods, feeds and high value bioactives. They can provide several types of renewable biofuels or methyl esters or bio-diesel or microalgal oil, which was demonstrated in 2000 by Belarbi et al. (2000) although the product was intended for pharmaceutical use. Unlike other crops, microalgae grow extremely rapidly and are exceedingly rich in oil. They can double their biomass within 24 h. Oil content can be between 20–50% by weight of dry biomass, even upto 80% has been reported (Spolaore, et al. 2006). Due to their higher photosynthetic efficiency, higher biomass productivities, a faster growth rate than higher plants, highest CO_2 fixation and O_2 production they can be grown in liquid medium which can be handled easily. They can be grown in variable climates and on non-arable land including marginal areas unsuitable for agriculture, in non potable water, so they do not displace food crop cultures. Their production is not seasonal and can be harvested daily (Chisti 2008). To utilize the algal resources effectively for production of bio-diesel on a commercial scale, research needs to be undertaken to screen the bio-diesel potential of existing strains of species native to a particular region, for example Kaur et al. (2009) have studied the algal strains *Ankistrodemus* spp, *Scenedesmus* spp, *Euglena* spp, *Chlorella* spp and *Chlorococcus* spp isolated from North East India and they were found to accumulate a high intracellular lipid content as their energy storage product.

The ability of algae to fix CO_2 can also be an interesting method of removing gases from power plants and thus they can be used to reduce greenhouse gases with a higher production microalgal biomass and consequently higher bio-diesel yield. Algal biomass production systems can be easily adapted to various levels of operational and technological skills. The physical and fuel properties of bio-diesel obtained from algae i.e. density, viscosity, acid value, heating value etc. are comparable to those of fuel diesel (Miao and Wu 2006). Bio-diesel may be produced from oleaginous crops like rapeseed, soybean, sunflower through a chemical transesterification process, but production yield from algae can be 10–20 times higher than the yield obtained from other sources Table 3 (Chisti 2007). Technical development is needed to test a pilot plant.

Microalgae can produce other different types of renewable biofuels. These include methane, produced by anaerobic digestion of the algal biomass (Spolaore et al. 2006), biodiesel from microalgal oil, photobiologically produced biohydrogen (Kapdanand Kargi 2006). The oil content of microalgae varies from species to species. 20–50% of oil level is very common. Oil productivity, that is the mass of oil produced per unit volume of the microalgal broth per day, depends on the algal growth rate and oil content of biomass (Table 4).

Micro-bio-diesel production from algae by heterotrophic fermentation of *Chlorella* was developed by Miao and Wu (2006), called algal fermentation based micro-bio-diesel (AFMD) production. This was found to be superior in comparison to classical photoautotrophic culture model, as it allows for much higher proportion of fatty acids. Another key issue in AFMD production is transesterification, which

Table 3 Comparison of some sources of biodiesel (Chisti 2007)

Crop	Oil yield (L ha^{-1})
Corn	172
Soybean	446
Canola	1,190
Jatropha	1,892
Coconut	2,689
Palm	5,950
Microalgae[a]	+36,900
Microalgae[b]	58,700

[a]70% oil (by weight) in biomass
[b]30% oil (by weight) in biomass

Table 4 Oil content of some microalgae (Chisti 2007)

Microalga	Oil content (% dry wt)
Botruococcus braunii	25–75
Chlorella sp.	28–32
Cylindrotheca	16–37
Dunaliella primolecta	23
Nannochloris sp.	20–35
Nannochloropsis sp.	31–68
Nitzschia sp.	45–47
Schizochytrium sp.	50–77
Tetraselmis sueica	15–23

can be catalyzed by acid, alkali or enzymes. Enzymatic methods are more environmentally friendly and the key factor of a successful enzymatic process is the lipase immobilization, (Shimmada et al. 1999). Growth media required for cultivation of algae is usually inexpensive. Growth medium must provide the essential elements like nitrogen (N), phosphorus (P), iron etc. which can be estimated using the formula $CO_{0.48}$, $H_{1.83}$, $N_{0.11}$, $P_{0.01}$ (Grohelaar 2004). *Neochloris deabundans* and *Nannochloropsis sp.* when grown in N-deficient medium shows a great increase in oil quantity i.e almost 56% in *Neochloris* in comparison to 29.0% in normal medium. Whereas *Scenedesmus obliquus* presents the most adequate fatty acid profile in terms of lindemic and polyunsaturated fatty acids (Gouveia and Oliveira 2008). Illman et al. (2000) reported that the limitation of nitrogen could increase oil content in all *Chlorella* strains. On the other hand Hossain et al. (2008) used common species *Oedogonium* and *Spirogyra* and found biomass (after oil extraction) was higher in *Spirogyra* and sediments (glycerine, water and pigments) were also higher in *Spirogyra*, but *Oedogonium* sp. has higher bio-diesel content.

4.1 Large Scale Cultivation

Currently, raceway ponds and tubular photobioreactors are usually adopted for large scale cultivation of autotrophic microalgae. Closed photobioreactors are also used since they could save water, energy and compared to some other open cultivation systems (Peer et al. 2008). A tubular bioreactor is another device used for scaling up of autotrophic algae. Micro algal broth is continuously pumped through the solar array where sunlight is absorbed. Generally speaking the scale up for atmospheric microalgae is more complicated since light is needed during cultivation. Heterotrophic microalgae also can accumulate oils with organic carbon sources instead of sunlight, eg. *Chlorella potothecoides* can use organic carbon sources for oil production and oil content obtained is about four times high than that in the corresponding autotrophic cells (Miao and Wu 2004). However Chisti, (2007) has concluded that photobioreactors provide much greater oil yield per hectare compared to raceway ponds, this is because the volumetric biomass productivity of photobioreactors is more than 13-fold greater in comparison with raceway ponds, but both are technically feasible.

5 Microbial Oils as a Source of Biofuels

In recent years, much attention has been paid to the exploration of microbial oils which might become one of the potential oil sources for biodiesel production in the future. This may be produced by yeast, fungi, bacteria and actinomycetes (Ma 2006). It has been demonstrated that such microbial oils can be used as feedstocks for bio-diesel production comparable to other vegetable oils and animal fats. Microbial oil production has several advantages over other bio-diesel feedstocks, like having a short life cycle, being less labour intensive, can be grown in a multiplicity of climates and locations and the processes are easy to scale up (Li and Wang 1997). However, this work is still at its infancy and in order to maximize production and efficiency further research is required.

5.1 Fungal Sources

Many species of yeast like *Crytococcus albidus, Lipomyces lipofera, Lipomyces starkeyi, Rhodospridium toruloides, Rhodotorula glutinis, Trichosporon pullulan* and *Yarronia lipolytica* can accumulate oils under cultivation conditions (Picture 6). Their lipid yield ranges from 5.9% to 37% g/l (Liang et al. 2006). The lipid component of yeast oils were found to be myristic acid, palmitic acid, stearic acid, oleic acid, linoleic acid and linolenic acid. It has been reported that such yeast oils can be used as oil feedstocks for bio-diesel production with the catalysis either by lipase or chemical catalyst (Li et al. 2007). Different cultivation conditions, C/N ratio,

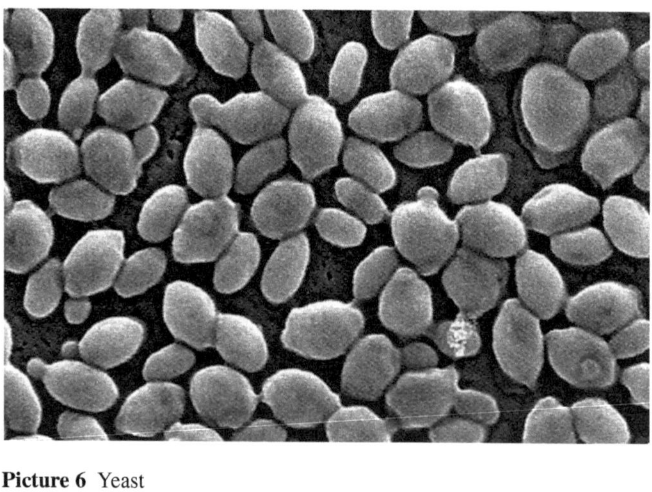

Picture 6 Yeast

nitrogen resources, temperature, pH, oxygen and concentration of trace elements and inorganic salt have varied influence on oil accumulation. It was reported (Mainul et al. 1996) that when C/N ratio increased from 25 to 70, oil content increased from 18% to 46%. Whereas Huang et al. (1998) reported that the inorganic nitrogen sources were good for cell growth but not suitable for oil production, while organic nitrogen sources such as peptone and broth were good for oil production but not suitable for cell growth. Exploration of cheap carbon sources for yeast oil accumulation is required to commercialize its production. Many workers have used a variety of sources like xylose, glycerol, sewage sludge, sweet potato starch processing waste etc. and have obtained up to 52.6% lipid content in *L. starkeyi* (Kong et al. 2007). In the future large scale bio-diesel productions shall be aimed to produce more of the by-product glycerol and this crude glycerol be utilized as a feedstock for yeast oil production.

Bio-ethanol is by far the largest scale microbial process used for both industrial ethanol and alcoholic beverage production and it is a nature industry. State of the art industrial ethanol production uses sugarcane molasses or enzymatically hydrolysed starch (from corn or other grains) and batch fermentation with yeast *Saccharomyces cerevisiae* to create ethanol and by products like CO_2 and small amounts of methanol, glycerol etc.. Inhibitor sensitivity, product tolerance, ethanol yield and specific ethanol productivity have been improved in modern industrial strains to the degree that up to 20% (w/v) of ethanol are produced in present day industrial yeast fermentation vessels from starch derived glucose (Antoni et al. 2007). Maize silage, which has a high glucan and total sugar content compared to other lignocellulosic materials, was studied by Thomsen et al. (2008). They concluded that the theoretical ethanol production of maize silage pretreated at 185°C for 15 min without oxygen using dry commercial yeast was 392 kg ethanol per ton of dry maize silage. On the other hand Roche et al. (2008) have optimized the enzymatic hydrolysis of the cellulose fraction of cashew apple bagasse and evaluated its fermentation to ethanol using *Saccharomyces cerevisiae.*

Although some types of fungi have the ability to produce oils most fungi are explored mainly for the production of special lipids such as DHA, GLA, EPA and ARA (Li et al. 2008). The reports of utilization of fungal oils for biodiesel production are very few (Table 5). Species of various filamentous fungi were tested for ability to produce lipids as a material for biodiesel. The mucoralean fungi *Cunninghanella japonica* was found to be most promising, producing over 7 g/l lipids (Sergeeva et al. 2008).

5.2 Bacterial Sources

Just like fungi, bacteria can also accumulate oil under special conditions. They produce polyunsaturated fatty acids and some branched chain fatty acids (Pathayak and Sree 2005). The fermentation of all sugars in cellulosic biomass

Table 5 Lipid accumulation by different yeasts and molds (Li et al. 2008)

Species	Lipid yield g/l
M. rouxii	1.0
C. ehinulata	8.0
M. mucedo	12.0
M. ranmamianna	31.3
M. isabellina	18.1
R. toruloides	13.8
L. starkeji	9.99
C. curvatus	37.1
R. glutinis	7.19
T. fermentans	5.32

can be performed by most bacteria although there is a problem of catabolite repression. *Zymomonas mobilis* is used for continuous ethanol fermentation a fluidized bed reactor (Weaster Botz 1993). Cheese whey has been used as a substrate for ethanol fermentation using *Streptococcus fragilis* in Gaerings industry process. *Kluyveromyces fragilis* is used in most commercial plants (Pesta et al. 2006). In future, cheese whey substrate will gain importance because of increasing cheese production and problems during disposal resulting form high organic matter content. Another industrial process based on the conversion of sugars (a CO_2/H_2 mixture from gasified biomass) to ethanol using *Carboxydocella gungdhlii* has been developed (Henstra et al. 2007). During this process municipal waste is gasified and cooled down to a fermentation temperature (while using the waste heat of produce electricity). This is blown into a fermenter and converted to ethanol.

A new concept of microbiological production of bio-diesel using metabolically engineered *E.coli* has been developed by Kalschener et al. (2006). *E.coli* cells were manipulated by introducing the pyruvate decarboxylase and alcohol dehydrogenase gene from *Zymomonas mobilis* for high ethanol production. Compared to others many expressions of genes in fatty acid synthesis are well understood in bacteria (Alexander et al. 2007). Therefore it is relatively easy to use biological engineering technology, genetic/metabolic engineering to modify bacteria performance to improve oil accumulation.

6 Agriwaste-Agriresidue as Source of Biofuels

To reduce the cost of biofuels, exploring sources other than glucose is very important. Many agricultural and industrial wastes could be used as carbon sources like xylose, starch hydrolysate, glycerol, sewage sludge, sweet potato starch processing waste etc. *C. potothecoides* could use starch hydrolysate as carbon source showing better accumulation of oil in comparison to glucose

(Han et al. 2006). Similarly yeast oil production was enhanced when industrial waste rich in sugar could be used (Du et al. 2007). However, if cellulose hydrolysate contain components like acetic acid, formic acid and furfural have a negative effect on cell growth and are toxic (Shen et al. 2007). Therefore, before using such cheap carbon sources for oil production detoxification becomes imperative.

7 Bio-Hydrogen

Hydrogen is also regarded as an ideal fuel for future transportation because it can be converted to electric energy in fuel cells or burnt and converted into mechanical energy without CO_2 production. Hydrogen can be produced biologically by algal and cyanobacterial biophotolysis of water or be photofermentation of organic substrates from photosynthetic bacteria. Production of hydrogen using miscenthus by *Thermotoga elfii* was studied by De Vrije et al. (2002) and biological hydrogen was produced from sweet sorghum using thermophilic bacteria (Claasen et al. 2004). Hydrogen production was carried out using glucose as substrate by *Enterobacter cloacae* in the dark fermentation. In the second stage fermentation *Rhodobacter sphaeroides* were used to photoproduce hydrogen. The overall hydrogen yield in this process was found to be higher than any other single stage process (Nath and Das 2006).

8 Bio-Gas

Bio-gas plants are another source of biofuels. They produce methane gas sustainably along with carbon dioxide from plant biomass, sewage, organic household or industrial waste. Substrate used can be very diverse from cow, pigs, chickens, horse etc. manure, fat from slaughter waste, frying oil, organic household or garden waste, municipal solid waste, waste from agriculture or food production. Energy crops such as maize, clover, grass, polar, willow etc. can be used and to ensure a homogenous substrate quality throughout the year the green plant material is usually stored as silage, preferably by a process favouring homofermentative lactobacilli to minimize carbon loss (Gassen 2005). Biogas formation is generally a three stage process involving hydrolysis of polysaccharides, acetogenesis (production of organic acids, carbon dioxide, hydrogen) and methanogenesis which produces upto 70% (v/v) CH_4, 30% CO_2 and the byproducts NH_3 and H_2S (Yadvika et al. 2004). Further development of biogas technology is expected to increase production efficiency which is possible at the hydrolysis stage.

9 New Developments

Recently several new approaches have been demonstrated like introduction by recombinant gene technology of the butanol metabolic pathway from one of the solvent producing *Clostridia* to another bacterium which is more tolerant of the products. This is combined with new and cheaper substrates derived from hydrolysis of lignocellulosic biomass as well as sterile fermentation and downstream processing technology. A biogas plant connected to the plant would further enhance energy conservation considerably (Antoni et al. 2007) (Table 6).

A recent concept is creating biorefinery schemes that utilizes all the components in cereal grains as well as associated residues to produce value added products, would lead to increase in validity of cereal based biorefinery and the displacement of fossil fuels. In USA cereal based biorefineries utilizing corn have been developed for production of fuel ethanol. In EU countries, wheat is one of the predominant feedstocks for the production of fuel ethanol in industrial operations (Koutinas et al. 2007). The wheat based biorefinery proposed by them has exploited each component of wheat for the production of high value (recombinant protein), intermediate value (platform chemicals) and low value (fuel ethanol) products leading to improved profitability and flexibility. Iogen has set up a US$22 million demonstration plant in Ontario which can convert 12,000–15,000 metric tons per year of wheat straw, barley straw and corn straw into 3–4 million liters per year of fuel grade ethanol (Bohlmann 2006). Here the process combines innovations in pretreatment, enzyme technology and advanced fermentation technology. The US ethanol industry is primarily based on processing of corn grain (i.e. starch) through either dry grind or wet milling processes. Projections for 2010 have indicated that 2.6 billion bushels will be required for ethanol- 1.2 billion bushels more than 2005

Table 6 Yields of biofuel per hectare (Antoni et al. 2007)

Biofuel	Country	Crop	GJ fuel/ha annum
Biodiesel	US	Soybean	16.3
	EU	Sunflower	32.5
		Barley	35.8
	Germany	Rapeseed	50.4
	Tanzania	Palm oil	186.0
Biobutanol	USA	Corn	66.0
	EU	Wheat	52.9
	EU	Sugarbeet	116.3
	Brazil	Sugarcane	137.5
	India	Sugarcane	112.1
	Germany	Wheat	54.1
	New Zealand	Switchgrass	84.6
Biomethane	Germany	Wheat	71.9
	Germany	Corn silage	163.4
	Germany	Energy crop residues	217.3
Plant oil		Rapeseed	50.8

(Fischer 2007) concerns have been raised over the availability and cost of corn grain for livestock faced globally. However, the question of agricultural capacity is difficult to answer and needs to be addressed on a regional basis. USA and EU have excess grain production, Latin America has a surplus of both oil and sugar, whereas Asia Pacific is a region in deficit with regard to both oil and grain/sugar and faces challenging issues in relation to both energy and food production (Bohlmann 2006).

Significant achievements are necessary for technologies breakthroughs. Genetically engineered crops and micro-organisms could boost crop production, yields and fermentation productivities, optimizes crop composition and biomass enzymatic hydrolysis minimize unwanted by products formation and maximized utilization of biomass hydrolysates.

10 Enzymes for Biofuel Production

Pure lipases extracted from different sources have been successfully used in the production of bio-diesel. Most lipases used as catalysts in organic synthesis are of microbial and fungal origin such as *Candida rugosa*, *Pseudomonas fluorescens*, *Rhizopus oryzae*, *Burkolderia cepacia*, *Aspergillus niger*, *Rhizomucor meihei* etc. A comparative study on the type of lipase powders from different sources (Iso et al. 2001) revealed that *Pseudomonas fluorescens* lipase showed the highest enzymatic activity. On the other hand Hama et al. (2004) successfully used the whole cell *R. oryzae* as biocatalyst and investigated the effect of cell membrane fatty acid composition on biodiesel fuel production. The cost of lipase is one of the obstacles facing full exploitation of its potential. The reuse of lipase has been achieved by immobilization, which has resulted in improved stability and activity. Immobilization of lipase from *B. cepacia* within a phyllosilicate solution gel matrix has been studied by Hsu et al. (2001). Membrane reactors with immobilized lipase in the form of flat sheet or hollow fibre form have also been proposed having the advantage of continuous reaction and separation (Hilal et al. 2004).

11 Conclusion

Transition from the inherently unsustainable dominant use of petroleum of the twentieth century to a more sustainable use of agricultural and forestry sources in combination with efficient use other biosources for the production of biofuels is one of the most challenging problems of the century. Public awareness/education is needed on biofuels, emphasizing potential environmental benefits, effects of reducing green house gas emissions and global warming. A comprehensive approach is needed involving plant breeders, agronomists, bioprocess engineers, biotechnologists and microbiologists. The major focus should also be directed on using traditional crops with objectives of optimization of crops and processes for improvements in yields, economics and efficiency. Bio-butanol has manifold advantages over bio-ethanol like greater energy content and direct use in cars without engine modification. However yield of butanol from biomass on an energy content basis will need to match that targeted for ethanol. Conversion of biomass to methane via anaerobic fermentation or gasification will provide another bio-based transportation fuel option.

The use of vegetable oils can be increased by developing high oil yield crops especially in non-agricultural areas. Plant oil content should be increased in vegetation as well as seed. Low input perennial herbaceous or woody species that use water efficiently like C4 species need to selected and improved genetically. High biomass yield and low cost of production are key. Development of micro-organisms for simultaneous fermentation of pentoses and hexoses will lead to sustainable long term production and profitability. Microbial oil is also one of the potential feedstocks for biofuel production. Yeast has the largest biomass and highest lipid content. Further optimizing and improving the ability of yeast to use cheap sources of carbon for oil accumulation is very important. Autotrophic algae have the advantage of using sunlight and carbon dioxide for oil accumulation and developing lower cost photobioreactor to bring down the costs also need to be explored further. Microbial oils have the advantage of renewability, fast growth rate and not taking arable land. Further modifications through genetic engineering and metabolic engineering have much potential for the performance improvement of microbes producing oils.

References

Alexander W, Trond EE, Hans KK, Sergey BZ, Mimmi TH (2007) Bacterial metabolism of long chain n-alkanes. Appl Microbiol Biotechnol 76:1209–1221

Antoni D, Zverlov VV, Schwarz WH (2007) Biofuels from microbes. Appl Microbiol Biotech 77:23–35

Belarbi EH, Molina GE, Chisti YA (2000) Process for high yield and scaleable recovery of high purity eichosapentaenoic acid esters form microalga and fish oilEnzyme Microbio. Enzyme Microbio Technol 26:516–529

Bevan MV, Franssen MCR (2006) Investing in green and white biotech. Nat Biotechnol 24:765–767

Blades T, Rudloff M, Schulze O (2006) Sustainable SunFuel from CHOREN's Carbo-V® Process. Presented at ISAF XV, San Diego – September 2005. www.choren.com/

Bohlmann GM (2006) Process economic consideration for production of ethanol from biomass feedstocks. Industrial Biotechnol 2(1(Spring)):14–20

Bricka RM (2007) Energy crop gasification. In: Eaglesham A, Hardy RWF (eds) Agriculture biofuels: technology, sustainability and profitability, pp 127–135

Chang HM, Liao HF, Lee CC, Sheih CJ (2005) Optimized synthesis of lipase catalyzed biodiesel by Novozym 435. J Chem Technol Biotechnol 80:307–312

Chisti Y (2007) Biodiesel from Microalgae. Biotechnol Advances 25:294–306

Chisti Y (2008) Biodiesel from Microalgae beats bioethanol. Trends Biotechnol 26:126–131. doi:10.1016/j.tibtech:2007.12.002

Claaswn PAM, De Vrije T, Budde MAW (2004) Biological hydrogen production from sweet sorghum by thermophilic bacteria. Proceeding of 2nd world conference on biomass of energy, Rome, pp 1522–1525

Clifton-Brown JC, Chiang YC, Hodkinson TR (2008) Miscanthus: genetic resources and breeding potential to enhance bioenergy production. In: Vermerris W (ed) Genetic improvement of bioenergy crops. Springer Science, New York, 273–294
Davenport R (2008) Chemicals and polymers from biomass. Industrial Biotechnol 4(1):59–94
De Vrije T, De Haas GG, Tan GB, Keijsers ERP, Cleaasen PAM (2002) Pretreatment of Miscanthus for hydrogen production by *Thermotoga elfii*. Int J Hydrogen Energy 27:1381–1390
Demain AL, New comb M, Wu JHD (2005) Cellulase, Clostridia and ethanol. Microbiol Mol Biol Rev 69:124–154
Desai S, Narayanaih Kumar CH, Reddy M, Gnenamianickam S, Rao G, Venkateshwarlu B (2007) Seed inoculation with *Bacillus* spp. Improves seedling vigour in oil seed plant *Jatropha curcas* L. Biol Fertil soils 44(1):229–234
Du J, Wang HX, Jin HL, Yang KL, Zhang XY (2007) Fatty acids production by fungi growing in sweeet potato starch processing waste water. Chin J Bioprocess Eng 5(1):33–36
EIA (2006) International energy: outlook, Energy Information administration office of integrated analysis and forecasting US Department of energy, Washington, DOE/EIA-0484
Fairless D (2007) The little shrub that could-may be. Nature 449(Oct.):652–655
Fischer F, Tropsch H (1930) Fischer-Tropsch synthesis. US Patent 1(246):464
Fischer JR (2007) Building a prosperous future in which agriculture uses and produces energy efficiently. In: Eaglesham A, Hardy WER (eds) NABC report IS agricultural biofuels, pp 27–40
Gapes JR (2000) The economics of acetone butanol fermentation: theoretical and market considerations. J Mol Microbiol Biotechnol 2:27–32
Gassen HG (2005) Ein Beitragzur ummelt freundlichen evergiehersorgung: Biogasanlagen. Biol in Unserer Zeif 6:384–392
Gibbons WR (2007) Challenges on the road to biofuels. In: Agricultural biofuels: technology, sustainability and profitability. NABC Report, pp 93–103
Goldemberg J (2000) World energy assessment preface. United Nations Development Programme, New York
Goldemberg J, Coelho ST, Guardabassi P (2008) The sustainability of ethanol production from sugarcane. Energy Pol 36:2086–2097
Gouveia L, Oliveira AC (2008) Microalgae as a new material for biofuels production. J Ind Microbiol Biotechnol. doi:10.1007/s 10295-008-0495-6
Grohelaar JV (2004) Algal nutrition. In: Richmond A (eds) Handbook of microalgal culture: biotechnol. Appl Phycol, pp 97–115. Blackwell
Grushcow J, Smith M (2006) Enginering high performance biolubricants in crop plants. Industrial Biotechnol 2(1):48–50
Hahn HB, Karhumaa K, Fonseca C, Spencer MI, Gorwa Grauslund MF (2007) Towards industrial pentose-fermenting yeast strains. Appl Microbiol Biotechnol 74:937–953
Hama S, Yamaji H, Kaiea M, Oda M, Kondo A, Fukuda H (2004) Effect of fatty acid membrane composition on whole cell biocatalysts for biodiesel fuel production. Biochem Eng J 21:155–160
Han X, Miao XL, Wu QY (2006) High quality biodiesel production from heterotropic growth of *Chlorella protothecoides* in fermenters using starch hydrolysate as organic carbon. J Biotechnol 124(4):499–507
Hastings AFStJ, Clifton-Brown JC, Wattenbach M, Mitchell CP, Smith P (2009a) Development of MISCANFOR a new Miscanthus crop growth model: towards more robust yield predictions under different soil and climatic conditions. Global Change Biology -Bioenergy, 2, early view online
Hastings AFStJ, Clifton-Brown JC, Wattenbach M, Stampfl P, Mitchell CP, Smith P (2009b) Future energy potential of Miscanthus in Europe. Global Change Biology- Bioenergy, 2, early view online
Henstra AM, Sipma J, Rinzema A, Stams AJM (2007) Microbiology of synthesis gas fermentation for biofuel production. Curr Opin Biotechnol 18:1–7
Hilal N, Kochkodan V, Nigmatullin R, Goncharuk V, Alklatib L (2004) Lipase immobilized biocatalytic membranes for enzymatic esterification. J Memb Sci 268:196–207

Hoffert MI, Calderia K, Benford G, Criswell DR, Green C, Herzog H, Jain AK, Kheshgi HS, Lackner KS, Lewis JS, Lightfoot HD, Manheimer W, Mankins JC, Manel ME, Perkins LJ, Schlesinger ME, Volk T, Wigley TML (2002) Advanced technology paths to global climate stability: energy for a greenhouse planet. Science 298(5595):981–987

Hossain ABMS, Salleh A, Boyce AN, Chowdhury P, Naqiuddin M (2008) Biodiesel fuel production from algae as renewable energy. Am J Biochem Biotechnol 4(3):250–254

Hsu AF, Jones K, Martner WN, Foglia TA (2001) Production of alkyl esters from fallow and grease using lipase immobilized in a phyllosilicate sol-gel. J Am Oil Chem Soc 78(6):585–588

Huang JZ, Shi QQ, Zhai XL, Lin YX, Xie BF, Wu SG (1998) Studies on the breeding of *Mortierella isabellina* mutant high producing lipid and its fermentation conditions. Microbiol 25(4):187–191

Illman AM, Scragg AH, Shales SW (2000) Increase in *Chlorella* strains calorific values when grown in low nitrogen medium. Enzyme Microbe Technol 27:631–635

IPCC (2007) Fourth Assessment Report (AR4) of Working Group 1: Chapter 2. The Physical science basis. http://www.ipcc.ch/pub/reports.htm

Iso M, Chen B, Eguchi M, Kudo T, Shrestha S (2001) Production of biodiesel fuel from triglycerides and alcohol using immobilized lipase. J Mol Catal B Enzymatic 16:53–58

Kalschener R, Stolting T, Steinbuschel A (2006) Microdiesel: *Escherichia coli* engineered for fuel production. Microbiology 152:2529–2536

Kapdan IK, Kargi F (2006) Biohydrogen production from waste materials. Enzyme Microbe Technol 38:569–582

Kaur S, Gogoi HK, Srivastava RB, Kalita MC (2009) Algal diversity as a renewable feedstock for biodiesel. Curr Sci 96(2):182

Klass LD (1998) Biomass for renewable energy fuels and chemicals. Academic Press, New York, pp 1–2

Kong XL, Lin B, Zhao ZB, Fen GB (2007) Microbiol production of lipids by co-fermentation of glucose and xylose with *Lipomyces starkeyiz*. Chin J Bioprocess Eng 5(2):36–41

Koutinas AA, Wang RH, Webb C (2007) The biochemurgist- bioconversion of agricultural raw materials for chemical production. Biofuel Bioprod Biorefine 1(1):24–38

Kumar GP, Yadav SK, Thawale PR, Singh SK, Jawarkar AA (2008) Growth of Jatropha curcas on heavy metal contaminated soil amended with industrial waste and *Azotobacter*: a green house study. Bioresour Technol 99(6):2078–2082

Kyoto Protocol (1998) unfccc.int/essential_background/**kyoto_protocol**/background/items/1351.php -

Lai CC, Zullaikah S, Vali SR, Ju YH (2005) Lipase catalyze production of biodiesel from rice bran oil. J Chem Technol Biotechnol 80:331–337

Lange JP (2007) Lignocellulosic conversion: an introduction to chemistry process and economics. Biofpr 1(1(Sept.)):39–48

Li Q, Wang MY (1997) Use food industry waste to produce microbial oil. Sci Technol FJood Indus 6:65–69

Li Q, Wei DU, Liu D (2008) Perspectives of microbial oils for biodiesel production. Appl Microbiol Biotechnol 80:749–756

Li W, Du W, Liu DH, Zhao ZB (2007) Enzymatic transesterification of yeast oil for biodiesel fuel production, Chin J. Process Eng 7(1):137–140

Liang XA, Dong WB, Miao XJ, Dai CJ (2006) Production technology and influencing factors of microorganism grease. Food Res Dev 27(3):46–47

Ma YL (2006) Microbial oils and its research advance. Chin J Bioprocess 4(4):7–11

Minaul H, Philippe JB, Louis MG, Alain P (1996) Influence of nitrogen and iron limitations on lipid production by Cryptococcus curvaters growth in batch and fedbatch culture. Process Biochem 31(4):355–361

Miao X, Wu Q (2006) Biodiesel production from heterotrophic microalgal oil. Bioresouce Technol 97:841–846. doi:10.1016/jbiotech:2005.04.008

Miao XL, Wu QY (2004) Bio-oil fuel production from Microalgae after heterotrophic growth. Renew Energy Resour 4(116):41–44

NABC Report (2007) Agriculture and forestry for energy, chemicals and materials. The Road Forward, 1–5 Feb

Nath K, Das D (2006) Amelioration of biohydrogen productions by a two stage fermentation process. Ind Biotechnol 2(1 Spring):44–47

Pathayak S, Sree A (2005) Screening of bacterial associates of marine sponges for single cell oil and PVFA. Lett Appl Microbiol 40:358–363

Patzek TW (2008) Thermodynamics of the corn-ethanol biofuel cycle. Critical Rev Plant Sci 23.6:519–567

Peer M, Skye R, Thomas H, Evan S, Ute C, Jam H, Clemens P, Olaf K, Ben H (2008) Second generation biofuels: high efficiency Microalgae for biodiesel production. Bioener Res 1(1):20–43. doi:10.1007/s12155-006-9008-8

Pesta G, Meyer-Pihroff R, Russ W (2006) Utilization of whey. In: Oreopoulou V, Russ W (eds) Utilization of byproducts and treatment of waste in the food industry. Springer, New York, pp 1–11

Pimental D, Patzek TW (2005) Ethanol production using corn, switchgrass and wood: biodiesel production using soybean and sunflower. Nat Resour Res 14(1):65–76

Quershi N, Blaschek HP (2005) Butanol production from agricultural biomass. In: Shetty K, Pometto A, Paliyath G (eds) Food biotechnology. Taylor and Fracis Group plc., Boca Raton, pp 525–551

Quershi N, Saha BC, Cotta MA (2007) Butanol production from wheat straw hydrolysate using Clostridium beijerinckii. Bioprocess Biocyst Eng 30:419–427

Ragauskas AJ, Nagy M, Kim DH, Eckert CA, Hallet JP, Liotta CL (2006) From wood to fuels, intergrating biofuels and pulp production. Industrial Biotechnol 2(1):55–65

Roche MVP, Rodrigues THS, Mecedo GR, De Goncalves LRB (2008) Enzymatic hydrolysis and fermentation of pretreated cashew apple bagasse with alkali and diluted sulphuric acid for bioethanol production. Appl Biochem Biotechnol. doi:10.1.807/s12010-008-8432-8, Published online 25 Nov 2008

Rosentrater KA (2007) Ethanol processing co-products: economics, impacts, sustainability, in: agricultural biofuels technology, sustainability and profitability. NABC 19:105–126

Sanchez F, Vasud evam PT (2006) Enzyme catalyzed production of biodiesel from olive oil. Appl Biochem Biotechnol 135:1–14

Schwarz WH, Gapes HG (2006) Butanol- rediscovering a renewable fuel. Bioworld Euro 01–2006:16–19

Searchinger T, Heimlich R, Houghton RA, Dong F, Elobeid A, Fabiosa J, Tokgoz S, Hayes D, Yu T-H (2008) Use of U.S. croplands for biofuels increases greenhouse gases through emissions from land use change. Science Express. www.sciencexpress.org/7 February 2008/Page 4/10.1126/science.1151861

Sergeeva YE, Galanina DA, Andrianova DA, Fedofilova EP (2008) Lipids of filamentous fungi as a material for producing biodiesel fuel. Appl Biochem Micobiol 44(5):523–527

Shen JJ, Li FC, Yang QL, Feng DW, Qin S, Zhao ZB (2007) Fermentation of Sportina anglica acid hydrolysate by *Trichosporum cutanarm* for microbial lipid production. Marine Sci 3(8):38–41

Shimmada Y, Watanbe Y, Samukama T (1999) Conversion of vegetable oil to biodiesel using immobilized Candida Antarctica lipase. J Am Oil Chem Soc 76:789–792

Smart LB, Camoron KD, Volk TA, Abrahamsm LP (2007) Breeding, selection and testing of shrub willow as a dedicated energy crop. In: Eaglesham A, Hardy RWF (eds) Agriculture biofuels: technology, sustainability and profitability, pp 85–92

Spolaore P, Joannis CC, Duran E, Isambert A (2006) Commercial applications of microalgae. J Biosci Bioeng 101:87–96

Thomsen MH, Nielsen JBH, Popiel PO, Thomsen AB (2008) Pretreatment of whole crop harvested ensiled maize for ethanol production. Appl Biochem Biotechnol 148:23–33

Tolan JS (2003) Conversion of cellulosic biomass to ethanol using enzymatic hydrolysis. In: 226th American Chemical Society National Meeting Abstracts, New York

Van de loo FJ, Broun P, Turner S, Somerville C (1995) An oleate 12- hydroxylase from castor (*Ricinus comminis*) a fatty acyl desaturase homologue. Proc Natl Acad Sci USA 92:6743–6747

Volk TA, Smart LB, Kimberly D, Abrahamson LP (2004) Growing fuel: a sustainability assessment of willow biomass crops frontiers. Ecol Environ 2:411–418

Webb C, Koutinas AA, Wang RH (2004) Developing a sustainable bioprocessing strategy based on a generic feedstock from wheat. Biotechnol Bioenergy 85:524–538

Wenster BD (1993) Continuous ethanol production by *Zymomonas mobilis* in a fluidized bed reactor I: Kinetic studies of immobilization in maroporous glass beads. Appl Microbiol Biotechnol 39:679–684

Wyse RE (2005) Biofuels are vital energy souces in growing Asian economies. Industrial Biotechnol 1(4):221

Yadvika, Santosh S, Sreekrishnan TR, Kolili S, Rana V (2004) Enhancement of biogas production from solid substrates using different techniques. A review. Bioresource Technol 95:1–10

Yan F, Bai F, Tian S, Zhang J, Zhang Z, Yang X (2008) Strain construction for ethanol production form dilute acid lignocellulosic hydrolysate. Appl Biochem Biotechnol. DOI: 10.1007/s 1210-00808343-8

Zhang Y, Dube MA, Mc Lean DD, Kates M (2003) Biodiesel production from waste cooking oil: 1. Process design and technological assessment. Bioresource Technol 89:1–16

Zuhair AS (2007) Production of biodiesel: possibilities and challenges. Biofuels Bioproduct Biorefin 1(1):57–66

Challenges and Opportunities of Soil Organic Carbon Sequestration in Croplands

Ilan Stavi and Rattan Lal

Abstract This chapter reviews the main farming practices related with soil organic carbon (SOC) sequestration in croplands, aimed, simultaneously, at improving soil quality and health, and reducing net emission of carbon dioxide into the atmosphere. The reviewed practices include conservation tillage, reduced tillage, no tillage, crop residues management, manuring and fertilizing, and cover cropping. In addition, this manuscript addresses prospects for future development by means of agroforestry, biochar application, and perennial grain crops, aimed at maintaining sustainability of agroecosystems, restoring degraded croplands, and increasing the SOC sink capacity. The data collected supports the conclusion that taking into consideration the prevailing physical conditions, wise integration of various conservation practices may increase the ability of croplands in sequestering SOC while maintaining ecosystem services and enhancing agronomic production.

Keywords Agro forestry • Brochar • C-sequestration • Compost • Cover crops • Crop productivity • Greenhouse gaseous emission • Land reclamation • Off-site water pollution • Soil erosion control

1 Introduction

World soils are of the largest stores of terrestrial carbon (C). Sequestration of SOC is naturally driven, encompasses humification of organic matter (OM), and formation of secondary carbonates which are stored in the soil profile (Lal et al. 2007). Landuse conversion from native- into arable- lands eliminates the natural plant communities that are integral to a range of ecosystem services (Cox et al. 2006),

I. Stavi (✉) and R. Lal
Carbon Management and Sequestration Center, The Ohio State University, Columbus, OH 43210, USA
e-mail: istavi@yahoo.com; Lal.1@osu.edu

E. Lichtfouse (ed.), *Biodiversity, Biofuels, Agroforestry and Conservation Agriculture*, Sustainable Agriculture Reviews 5, DOI 10.1007/978-90-481-9513-8_5,

with the related negative impact on SOC pools. Thus, SOC sequestration becomes crucially important due to the growing need to offset the increased atmospheric concentration of carbon dioxide (CO_2), along with the attendant increase in climatic changes and global warming (Lehmann 2007).

SOC comprises of a great variety of organic compounds. However, it is usually classified into three fractions, according to their rate of mineralization or turnover. Liable/active or easily mineralizable compounds, such as simple sugars along with microbial and fungal biomass, comprise 5–15% of the total SOC pool. Turnover rate of this fraction ranges between months and years. The intermediate fraction comprises 20–40% of the total SOC pool, and its turnover rate spans over several decades. The stable or recalcitrant fraction comprises the remaining 60–70% of the total SOC pool, with turnover time of hundreds to thousands of years (Lichtfouse 1997; Rice 2002).

The SOC pool is determined by chemical, physical, and hydrological characteristics of the soil. For example, a strong relationship exists between soil organic matter (SOM) and nutrients storage and cycling (Lal et al. 2007). Since SOC is mainly associated with fine soil particles, especially the clay fraction (Greene and Tongway 1989), it modifies the soil structure through decrease in soil bulk density, and increase in proportion of macroaggregates (Oades and Waters 1991) and their stability, and macroporosity. Thus, SOC pool also enhances the soil's hydraulic conductivity, and water infiltrability and holding capacity. It is also a source of energy for soil fauna, increases soil biodiversity, purifies water, and neutralizes pollutants (Lal 2007). There are several factors which moderate the rates of SOC decomposition. The latter is determined by the C:N ratio; >30 slows decomposition, and <20 enhances it. Also soil temperature regime controls decomposition; as temperature increases the microbial activity increases and vice versa. Generally, for every 10°C increase in temperature, the microbial activity doubles. Soil moisture content is also an important determinant of the decomposition rate. Optimal microbial activity occurs at near field capacity, when ~60% of the soil pores are filled with water (Rice 2002).

Depletion of the SOC pool in cultivated soils is attributed to higher oxidation rates, and lower input of biomass (Six et al. 2000). Conventional inversion-tillage reduces soil structural stability, increases bulk density, and decreases water holding capacity of the surface layer (Moebius-Clune et al. 2008). Incorporation of crop residues into the soil by inversion tillage, increases raindrop splash, enhances aggregate slaking and dispersion, exacerbates SOC decomposition (Polyakov and Lal 2004), increases CO_2 flux, and decreases SOC pool (Lal 2003). The rate of CO_2 emission varies widely depending on ambient and soil temperatures, and soil moisture content (Ussiri and Lal 2009). In addition to CO_2, tillage also increases nitrous oxide (N_2O) efflux from the soil. The rate of N_2O emission varies considerably with time, and highest rates are generally associated with periods of high soil moisture content, high temperature, and low soil NO_3–N concentration (Perdomo et al. 2009). Within the total SOC pool, the particulate organic carbon (POC) is the most oxidative fraction. The POC is a sand-sized fraction (50–2,000 μm) of OM derived from semi decomposed above- or below- ground residues. Its concentration is more near the soil surface than at deeper depths (Franzleubbers 2009). This fraction is highly vulnerable to disturbance by tillage. Tillage also increases the breakdown of macroaggregates, which protect SOC from consumption by soil fauna and microorganisms (Six et al. 2000). Mineralization of the formerly macroaggregate-associated SOC pool is rapid, suggesting that this fraction is also highly prone to depletion following tillage. Tillage also stimulates soil microbial activity. The biologically-active fraction of SOC moderates nutrient cycling and decomposition of the vegetative material. It includes microbial biomass and readily mineralizable carbon (C) and nitrogen (N), has a short turnover time, and easily becomes a part of the active SOC pool. Hence, inversion tillage leads to an exhaustion of the mineralizable C and N, which would otherwise contribute to their pools in the soil (Franzleubbers 2009).

The combined effect of the soil inversion and crop residues removal increases susceptibility of the soil to crusting under heavy rainstorms, and enhances overland flow generation, soil erosion, and SOC depletion (Lal et al. 2004). These processes indicate decreased functionality of the agroecosystem, i.e., reduced ability of the system to retain vital resources within its boundaries (Tongway and Ludwig 2003). Throughout time, the increased dysfunctionality reduces the productive potential of the agroecosystem, forming a positive feedback between functionality and productivity, which further weaken each other (Stavi et al. 2009). These processes result in land degradation, with the attendant reduction in crop yields (Lal et al. 2007).

Since the dawn of agriculture about 10,000 years ago, arable lands have been a source of atmospheric CO_2 (Ruddiman 2003). Most agricultural soils have lost a large portion of their SOC pool, resulting in much lower SOC pool than their potential capacity (Hutchinson et al. 2007). The reduction in SOC concentration of the uppermost soil layers decreases soil fertility and crop productivity. If not replenished, the long-term consequence of SOC reduction may cause land degradation and accelerated soil erosion, jeopardizing global food security (Lal et al. 2007). Improved land management involves an increasing input and decreasing output of C in the topsoil (Hutchinson et al. 2007). Currently, agricultural practices account for about 25% of the CO_2, 50% of the methane (CH_4), and 70% of the N_2O emissions globally. Under the terms of The Kyoto Protocol (1998), several countries are committed to reducing their greenhouse gas (GHG) emissions in the first commitment period (2008–2012) below the 1990 baseline. The agricultural sector can help realize this target, by reducing GHG emissions, and increasing C sequestration in terrestrial agroecosystems (Hutchinson et al. 2007).

The main objectives of this review are to: (1) describe the most prevalent agricultural practices related to SOC management and sequestration, (2) highlight the opportunities and challenges to adoption of these practices, and (3) address prospects for research and development in related agronomic issues.

2 Conservation/Reduced Tillage Systems

Conservation tillage (CT) encompasses a range of techniques aimed at reducing soil disturbance as caused by conventional practices. The concept of CT is based on loosening and aerating the root zone, without inversion of the surface soil and mixing it with the sub-soil (Franzleubbers 2009). Thus, CT reduces adverse impacts on soil physical and hydrological characteristics, and limits disturbance of crop residues on the soil surface (Lal et al. 2007).The reduced disturbance of the soil decreases rates of SOC mineralization and CO_2 emission (Dendoncker et al. 2004; Al-Kaisi and Yin 2005). Avoiding soil disturbance is especially important for accumulation of POC, macro-aggregate associated SOC, and biologically active C. Within the latter fraction, the most visible effect of tillage is on earthworms, as they require a moist environment with adequate OM, which are provided by CT (Franzleubbers 2009).

Chisel plowing (Table 1) is widely used, and is an alternative to moldboard plow. It is aimed at maintaining soil structure and fertility. Tillage depth of the chisel may be similar to that of moldboard, but the degree of soil inversion is much lesser (Franzleubbers 2009). This apparatus is often equipped with forward slanted legs of various shapes. A range of scarifiers may be attached to the legs, depending on tillage goals, soil texture, and structure. Since the soil is lifted as it flows over the legs, its loosening occurs along natural cracks, with limited disturbance to the soil structure. However, despite its positive impact on soil physical quality as compared with conventional tillage (Gomez et al. 2001), some studies have indicated no differences in SOC pools under chisel- and moldboard- plowing (Alvarez 2005; Ussiri and Lal 2009).

Ripper/paraplow rather than moldboard- or disk-plows is widely used to maintain soil structural stability and for SOC sequestration (Table 2). Deep ripping (up to a depth of ~40 cm) is an efficient practice to loosen the soil without excessive incorporation of crop residues (Franzleubbers et al. 2007). The paraplow's legs are

Table 1 Impact of chisel tillage on total soil organic carbon (SOC) sequestration

Location	Climate	Soil type or texture	Crop	Soil depth	SOC sequestration	Crop yield	Reference
Various [b]	Various	Various	Various	Various	Not differed from moldboard tillage	Various	Alvarez 2005
USA	Temperate-humid	Fine-silty	Corn/soybean	0–75 cm	Higher than in moldboard tillage	[a]	Olson et al. 2005
Argentina	Semi-arid	Silt loam	Corn/soybean	0–40 cm	Higher than in moldboard tillage	[a]	Apezteguía et al. 2009
Spain	Semi-arid	Calcic Luvisol	[a]	0–30 cm	Higher than in moldboard tillage	[a]	López-Fando and Pardo 2009
USA	Temperate-humid	Silt-loam	Corn	0–30 cm	Not differed from moldboard tillage	[a]	Ussiri and Lal 2009

[a] Not available/not measured
[b] Synthesis/review article

Table 2 Impact of ripping/paraplow tillage on total soil organic carbon (SOC) sequestration

Location	Climate	Soil type or texture	Crop	Soil depth	SOC sequestration	Crop yield	Reference
USA	Humid subtropical	Sandy loam/ sandy clay loam	Various	0–20 cm	Lower than in NT at 0–3 cm and higher than in NT at 3–20 cm	[a]	Franzleubbers et al. 2007
Spain	Semi-arid	Loamy sand	Grey pea/ barley	0–45 cm	Higher than in moldboard tillage	Higher than in moldboard tillage	López-Fando et al. 2007
Spain	Semi-arid	Calcic Luvisol	[a]	0–30 cm	Higher than in moldboard tillage	[a]	López-Fando and Pardo 2009

[a] Not available/not measured

slanted both forward and at the lateral axis, at an angle of about 45°. Adjustable shatterplates behind the legs cause disturbance of the soil but this can be controlled. Similar to chisel plowing, the soil is loosened as it is lifted over the legs and plates, but not compressed or mixed as under conventional tillage. This practice is mostly used to loosen compacted soils (López-Fando et al. 2007). Soil is ripped under dry conditions, to alleviate compaction (Franzleubbers et al. 2007).

Strip tillage is another CT practice, used as an alternative to conventional full-width tillage in row crops (Table 3). It is designed to loosen the soil only in the row zone (strips of 15–20 cm width), creates disturbance levels similar to that of paraplow, and soil properties in the interrow remain undisturbed (Overstreet and Hoyt 2008). It is most suited in soils prone to erosion, drought, compaction, or plough pans (Overstreet 2009). Strip tillage reduces soil disturbance minimizes CO_2 emission (Al-Kaisi and Yin 2005). Another benefit of this practice, along with economic and environmental aspects, is the reduced energy consumption compared with full-width tillage (Overstreet 2009), without any reduction in crop yields (Al-Kaisi et al. 2005).

No-till (NT) is widely used in conservation farming (Table 4). It reduces the adverse impacts of tillage by maintaining the soil structure and stability (Moebius-Clune et al. 2008). NT systems also enhance nutrient uptake and utilization by plants. Thiagalingam et al. (1996) reported that in Queensland, Australia, uptake rates of N, phosphorous (P), potassium (K), sulfur (S), and zinc (Zn) by corn (*Zea mays* L.) and soybean (*Glycine max*) were higher under NT compared with conventional tillage. Continuous use of NT increases the SOC concentration (e.g., Thiagalingam et al. 1996; Dendoncker et al. 2004; Olson et al. 2005), mainly in the uppermost soil layer. Smith et al. (2001) calculated that 100% conversion of arable lands to NT farming would sequester about 23 Tg C year^{-1} in the European Union, and about 43 Tg C year^{-1} in the wider Europe (excluding the former Soviet Union). Vågen et al. (2005) estimated that attainable SOC sequestration rates in Sub-Sahran Africa under NT systems range from 0.2 to 1.5 Tg

Table 3 Impact of strip-tillage on total soil organic carbon (SOC) sequestration

Location	Climate	Soil type or texture	Crop	Soil depth	SOC sequestration	Crop yield	Reference
Nigeria	Humid	Alfisol	[a]	0–15 cm	Higher than in conventional tillage	[a]	Agele et al. 2005
USA	Temperate-humid	Various	Various	0–30 cm	Various	[a]	Al-Kaisi and Yin 2005
USA	Temperate-humid	Various	Corn/soybean	0–30 cm	Higher than in conventional tillage	Not affected by tillage system	Al-Kaisi et al. 2005
USA	Humid sub-tropical	Sandy loam	Various vegetables	0–25 cm	Not differed between rows and interrow spaces	[a]	Overstreet and Hoyt 2008
Various[b]	Various	Various	Sugar beet	Various	Higher than in conventional tillage	Various	Overstreet 2009

[a] Not available/not measured
[b] Synthesis/review article

C year^{-1}. However, other studies found no positive effect of NT on SOC pool. In the Midwest USA, Chatterjee and Lal (2009), investigated in five sites, the impact of conventional tillage and NT on SOC pool in the 0–60 cm depth. They reported that SOC pool in the uppermost 5-cm was larger under NT, but that the total SOC pool for the whole soil profile was not different between the tillage treatments. Similar finding were reported by Christopher et al. (2009), who compared the effect of tillage in five sites at the same region. In some of the sites Christopher and colleagues even reported negative SOC sequestration rates following conversion from conventional tillage to NT. It seems that the impact of tillage *vs.* NT on SOC pool is site-specific, and determined by a range of natural conditions and anthropogenic effects.

Despite its advantageous effect on the soil structure, SOC sequestration (in some soils), and nutrient cycling, NT is not applicable in soils with poor drainage (Vetsch et al. 2007), and low temperatures early in the growing season. These conditions decrease biological activity, inhibit seed germination, and retard seedlings development (da Silva et al. 2006). Reduced biological activity decreases soil health, with the attendant increase in pathogens infestation (Paulitz et al. 2002; Govaerts et al. 2007a, b). Among the major challenges involved with adoption of reduced tillage systems, and especially under NT farming, is the problem of weed infestation. To control weeds these systems rely on herbicides application (Matsumoto et al. 2008; Franzleubbers 2009). Application of herbicides may adversely impact soil food chain, water quality, and human health. Thus, rotational/occasional tillage in long-term NT system may be useful to reducing pest infestation (Pierce et al. 1994) or the spreading of weeds, while maintaining many of the soil quality benefits of CT systems (Kettler et al. 2000). However, Lal et al. (2007) reported that occasional tillage following prolonged NT practice could reduce easily the SOC pool that was accumulated during several

Table 4 Impact of no tillage on total soil organic carbon (SOC) sequestration

Location	Climate	Soil type or texture	NT duration	Crop	Soil depth	SOC sequestration	Crop yield	Reference
Australia	Semi-arid	Loamy/sandy	4–8 years	Corn/soybean	0–30 cm	Higher than in disk plow followed by chisel plow	Higher under NT	Thiagalingam et al. 1996
Europe[b]	Various	Various	Various	Various	Various	Higher than in conventional tillage	[a]	Smith et al. 2001
Argentina	[a]	Chernozemic clay loam	5 years	Corn/wheat/soybean	0–7.5 cm	Higher than in chisel tillage	[a]	Gomez et al. 2001
Brazil/USA[b]	Tropical/temperate	Various	Various	Various	Various	Higher than in conventional tillage	[a]	Six et al. 2002
Belgium[b]	Temperate	Various	Various	Various	Various	Higher than in conventional tillage	[a]	Dendoncker et al. 2004
Various[b]	Various	Various	Various	Various	Various	Various	[a]	Alvarez 2005
Sub-Saharan Africa[b]	Humid/sub-humid/semi-arid	Various	Various	Various	Various	Various	[a]	Vågen et al. 2005
SE Asia/Mexico[b]	Various	Various	Various	Various	Various	Various	[a]	Hobbs et al. 2008
USA	Temperate humid	Various	Various	Various	0–60 cm	Lower than in conventional tillage	[a]	Chatterjee and Lal 2009
USA	Temperate humid	Various	Various	Various	0–60 cm	Lower than in Conventional tillage	[a]	Christopher et al. 2009

[a] Not available/not measured
[b] Synthesis/review article

years. But, other studies have reported no changes in SOC pool in the tilled layer following occasional tillage. Instead, occasional inversion tillage in prolonged NT systems redistributes the SOC pool throughout the tilled zone which is otherwise concentrated in the surface layer (Pierce et al. 1994; Kettler et al. 2000; Quincke et al. 2007). Insects are a useful biological means to control pests, which attack old weeds even at the flowering and seed-setting stage. These biocontrol agents can be used as a complementary practice to mechanical control (Hatcher and Melander 2003). Another means include pathogenic fungi, which may be used to control weed emergence by incorporating them into the soil (Jackson et al. 1996). Flaming is another organic alternative weed control, though involving some limitations, such as non-selectivness, heat tolerance of some weeds, and smaller efficiency against broadleaf as compared with grasses. Also, the phenological stage of the weeds influences its effectiveness; usually, the earlier the growth stage, the more efficient the flaming treatment (Cisneros and Zandstra 2008). In addition to weed, this method is also useful for insect control, with the degree of decline depending on magnitude of the fire and mobility of the insect (Hatcher and Melander 2003). Indeed, several technical and agronomic barriers prevent the wide adoption of biological and other environment-friendly methods of weed and pathogens control. Special attention must be given to research and development of new means, and to increase the environmental sustainability of reduced tillage farming systems.

For increasing SOC pool and improving related soil properties, conversion to NT or any reduced tillage method alone may not be suitable for all soils and ecoregions. Yet, its benefits to erosion control, water conservation, and saving in energy cannot be overemphasized. Thus, considerable improvements may be achieved by a judicious coupling of CT practices with other, complementary, conservation means.

3 Crop Residue Management

Mulching with crop residues has numerous beneficial effects on soil quality. Residues protect the soil against raindrop impact, and reduce direct soil losses through splash (Kladivko 1994). The plant litter also reduces indirect soil losses by limiting aggregate breakdown and decreasing crust formation, leading to reduced overland flow generation (Alberts and Neibling 1994), and soil erosion. Residue cover also decreases radiation intensity at the soil surface, thus, reducing soil temperature fluctuations and evaporation rates. The more stabilized conditions of moisture and temperature enhance activity of micro- and meso- fauna, which use the plant litter as a source of energy (Tisdall and Oades 1979). In addition, there is an increase in detritivore abundance under larger litter cover and SOC concentration (Tisdall et al. 1978; Franzleubbers 2009). The faunal activity enhances soil stabilization by the decomposition of litter and incorporation of the resulting humus into the soil, as well as by the incorporation of dead faunal cells, which bond soil particles (Tisdall and Oades 1979; Tisdall et al. 1978). Also fungal activity is augmented with litter cover, and improves macroaggregate stability through the fungal hyphae, which act as a physical bonding agent (Oades and Waters 1991). In addition, in cases of dense residue cover, the shade provided by the residue reduces weed emergence (Dabney et al. 2001). Residue retention also increases fertilizer-N availability for plants by reducing its loss through leaching (Gentile et al. 2009) and surface run-off.

Incorporation of the crop residues into the soil by tillage increases its mineralization, with the attendant loss of C (Govaerts et al. 2007b) and N (Chivenge et al. 2007). Rate of decomposition of plant residues, on- or below- the soil surface, depends on moisture availability (Adiku et al. 2008), and is negatively related to the residue C:N ratio (Blanco-Canqui and Lal 2009). Residue management comprises both above- and below- ground biomass which remain in the field after harvest. Retention of crop residue on the soil surface reduces its decomposition rate, with increasing the attendant beneficial impact on SOC pool, soil quality (Blanco-Canqui et al. 2006), and crop productivity (Govaerts et al. 2007a) (Table 5). Accumulation of SOC occurs only when residue-C inputs exceeds residue-C outputs, and soil disturbance is minimized. Thus, residue management is usually combined with reduced tillage or NT, to maximize SOC sequestration (Govaerts et al. 2007b; Blanco-Canqui and Lal 2009). In the Corn Belt of the Midwest USA, Blanco-Canqui et al. (2006) reported a positive relationship between residue quantity and SOC pool. In the dryland farming region of northern China, Wang et al. (2005) calculated that on an average, at least 50% of crop residues must be returned into the soil annually in order to maintain a favorable SOC balance.

Retention of crop residues on the field surface also impacts soil biological activity. Crop residues improve conditions for microbial reproduction and diversity, favoring development of natural predators, and strengthening ecological stability (Govaerts et al. 2007a). Govaerts et al. (2007b) reported that in a wheat (*Triticum aestivum*)-corn agroecosystem in the subtropical highlands of central Mexico, crop residue retention increased soil microbial biomass and micro-flora activity. They suggested that cropping systems that include NT, crop rotation, and crop residue retention can increase overall biomass and micro-flora activity and diversity. In contrast, NT without residue retention is an unsustainable practice that leads to poor soil health in the long run. However, under some circumstances, crop residues may increase pathogen infestation, and root and foliar diseases (Paulitz et al. 2002). For example, Govaerts et al. (2007a) observed that in central Mexico, residue retention decreased incidence of root rot (soil-borne fungus) in wheat, but increased that in corn. Compared with monoculture systems, rotation of corn and wheat decreased the incidence of corn root rot but increased that in wheat. Govaerts and colleagues also observed that density of plant parasitic nematodes was lower for treatments with continuous corn compared to those where corn was rotated with wheat, irrespective of residue management. It seems that the effect of crop residue retention on pathogens infestation is site-specific. Further studies are needed in assessing the impact of crop residue management on soil's biological activity and health under various natural conditions and management practices.

In many parts of the world, a competing factor for crop residues is its usage as feed for livestock; as off-site fodder or grazed in situ (Chivenge et al. 2007). If the OM is not replenished, both of these practices cause nutrient depletion, land

Table 5 Impact of crop residue retention on total soil organic carbon (SOC) sequestration

Location	Climate	Soil type or texture	Crop residues' origin	Soil depth	SOC sequestration	Crop yield (in the following season)	Reference
Western Sahel[b]	Semi-arid/ sub-tropical	Various	Various	Various	Higher with than without residue retention	Various	Bationo and Buerkert 2001
USA	Temperate-humid	Silt loam	Corn	0–30 cm	Higher with than without residue retention	[a]	Allmaras et al. 2004
China[b]	Arid	Various	Various	Various	Higher with than without residue retention	[a]	Wang et al. 2005
USA	Temperate-humid	Silt loam/clay loam	Corn	0–10 cm	Higher with than without residue retention in the 0–5 cm; no effect at 5–10 cm	[a]	Blanco-Canqui et al. 2006
Zimbabwe	Tropical-humid	Sandy/red clay	[a]	0–30 cm	Higher with than without residue retention in a sandy soil; no effect in a red clay soil	[a]	Chivenge et al. 2007
Mexico	Subtropical	Cumulic Phaeozem	Corn/wheat	0–15 cm	Higher with than without residue etention	Higher with than without residue retention	Govaerts et al. 2007a
Various[b]	Various	Various	Various	Various	Higher with than without residue retention	Various	Blanco-Canqui and Lal 2009

[a]Not available/not measured
[b]Synthesis/review article

degradation, and accelerated soil erosion. Another competing usage is the emerging agricultural sector of biofuel feedstock production as an alternative source of energy. The economic growth of this sector encourages many farmers to consider crop residues as legitimate substances for biofuel production. This practice is often encouraged by policy-makers, with the objective of reducing dependence on fossil fuels. As a result, a large percentage of the residue is harvested as a ligno-cellulosic biomass for the ethanol industry. Since agricultural practices should sustain food (and fuel) production, and simultaneously, maintain soil quality (Lal 2007), bio-ethanol farmers have to use organic amendments (e.g., manure, compost, or cover crops) to replace C removed with the harvested stover (Fronning et al. 2008). A more sustainable solution is the judicious usage of marginal lands, which are not suited for grain crops production, for growing alternative crops, such as perennial grasses (Fargione et al. 2008), or sustainably managed forests (Sedjo 2008) for biofuel production. Even horticultural co-products from urban areas may be used as an alternative source for feedstock production (Koh et al. 2008), without interfering with soil quality or competing with food production (Fargione et al. 2008).

4 Manuring/Composting and Nutrient Management

Among plant nutrients, N is the most important for vegetative production. The effect of N fertilizer on SOC pool has been widely studied. Alvarez (2005) demonstrated that N fertilizer increases SOC pool but only when crop residues are returned to the soil. He developed a model, in which cumulative N fertilizer rate, rainfall, temperature, soil texture, and cropping intensity index were included, to calculate the anticipated SOC pool. Wang et al. (2006) reported that the net impact of NT and reduced tillage on N concentration includes greater rates of immobilization and denitrification, and lower rates of mineralization and nitrification, with more OM accumulation and a less oxidative biochemical environment. Wang and colleagues concluded that optimal N fertilization is more critical with NT and CT systems as compared with conventional tillage. Christopher and Lal (2007) observed that farming practices that enhance nutrient use, reduce or eliminate tillage, and increase crop rotation, strengthen N availability to plants, and increase SOC sequestration. However, they stressed that the sequestered C may be negated by emission of CO_2 associated with various farming practices, and N_2O, which is associated with application of fertilizers. Application of chemical fertilizers has large environmental impacts, of these, the most important are the emissions of N_2O, which constitute a potent GHG (Aneja et al. 2009), and eutrophication of above- and below- ground water sources (Isermann 1990). As the ability of soils to retain cations in an exchangeable (plant-available) form increases in proportion to the amount of SOM (Lehmann 2007), an increased SOC concentration reduces N fertilizer requirement.

Lack of C input to feed the heterotrophic community of the soil organisms will lead to a reduction in the biologically active SOC pool (Franzleubbers 2009), and to an attendant reduced soil fertility and crop yield (Matsumoto et al. 2008). A very common practice throughout the world is the use of organic substances such as (unprocessed) manure or (processed) compost, aimed at returning what has been removed: mainly C (Table 6), N, P, and K (Su et al. 2006). In a 3-year study in northeast Thailand, Matsumoto et al. (2008) examined the impact of tillage treatment and manure application on corn yields. They reported that during the study period, yields under tillage with manuring increased (4.2–6.8 Mg ha^{-1}), while those under tillage without manuring remained relatively constant (3.3–3.5 Mg ha^{-1}), and those under NT without manuring, decreased (1.9–0.9 Mg ha^{-1}). In a corn agroecosystem in Michigan, USA, Fronning et al. (2008) reported that compost and manure application increased SOC concentration in the 0–25 cm depth by 41% and 25%, respectively, compared with no additives. Manure or compost can be applied in conjunction with chemical fertilizers, in order to increase nutrient uptake efficiency, soil microbial activity, SOC concentration, and crop yields. For example, Ogunwole

Table 6 Impact of manuring/composting on total soil organic carbon (SOC) sequestration

Location	Climate	Soil type or texture	Crop	Soil depth	SOC sequestration	Crop yield	Reference
China	Dry	Silt loam	Wheat/corn	0–20 cm	Highest under combined treatment (chemical fertilizer and manure)	[a]	Su et al. 2006
China	[a]	Loam	Wheat/corn	0–20 cm	Higher than under chemical fertilizer	Various	Ding et al. 2007
USA	Temperate-humid	Loam/sandy loam	Corn	0–25 cm	Higher than without manuring/composting	[a]	Fronning et al. 2008
Thailand	Tropical-humid	Sandy	Corn	0–50 cm	Higher than without manuring	Higher than without manuring	Matsumoto et al. 2008
Nigeria[b]	Tropical-humid	Sandy loam	Various	0–15 cm	Highest under combined treatment (chemical fertilizer and manure)	[a]	Ogunwole 2008
Hawaii	Tropical humid	Waialua gravelly clay	Corn	0–10 cm	No difference between poultry and cattle manure	[a]	Abbas and Fares 2009

[a] Not available/not measured
[b] Synthesis/review article

(2008) reported that in Nigeria, 45 years of annual application with combined farmyard manure and NPK chemical fertilizer resulted in larger SOC sequestration than farmyard manure alone. In contrast, Ogunwole reported that SOC concentration under chemical fertilizer alone was similar to that of soils without receiving any amendment. Su et al. (2006) reported that in northwest China, application of inorganic fertilizers even reduced SOC concentration on a long-term (22 years) basis. At the same time, combined application of manure and inorganic fertilizers, increased the soil concentrations of C, N, P, and K.

Despite the beneficial effects on soil quality, SOC pool, and crop productivity, manure and compost application also increases rates of CO_2 emission from the agroecosystem (Abbas and Fares 2009). Based on a 3-year study in Thailand, Matsumoto et al. (2008) observed that application of cattle manure caused CO_2 fluxes at rates of 10.5, 13.3, and 15.3 Mg ha^{-1} $year^{-1}$, during the first, second, and third year, respectively. These rates were larger than in conventional tillage (10.1–11.5 Mg ha^{-1} $year^{-1}$) and NT (8.5–9.4 Mg ha^{-1} $year^{-1}$) without manuring. In China, Ding et al. (2007) observed that CO_2 emission rates were considerably higher under manuring than under chemical fertilizer application. The emission rates were 228 versus 188, 132 versus 123, and 401 versus 346 g C m^{-2} in the first, second, and third year, respectively. To reduce decomposition rates, manure and compost may be retained on the soil surface (as opposed to their incorporation into the soil by inversion tillage). However, Franzleubbers et al. (2007) stressed that surface application of manure without incorporation into the soil may potentially cause undesired nutrient enrichment in surface water runoff. Further research is needed in addressing integrative manuring-fertilizing systems, aimed at increasing crop yields, maintaining SOC pools, and reducing water sources contamination and GHG emission.

5 Cover Crops

Cover crops may be incorporated into the soil by tillage ("green manure"), or left undisturbed on the soil surface. In the latter, planting of a subsequent crop is done through the dead or living biomass. Cover crops maintain soil fertility and crop productivity, and minimize agricultural impact on the environment (Reicosky and Forcella 1998). Cover crops are usually planted at the end of the "conventional" growing season, to cover the soil surface during the off-season. They reinforce soil physical structure and quality (Reicosky and Forcella 1998), increase water infiltration capacity (Dabney 1998), provide protection to the soil surface from erosion (Unger and Vigil 1998), increase SOC concentration (Table 7) (Dabney 1998; Reicosky and Forcella 1998; Unger and Vigil 1998), and suppress weed infestation (Dabney et al. 2001).

Cover crops encompass a range of species and varieties, mainly gramineae, such as winter wheat (*Triticum aestivum*), cereal rye (*Secale cereale*), and oats (*Avena sative*), and leguminosae, such as red clover (*Trifolium pratense*) (Unger and Vigil 1998; Singer 2008), hairy vetch (*Vicia villosa*) (Fortuna et al. 2008), and Austrian winter peas (*Pisum sativum*). The latter is also efficient in controlling pathogen infestation (Mahler

Table 7 Impact of cover cropping on total soil organic carbon (SOC) sequestration

Location	Climate	Soil type or texture	Cover crop	Main crop	SOC sequestration	Crop yield	Reference
Various[b]	Various	Various	Various	Various	Higher than without cover crop	Various	Unger and Vigil 1998
Various[b]	Various	Various	Various	Various	Higher than without cover crop	Various	Wagger et al. 1998
Various[b]	Various	Various	Various	Various	Higher than without cover crop	Various	Dabney et al. 2001
Belgium[b]	Temperate	Various	Various	Various	Higher than without cover crop	[a]	Dendoncker et al. 2004
India	Semi-arid tropical	Alfisol	Horsegram (*Macrotyloma uniflorum*)	Sorghum/ sunflower	Higher than without cover crop	Higher than without cover crop	Venkateswarlu et al. 2007
USA	Humid sub-tropical	Silt loam	Hairy vetch	Corn/ soybean	Higher than without cover crop	Higher than without cover crop	Fortuna et al. 2008

[a] Not available/not measured

[b] Synthesis/review article

and Auld 1989). Leguminous cover crops can increase considerably soil N concentration (Unger and Vigil 1998; Singer 2008). Radish (*Raphanus sativus*) is a relatively new cover crop, with potential to control weeds in early sown spring crops (Lawley et al. 2007). Mixed cover crops, such as combining legumes with cereals further reduce the need of chemical fertilizers during the subsequent growing season (Kuo and Sainju 1998).

In the semi-arid tropical region of southern India, Venkateswarlu et al. (2007) studied the impact of legume horsegram (*Macrotyloma uniflorum*) cover crop in long term sorghum (*Sorghum bicolor*) and sunflower (*Helianthus annus*) agroecosystems. They reported that annual incorporation of the horsegram into the soil improved its fertility, increased crop yields by 28% and 18% for sorghum and sunflower, and enhanced SOC concentration by 24%. Impact of cover crop was even larger in plots receiving chemical fertilizers. In a corn-soybean agroecosystem in a humid subtropical region of Kentucky, USA, Fortuna et al. (2008) studied the effects of tillage and hairy vetch cover crop, on C and N cycles and crop yields. Fortuna and colleagues reported that this cover crop in conjunction with NT produced yields equivalent to those of moldboard-plow and N- fertilized system. They concluded that hairy vetch cover crop reduced the N fertilizer requirement and increased total soil N and C levels. Venkateswarlu et al. (2007) stressed that the low-cost and simplicity make this practice applicable even in small and marginal farms.

It seems that cover cropping, with or without the implementation of other conservation farming practices, has numerous environmental benefits. However, numerous management problems have resulted in inadequate adoption of these practices (Johnson et al. 1998). For example, in non-humid regions, despite increasing infiltration and water holding capacity, the usage of water by cover crops for self-growth, may reduce water availability for the subsequent main crops. Another problem is the delay and reduction of soil surface warming in cold or temperate regions. If incorporated into the soil by tillage, cover crops have only small effect on soil temperature. However, living cover crops or dry residues left on the soil surface, can considerably alter soil temperature. The cooler soil temperatures may retard emergence and development of subsequent crops (Dabney et al. 2001). On the contrary, reduced maximum temperatures created by the cover crop can enhance crop production in the tropics (Thiagalingam et al. 1996). Improved understanding of cover crop impact on various agronomic aspects is needed to enhance their adoption in a wide range of eco-regions.

6 Agroforestry

On degraded or eroded agroecosystems, where loss of water, soil and nutrients is associated with reduced SOC pools, agroforestry may be considered as an efficient reclamative means. In agroforestry systems, trees and shrubs are planted according to a certain spatial pattern, reducing the landform homogeneity while increasing vegetative patchiness (Theng 1991; Tenbergen et al. 1995). In cases of planting leguminous trees or shrubs, they also enrich the soil with N, which become available to herbaceous vegetation in the inter-patch spaces (Theng 1991). Agroforestry systems can also combine food crops with energy plantations (Lal 2007), to increase economic viability.

Agroforestry systems imitate natural ecosystems, which are comprised of woody vegetation patches and herbaceous vegetation in the inter-patch spaces. These natural systems maintain primary productivity while reducing loss of resources through water overland flow and soil erosion (e.g. Bromley et al. 1997; Tongway and Ludwig 2003; Stavi et al. 2009). As woody vegetation tends to form "fertility islands", they increase also patchiness of the soil, as determined by a range of physical and chemical soil features (Garner and Steinberger 1989). Throughout time, increased productivity of the woody vegetation patches and their improved soil characteristics reinforce each other by a positive feedback, with an attendant increase in the SOC pool (Stavi et al. 2008a). The reduced spatial consecutiveness of the surface increases the retaining of water and soil resources, results in an attendant improvement of the soil characteristics and herbaceous growth in the inter-patch spaces (Stavi et al. 2008b).

To increase the conservation efficiency of soil and water resources in agroforestry systems, planting trees and shrubs may be accompanied by the construction of structures such as contour ridges, checkdams, or bench terraces. In such cases, the trees or shrubs are planted in upslope proximity of the structures, taking advantage of the accumulated run-on water and its associated dissolved nutrients and suspended materials. Theng (1991) reported that such structures are efficient in soil erosion control even on steep terrains of 20–30°. Checkdams and bench terraces have been constructed in upland areas of the African tropics and sub-tropics. However, the high costs involved with their construction and maintenance limit widespread adoption (Theng 1991). Slow-forming terraces, created via soil translocation by tillage, and comprised of contour grass barrier strips, are widespread in the Andes region of Ecuador as means to control soil erosion and increase crop yields. However, soil fertility in these systems reported to be highly spatially heterogeneous, increases from the upper to the lower edge of the terraces. Nevertheless, the SOC pool is almost evenly distributed widthwise in the terraces (Dercon et al. 2003). In the semi-arid hilly region of Israel, a range of trees species have been planted in mini-catchment structures, enabling larger primary production of both the introduced trees and their surrounding native vegetation (Tenbergen et al. 1995).

The potential of agroforestry in mitigating land degradation, soil erosion, and SOC depletion in croplands agroecosystems is large. In the tropics, agroforestry remains the primary method by which SOC sequestration rates may be considerably increased (Theng 1991; Hutchinson et al. 2007). Yet, a wide range of SOC sequestration capacity is reported in the literature. Hutchinson et al. (2007) estimated that total C accumulation (above- and below-ground) in tropical agroforestry systems range from 4 to 9 Mg C ha^{-1} $year^{-1}$. Hutchinson and colleagues added that over a period of 20–25 years, C accumulation in above-ground plant biomass can be as high as 50 Mg C ha^{-1}, whereas another 50 Mg C ha^{-1} can be sequestered in the soil. Vågen et al. (2005) calculated that agroforestry in the African Sub-Sahara has the potential for SOC sequestration of 28.5 Tg C $year^{-1}$. In the Indian subtropics, Purakayastha et al. (2007) studied the impact of conversion of arable lands into agro-forestry systems on different SOC fractions. Compared with a range of agricultural landuses, soils under agro-forestry contained the highest pools of total SOC (33.7 Mg ha^{-1}), POC (3.58 Mg ha^{-1}), and microbial biomass C (0.81 Mg ha^{-1}). At the same time, the relative amount of SOC mineralized in the agro-forestry soils was lower than in other landuses, suggesting that agroforestry systems could support a more stable SOC pool. Purakayastha and colleagues concluded that agroforestry may be an important management practice for reclaiming degraded arable lands, while sequestering large amounts of SOC. Although many studies have assessed the SOC sequestration capacity by agroforestry systems under tropical and sub-tropical regions, few if any have been conducted in other climatic regions. As degradation processes including accelerated soil erosion and SOC loss are widespread in arable agroecosystems, agroforestry may be an important reclamation strategy. Additional research is needed in a range of climatic, topographic, lithologic, and pedogenic conditions, to enable the adoption of the most appropriate agroforestry systems under various site-specific scenarios.

7 Biochar

A promising approach in lowering atmospheric CO_2 concentration while producing highly valuable soil amendment is low-temperature (400–550°C) pyrolysis technology (combustion under complete or partial exclusion of oxygen) to produce biofuels, with the attendant production of biochar as a co-product. This technology relies on capturing the off-gases from thermal decomposition of vegetative materials. This process produces three to nine times more energy than is used in generating it, and at the same time, about half of the C can be sequestered in soil (Lehmann 2007). Rates of sequestration by this technology range between ~4.3 and 10.9 Mg CO_2 ha^{-1} year^{-1} (Gaunt and Lehmann 2008). A similar, "low-tech" scenario, to that of biochar application in agricultural soils is the anthropogenic black soils of the *terra preta do índio* in the Brazilian Amazon, which contain much larger SOC pool as compared with soils in the proximity. The char particles of the *terra preta* absorb nutrients and water which may otherwise be leached below the reach of roots, and used as a habitat for microorganisms which turn the soil into a spongy, fragrant, dark material (Marris 2006).

The global use of biochar as soil amendment is gradually increasing, to improve soil quality and crop yields (Steiner et al. 2007). To be effective, biochar must be incorporated into the soil by inversion tillage. As opposed to "conventional" organic substances, characterized with rapid turnover especially under warm-humid conditions (Glaser et al. 2002), biochar is highly recalcitrant to microbial decomposition, and retained in the soil for the long term (Steiner et al. 2007). The residence time of biochar in the soil is estimated between decades and millennia, depending on the biochar's origin and quality, as well as on the prevailing climatic and pedogenic conditions (Lehmann 2007). The very high C content of biochar results in its very low bulk density. Hence, application of biochar increases the soil's saturated hydraulic conductivity, and its water permeability and holding capacity (Asai et al. 2009). In addition, biochar has an increased capability to retain nutrients compared with other forms of OM. In contrast to other organic substances, biochar also strongly adsorbs phosphate (Lehmann 2007). As a result, fertilizers efficiency and crop yields are increased, whereas water overland flow, soil erosion, and N_2O emission are decreased (Gaunt and Lehmann 2008).

Despite the positive effects on soil quality, the impact of biochar application on crop yields varies greatly (Asai et al. 2009). In a study in northern Laos, Asai et al. (2009) reported that biochar application increased rice (*Oryza sativa*) yields in sites with low P availability and improved crop response to fertilizer. However, biochar reduced rice yields in relatively P-rich sites. Furthermore, biochar application without additional N also reduced rice yields in soils with low antecedent N concentration. Thus, the impact of biochar application on crop yields depends on antecedent soil fertility and fertilizer management (Asai et al. 2009). In New South Wales, Australia, Chan et al. (2007) also observed that application of biochar without N fertilizer, did not increase radish (*Raphanus sativus*) yields even at a rate of 100 Mg ha^{-1}. Higher radish yields were associated with larger rates of biochar application only in the presence of N fertilizer. Chan et al. (2007) used a mixture of grass clippings, cotton (*Gossypium hirsutum*) trash, and plant prunings as a biochar. There was a positive relationship between biochar application rates and the resultant SOC concentration: 0, 10, 50, and 100 Mg biochar ha^{-1} yielded 21.6, 27.0, 43.4, and 64.6 g C kg^{-1}, respectively in the 0–10 cm depth.

The basic aspects of biochar production are well-known and the required tools and resources are readily available. Hence, biochar can be easily produced locally by farmers, including those with low economic power (Glaser et al. 2002). Yet, there are gaps in understanding of several aspects related with biochar production and application. For example, in their review about charcoal application in tropical regions, Glaser et al. (2002) reported a wide range of biochar application rates (0.5–135 Mg ha^{-1}), and an equally very large range of plant responses (between −29 and +324%). It is, therefore, important that properties of biochar are duly reported. Thus, standardization of biochar is crucial to validation and comparison among a range of production methods of diverse biochar products. Also, increased understanding is essential for development of agricultural markets for biochars, as well as for further development in technologies to produce high quality biochars (Chan et al. 2007). Additional research is needed to explore several post-application aspects, such as the role of microorganisms in oxidizing charcoal and releasing nutrients from its surface (Glaser et al. 2002). The cation retention of fresh biochar is relatively low compared to aged biochar, and it is not clear under what conditions, and over what period of time, biochar develops its adsorbing properties. For example, the production method that would attain high cation exchange capacity (CEC) in soils of cold climatic regions is not currently known. Research is also needed in maximizing the favorable attributes of biochar while evaluating the associated environmental risks. For example, the effects of biochar on the soil N cycle, the emissions of CH_4, as well as various human health considerations associated with the pyrolysis process, require more attention (Lehmann 2007). A considerable obstacle in the development of biochar markets is that its purported benefits do not slot easily into the framework of the Kyoto Protocol (Marris 2006). As biochar has the potential to provide an important C sink, to improve soil fertility and crop yields, and to reduce environmental pollution by fertilizers (Lehmann 2007), an intervention at the inter-governmental level is needed in legislation, leading to the acknowledgement of biochar as a tradable merchandise.

8 Perennial Grain Crops

Most of humanity's food comes directly or indirectly (as animal feed) from cereal grains, legumes and oilseed crops, which comprise together ~70% of global agricultural land. All these crops are annuals (Cox et al. 2006), which require frequent and expensive care to remain productive. The annuals have relatively shallow roots, most of which grow in the top 0.3 m of soil. Coupled with the short life cycle of the root system, this leads to depletion of soil fertility, land degradation, and off-site water contamination (Glover et al. 2007). Yet, annual grain crops produce high and profitable yields despite being environmentally unsound. In hilly or marginal lands, such crops cause severe soil erosion and degradation, and produce low yields (Wagoner 1989; Cox et al. 2006). Consequently, a growing number of human population faces food insecurity (Glover 2005).

Perennial grain crop breeders use two methods; domestication of wild plants, and hybridization of existing annual crop plants with their wild relatives. Domestication is the more straightforward approach in creating perennial crops. Relying on selection of superior individual plants, breeders seek to increase the frequency of genes for desirable traits, such as easy separation of seed from husk, a nonshattering seed, large seed size, synchronous maturity, palatability, strong stems, and high seed yield. Domestication programs are currently focused on intermediate wheatgrass (*Thinopyrum intermedium*), Maximilian sunflower (*Helianthus maximiliani*), Illinois bundleflower (*Desmanthus illinoensis*) and flax (a perennial species of the *Linum* genus) (Cox et al. 2006; Glover et al. 2007). Compared with domestication, hybridization is a potentially faster means to create a perennial crop plant, although more technology is required to overcome genetic incompatibilities between the parent plants. This process encompasses hybridization of two plant species, and can bring together the best qualities of the domesticated annual and its wild perennial relative (Cox et al. 2006). Of the most widely grown grain and oilseed crops, wheat, rice, corn, sorghum, flax, and oilseed sunflower are capable of hybridization with perennial relatives (Glover et al. 2007).

The concept in perennial grain-cropping systems is that they will function similarly to the natural ecosystems displaced by agriculture (Glover 2005; Glover et al. 2007). Whereas in annual plants a large part of their energy is directed to seed production, perennials spend a large portion of their energy on developing vigorous root systems. Thus, perennials produce lower grain yields but increase their capability in extracting resources from the soil (Scheinost et al. 2001). Well developed root systems, often to 2-m depth, enable perennial grain crops the access to water and nutrients in larger volumes of the soil as compared with annuals (Glover et al. 2007). In turn, perennials' roots support biological activity, soil structure formation, water conservation, nutrient cycling, and soil erosion control (Wagoner 1989; Cox et al. 2006). Also the surface cover by the perennials' shoot provides an efficient protection from water and wind erosion. Post harvest and following winter dormancy, re-growth is initiated from the rhizomes in the early spring, allowing a crop to be re-harvested for several years (Scheinost et al. 2001).

In economic terms, mechanical field operations under perennial crops are considerably reduced as the soil would not be re-worked each year, thus, saving on fuel and labor costs (Wagoner 1989). Also herbicide costs for perennial crop production may be 4–8.5 times less than that for annuals (Glover et al. 2007). The deep and dense root system also suppresses weeds, reducing the need for herbicide use. In addition, greater root depth and longer growing season let perennials boost their SOC sequestration by 50% or more as compared with annuals (Glover et al. 2007). Glover et al. (2007) stressed that perennial grain crops production involves a negative compared with a positive C balance in annual crops. Potential SOC sequestration by perennial grain crops ranges between 300 and 1,100 kg ha^{-1} $year^{-1}$, as compared with 0–450 kg ha^{-1} $year^{-1}$ in annual grain crops. An estimated impact of 3–8°C temperature increase may be positive on yields of perennial grain crops; ~+5 Mg ha^{-1}, and negative on annual crops; −1.5 to −0.5 Mg ha^{-1} (Glover et al. 2007).

Mixtures of perennial crops at the same stand may increase crop yields by reducing vulnerability to biotic and abiotic stress conditions, through exploiting resources in different soil layers, and by providing an efficient weed control (Weik et al. 2002). However, in south-west Germany, Weik et al. (2002) observed no clear effect of mixing perennial grain crops on their yields. They tested various combinations of perennial grain crops, including rye (*S. cereale*, *S. montanum*), intermediate wheatgrass, lupin (*Lupinus polyphyllus*), and linseed (*Linum perenne*). In some stands, white clover (*T. repens*) was included as an undersown intercrop for improved N supply. Maturation of the species differed by up to 6 weeks, and consequently the yields were not satisfactory (Weik et al. 2002).

Much more research is needed in developing agronomic and economic feasible varieties which considerably enhance the attractiveness of perennial grain crop production worldwide, and not only on erodible or marginal lands. Glover et al. (2007) anticipated that large-scale development of high-yield perennial grain crops may become feasible within the next 25–50 years.

9 General Discussion

Most agricultural soils throughout the world have lost 25–75% of their antecedent SOC pools. Soil with severely depleted SOC pool have lost as much as 30–40 Mg C ha^{-1} (Lal 2007; Lal et al. 2007). On a cumulative global scale, soils have lost ~78 Pg C (Lal et al. 2004), and current rate of emission from agricultural soils is estimated at 0.5–2.5 Pg $year^{-1}$. Thus, the SOC sequestration potential in the world croplands is far beyond the actual levels of SOC pool, and through adoption of conservation farming systems, is estimated at 0.6–1.2 Pg C $year^{-1}$ (Lal et al. 2007).

Integrative agroecosystems which involve various practices, such as conservation tillage, residue management, manuring/composting, and cover cropping, are crucial to enhancing SOC pool, improving nutrient uptake, reducing fertilizers rates, controlling soil erosion, and decreasing CO_2 emissions (Table 8). At the same time, CH_4 emission can be reduced by shifting anaerobic paddy fields to aerobic and sustainable managed agroecosystems (Hobbs et al. 2008), whereas N_2O emissions can be reduced considerably by increasing efficiency of N uptake and decreasing the fertilizers rate. Further efforts should be directed to research and development of agroforestry means, biochar technologies, and perennial grain crop production, which have the capability to considerably enhance the SOC sequestration in cropland agroecosystems. Application of the recommended management practices in severely degraded or eroded lands would considerably recuperate their carrying capacity and productivity. Over the long-term, the enhanced SOC pool, coupled with reduced emissions of GHG, is anticipated to initiate a positive feedback, in which lower gaseous concentrations reduce global surface temperature, and increase soil C sink capacity.

Table 8 Recommended management practices, their applicability, and co-benefits

Recommended management practice	Applicability	Co-benefits
1. Conservation tillage (CT)	Limited efficiency under poor drainage conditions or very low temperatures	Soil erosion control; reduced off-site water sources pollution; reduced fiscal and C inputs associated with conventional tillage
2. Crop residue management	In conjunction with conservation or conventional tillage systems	Soil erosion control; reduced off-site water sources pollution; weed control; increased agroecosystem health
3. Manuring/composting	In conjunction with conservation or conventional tillage systems	Reduced off-site water sources pollution in proximity with accumulation sites, e.g., dairy or poultry farms
4. Cover cropping	Not under: (i) poor drainage conditions, (ii) very low temperatures, (iii) very dry conditions	Soil erosion control; reduced off-site water sources pollution; weed control; pathogen control (in certain cover crops); reduced requirement in fertilizer input (in leguminous cover crops)
5. Agroforestry	Poorly arable to highly productive lands; leveled to steep lands	Soil erosion control at the field- to the hillslope- scale; increased agroecosystem diversity
6. Biochar	Wherever technology is available and economic sounds	Reduced amounts of (urban- or horticultural-) woody waste
7. Perennial grain crops	Steep-, degraded-, or marginal- lands; also in productive lands considering development of improved hybrids	Soil erosion control; reduced off-site water sources pollution; weed control; increased agroecosystem diversity; reduced fiscal and C inputs associated with annual grain crops
8. Aerobic rice	Lowlands and uplands in tropical regions	Increased water use efficiency; decreased labor costs

Yet, emphasizing the importance of increased SOC sequestration in arable lands while ignoring adverse impact of related agricultural practices, is environmentally unsound. Much remains to be done in research and development of complementary agro-technical means which would enhance the sustainability of these agroecosystems. Crucially urgent is the need of further development and adoption of non-chemical practices to control pests. Such means would manifest advantages for ecosystem services, mainly decreased contamination of water and soil sources, with the associated increased environmental- and human-health.

To meet the growing population demands, agriculture will have to produce more food from less available land through more efficient use of natural resources, and with minimal environmental impact (Glaser et al. 2002; Hobbs et al. 2008). Due to the high environmental efficiency and the relative ease and low-cost of conservation agricultural practices, both farmers and policy-makers cannot afford to ignore them (Lal 2007). As energy prices soar and the costs of environmental degradation are increasingly acknowledged, budgeting public money for long-term projects which reduce resource consumption and land depletion will become increasingly politically correct (Glover et al. 2007). A new approach, emphasizing agricultural multifunctionality, would have to support production of standard commodities (food or fiber) and at the same time, maintaining ecosystem services. Such multifunctional production systems can be feasible if managed properly by stakeholders, and supported by interest groups or governmental agencies (Jordan et al. 2007).

Acknowledgments We gratefully acknowledge the Midwest Regional Carbon Sequestration Partnership (MRCSP) for funding the study.

References

Abbas F, Fares A (2009) Soil organic carbon and carbon dioxide emission from on organically amended Hawaiian tropical soil. Soil Sci Soc Am J 73:995–1003

Adiku SGK, Narh S, Jones JW, Laryea KB, Dowuona GN (2008) Short-term effects of crop rotation, residue management and soil water on carbon mineralization in a tropical cropping system. Plant Soil 311:29–38

Agele SO, Ewulo BS, Oyewusi IK (2005) Effects of some soil management systems on soil physical properties, microbial biomass and nutrient distribution under rainfed maize production in a humid rainforest Alfisol. Nutr Cycl Agroecosyst 72:121–134

Al-Kaisi MM, Yin XH (2005) Tillage and crop residue effects on soil carbon and carbon dioxide emission in corn-soybean rotations. J Environ Qual 34:437–445

Al-Kaisi MM, Yin XH, Licht MA (2005) Soil carbon changes as affected by tillage system and crop biomass in a corn-soybean rotation. Appl Soil Ecol 30:174–191

Alberts EE, Neibling WH (1994) Influence of crop residues on water erosion. In: Unger PW (ed) Managing agricultural residues. Lewis Publishers, Boca Raton, FL, pp 19–39

Allmaras RR, Linden DR, Clapp CE (2004) Corn-residue transformations into root and soil carbon as related to nitrogen, tillage, and stover management. Soil Sci Soc Am J 68:1366–1375

Alvarez R (2005) A review of nitrogen fertilizer and conservation tillage effects on soil organic carbon storage. Soil Use Manage 21:38–52

Aneja VP, Schlesinger WH, Erisman JW (2009) Effects of agriculture upon the air quality and climate: research, policy, and regulations. Environ Sci Technol 43:4234–4240

Apezteguía HP, Izaurralde RC, Sereno R (2009) Simulation study of soil organic matter dynamics as affected by land use and agricultural practices in semiarid Córdoba, Argentina. Soil Till Res 102:101–108

Asai H, Samson BK, Stephan HM, Songyikhangsuthor K, Homma K, Kiyono Y, Inoue Y, Shiraiwa T, Horie T (2009) Biochar amendment techniques for upland rice production in Northern Laos. 1. Soil physical properties, leaf SPAD and grain yield. Field Crop Res 111:81–84

Bationo A, Buerkert A (2001) Soil organic carbon management for sustainable land use in Sudano-Sahelian west Africa. Nutr Cycl Agroecosyst 61:131–142

Blanco-Canqui H, Lal R, Post WM, Izaurralde RC, Owens LB (2006) Rapid changes in soil carbon and structural properties due to stover removal from no-till corn plots. Soil Sci 171:468–482

Blanco-Canqui H, Lal R (2009) Crop residue management and soil carbon dynamics. In: Lal R, Follett RF (eds) Soil carbon sequestration and the greenhouse effect, Soil Science Society of America Special Publication 57, 2nd edn. Soil Science Society of America, Madison, WI, pp 291–309

Bromley J, Brouwer J, Barker AP, Gaze SR, Valentin C (1997) The role of surface water redistribution in an area of patterned vegetation in a semi-arid environment, south-west Niger. J Hydrol 198:1–29
Chan KY, Van Zweiten L, Meszarosa I, Downiec A, Joseph S (2007) Agronomic values of greenwaste biochar as a soil amendment. Aust J Soil Res 45:629–634
Chivenge PP, Murmira HK, Giller KE, Mapfumo P, Six J (2007) Long-term impact of reduced tillage and residue management on soil carbon stabilization: implication for conservation agriculture on contrasting soils. Soil Till Res 94:328–337
Chatterjee A, Lal R (2009) On farm assessment of tillage impact on soil carbon and associated soil quality parameters. Soil Till Res 104:270–277
Christopher SF, Lal R (2007) Nitrogen management affects carbon sequestration in North American cropland soils. Crit Rev Plant Sci 26:45–64
Christopher SF, Lal R, Mishra U (2009) Regional study of no-till effects on carbon sequestration in the Midwestern United States. Soil Sci Soc Am J 73:207–216
Cisneros JJ, Zandstra BH (2008) Flame weeding effects on several weed species. Weed Technol 22:290–295
Cox TS, Glover JD, Van Tassel DL, Cox CM, DeHaan LR (2006) Prospects for developing perennial grain crops. BioScience 56:649–659
da Silva VR, Reichert JM, Reinert DJ (2006) Soil temperature variation in three different systems of soil management in blackbeans crop. RevBras Ciênc Solo 30:391–399
Dabney SM (1998) Cover crops impacts on watershed hydrology. J Soil Water Conserv 53:207–213
Dabney SM, Delgado JA, Reeves DW (2001) Using winter cover crops to improve soil and water quality. Commun Soil Sci Plan 32:1221–1250
Dendoncker N, Wesemael BV, Rounsevell MDA, Roelandt C, Lettens S (2004) Belgium's CO2 mitigation potential under improved cropland management. Agric Ecosyst Environ 103:101–116
Dercon G, Deckers J, Govers G, Poesen J, Sánchez H, Vanegas R, Ramírez M, Loaiza G (2003) Spatial variability in soil properties on slow-forming terraces in the Andes region of Ecuador. Soil Till Res 72:31–41
Ding WX, Meng L, Yin YF, Cai ZC, Zheng XH (2007) CO_2 emission in an intensively cultivated loam as affected by long-term application of organic manure and nitrogen fertilizer. Soil Biol Biochem 39:669–679
Fargione J, Hill J, Tilman D, Polasky S, Hawthorne P (2008) Land clearing and the biofuel carbon debt. Science 319:1235–1238
Fortuna A, Blevines R, Frye WW, Grove J, Cornelius P (2008) Sustaining soil quality with legumes in no-tillage systems. Commun Soil Sci Plan 39:1680–1699
Franzleubbers AJ (2009) Linking soil organic carbon and environmental quality through conservation tillage and residue management. In: Lal R, Follett RF (eds) Soil carbon sequestration and the greenhouse effect, Soil Science Society of America Special Publication 57, 2nd edn. Soil Science Society of America, Wisconsin, pp 263–289
Franzleubbers AJ, Schomberg HH, Endale DM (2007) Surface-soil responses to paraplowing of long-term no-tillage cropland in the Southern Piedmont USA. Soil Till Res 96:303–315
Fronning BE, Thelen KD, Min DH (2008) Use of manure, compost, and cover crops to supplant crop residue carbon in corn stover removed cropping sytems. Agron J 100:1703–1710
Garner W, Steinberger Y (1989) A proposed mechanism for the formation of 'fertile islands' in the desert ecosystem. J Arid Environ 16:257–262
Gaunt JL, Lehmann J (2008) Energy balance and emissions associated with biochar sequestration and pyrolysis bioenergy production. Environ Sci Technol 42:4152–4158
Gentile R, Vanlauwe B, van Kessel C, Six J (2009) Managing N availability and losses by combibing N-fertilizer with different quality residues in Kenya. Agric Ecosyst Environ 131:308–314
Glaser B, Lehmann J, Zech W (2002) Ameliorating physical and chemical properties of highly weathered soils in the tropics with charcoal – a review. Biol Fert Soils 35:219–230
Glover J (2005) The necessity and possibility of perennial grain production systems. Renew Agric Food Syst 20:1–4
Glover JD, Cox CM, Reganold JP (2007) Future farming: a return to roots? Sci Am 297:82–89

Gomez E, Ferreras L, Toresani S, Ausilio A, Bisaro V (2001) Changes in some soil properties in a Vertic Argiudoll under short-term conservation tillage. Soil Till Res 61:179–186
Govaerts B, Fuentes M, Mezzalama M, Nicol JM, Deckers J, Etchevers JD, Figueroa-Sandoval B, Sayre KD (2007a) Infiltration, soil moisture, root rot and nematode populations after 12 years of different tillage, residue and crop rotation managements. Soil Till Res 94:209–219
Govaerts B, Mezzalama M, Unno Y, Sayre KD, Luna-Guido M, Vanherck K, Dendooven L, Deckers J (2007b) Influence of tillage, residue management, and crop rotation on soil microbial biomass and catabolic diversity. Appl Soil Ecol 37:18–30
Greene RSB, Tongway DJ (1989) The significance of (surface) physical and chemical properties in determining soil surface condition of red earths in rangelands. Aust J Soil Res 27:213–225
Hatcher PE, Melander B (2003) Combining physical, cultural and biological methods: prospects for integrated non-chemical weed management strategies. Weed Res 43:303–322
Hobbs PR, Sayre K, Gupta R (2008) The role of conservation agriculture in sustainable agriculture. Philos Trans R Soc 363:543–555
Hutchinson JJ, Campbell CA, Desjardins RL (2007) Some perspectives on carbon sequestration in agriculture. Agric Forest Meteorol 142:288–302
Isermann K (1990) Share of agriculture in nitrogen and phosphorous emissions into the surface waters of Western-Europe against the background of their eutrophication. Fert Res 26:253–269
Jackson MA, Baruch SS, Schisler DA (1996) Formulation of *Colletotrichum truncatum* microsclerotia for improved bioconrol of the weed hemp sesbania (*Sesbania exaltata*). Biol Control 7:107–113
Johnson TJ, Kaspar TC, Koheler KA, Corak SJ, Logsdon SD (1998) Oat and rye overseeded into soybean as fall cover crops in the upper Midwest. J Soil Water Conserv 53:276–279
Jordan N, Boody G, Broussard W, Glover JD, Keeney D, McCown BH, McIsaac G, Muller M, Murray H, Neal J, Pansing C, Turner RE, Warner K, Wyse D (2007) Sustainable development of the agricultural bio-economy. Science 316:1570–1571
Kettler TA, Lyon DJ, Doran JW, Powers WA, Stroup WW (2000) Soil quality assessment after weed-control tillage in a no-till wheat-fallow cropping system. Soil Sci Soc Am J 64:339–346
Kladivko EJ (1994) Residue effects on soil physical properties. In: Unger PW (ed) Managing agricultural residues. Lewis, Boca Raton, FL, pp 123–141
Koh LP, Tan HTW, Sodhi NS (2008) Biofuels: waste not, want not. Science 320:1419
Kuo S, Sainju UM (1998) Nitrogen mineralization and availability of mixed leguminous and non-leguminous cover crop residues in soil. Biol Fert Soils 26:346–353
Kyoto Protocol to the United Nations Framework Convention on Climate Change (1998) United Nations
Lal R (2003) Soil erosion and the global carbon budget. Environ Int 29:437–450
Lal R, Griffin M, Apt J, Lave L, Morgan GM (2004) Managing soil carbon. Science 304:393
Lal R (2007) Farming carbon. Soil Till Res 96:1–5
Lal R, Follett RF, Stewart BA, Kimble JM (2007) Soil carbon sequestration to mitigate climate change and advance food security. Soil Sci 172:943–956
Lawley Y, Weil R, Teasdale J (2007) Forage radish winter cover crop as an integrated weed management tool in the Mid-Atlantic region. International Annual Meeting of the ASA-CSSA-SSSA, New Orleans, LA
Lehmann J (2007) Bio-energy in the black. Front Ecol Environ 5:381–387
Lichtfouse E (1997) Heterogeneous turnover of molecular organic substances from crop soils as revealed by 13C labeling at natural abundance with Zea mays. Naturwissenschaften 84:22–23. doi: 10.1007/s001140050342
López-Fando C, Dorado J, Pardo MT (2007) Effects of zone-tillage in rotation with no-tillage on soil properties and crop yields in a semi-arid soil from central Spain. Soil Till Res 95:266–276
López-Fando C, Pardo MT (2009) Changes in soil chemical characteristics with different tillage practices in a semi-arid environment. Soil Till Res 104:278–284
Mahler RL, Auld DL (1989) Evaluation of green manure potential of Austrian winter peas in northern Idaho. Agron J 81:258–264
Marris E (2006) Black is the new green. Nature 442:624–626

Matsumoto N, Paisancharoen K, Hakamata T (2008) Carbon balance in maize fields under cattle manure application and no-tillage cultivation in northeast Thailand. Soil Sci Plant Nutr 54:277–288

Moebius-Clune BN, van Es HM, Idowu OJ, Schindelbeck RR, Moebius-Clune DJ, Wolfe DW, Abawi GS, Thies JE, Gugino BK, Lucey R (2008) Long-term effects of harvesting maize stover and tillage on soil quality. Soil Sci Soc Am J 72:960–969

Oades JM, Waters AG (1991) Aggregate hierarchy in soils. Aust J Soil Res 29:815–828

Ogunwole JO (2008) Soil aggregate characteristics and organic carbon concentration after 45 annual application of manure and inorganic fertilizer. Biol Agric Hortic 25:223–233

Olson KR, Lang JM, Ebelhar SA (2005) Soil organic carbon changes after 12 years of no-tillage and tillage of Grantsburg in southern Illinois. Soil Till Res 81:217–225

Overstreet LF, Hoyt GD (2008) Effects of strip tillage on soil biology across a spatial gradient. Soil Sci Soc Am J 72:1454–1463

Overstreet LF (2009) Strip tillage for sugarbeet production. Int Sugar J 111:292–+

Paulitz TC, Smiley RW, Cook RJ (2002) Insights into the prevalence and management of soilborne cereal pathogens under direct seeding in the Pacific Northwest USA. Can J Plant Pathol 24:416–428

Perdomo C, Irisarri P, Ernest O (2009) Nitrous oxide emission from an Uruguayan argiudoll under different tillage and rotation treatments. Nutr Cycl Agroecosyst 84:119–128

Pierce FJ, Fortin MC, Staton MJ (1994) Periodic plowing effects on soil properties in a no-till farming system. Soil Sci Soc Am J 58:1782–1787

Polyakov V, Lal R (2004) Modeling soil organic matter dynamics as affected by soil water erosion. Environ Int 30:547–556

Purakayastha TJ, Chhonkar PK, Bhadraray S, Patra AK, Verma V, Khan MA (2007) Long-term effects of different land use and soil management on various organic carbon fractions in an Inceptisol of subtropical India. Aust J Soil Res 45:33–40

Quincke JA, Wortmann CS, Mamo M, Franti T, Drijber RA (2007) Occasional tillage of no-till systems: carbon dioxide flux and changes in total and labile soil organic carbon. Agron J 99:1158–1168

Reicosky DC, Forcella F (1998) Cover crop and soil quality interactions in agroecosystems. J Soil Water Conserv 53:224–229

Rice CW (2002) Organic matter and nutrient dynamics. In: Lal R (ed) The encyclopedia of soil science. Marcel Dekker, New York, pp 925–928

Ruddiman WF (2003) The anthropogenic greenhouse era began thousands of years ago. Clim Change 61:261–293

Sedjo RA (2008) Boifuels: think outside the cornfield. Science 320:1420–1421

Smith P, Goulding KW, Smith KA, Powlson DS, Smith JU, Falloon P, Coleman K (2001) Enhancing the carbon sink in European agricultural soils: including trace gas fluxes in estimates of carbon mitigation potential. Nutr Cycl Agroecosyst 60:237–252

Singer JW (2008) Corn Belt assessment of cover crop management and preferences. Agron J 100:1670–1672

Six J, Elliot ET, Paustin K (2000) Soil macroaggregate turnover and microaggregate formation: a mechanism for C sequestration under no-tillage agriculture. Soil Biol Biochem 32:2099–2103

Six J, Feller C, Denef K, Ogle SM, de Moraes Sa JC, Albrecht A (2002) Soil organic matter, biota and aggregation in temperate and tropical soils – effects of no-tillage. Agronomie 22:755–775

Scheinost PL, Lammer DL, Cai X, Murray TD, Jones SS (2001) Perennial wheat: the development of a sustainable cropping system for the U.S. Pacific Northwest. Am J Altern Agric 16:147–151

Stavi I, Ungar ED, Lavee H, Sarah P (2008a) Grazing-induced spatial variability of soil bulk density and content of moisture, organic carbon and calcium carbonate in a semi-arid rangeland. Catena 75:288–296

Stavi I, Ungar ED, Lavee H, Sarah P (2008b) Surface microtopography and soil penetration resistance associated with shrub patches in a semiarid rangeland. Geomorphology 94:69–78

Stavi I, Lavee H, Ungar ED, Sarah P (2009) Eco-geomorphic feedbacks in semi-arid rangelands: a review. Pedosphere 19:217–229

Steiner C, Teixeira WG, Lehmann J, Nehls T, de Macedo JLV, Blum WEH, Zech W (2007) Long term effects of manure, charcoal and mineral fertilization on crop production and fertility on a highly weathered Central Amazonian upland soil. Plant Soil 291:275–290

Su YZ, Wang F, Sou DR, Zhang ZH, Du MW (2006) Long-term effect of fertilizer and manure application on soil-carbon sequestration and soil fertility under the wheat-wheat-maize cropping system in northwest China. Nutr Cycl Agroecosyst 75:285–295

Tenbergen B, Günster A, Schreiber KF (1995) Harvesting runoff - the minicatchment technique – an alternative to irrigated tree plantations in semiarid regions. Ambio 24:72–75

Thiagalingam K, Dalgliesh NP, Gould NS, McCown RL, Cogle AL, Chapman AL (1996) Comparison of no-tillage and conventional tillage in the development of sustainable farming systems in the semi-arid tropics. Aust J Exp Agric 36:995–1002

Theng BKG (1991) Soil science in the tropics – the next 75 years. Soil Sci 151:76–90

Tisdall JM, Cockroft B, Uren NC (1978) The stability of soil aggregates as affected by organic materials, microbial activity and physical disruption. Aust J Soil Res 16:9–17

Tisdall JM, Oades JM (1979) Stabilization of soil aggregates by the root systems of ryegrass. Aust J Soil Res 7:429–441

Tongway DJ, Ludwig JA (2003) The nature of landscape dysfunction in rangelands. In: Ludwig JA, Tongway DJ, Freudenberger D, Noble J, Hodgkinson K (eds) Landscape ecology function and management. CSIRO Publishing, Canberra, Australia, pp 49–61

Unger PW, Vigil MF (1998) Crop cover effects on soil water relationships. J Soil Water Conserv 53:200–206

Ussiri DAN, Lal R (2009) Long-term tillage effects on soil carbon storage and carbon dioxide emissions in continuous corn cropping system from an alfisol in Ohio. Soil Till Res 104:39–47

Vågen TG, Lal R, Singh BR (2005) Soil carbon sequestration in Sub-Saharan Africa: a review. Land Degrad Dev 16:53–71

Venkateswarlu B, Srinivasarao C, Ramesh G, Venkateswarlu S, Katyal JC (2007) Effects of long-term legume cover crop incorporation on soil organic carbon, microbial biomass, nutrient build-up and grain yields of sorghum/sunflower under rain-fed conditions. Soil Use Manage 23:100–107

Vetsch JA, Randall GW, Lamb JA (2007) Corn and soybean production as affected by tillage systems. Agron J 99:952–959

Wagger MG, Cabrera ML, Ranells NN (1998) Nitrogen and carbon cycling in relation to cover crop residue quality. J Soil Water Conserv 53:214–218

Wagoner P (1989) Philosophy and theory: why the need for perennial grain crops, vol 108, North Dakota Research Report. North Dakota State University, Fargo, ND, pp 2–4

Wang XB, Cai DX, Hoogmoed WB, Oenema O, Perdok UD (2005) Scenario analysis of tillage, residue and fertilization management effects on soil organic carbon dynamics. Pedosphere 15:473–483

Wang XB, Cai DX, Hoogmoed WB, Oenema O, Perdok UD (2006) Potential effect of conservation tillage on sustainable land use: a review of global long-term studies. Pedosphere 16:587–595

Weik L, Kaul HP, Kübler E, Aufhammer W (2002) Grain yields of perennial grain crops in pure and mixed stands. J Agron Crop Sci 188:342–349

Conservation Agriculture Under Mediterranean Conditions in Spain

F. Moreno, J.L. Arrúe, C. Cantero-Martínez, M.V. López, J.M. Murillo, A. Sombrero, R. López-Garrido, E. Madejón, D. Moret, and J. Álvaro-Fuentes

Abstract Intensive agriculture with deep tillage and soil inversion causes rapid soil deterioration with loss of soil organic matter content. This practice leads to a decrease of soil biological activity, a damage of the physical properties and a reduction of crop yields. Conservation agriculture aims to achieve sustainable and profitable agriculture through the application of three basic principles: minimal soil disturbance by conservation tillage, permanent soil cover and crop rotations. Any practice of conservation agriculture must maintain on the soil enough surface residues throughout the year. Conservation tillage is thus any tillage and planting system that maintains at least 30% of the soil surface covered by residues after planting to reduce soil erosion by water. Here we review the main advances about the adoption of conservation agriculture under Mediterranean conditions in Spain. There are major cost savings, e.g. fuel and fertilizer costs, compared with conventional agriculture. Conservation tillage has been proven to be highly efficient for water storage, to increase moderately the organic matter in the soil top layer, and to improve soil physical properties and aggregation. However, no tillage may induce greater soil compaction in some cases. In this case, an occasional tillage is advised. Furthermore, conservation tillage can reduce soil CO_2 emissions, mobility and persistence of herbicides. In general, conservation tillage enhances biodiversity compared to conventional tillage. Crop yields under conservation tillage are similar or even greater than yields of traditional tillage. All these benefits

F. Moreno (✉), J.M Murillo, R. López-Garrido, and E. Madejón
Consejo Superior de Investigaciones Científicas (CSIC), Instituto de Recursos Naturales y Agrobiología de Sevilla (IRNAS), Avda. de Reina Mercedes 10, 41012 Sevilla Spain
e-mail: fmoreno@irnase.csic.es

J.L. Arrúe, M.V. López, D. Moret, and J. Alvaro-Fuentes
Consejo Superior de Investigaciones Científicas (CSIC), Estación Experimental de Aula Dei (EEAD), Avda. de Montañana 1005, 50059 Zaragoza, Spain

C. Cantero-Martínez
Universidad de Lleida (UdL), Avda. Rovira Roure 191, 25198 Lleida, Spain

A. Sombrero
Instituto Tecnológico Agrario de la Junta de Castilla y León (ITACyL), Ctra. de Burgos, Km. 118, 47071, Valladolid, Spain

E. Lichtfouse (ed.), *Biodiversity, Biofuels, Agroforestry and Conservation Agriculture*, Sustainable Agriculture Reviews 5, DOI 10.1007/978-90-481-9513-8_6,

show that conservation agriculture in Spain is a more sustainable alternative than conventional agriculture. Nonetheless, we have found from the literature analysis some constraints for its adoption, mainly due to inadequate extension and technology transfer systems and lack of access to specific inputs.

Keywords Conservation tillage • Soil quality • Water storage • Environmental impact • Socio-economic impact • Soil carbon sequestration • CO_2

1 Introduction

Intensive agriculture causes rapid soil deterioration, with loss of soil organic matter content, leading to a decrease of soil biological activity, a damage of the physical properties and a reduction of crop productivity. Losses of organic matter derive from the soil inversion by tillage that characterizes other kind of agriculture, such as organic farming and integrated agriculture. Conservation agriculture aims to achieve sustainable and profitable agriculture through the application of three basic principles: minimal soil disturbance (conservation tillage), permanent soil cover and crop rotations. In general, conservation agriculture includes any practice which reduces changes or eliminates soil tillage and avoids residues burning to maintain enough surface residues throughout the year. Soil is protected from rainfall erosion and water runoff; soil aggregates are stabilised, organic matter and the fertility level naturally increase, and less surface soil compaction occurs. Furthermore, the contamination of surface water and the emissions of CO_2 to the atmosphere are reduced, and biodiversity increases (ECAF 1999). The efficiency of conservation agriculture to reduce soil erosion and to improve the organic content and water storage is universally recognized. This is particularly important in arid and semi-arid zones, in which soil organic matter content is very low and the climatic conditions leads to continuous losses. In these conditions, water is the limiting factor for crop development under rainfed agriculture and the management of crop residues is of prime importance to obtain sustainable crop productions (Du Preez et al. 2001).

Basically, this is the general picture in Spain, where about 80% of its surface is devoted to extensive agriculture, mostly under dryland conditions. In general, soils are basic and calcareous in Central and Eastern Spain and acid in North and North-Western Spain, most of them with a low soil organic matter content, due, among other factors, to more than 2,000 years of continuous cultivation and to a low development of natural vegetation under adverse climatic conditions in dry and semi-arid areas, very frequent in Spain.

Besides limited water availability for agriculture and other uses, the worst environmental issue facing the Spanish agriculture is soil erosion. The average soil loss by water impact in Spain has been estimated in about 34 t ha^{-1} $year^{-1}$, with low rates in North-Western areas and high rates in Eastern and Southern Spain, especially in Andalusia, where annual soil losses can reach 60–80 t ha^{-1}. On the basis of the knowledge available, conservation agriculture appears to be the most important sustainable alternative system to conventional agriculture based on intensive tillage to cope with negative agro-environmental problems like the loss of fertile soil in areas prone to erosion processes (Photos 1 and 2).

Photo 1 Conventional tillage using mouldboard ploughing in semi-arid conditions

Photo 2 Direct drilling in semi-arid conditions

Any practice of conservation agriculture must maintain on the soil enough surface residues throughout the year, an important aspect when considering tillage. As pointed out by Gajri et al. (2002), initially, the concept of minimum tillage was aimed at reducing the number of tillage trips across the field. Later, the emphasis was put on leaving the soil surface covered with residues, rather than merely reducing the number of operations, and the term *conservation tillage* was introduced (Hill 1996). Sometimes, no distinction between conservation tillage, minimum tillage or reduced tillage is made (Bradford and Peterson 2000). Types of conservation tillage include no-tillage, ridge tillage, mulch tillage and zone tillage (see Hill 1996 for definitions). Designs for tillage experimentation have been reviewed by López and Arrúe (1995).

The most commonly definition used for conservation tillage is any tillage and planting system that maintains at least 30% of the soil surface covered by residues after planting to reduce soil erosion by water. Where soil erosion by wind is a primary concern, the system must maintain a 1.1 Mg ha^{-1} flat small grain residue equivalent on the surface during the critical wind erosion period. Basically, is a year-round conservation system that usually involves a reduction in the number of passes over the field and/or in the intensity of tillage, avoiding ploughing (soil inversion) (Gajri et al. 2002), that would incorporate residues into the soil mass.

Permanent soil cover (cover crops) imply sowing of appropriates species, or growing spontaneous vegetation, in between rows of trees, or in the period of time in between successive annual crops, as a measure to prevent soil erosion and to control weeds. Cover crops are generally managed with herbicides with a minimal environmental impact (ECAF 1999). Steep gradients of the terrain with a lack of soil cover (frequent in olive groves) under Mediterranean climate, which typically has long periods of drought followed by torrential storms, conduces to a serious loss of soil and fertility irrespective of the tillage system applied. Alternative cover olive-grove systems have recently been introduced for these situations (Castro et al. 2008). The importance of crop rotations can be seen in Karlen et al. (1994), however not always is possible to use a wide range of crops in such as dryland conditions and a choose of winter cereals could be used (Alvaro et al. 2009).

2 Conservation Agriculture in Spain: Main Research Topics

According to Fernández-Quintanilla (1997), it was in the 1970s when the concepts of tillage reduction and the use of conservation agriculture practices for annual and perennial crops were first introduced in Spain mainly through knowledge gathered in the USA. The release on the market of new herbicides, as paraquat and gliphosate, for a full control of volunteers and weeds before sowing was definitely a key factor.

Since mid 1970s, in perennial crops, and the late 1970s in annual crops, a large number of conservation agriculture field studies have been carried out across Spain, implemented as a farmer's initiative in some cases. Most of these studies were based on the comparison of conventional primary tillage (i.e., mouldboard ploughing with soil inversion) with two forms of conservation agriculture, (i) minimum or reduced tillage, in which the conventional primary tillage is replaced by a vertical or surface tillage with different ploughs (e.g., chisel or cultivator) and (ii) no-tillage or direct drilling. The main research topics considered by different Spanish research groups could be classified according four main knowledge areas: (1) socio-economic aspects: energy use and consumption, (2) soil quality and water saving, (3) environmental issues, and (4) crops and crops protection.

2.1 *Socio-economic Aspects: Energy Use and Consumption*

At the beginning, the main driving forces for conservation agriculture development in Spain were based on labour simplification and savings of fuel and costs for machinery required for tillage and other kind of inputs. Later, the advantageous agronomic and environmental aspects of conservation agriculture practices (soil water conservation, soil protection, and increase of soil organic carbon and soil biological activity) were recognised by farmers. Important cost savings have been reported for minimum tillage and zero tillage in Spain, compared with conventional tillage. Reduction in fuel consumption can range between 30% (minimum tillage) and 60% (no-tillage); time saving for tillage operations, derived from reduction of the number of labours, can reach up to 45% (no-tillage) (Hernanz et al. 1995; Sombrero et al. 2001b; Sánchez-Girón et al. 2004) (Table 1).

However, the acceptance of conservation agriculture technologies in Spain is still low, especially in those areas where these technologies were not initially well introduced. As pointed out by Cantero-Martínez and Gabiña (2004) and Angás et al. (2004), this low degree of adoption is a consequence of inadequate extension and technology transfer systems and lack of access to specific inputs, machinery and

Table 1 Net margin (Euros/ha) of a economical study conducted in the Ebro Valley in 2006, comparing different tillage system in three different areas (arid, semiarid and subhumid) for three different farm sizes

Zone/farm ha	Intensive	Vertical	Minimum	No tillage	[a]No tillage
Arid/300	257.11	268.54	316.94	305.62	302.15
Arid/150	256.08	266.47	310.64	297.06	295.56
Arid/75	242.98	246.75	289.59	268.96	280.66
Semiarid/300	203.51	241.09	246.69	265.16	261.69
Semiarid/150	202.40	239.00	240.30	256.60	255.10
Semiarid/75	189.38	219.30	219.34	228.50	240.20
Semihumid/300	365.04	376.47	403.27	401.97	398.50
Semihumid/150	364.01	374.40	396.97	393.41	391.91
Semihumid/75	350.91	354.68	375.92	365.31	377.01

[a] No tillage is used but planting machine is rented for sowing operations (Cantero-Martínez et al. 2009)

Table 2 Main driving forces and constraints for conservation agriculture in Spain

Driving forces
Better economy (labour simplification; less time requirements for tillage operations; less fuel consumption; less machinery required for tillage; less power machinery)
Flexible sowing time. Double crop possibilities in some areas
Better water economy and soil protection. Better soil quality
Greater nutrient-use efficiency. Less use of fertilizers. Faster crop establishment and development. Same yield or slight yield increases (10–15%)
Potential constraints
Economic reasons. Farmer's reluctance and fear to acquire new and expensive specific machinery or to higher herbicide costs
Occasional soil deterioration (compaction, poor aeration, waterlogging)
Crop residue management difficulties. Occasional allelopathic problems
Occasional higher incidence of weeds, pests and diseases (the reverse situation can be a driving force)
Irregular incidence of rodents and slugs
Poor crop development under particular conditions (lower soil temperatures under irrigated spring crops)
Insufficient information and technical support
Social relationships among farmers (criticisms discouraging hesitant farmers)

equipment. Table 2 shows the main driving forces and constraints for conservation agriculture in Spain, information derived from different farmer surveys carried out along this country.

In summary, important cost savings (fuel, fertilizers) have been reported for conservation agriculture in Spain compared with conventional tillage. However, its adoption is still low mainly due to inadequate extension and technology transfer systems and lack of access to specific inputs, machinery and equipment. Crop residue management difficulties and occasional higher incidence of weeds, pests and diseases, besides social relationships among farmers (criticisms) may also difficult the establishment of conservation agriculture in local scenarios.

2.2 *Soil Quality and Water Storage*

The efficiency of conservation tillage to improve the water storage in soil is universally recognized. This is very important in arid and semi-arid zones, where management of crop residues is of prime importance to obtain sustainable crop productions (Du Preez et al. 2001; Lampurlanés and Cantero-Martínez 2006; Moreno et al. 1997; Pelegrín et al. 1990). Organic matter is highly related to the soil capacity for water storage; most soils under Mediterranean semi-arid conditions are rather low in organic matter. Increases in soil organic carbon under conservation tillage (reduced tillage, minimum tillage, no-tillage) have been reported by different authors in Spain (Álvaro-Fuentes et al. 2008a; Bescansa et al. 2006; Bravo et al. 2003; De Santiago et al. 2008; Hernanz et al. 2002; López-Bellido et al. 1997; López-Fando and Pardo 2001; Ordóñez Fernández et al. 2003, 2007; Murillo et al. 1998).

However, semi-arid Mediterranean conditions may suppose a limiting factor for the accumulation of organic carbon in the top soil layers. Thus, the simple determination of the total content of organic carbon can not be the best indicator of the improvement caused by conservation tillage. Under semi-arid conditions could be more interesting the knowledge of the stratification ratio of soil organic carbon, defined as the content at the surface layer (e.g., 0–5 cm) divided by the content at deeper layers (Franzluebbers 2002; Murillo et al. 2004; Moreno et al. 2006; López-Fando et al. 2007). As reported by Franzluebbers (2002), stratification ratio greater than 2 are not frequent in degraded soils. As pointed out by Franzluebbers (2004), soils with low inherent levels of organic matter could be the most functionally improved with conservation tillage, despite modest or no change in total standing stock of soil organic carbon within the rooting zone. Stratification of soil organic matter pools with depth under conservation tillage systems has consequences on soil functions beyond that of potentially sequestering more carbon in soil. The "more is better" argument referred to soil organic carbon is weaker when applied to agricultural productivity, where the benefits of greater soil organic matter contents on intensively managed arable soils are sometimes obscure (Sojka and Upchurch 1999). In the absence of a clear critical point and demonstrable ecological consequence, the setting of soil quality targets within a continuum requires human value judgments (Sparling et al. 2003).

Although under semi-arid climate there could not be a great enrichment of soil organic carbon at surface in conservation tillage, slight increases could have created particular conditions for the physico-chemical and biological soil dynamics. It has been reported that despite the stratification ratio of the total soil organic carbon may only slightly increase under conservation tillage, other variables related to the biological dynamics of the soil, such as microbial biomass carbon and some enzyme activities (and their stratification ratios) may increase to a greater extent (Madejón et al. 2007; Murillo et al. 2006). Nutrients, and general soil fertility, are also positively affected by the soil organic matter increase derived from the conservation tillage establishment (Bravo et al. 2006, 2007; de Santiago et al. 2008; Ordóñez Fernández et al. 2003; López-Fando and Almendro 1995; Moreno et al. 2006, Martín-Rueda et al. 2007; Saavedra et al. 2007).

Short- and long-term experiments under conservation and conventional tillage have shown that organic matter also influences soil physical properties and aggregation (Álvaro-Fuentes et al. 2007c, 2008b; Lampurlanés and Cantero-Martínez 2003; López et al. 1996; Moreno et al. 1997, 2000, 2001; Moret and Arrúe 2007a, b). In general, aggregate stability was greater under conservation tillage, especially under no-tillage, than under conventional tillage, minimum tillage or reduced tillage. At long-term, higher soil bulk density and compaction under no-tillage than under conventional tillage (and reduced tillage) have been reported (Álvaro-Fuentes et al. 2008a; Moret and Arrúe 2007a, b). The hydraulic conductivity was significantly lower under no-tillage than under the tilled treatments due to a lower number of water transmitting pores per unit of area. The effects of subsoil compaction on soil properties and the use of models to simulate the subsoil compaction process have been investigated by Coelho et al. (2000); Moreno et al. (2003); Perea et al. (2003). Soil density tended to be greater under reduced tillage in dry years, although there were no differences in years with high rainfall (Moreno et al. 2001). However, in many cases those situations do not creates a depression on the crop productivity.

In relation to water storage, conservation tillage is frequently highly efficient to reduce soil water losses, especially in years with rainfall lower than the average (Moreno et al. 2000). The use of cover crops in olive orchards in Andalusia has been carried out by Pelegrín et al. (2001) who specifically investigated cover crop systems for soil and water conservation and designed a seed driller for cover crop sowing under no-tillage management conditions.

Results obtained by López et al. (1996) suggested that reduced tillage could replace the conventional tillage without adverse effects on soil water content and storage. However, no-tillage was not a viable alternative for extremely arid zones of NE Spain. On the contrary, Lampurlanés et al. (2001, 2002, 2003), Lampurlanés and Cantero-Martínez (2004) and Cantero-Martínez et al. (2003) showed that no-tillage resulted potentially better for semi-arid regions of NE Spain because it maintains greater water content in the soil and promotes root growth in the surface soil layers and, in some cases, deep in the soil profile also, especially in years of low rainfall. In shallow soils because of the low soil water-holding capacity, no-tillage proved to be better under low amount and frequent events of rainfall in spring that matching with the filling grain period of the crop. Moret et al. (2006, 2007a) quantified the efficiency of long fallow and tillage for soil water storage in NE Spain showing that conservation tillage systems could replace conventional tillage for soil management during fallow without adverse effects on soil water conservation. Tillage effects on water storage during fallow in NE Spain have also been exhausted discussed by Lampurlanés et al. (2002).

Water storage is related, among other factors, to residues management and evolution. López et al. (2003, 2004, 2005b) have studied the evolution of barley residues during four long fallow periods under conservation tillage, reduced tillage and no-tillage, and under both continuous cropping and cereal-fallow rotation. The lack of residue-disturbing operations in no-tillage makes this practice the best strategy for fallow management. Under no-tillage, the soil surface still conserved between 10% and 15% of residue cover after long-fallowing and percentages of standing residues ranging from 20% to 40% of the total mass after the first 11–12 months.

In summary, increases in soil organic carbon under conservation tillage (reduced tillage, minimum tillage, no-tillage) have been reported by different authors in Spain, although under semi-arid Mediterranean conditions there could not be a great enrichment of soil organic carbon at surface. However, slight increases of organic matter have created particular conditions for the physico-chemical and biological soil dynamics increasing soil quality. In general, better soil physical properties and aggregation under conservation agriculture have been reported. However, greater compaction under no-tillage can result at long-term in particular scenarios, which could make advisable an occasional tillage labour.

Conservation tillage has been proved to be highly efficient to water storage, especially in years with rainfall lower than the average, and can replace conventional tillage for soil management during fallow, due to the lack of residue-disturbing operations.

2.3 Environmental Factors. Soil Erosion and Biodiversity

Several studies have been focused on the development of simulation models and expert systems to predict the effects of tillage systems on water erosion under different climatic conditions and to design site-specific agricultural machinery (Simota et al. 2005; de la Rosa et al. 2005; Dexter et al. 2005; Horn et al. 2005; Díaz-Pereira et al. 2002; Gómez et al. 2002). These models provide a tool for recommendations for site-specific land use and management strategies. In order to obtain new knowledge for a better management of the olive crop, Gómez et al. (1999, 2004, 2005) have studied the effects of different tillage systems on soil physical properties, infiltration, water erosion and yield of olive orchards.

In areas where strong and dry winds are frequent (such as *Aragon*, NE Spain), fallow lands are susceptible to wind erosion due to insufficient crop residues on the surface (López et al. 2001). Results obtained by López (1998), López and Arrúe (1997) and López et al. (2000, 2003, 1998, 2003, 2005a, b) indicated that reduced tillage, and especially no-tillage, could be considered as a viable alternative to conventional tillage for wind erosion control during the fallow period in these areas. Consequently, no significant dust emission and saltation transport was observed in conservation tillage plots (Sterk et al. 1999; Gomes et al. 2003). Reduced tillage provides higher soil protection than conventional tillage through a lower wind-erodible fraction (aggregates <0.84 mm in diameter) of soil surface (on average, 10% less) and a significantly higher percentage of soil cover with crop residues and clods (30% higher).

Adoption of conservation tillage can reduce soil CO_2 emissions to the atmosphere thus minimising soil organic carbon losses and mitigating the greenhouse effect (Arrúe 1997). Álvaro-Fuentes et al. (2004, 2007a, b, 2008a) have evaluated the influence of conventional tillage and conservation tillage (reduced tillage and no-tillage) on short- and long-term CO_2 fluxes at NE Spain. Soil CO_2 emissions just after tillage were 40% higher under conventional tillage than under no-tillage as the CO_2 accumulated on soil pores was released to the atmosphere after the tillage event (Reicosky et al. 1997). At the same time, tillage has an effect during the whole growing season increasing microbial decomposition resulting in a 20% higher soil CO_2 emissions under no-tillage than under no-tillage during the whole growing season. Reduction of CO_2 fluxes under reduced tillage respect conventional tillage have also been reported by Sánchez et al. (2002, 2003) in the Spanish plateau. The influence of N fertilization on N_2O and CO_2 emissions under rainfed Mediterranean conditions have recently been studied by Menéndez et al. (2008) and Morell et al. (2007).

The transport and persistence of herbicides in soils under conservation tillage (reduced tillage) has also been studied in Spain (Cox et al. 1996, 1999; Cuevas et al. 2001). Results from these studies showed that the mobility and persistence of herbicides (e.g., trifluralin and metmitron) were lower under conservation tillage than under conventional tillage.

In general, conservation tillage enhances biodiversity. The effect of tillage on nematode populations was early studied by López-Fando and Bello (1995). The effect of conservation tillage systems on earthworm activity as a biological indicator has been studied in NE Spain (Cantero-Martínez et al. 2004). The most important finding never described in the Iberian Peninsula was the higher earthworm population and activity measured under no-tillage compared to conventional tillage. Soil moisture conditions, as influenced by the climatic conditions of the year, was a determinant factor for the number of the earthworms during and between years. Tillage system influences greatly the earthworm population in the long term experiments and much higher populations were found during several years under no-tillage. Despite the number of earthworm adults and eggs were influenced by the water regime of the year, and in drier years the level of adults was always higher under conservation tillage systems in the first 30 cm of the soil profile.

In summary, conservation tillage could be considered as a viable alternative to conventional tillage for wind and water erosion control during the fallow period. Furthermore, adoption of conservation tillage can reduce soil CO_2 emissions to the atmosphere thus minimising soil organic carbon losses (that can enhance soil erosion) and mitigating the greenhouse effect. Soil CO_2 emissions just after tillage were 40% higher under conventional tillage than under no-tillage. Conventional tillage also had an effect during the whole growing season increasing microbial decomposition, resulting in a 20% higher soil CO_2 emissions than in conservation tillage. Mobility and persistence of herbicides were lower under conservation tillage than under conventional tillage. Moreover, conservation tillage enhances biodiversity, and for example, a higher earthworm population and activity were measured under no-tillage compared to conventional tillage.

2.4 Crops and Crops Protection

Conservation tillage is also especially important to achieve sustainable yields in semi-arid climate regions. However, the implementation of no tillage systems has occasionally caused yield losses, especially in humid and subhumid regions due to cooler and wetter soil conditions, inadequate physical properties, thermal and aeration regimes, root growth increased grassy weeds and residue problems during seeding (Kirkegaard et al. 1995; Gajri et al. 2002).

Nevertheless, with correct management, the global experience with conservation tillage does not result in smaller harvests than conventional tillage (Warkentin 2001; Gajri et al. 2002). In water-limiting environments, no till and other moisture conservation practices can increase crop yields (Gajri et al. 2002). This is corroborated by global results related to crop response in Spain, one of the major research subjects of the Spanish groups.

Conservation Agriculture has been developed in Spain since late seventies and has been focused mainly in field crops under dryland conditions as winter cereals. Less attention has been paid to field crops under irrigation. Only some studies have been conducted in orchards as olive, vineyards or almond under dryland condition or deficit irrigation. No studies have been done in horticultural crops.

In field crops, many studies at the dry conditions of SW Spain have shown that crop yields under conservation tillage (reduced tillage, no-tillage) were similar to or even greater than those in conventional tillage. A significant number of studies have dealt with the effects of these tillage systems on crop yield under different crop rotations and N fertiliser rates in rainfed conditions, with particular attention to no-tillage (Bravo et al. 2003; González et al. 2003; López-Bellido and López-Bellido 2001; López-Bellido et al. 1996, 1997, 1998, 2000, 2001, 2002, 2003a, b, 2004a, b; Moreno et al. 1997; Murillo et al. 1998). Murillo et al. (1998, 2000) reported data about the nutritional status of the crop, and showed the differences in crop development, at the earlier stages, frequently slightly better under conventional tillage although the difference disappeared at more advanced stage of growth.

In semi-arid central Spain, Herranz et al. (2002) and Sombrero et al. (1998, 2001a, 2004) compared different conservation tillage systems (minimum tillage, reduced tillage, no-tillage) to conventional tillage and showed that in general there not were differences in the crop yields. As a conclusion, the use of conservation tillage, especially no-tillage, was recommended as a viable management practice for cereal production in those areas. Total number of weeds was significantly lower in no-tillage than in other conservation tillage systems, although greater than in conventional tillage (Sombrero et al. 2001a, 2004). In NE Spain barren brome (Bromus sterilis L.) is the most difficult weed to control in cereal cropping systems under conservation tillage although, in general, tillage reduction is not detrimental for weed control in any crop (Catalán et al. 2003). Crop yield was also higher under conservation tillage in North Spain, except for subhumid areas. In Northern areas of Spain delay of planting under no-tillage has proven to be an effective method for brome control and effective reduction of some pest to be Hessian fly (Mayetiola destructor) and diseases as Helmintosporium and Oidium.

Studies by Angás and Cantero-Martínez (2000), Gabrielle et al. (2002) and Cantero-Martínez et al. (2003, 2007) in NE Spain were aimed to establish the optimal nitrogen (N) fertilisation for different tillage systems. Two models (CERES and CROPsyst) were tested as support decision tools for agronomic recommendations. Conservation tillage improved the yield and water-use efficiency (WUE) of barley and proved to be a valuable system, especially under dry conditions, providing greater water storage in the recharge period October–January. Only in wet years, higher yields were obtained in no-tillage when some N fertilizer was applied (Angás et al. 2006). However, in dry years with scarce rainfall during autumn, N should not be applied in any tillage system. In the same area, Lampurlanés et al. (1997, 2001, 2002), Lampurlanés and Cantero (2003) and Cantero-Martínez et al. (2004) evaluated the use of conservation tillage in the long term. No-tillage resulted potentially better for semi-arid regions. In these soils, yield depends on favourable rainfall distribution throughout the growing season, including the grain filling period. Nonetheless, in particular areas (e.g. semi-arid Aragón, NE Spain) no-tillage reduced barley growth, yield and water-use efficiency when compared with reduced tillage and conventional tillage. On average, winter cereal yield under no-tillage was about 9% lower than in conventional tillage; in these conditions, reduced tillage (chiselling) is recommended as a suitable alternative to conventional tillage (mouldboard ploughing) without detrimental effect on crop yield (López and Arrúe 1997; Moret et al. 2007a, b).

Under irrigated agriculture soil compaction and crop residue accumulation were sometimes constraints under no-tillage, being minimum tillage the best soil management option under this system (Santiveri et al. 2002; Berenguer et al. 2004). However, a recent study by Muñoz et al. (2007) showed a consistent improvement of soil quality by direct seeding under irrigation. Conservation tillage is specific to site and soil conditions (Lal 1989), which makes necessary experimentation for each local scenario. For that reason, more studies are needed in Spain on this respect.

Conservation agriculture in Spain has been focused mainly in field crops under rainfed conditions as winter cereals; crop yields were similar to or even greater than those in conventional tillage. At the earlier stages, frequently slightly better plant growth has been reported under conventional tillage, although the difference disappeared at a more advanced stage of growth. Crop yield was also higher under conservation tillage in humid areas. When adequately managed tillage reduction is not detrimental for weed control in any crop. Different studies were aimed to establish the optimal nitrogen (N) fertilisation for different tillage systems. Only in wet years, higher yields were obtained in no-tillage when some N fertilizer was applied. However, in dry years with scarce rainfall during autumn, N should not be applied in any tillage system.

3 Conclusion

From this literature analysis we conclude that field experiments carried out all over the Spanish geography give wide experience and knowledge on Conservation Agriculture. The major findings show that there are important cost savings (fuel, fertilizers) compared with conventional agriculture. Conservation tillage has proved to be highly efficient to water storage, especially in years with rainfall lower than average. Under our Mediterranean conditions moderate increases of organic matter have been observed in the soil top layer. However, this moderate increase of organic matter creates particular conditions for the physico-chemical and biological soil dynamics increasing soil quality. In general, better soil physical properties and aggregation under Conservation Agriculture have been reported. However, greater compaction under no-tillage can result at long-term in particular scenarios, which could make advisable an occasional tillage labour. Furthermore, adoption of conservation tillage can reduce soil CO_2 emissions to the atmosphere thus minimising soil organic carbon losses. Mobility and persistence of herbicides were lower under conservation tillage than under conventional tillage. In general, conservation tillage enhances biodiversity compared to conventional tillage. Crop yields were similar to or even greater than those in traditional tillage, although at the earlier stages, frequently slightly better plant growth has been reported under conventional tillage, the difference disappearing at a more advanced stage of growth. All these advantages demonstrate that Conservation Agriculture in Spain can be a more sustainable alternative than the conventional agriculture. However, we have found from the literature analysis some constraints for its adoption, mainly due to inadequate extension and technology transfer systems and lack of access to specific inputs, machinery and equipment. Crop residue management difficulties and occasional higher incidence of weeds, pests and diseases, besides social relationships among farmers (criticisms) may also difficult the establishment of Conservation Agriculture in local scenarios.

Despite of all these findings we have detected from the literature analysis that would be desirable more integrated studies on the suitability of annual and perennial crops for Conservation Agriculture techniques under both rainfed and irrigated conditions, as well as on the adoption of crop rotations and cover crops adapted to those technologies.

Acknowledgements This work was carried out with the support of the projects GOCE-CT-2004–505582–KASSA (EU), AGL2004–07763–CO2–01–AGR, AGL2007–66320–CO2–01–AGR and AGL2008–00424 from the Spanish CICYT, and the Research Group AGR 151 (Junta de Andalucía).

References

Álvaro-Fuentes J, López MV, Gracia R, Arrúe JL (2004) Effect of tillage on short-term CO_2 emissions from a loam soil in semi-arid Aragón (NE Spain). Options Méditerranéennes, Serie A 60:51–54

Álvaro-Fuentes J, Cantero-Martínez C, López MV, Arrúe JL (2007a) Soil carbon dioxide fluxes following tillage in semiarid Mediterranean agroecosystems. Res Soil Till 96:331–341

Álvaro-Fuentes J, López MV, Arrúe JL, Cantero-Martínez C (2007b) Management effects on soil carbon dioxide fluxes under semiarid Mediterranean conditions. Soil Sci Soc Am J 72:194–200

Álvaro-Fuentes J, Arrúe JL, Gracia R, López MV (2007c) Soil management effects on aggregate dynamics in semiarid Aragon (NE Spain). Sci Total Environ 378:179–182

Álvaro-Fuentes J, López MV, Cantero-Martínez C, Arrúe JL (2008a) Tillage effects on soil organic carbon fractions in Mediterranean dryland agroecosystems. Soil Sci Soc Am J 72:541–547

Álvaro-Fuentes J, Arrúe JL, Gracia R, López MV (2008b) Tillage and cropping intensification effects on soil aggregation: temporal dynamics and controlling factors under semiarid conditions. Geoderma 145:390–396

Álvaro-Fuentes J, Lampurlanes J, Cantero-Martínez C (2009) No-tillage rotations under Mediterranean rainfed conditions: I. Biomass, grain yield and water-use efficiency. Agron J 101:1227–1234

Angás P, Cantero-Martínez C (2000) Trends of soil nitrate in semi-arid areas of Ebro Valley: tillage and N-fertilization effects. In: Proceedings of III international crop science congress – VI congress European society agronomy, Hamburg, Germany, p 46

Angás P, Cantero-Martínez C, Karrou M, Benbelkacem A, Avci M (2004) Crop production technologies in Mediterranean region. In: Cantero C, Gabiña D (eds) Mediterranean rainfed agriculture: strategies for sustainability. Options Méditerranéennes 60:99–126

Angás P, Lampurlanés J, Cantero-Martínez C (2006) Tillage and N fertilization effects on N dynamics and barley yield under semiarid Mediterranean conditions. Soil Till Res 87:59–71

Arrúe JL (1997) Impacto potencial del laboreo de conservación sobre el suelo como sumidero de carbono atmosférico. In: García-Torres L, González P (eds) Agricultura de conservación: fundamentos agronómicos medioambientales y económicos. Asociación Española Laboreo de Conservación/Suelos Vivos, Córdoba, Spain, pp 189–199

Berenguer P, Borras G, Santiveri F, Cantero-Martínez C, Lloveras J (2004) Yield response of maize to nitrogen fertilisation in irrigated areas of the Ebro valley (Spain). In: Proceedings of VIII congress European society for agronomy, Copenhagen, Denmark, pp 365–366

Bescansa P, Imaz MJ, Virto MJ, Enrique A, Hoogmoed WB (2006) Soil water retention as affected by tillage and residue management in semiarid Spain. Soil Till Res 87:19–27

Bradford JM, Peterson GA (2000) Conservation tillage. In: Sumner ME (ed) Handbook of soil science. CRC Press, Boca Raton, FL, pp G247–G269

Bravo C, Giráldez JV, Ordóñez R, González P, Perea F (2003) Surface impact of conservation agriculture on pea yield and soil fertility in a heavy clay soil of southern Spain. In: Proceedings of II world congress on Conservation agriculture, Iguassu Falls, Parana-Brasil, pp 363–366

Bravo C, Torrent J, Giráldez JV, González P, Ordóñez R (2006) Long-term effect of tillage on phosphorus forms and sorption in a vertisol of southern Spain. Eur J Agron 25:264–269

Bravo C, Giráldez JV, Ordóñez R, González P, Torres F, Perea Torres F (2007) Long-term influence of conservation tillage on chemical properties of surface horizon and legume crops yield in a vertisol of southern Spain. Soil Sci 172:141–148

Cantero-Martínez C, Angás P, Lampurlanés J (2003) Growth, yield and water productivity of barley (*Hordeum vulgare* L.) affected by tillage and N fertilization in Mediterranean semiarid, rainfed conditions of Spain. Field Crop Res 84:341–357

Cantero-Martínez C, Angás P, Ameziane T, Pisante M (2004) Land and water management technoogies in the Mediterranean region. In: Cantero C, Gabiña D (eds) Mediterranean rainfed agriculture: strategies for sustainability. Options Méditerranéennes 60:35–50

Cantero-Martínez C, Gabiña D (2004) Evaluation of agricultural practices to improve efficiency and environment conservation in Mediterranean arid and semi-arid production systems. MEDRATE project. In: Cantero C, Gabiña D (eds) Mediterranean rainfed agriculture: strategies for sustainability. Options Méditerranéennes 60:21–34

Cantero-Martínez C, Angás P, Lampurlanés J (2007) Long-term yield and water use efficiency under various tillage systems in Mediterranean rainfed conditions. Ann Appl Biol 150:293–305

Castro J, Fernández-Ondoño E, Rodríguez C, Lallena AM, Sierra M, Aguilar J (2008) Effects of different olive-grove management systems on the organic carbon and nitrogen content of the soil in Jaén (Spain). Soil Till Res 98:56–67

Catalán G, Hervella A, de Andrés EF, Tenorio JL (2003) Study of the main weeds in three tilling systems in cold semi-arid Spain. Proceedings of II world congress on conservation agriculture, Iguassu Falls, Parana-Brasil, pp 532–534

Coelho MB, Mateos L, Villalobos FJ (2000) Influence of a compacted loam subsoil layer on growth and yield of irrigated cotton in Southern Spain. Soil Till Res 57:129–142

Cox L, Calderón MJ, Celis R, Hermosín MC, Moreno F, Cornejo J (1996) Mobility of metamitron in soils under conventional and reduced tillage. Fresen Environ Bull 5:528–533

Cox L, Calderón MJ, Hermosín MC, Cornejo J (1999) Leaching of clopyralid and metamitron under conventional and reduced tillage systems. J Environ Qual 28:605–610

Cuevas MV, Calderón MJ, Fernández JE, Hermosín MC, Moreno F, Cornejo J (2001) Assessing herbicide leaching from field measurements and laboratory experiments. Acta Agrophys 57:15–25

De la Rosa D, Díaz-Pereira E, Mayol F, Czyz EA, Dexter AR, Dumitru E, Enache R, Fleige H, Horn R, Rajkay K, Simota C (2005) SIDASS project Part 2. Soil erosion as a function of soil type and agricultural management in a Sevilla olive area, southern Spain. Soil Till Res 82:19–28

De Santiago A, Quintero JM, Delgado A (2008) Long-term effects of tillage on the availability of iron, copper, manganese, and zinc in a Spanish Vertisol. Soil Till Res 98:200–207

Dexter AR, Czyz EA, Birkás M, Diaz-Pereira E, Dumitru E, Enarche R, Fleige H, Horn R, Rajkaj K, de la Rosa D, Simota C (2005) SIDASS project Part 3. The optimum and the range of water content for tillage – further developments. Soil Till Res 82:29–37

Díaz-Pereira E, Prange N, Fernández M, de la Rosa D, Moreno F. (2002) Predicting soil water erosion using the ImpelERO model and a mapped reference area in the Sevilla province (Spain). Adv GeoEcol 35:533–542. Catena-Verlag, Reiskirchen, Germany

Du Preez CC, Steyn JT, Kotze E (2001) Long-term effects of wheat residue management on some fertility indicators of a semi-arid plinthosol. Soil Till Res 63:25–33

ECAF (European Conservation Agricultural Federation) (1999) Conservation agriculture in Europe: environmental, economic and EU Policy perspectives. Brussels, p 23

Fernández-Quintanilla C (1997) Historia y evolución de los sistemas de laboreo. El laboreo de conservación. In: García Torres L, González Fernández P (eds) Agricultura de Conservación. Fundamentos Agronómicos, Medioambientales y Económicos. Asociación Española de Laboreo de Conservación, Córdoba, Spain, pp 3–12

Franzluebbers AJ (2002) Soil organic matter stratification ratio as an indicator of soil quality. Soil Till Res 66:95–106

Franzluebbers AJ (2004) Tillage and residue management effects on soil organic matter. In: Magdoff F, Weil RR (eds) Soil organic matter in sustainable agriculture. CRC Press, Boca Raton, FL, pp 227–268

Gabrielle B, Roche R, Angás P, Cantero-Martínez C, Cosentino L, Mantineo M, Langensiepen M, Hénault C, Laville P, Nicoullau D (2002) A priori parameterization of the CERES soil-crop models and test against several European data sets. Agronomie 22:119–132

Gajri PR, Arora VK, Prihar SS (2002) Tillage for sustainable cropping. International Book Distributing Company, Lucknow, India

Gomes L, Arrúe JL, López MV, Sterk G, Richard D, Gracia R, Sabre M, Gaudichet A, Frangi JP (2003) Wind erosion in a semi-arid agricultural area of Spain: the WELSONS project. Catena 52:235–256

Gómez JA, Giráldez JV, Pastor M, Fereres E (1999) Effects of tillage method on soil physical properties, infiltration and yield in an olive orchard. Soil Till Res 52:167–175

Gómez JA, Orgaz F, Villalobos FJ, Fereres E (2002) Analysis of the effects of soil management on runoff generation in olive orchards using a physically based model. Soil Use Manag 18:191–198

Gómez JA, Romero P, Giráldez JV, Fereres E (2004) Experimental assessment of runoff and soil erosion in an olive grove on a Vertic soil in southern Spain as affected by soil management. Soil Use Manag 20:426–431

Gómez JA, Giráldez JV, Fereres E (2005) Water erosion in olive orchards in Andalusia (Southern Spain): a review. Geophys Res Abstr 7:08406

González P, Giráldez JV, Ordóñez R, Perea F (2003) Yields in a 20-year soil management experiment in a vertisol of southern Spain. In: Proceedings of II world congress for conservation agriculture, Iguassu Falls, Parana-Brasil, pp 359–362

Hernanz JL, Sánchez-Girón V, Cerisola C (1995) Long term energy use and economic evaluation of three tillage systems for cereal and legume production in Central Spain. Soil Till Res 35:183–198

Hernanz JL, López R, Navarrete L, Sánchez-Girón V (2002) Long-term effects of tillage systems and rotations on soil structural stability and organic carbon stratification in semi-arid central SpainSoil Till. Soil Till Res 66:129–141

Hill PR (1996) Conservation tillage: a checklist for U.S. farmers. Conservation Technology Information Center, West Lafayette, IN, p 35

Horn R, Fleige H, Richter FH, Czyz EA, Dexter A, Diaz-Pereira E, Dumitru E, Enarche R, Mayol F, Rajkaj K, de la Rosa D, Simota C (2005) SIDASS project Part 5. Prediction of mechanical strength of arable soils and its effects on physical properties at various map scales. Soil Till Res 82:47–56

Karlen DL, Varvel GE, Bullock DG, Cruse RM (1994) Crop rotations for the 21st century. Adv Agron 53:1–45

Kirkegaard JA, Munns R, James RA, Gardner PA, Angus JF (1995) Reduced growth and yield of wheat with conservation cropping. II. Soil biological factors limit growth under direct drilling. Aust J Agric Res 46:75–88

Lal R (1989) Conservation tillage for sustainable agriculture: tropics versus temperate environments. Adv Agron 42:85–197

Lampurlanés XS, Espadalé RMA, Carrera EG (1997) Plantas de compostaje para el tratamiento de residuos: riesgos higiénicos. NTP 597, comisionnacional.insht.es. http://comisionnacional.insht.es/InshtWeb/Contenidos/Documentacion/FichasTecnicas/NTP/Ficheros/501a600/ntp_597.pdf
Lampurlanés J, Angás P, Cantero-Martínez C (2001) Root growth, soil water content and yield of barley under different tillage systems on two soils in semi-arid conditions. Field Crop Res 6:27–40
Lampurlanés J, Angás P, Cantero-Martínez C (2002) Tillage effect on water storage efficiency during fallow, and soil water content, root growth and yield of the following barley crop on two different soils in semi-arid conditions. Soil Till Res 65:207–220
Lampurlanés J, Cantero-Martínez C (2003) Soil bulk density and penetration resistance under different tillage and crop management systems and their relationship with barley root growth. Agron J 95:526–536
Lampurlanés J, Cantero-Martínez C (2006) Hydraulic conductivity, residue cover and soil surface roughness under different tillage systems in semiarid conditions. Soil and Tillage Research 85:13–26. doi:10.1016/j.still.2004.11.006
López MV, Arrúe JL (1995) Efficiency of an incomplete block design based on geostatistics for tillage experiments. Soil Sci. Soil Sci Soc Am J 59:1104–1111
López MV, Arrúe JL, Sánchez-Girón V (1996) A comparison between seasonal changes in soil water storage and penetration resistance under conventional and conservation tillage systems in Aragón. Soil Till Res 37:251–271
López MV, Arrúe JL (1997) Growth, yield and water use efficiency of winter barley in response to conservation tillage in a semi-arid region of Spain. Soil Till Res 44:35–54
López MV (1998) Wind erosion in agricultural soils: an example of limited supply of particles available for erosion. Catena 33:17–28
López MV, Sabre M, Gracia R, Arrúe JL, Gomes L (1998) Tillage effects on soil surface conditions and dust emission by wind erosion in semi-arid Aragon (NE Spain). Soil Till Res 45:91–105
López MV, Gracia R, Arrúe JL (2000) Effects of reduced tillage on soil surface properties affecting wind erosion in semi-arid fallow lands of Central Aragon. Eur J Agron 12:191–199
López MV, Gracia R, Arrúe JL (2001) An evaluation of wind erosion hazard in fallow lands of semi-arid Aragón (NE Spain). J Soil Water Conserv 56:212–219
López MV, Moret D, Gracia R, Arrúe JL (2003) Tillage effects on barley residue cover during fallow in semi-arid Aragón. Soil Till Res 72:53–64
López MV, Arrúe JL, Álvaro-Fuentes J, Moret D (2004) Cereal residue management through conservation tillage in semi-arid Aragon (NE Spain). In: Proceedings of the 8th congress European society for agronomy, Copenhagen, Denmark, pp 625–626
López MV, Arrúe JL (2005a) Soil tillage and wind erosion in fallow lands of Central Aragón (Spain): an overview. In: Faz-Cano A, Ortíz R, Mermut AR (eds) Sustainable use and management of soils – arid and semiarid regions. Adv GeoEcol 36:93–102. Catena-Verlag, Reiskirchen, Germany
López MV, Arrúe JL, Álvaro-Fuentes J, Moret D (2005b) Dynamics of surface barley residues during fallow as affected by tillage and decomposition in semiarid Aragón (NE Spain). Eur J Agron 23:26–36
López-Bellido L, Fuentes M, Castillo JE, López-Garrido FJ, Fernández EJ (1996) Long-term tillage, crop rotation, and nitrogen fertilizer effects on wheat yield under rainfed Mediterranean conditions. Agron J 88:783–791
López-Bellido L, López-Garrido FJ, Fuentes M, Castillo JE, Fernández EJ (1997) Influence of tillage, crop rotation and nitrogen fertilization on soil organic matter and nitrogen under rainfed Mediterranean conditions. Soil Till Res 43:277–293
López-Bellido L, Fuentes M, Castillo JE, López-Garrido FJ (1998) Effects of tillage, crop rotation and nitrogen fertilization on wheat-grain quality grown under rainfed Mediterranean conditions. Field Crop Res 57:265–276

López-Bellido L, López-Bellido RJ, Castillo JE, López-Bellido FJ (2000) Effects of tillage, crop rotation, and nitrogen fertilization on wheat under rainfed Mediterranean conditions. Agron J 92:1054–1063
López-Bellido RJ, López-Bellido L (2001) Efficiency of nitrogen in wheat under Mediterranean conditions: effect of tillage, crop rotation and N fertilization. Field Crop Res 71:31–46
López-Bellido L, López-Bellido RJ, Castillo JE, López-Bellido FJ (2001) Effects of long-term tillage, crop rotation and nitrogen fertilization on bread-making quality of hard red spring wheat. Field Crop Res 72:197–210
López-Bellido RJ, López-Bellido L, Castillo JE, López-Bellido FJ (2002) Sunflower response to tillage and soil residual nitrogen in a wheat-sunflower rotation under rainfed Mediterranean conditions. Aust J Agric Res 53:1027–1033
López-Bellido RJ, López-Bellido L, Castillo JE, López-Bellido FJ (2003a) Nitrogen uptake by sunflower as affected by tillage and soil residual nitrogen in a wheat-sunflower rotation under rainfed Mediterranean conditions. Soil Till Res 72:43–51
López-Bellido RJ, López-Bellido L, López-Bellido FJ, Castillo JE (2003b) Faba Bean (*Vicia faba* L.) response to tillage and soil residual nitrogen in a continuous rotation with wheat (*Triticum aestivum* L.) under rainfed Mediterranean conditions. Agron J 95:1253–1261
López-Bellido RJ, López-Bellido L, Castillo JE, López-Bellido FJ (2004a) Chickpea response to tillage and soil residual nitrogen in a continuous rotation with wheat. II. Soil nitrate, N uptake and influence on wheat yield. Field Crop Res 88:201–210
López-Bellido L, López-Bellido RJ, Castillo JE, López-Bellido FJ (2004b) Chickpea response to tillage and soil residual nitrogen in a continuous rotation with wheat. I. Biomass and seed yield. Field Crop Res 88:191–200
López-Fando C, Almendros G (1995) Interactive effects of tillage and crop rotations on yield and chemical properties of soils in semi-arid central Spain. Soil Till Res 36:45–57
López-Fando C, Bello A (1995) Variability in soil nematode populations due to tillage and crop rotation in semi-arid Mediterranean agrosystems. Soil Till Res 36:59–72
López-Fando C, Pardo MT (2001) The impact of tillage systems and crop rotation on carbon sequestration in a Calcic Luvisol of Central Spain. In: Proceedings of I world congress on conservation agriculture. Madrid, Spain, pp 135–139
López-Fando C, Dorado J, Pardo MT (2007) Effects of zone-tillage in rotation with no-tillage on soil properties and crop yields in a semi-arid soil from central Spain. Soil Till Res 95:266–276
Madejón E, Moreno F, Murillo JM, Pelegrín F (2007) Soil biochemical response to long-term conservation tillage under semi-arid Mediterranean conditions. Soil Till Res 94:346–352
Martín-Rueda I, Muñóz-Guerra LM, Yunta F, Esteban E, Tenorio JL, Lucena JJ (2007) Tillage and crop rotation effects on barley yield and soil nutrients on a calciortidic Haploxeralf. Soil Till Res 92:1–9
Menéndez S, López-Bellido RJ, Benítez-Vega J, González-Murua C, López-Bellido L, Estavillo JM (2008) Long-term of tillage, crop rotation and N fertilization to wheat on gaseous emissions under rainfed Mediterranean conditions. Eur J Agron 28:559–569
Morell FJ, Álvaro-Fuentes J, Capell A, Cantero-Martínez C (2007) Short-term soil CO_2 efflux following tillage after ten years of mineral N fertilization in a Mediterranean semiarid agro-ecosystems. In: International symposium on Organic matter dynamics in agro-ecosystems. University of Poitiers-INRA, Poitiers, France, pp 169–170
Moreno F, Pelegrín F, Fernández JE, Murillo JM (1997) Soil physical properties, water depletion and crop development under conventional and conservation tillage in southern Spain. Soil Till Res 41:25–42
Moreno F, Pelegrín F, Fernández JE, Murillo JM, Girón IF (2000) Effects of conventional and conservation tillage on soil physical properties, water depletion and crop development in southern Spain. In: Proceedings of 15th conference Int. Soil Till. Res. Org., Fort Worth, Texas, USA, pp. 2–7.
Moreno F, Murillo JM, Girón IF, Fernández JE, Pelegrín F (2001) Conservation and conventional tillage in years with lower and higher precipitation than the average (south-west Spain).

In: García-Torres L, Benites J, Martínez-Vilela A (eds) Proceedings of the I world congress for conservation agronomy, Madrid, pp 591–595
Moreno F, Murer EJ, Stenitzer E, Fernández JE, Girón IF (2003) Simulation of the impact of subsoil compactation on soil water balance and crop yield of irrigated maize on a loamy sand soil in SW Spain. Soil Tillage Research 73: 31–41. doi:10.1016/S0167-1987(03)00097-7
Moreno F, Murillo JM, Pelegrín F, Girón IF (2006) Long-term impact of conservation tillage on stratification ratio of soil organic carbon and loss of total and active $CaCO_3$. Soil and Tillage Research 85:86–93. doi:10.1016/j.still.2004.12.001
Moret D, Arrúe JL (2007a) Dynamics of soil hydraulic properties during fallow as affected by tillage. Soil Till Res 96:103–113
Moret D, Arrúe JL (2007b) Characterizing soil water-conducting macro- and mesoporosity as influenced by tillage using tension inflitrometry. Soil Sci Soc Am J 71:500–506
Moret D, Arrúe JL, López MV, Gracia R (2006) Influence of fallowing practices on soil water and precipitation storage efficiency in semiarid Aragón (NE Spain). Agric Water Manag 82:161–167
Moret D, Braud I, Arrúe JL (2007a) Water balance simulation of a dryland soil during fallow under conventional and conservation tillage in semi-arid Aragón, Northern Spain. Soil Till Res 92:251–263
Moret D, Arrúe JL, López MV, Gracia R (2007b) Winter barley performance under different cropping and tillage systems in semiarid Aragon (NE Spain). Eur J Agron 26:54–63
Muñoz A, López-Piñeiro A, Ramírez M (2007) Soil quality attributes of conservation management regimes in a semi-arid region of south western Spain. Soil Till Res 95:255–265
Murillo JM, Moreno F, Pelegrín F, Fernández JE (1998) Responses of sunflower to conventional and conservation tillage under rainfed conditions in southern Spain. Soil Till Res 49:233–241
Murillo JM, Moreno F, Girón IF, Oblitas MI (2004) Conservation tillage: long term effect on soil and crops under rainfed conditions in south-west Spain (Western Andalusia). Span J Agric Res 2:35–43
Murillo JM, Moreno F, Madejón E, Girón I, Pelegrín F (2006) Improving soil surface properties: a driving force for conservation tillage under semi-arid conditions. Span J Agric Res 4:97–104
Ordóñez Fernández R, González Fernández P, Giraldez Cervera JV, Perea Torres F (2003) Effects of conservation agriculture on the fertility level of heavy clay soils under dry farming. In: Proceedings of II world congress for conservation agriculture, Iguassu Falls, Parana, Brasil, pp 405–407
Ordóñez FR, González FP, Giraldez Cervera JV, Perea Torres F (2007) Soil properties and crop yields after 21 years of direct drilling trials in southern Spain. Soil Till Res 94:47–54
Pelegrín F, Moreno F, Martín-Aranda J, Camps M (1990) The influence of tillage methods on soil physical properties and water balance for a typical crop rotation in SW Spain. Soil Till Res 16:345–358
Pelegrín F, Moreno F, Madueño A, Franco A, Girón IF, Fernández JE (2001) The use of green covers to conserve soil and water in a water harvesting systems within an olive orchard. In: García-Torres L, Benites J, Martínez-Vilela A (eds) Proceedings of I world congress for conservation agronomy, Madrid, pp 401–407
Perea Torres F, Giráldez Cervera JV, González Fernández P, Gil J, Ordóñez Fernández R (2003) Influence of soil moisture on the compaction of clay soils under conservation agriculture in Mediterranean environments. In: Proceedings of II world congress for conservation agriculture, Iguassu Falls, Parana, Brasil, pp 498–501
Reicosky DC, Dugas WA, Torbert HA (1997) Tillage-induced soil carbon dioxide loss from different cropping systems. Soil Till Res 41:105–118
Saavedra C, Velasco J, Pajuelo P, Perea F, Delgado A (2007) Effects of tillage on phosphorus release potential in a Spanish vertisol. Soil Sci Soc Am J 71:56–63
Sánchez-Girón V, Serrano A, Hernanz JL, Navarrete L (2004) Economics assessment of three long-term tillage systems for rainfed cereal and legume production in semi-arid central Spain. Soil Till Res 78:35–44

Sánchez ML, Ozores MI, Colle R, López MJ, De Torre B, García MA, Pérez I (2002) Soil CO_2 fluxes in cereal land use of the Spanish plateau: influence of conventional and reduced tillage practices. Chemosphere 47:837–844

Sánchez ML, Ozores MI, López MJ, Colle R, De Torre B, García MA, Pérez I (2003) Soil CO_2 fluxes beneath barley on the central Spanish plateau. Agric For Meteorol 118:85–95

Santiveri P, Berengué P, Lloveras J, Cantero-Martínez C (2002) Effect of tillage and residual nitrogen on yield in irrigated wheat. In: Proceedings of VII European society for agronomy congress, Córdoba, Spain, pp 537–538

Simota C, Horn R, Fleige H, Dexter A, Czyz EA, Díaz-Pereira E, Mayol F, Rajkai K, de la Rosa D (2005) SIDASS project Part 1: a spatial distributed simulation model predicting the dynamics of agro-physical soil state for selection of management practices to prevent soil erosion. Soil Till Res 82:15–18

Sojka RE, Upchurch DR (1999) Reservations regarding the soil quality concept. Soil Sci Soc Am J 63:1039–1054

Sombrero A, De Benito A, Escribano C (1998) Tillage systems and crop rotations effect on growth and yield barley in semi-arid areas. In: Proceedings of V European society for agronomy congress, Nitra, Eslovequia, pp 59–60

Sombrero A, De Benito A, Nieto M (2001a) Effects of tillage systems and crop rotations on the weed population and cereal yield in a semi-arid area. In: Proceedings of VII European society for agronomy congress, Córdoba, Spain, pp 541–542

Sombrero A, Benito A, Escribano C, Tenorio JL, Catalán G, Pérez de Ciriza JJ, Irañeta J (2001b) Profitability of cereal crop under three tillage systems in semi-arid zone of North Spain. In: Proceedings of I world congress for conservation agriculture, Madrid, Spain, pp 795–799

Sombrero A, De Benito A, Nieto M (2004) Rotation effects on barren brome (Bromus sterilis L.) control in conservation tillage. In: Proceedings of VIII European society for agronomy congress, Copenhagen, Denmark, pp 669–670

Sparling G, Parfitt RL, Hewitt AE, Schipper LA (2003) Three approaches to define desired soil organic matter contents. J Environ Qual 32:760–766

Sterk G, López MV, Arrúe JL (1999) Saltation transport on a silt loam soil in northeast Spain. Land Degrad Dev 10:545–554

Warkentin BP (2001) The tillage effect in sustaining soil functions. J Plant Nutr Soil Sci 164:345–350

Conservation Tillage and Sustainable Agriculture in Semi-arid Dryland Farming

Mohammad J. Zarea

Abstract One major crop production type in semi-arid area is dryland farming dependent on rainfall. The major factors that constrain semi-arid soil fertility and sustainable agriculture are low rainfall, low nutrient capital, moisture stress, soil erosion, high P fixation, high alkalinity, and low soil biodiversity. The water stress, low rainfall and shallow depth of many semi-arid soils limit food production in annual cropping systems. The management of beneficial microorganisms in the rhizosphere has emerged as an alternative to chemical fertilizers to increase soil fertility and crop production in sustainable agroecosystems; but it seems that major agricultural practices that strongly affect every approach to sustainable dryland farming in this area are affected by the choice of soil tillage practices. Crop response to tillage systems is diverse due to the complex interactions between tillage-induced soil, edaphic crop requirements and weather. The use of crop rotation, earthworms and mycorrhizae give several benefits in this area, and could be improved by adopting the best soil tillage system. This review treats the role of conservation tillage practices in enhancing soil water retention and infiltration, as well as physical, chemical and biological soil quality.

Keywords Arbuscular mycorrhiza • Persian plough • Weeds • Earthworms • Soil organic carbon • Soil enzyme

1 Introduction

The semi-arid region encompasses a wide variety of agricultural systems where water is probably one of the main keys to productivity. Water frequently limits rainfed crop production in this area because of low annual precipitation (<450 mm) and an uneven interannual distribution.

M.J. Zarea (✉)
Department of Agronomy and Breeding, Faculty of Agriculture, Ilam University, Ilam, Iran
e-mail: mj.zarea@ilam.ac.ir

E. Lichtfouse (ed.), *Biodiversity, Biofuels, Agroforestry and Conservation Agriculture*, Sustainable Agriculture Reviews 5, DOI 10.1007/978-90-481-9513-8_7,

The adoption of conservation tillage (CT) has increased worldwide over the past few decades. Conservation tillage has considerable potential for stabilizing production in semi-arid zones.

Semi-arid soils under conservation tillage generally contain greater concentrations of organic C (Franzluebbers et al. 1996; Zibilske et al. 2002; Fernández-Ugalde et al. 2009), microbial biomass (Bescansa et al. 2006), N mineralization (Franzluebbers et al. 1996), aggregate stability (Fernández-Ugalde et al. 2009; Muñoz et al. 2007), steady infiltration rates (Gicheru et al. 2004), water availability (Fernández-Ugalde et al. 2009), predator groups density and diversity (López-Fando and Bello 1995) and enzymatic activities (Madejón et al. 2007), especially in the upper layer. Conservation tillage decreases runoff (Dimanche and Hoogmoed 2002) and wind erosion (Buschiazzo et al. 2007) and is an alternative for reducing costs relative to conventional tillage for this area (Dimanche and Hoogmoed 2002).

Weeds are a major problem in no-tillage areas (Rapp et al. 2004; Walters et al. 2008; Bachthaler 1974; Schwerdtle 1977; Nielsen and Pinnerup 1982). While the use of effective herbicides in combination with cover crops integrated into no-tillage planting systems may provide a feasible option for enhancing weed control, these compounds tend to accumulate in the soil and subsurface water (Jacobson et al. 2005) where they can become a toxicological risk for invertebrates such as arbuscular mycorrhiza (AM) fungi and earthworms that ingest large amounts of soil and play a key role in soil biology.

The major semi-arid dryland crop is usually cereal grain, whose response to conservation tillage practices is variable (Rao and Dao 1996). The conservation tillage systems have been reported to yield equal (Bescansa et al. 2006; Aboudrare et al. 2006), or better (Blaise and Ravindran 2003; Ozpinar and Cay 2006; Muñoz et al. 2007; Gicheru et al. 2004; Ozpinar and Baytekin 2006) or lower (López-Fando et al. 2007; De Vita et al. 2007) than CT systems in semi-arid areas.

Many researchers have reported greater bulk density (Fernández-Ugalde et al. 2009; Ozpinar and Cay 2006), soil penetration resistance (Muñoz et al. 2007) and lower total porosity in no-till (NT) compared with mouldboard ploughing and chisel ploughing in semi-arid areas. Use of biological resources of the soil, such as earthworms, can counter these disadvantages.

The soil structural stability can provide better pore connectivity and the channels formed by earthworms and decayed roots provide higher hydraulic conductivity even for high bulk density soils when compared to other tillage methods (Ehlers et al. 1983; Wang et al. 1986; Osunbitan et al. 2005).

The symbiotic relationship between AM and the roots of higher plants contributes significantly to plant nutrition and growth (Augé 2001), and has been shown to increase the productivity of a variety of agronomic crops. These positive responses in productivity to AM colonization have mainly been attributed to the enhanced uptake by AM of relatively immobile soil ions (Marschner and Dell 1994; Marschner 1995; Liu et al. 2000a, b, 2007), but also involve the enhanced uptake and transport of far more mobile nitrogen (N) ions, particularly under drought conditions (Tobar et al. 1994; Liu et al. 2007).

For better nutrient management in semi-arid areas, an increased use of the biological potential is important. Many keys to agricultural success in semi-arid areas are to use proper tillage practices, adequate plant species and to use the soil biology potential to maintain soil fertility, and to guard against erosion and water limiting. Using earthworms can benefit these areas, providing increased soil porosity, soil aggregation, enhanced soil organics and increased soil filtration, particularly under no-tillage with increased soil-penetration resistance. Use of AM is important for producing several benefits of plant symbiosis under drought stress and limited water. Since nutrient mobility is limited in drought conditions, AM may have a larger impact on overall plant growth and development in dry conditions compared to well-watered conditions (Sánchez-Díaz and Honrubia 1994); but the efficiency of this biological potential is strongly dependent on tillage practices.

The purpose of this review is to outline the current state of knowledge about the effect of conservation tillage on crop yield, especially in semi-arid areas. The potential influence of tillage practices on soil properties, crop production, water-use efficiency, mycorrhizal fungi, earthworms, soil nutrient, chemical soil quality and soil enzyme activities are also discussed. The review focuses on interactions between tillage practice and soil properties, including a brief discussion on how this knowledge is currently being used and how an understanding of this could prove to be important for sustainable agriculture in the future. It is not meant to be a complete review of tillage practices on semi-arid farming (dryland farming), but rather concentrates on aspects of those interactions between soil tillage and soil properties which may have practical applications. The paper outlines the effect of conservation tillage on yield and soil biochemistry and biology in semi-arid areas. Methods of applying rhizosphere communities, such as earthworms and mycorrhizae, that can overcome the challenges of semi-arid crop production, are presented.

2 Towards Conservation Tillage

The term "tillage" is a generic term and is used broadly. Tillage embraces all operations of seedbed preparation that optimize soil and environmental conditions for seed germination, seedling establishment and crop growth. It includes: mechanical methods based on conventional techniques of ploughing and harrowing; weed control using chemical herbicides and growth regulators; and fallowing with an aggressive cover crop that can be easily controlled, for direct seeding through its residue mulch. There is a wide range of tillage systems used in the arid and semi-arid regions.

Last century, interest in avoiding unnecessary tillage for energy, moisture, labor and soil conservation benefits increased tremendously. Reduced tillage, including a variety of tillage practices that conserve soil and water by leaving at least 30% of the soil surface covered by residue, have been developed in recent decades. Conservation tillage, which includes any form of minimum or reduced tillage, has greatly improved our ability to capture and retain precipitation in the soil during the non-crop periods of the cropping cycle, and has made it possible to reduce fallow frequency and intensify cropping systems. Minimum or reduced tillage involves a cultivation operation whereby the soil is disturbed as little as possible to produce a crop. No-till or zero tillage, a form of minimum tillage where a slot is opened in the soil only sufficiently deep and wide to properly deposit and cover seeds, is increasingly becoming a way of management for agricultural semi-arid area soils.

2.1 Early Plough

The plough has been developed in early days of agriculture and was first pulled by humans and later by animals. One of the early types of plough that had been used for thousands of years in the Persian area for dryland semi-arid cropping shown in Fig. 1, was pulled by humans and later by animals (especially cows and horses). Farmers in Ethiopia have used the Maresha plough (Fig. 1) for thousands of years (Goe 1987; Gebregziabher et al. 2006). Both the Persian plough and Maresha plough are very simple, light in weight, cheap, and locally made (Sime 1986).

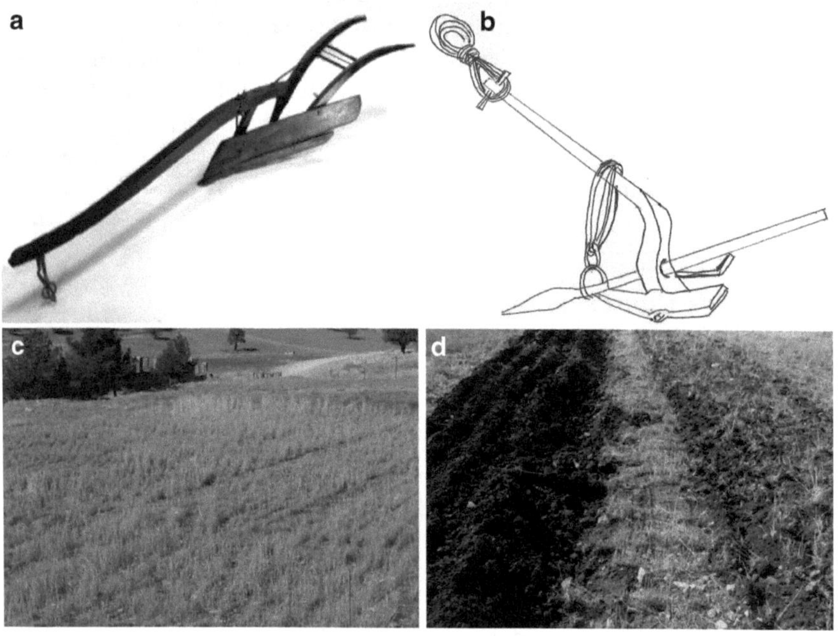

Fig. 1 (**a**) Early Persian ploughing that was commonly used in some Iran semiarid region for thousands years. This tillage has been used since ancient times by indigenous cultures. The plough has been developed in early days of agriculture and was first pulled by man and later by animals. (**b**) Maresha plow have used for thousands of years (Adapted from Temesgen et al. 2009). Bare soil (**c**) after two tillage practices (conventional tillage and minimum tillage) result in different crop residue (**d**)

2.2 Conventional Tillage

The most commonly applied conventional tillage system consists of a ploughing operation with mouldboard plough followed by one or more passes with a disc harrow. The status and performance of agricultural mechanisation are important factors for seedbed preparation and planting, and the application of fertilizer and pesticides.

The use of the plough is often mentioned in the Bible, one of the best-known quotations being "they shall beat their swords into plough shares" (Isaiah 2.4.) (Derpsch 1999). But the plough of biblical times had nothing to do with modern ploughs of the nineteenth century. In those days a plough was nothing more than a branch from a tree that scratched or scarified the soil surface without mixing the soil layers. Ploughs that inverted the soil layers and thus gave a better weed control were not developed until the seventeenth century. Not until the eighteenth and nineteenth centuries did ploughs become increasingly sophisticated (Derpsch 1999). Towards the end of the eighteenth century, German, Dutch and British developments of this tool led to an almost perfect mouldboard shape that turned the soil by 135° and was very efficient in weed control.

Conventional tillage can result in soil erosion, loss of organic matter, decreased water infiltration, loss of soil structure, decreased soil fertility and a reduction in overall soil quality due to the destruction of soil aggregates and structure (Parr et al. 1992; Paustian et al. 1997; Allmaras et al. 2000; Nyakatawa et al. 2000), all of which are beneficial under dryland farming. The natural roles mycorrhizosphere organisms may have been marginalised in intensive agriculture, since microbial communities in conventional farming systems have been modified due to tillage (Sturz et al. 1997; McGonigle and Miller 1996). Different management strategies introduce different types of disturbances, which may influence microbial communities in various ways (Johansson et al. 2004). Excessive soil compaction can lead to reduced soil microbial biomass and enzyme activity with adverse implications for long-term soil health It would be worth rewriting the above as it resembles that from Johansson et al. 2004 (Frankenberger and Dick 1983).

2.3 Conservation Tillage

Conservation tillage [no-tillage or zero tillage (NT, ZT)] and reduced tillage or minimum tillage (RT/MT)] brings many benefits with respect to soil fertility and energy use (e.g. Rapp et al. 2004). Agriculture in semi-arid areas suffers from strong annual variations both in crop yield and profitability. Implementation of conservation tillage in semi-arid areas began with the aim of improving retention of water in soil, and reducing erosion. Lower energy input, which reduces cropping costs and increases the profitability of agriculture, has facilitated rapid dissemination of conservation tillage in these areas, where economic efficiency of crops is close to marginal (Bescansa et al. 2006) Other factors, such as sustainable utilisation of soil and the environment, have also been influential (Bescansa et al. 2006; Tebrügge and Böhrnsen 2001). Numerous studies in several semi-arid areas have discovered the advantages of conservation tillage compared to conventional tillage (Table 1). It is widely documented that conservation tillage has significant and, in general, positive effects on several chemical soil properties, including organic matter and nutrient status (Bescansa et al. 2006) (Table 1). Conservation tillage also leads to positive changes in some physical properties of soil, such as aggregation, aggregate stability and soil water content (Table 1). One of the major advantages of conservation tillage is the greater availability of soil water which, especially in years of low rainfall, is very important.

Semi-arid areas are prone to soil erosion. Conservation tillage, using either no tillage or reduced tillage to provide year-round cover, is the best management practice for the control of wind erosion in the low-precipitation zone (Papendick 2004). The amount of organic material in the soil, and thus the potential fertility, is likely to be high in semi-arid zones under NT and reduced tillage (Table 1). Use of conservation tillage improves biological activity, important for sustainable agriculture (Table 1). Research has found organisms such as earthworms, fungi, bacteria and other microorganisms to be beneficial for crop production and agroecosystem management, but modern agricultural practice has reduced roles of these organisms to the detriment of the natural ecosystem.

Reduced tillage has been adopted more often in order to conserve soil water and reduce erosion and soil compaction caused by conventional intensive management. The terms "semi-arid" and "arid" region implies prolonged dryness, being used with respect to the climate itself, and to the land of the region. The ability to produce agricultural crops is restricted in such regions. Usually on arid lands evaporation is high, the soil is prone to erosion and a large proportion of the phosphorus in the soil is found as precipitated calcium-phosphate minerals, which are insoluble and unavailable to plants in the short term. It is possible to increase soil water content by choosing to adopt proper tillage practices in this area. Soil water content under conventional tillage can be 10–20% lower than under zero-tillage systems (Blevins et al. 1971; Ghaffarzadeh et al. 1994).

Tillage increases decomposition of crop residues and changes the structure of the soil web by relocating food resources and exposing protected carbon (Hendrix et al. 1986; Moore and de Ruiter 1991; Beare et al. 1992; Wardle 1995; Six et al. 2002). Conservation tillage is used to conserve soil nutrients and structure, increase sequestration of soil carbon, and to provide habitat and substrate for biota (Hendrix et al. 1998; Lal et al. 1998; Paustian et al. 2000; Holland 2004; Simmons and Coleman 2008). However, a deterrent for growers considering the transition to conservation tillage is the slow soil response to a cessation of tillage (e.g. increased soil carbon, efficient nutrient cycling, impacts on yield) and the consequent equilibration of the soil food web (Phatak et al. 1999; Simmons and Coleman 2008). However, the subsoil responds more quickly than the surface soil (Simmons and Coleman 2008) and there is evidence that below-ground food webs respond quickly to a cessation in tillage, suggesting that the delay in soil response may be

Table 1 Summary of conservation tillage on soil biochemical, nitrogen, phosphorus available, soil nematode population, cation exchange capacity in various semi arid regions

Conservation tillage	Parameter(s) affected	Reference
No tillage (NT)	>Soil organic carbon	Franzluebbers et al. 1996, Zibilske et al. 2002 and Fernández-Ugalde et al. 2009
No tillage (NT)	<Basa soil respiration	Franzluebbers et al. 1996
No tillage (NT)	>N mineralization	Franzluebbers et al. 1996
No tillage (NT)	>Soil denser	Franzluebbers et al. 1996
No tillage (NT)	>Aggregate stability	Fernández-Ugalde et al. 2009
No tillage (NT)	>Penetration resistance	Fernández-Ugalde et al. 2009
No tillage (NT)	>Water availability	Fernández-Ugalde et al. 2009
No tillage (NT)	>Soil water content	Fernández-Ugalde et al. 2009
No tillage (NT)	>Soil bulk density	Fernández-Ugalde et al. 2009
No tillage (NT)[c]	<Evaporative losses	Tessier et al. 1990
No tillage (NT)	>Organic carbon	Thomas et al. 2007 and Ouèdraogo et al. 2006
No tillage (NT)	>Total soil N concentration	Zibilske et al. 2002 and Thomas et al. 2007
No tillage (NT)	>Total soil P	Selles et al. 1999
No-till[e]	<Crop N uptake	Ouèdraogo et al. 2006
No tillage (NT)	>Exchangeable K	Thomas et al. 2007
No tillage (NT)	>Biocarbonate extractable P concentration	Thomas et al. 2007
No tillage (NT)	>Bacterial – feeding, fungivorous omnivorous predator groups density and diversity	López-Fando and Bello 1995
No tillage (NT)[d]	>Soil organic carbon, >soil water retention, >available soil water capacity	Bescansa et al. 2006
No tillage	Loss in total soil organic carbon	Ouèdraogo et al. 2006
No tillage	Improved surface accumulation of soil organic carbon (SOC), total N and available P and K	López-Fando et al. 2007
No tillage	No difference on Stratification ratio of SOC	López-Fando et al. 2007
No tillage	>Soil N stratification	López-Fando et al. 2007
Direct seeding, direct seeding with a winter crop cover	>Soil water content, >organic C, >nitrogen, >aggregate stability soil, >penetration resistance	Muñoz et al. 2007
Conservation tillage (CT)[a]	>Organic matter content	Madejón et al. 2007
Conservation tillage (CT)[a]	>Microbial biomass carbon	Madejón et al. 2007
Conservation tillage (CT)[a]	>Enzymatic activities	Madejón et al. 2007
Reduced tillage (RT)[b]	<Runoff	Dimanche and Hoogmoed 2002

(continued)

Table 1 (continued)

Conservation tillage	Parameter(s) affected	Reference
Reduced tillage (RT)[f]	>Total N in the soil, >penetration resistance, >organic carbon, > bulk density	Ozpinar and Cay 2006
Minimum tillage[g]	>Steady infiltration rates, >soil water stored, >drainage	Gicheru et al. 2004
Minimum tillage with chisel plow (MT)	>Soil N stratification	López-Fando et al. 2007
Reduced tillage using a spring tined cultivator	Less costly, lower losses of water by runoff, >water available for the crop	Dimanche and Hoogmoed 2002
Shallow tillage with rototiller (ST) and double disc tillage (DD)	<Plant biomass (common vetch) nitrogen content	Ozpinar and Baytekin 2006
Double disc tillage (DD)	>Grain soil nitrogen, improved organic carbon	Ozpinar and Baytekin 2006
No-till (NT)	<Wind erosion	Buschiazzo et al. 2007

> – Increased; < – decreased

[a] Mouldboard ploughing, by reduction of the number of tillage operations and leaving the crop residues on the surface

[b] Espiring tined cultivator

[c] Zero tillage fallow compared with conventional fallow

[d] No tillage with and without stubble burning

[e] Animal power was used for the tillage (12 cm depth)

[f] Rototilling followed by one discing

[g] With manure and mulching

due more to the time necessary for building organic matter than to a slow response by the biota (Simmons and Coleman 2008). One major constraint stopping farmers adopting RT or NT practices, however, is the weed problem.

3 Weed Problems

Weeds, and especially grasses and perennial weeds, may become a problem, both in no-tillage systems (Bachthaler 1974; Schwerdtle 1977; Nielsen and Pinnerup 1982) and in reduced tillage systems (Pleasant et al. 1990). The major limitation to widespread adoption of the no-tillage crop production system is the lack of effective weed control strategies in the absence of tillage (Rapp et al. 2004; Walters et al. 2008).

Under no-tillage, weed seeds are no longer distributed throughout the soil profile but tend to accumulate in the topsoil layer. Densities of weed populations may increase because most weed seeds are in a condition favoring germination. Experiments conducted over many years have shown that weed seed density is greatest in a no-tillage system, and declines with tillage intensity (Cardina et al. 1991, 2002; Feldman et al. 1997). Tillage practices influence the vertical distribution of weed seeds in soil in addition to the rate of seedbank decline (Ball 1992; Barberi and Cascio 2000; Buhler et al. 1996, 1997; Cardina et al. 1991). Studies of weed communities under reduced tillage have shown that such systems favor annual grasses, perennial species disseminated by wind, and volunteer crops. By contrast, annual broad-leaved weed species tend to thrive under conventional tillage systems (Moyer et al. 1994; Streit et al. 2002).

In Norway, Tørresen and Skuterud (2002) showed that both post-emergence herbicide and glyphosate application were necessary to control different weed groups when tillage was reduced. The review of Belde et al. (2000) describes long-term impacts (4–25 years) of reduced-tillage systems on the flora with traditional selective herbicides; in four investigations the abundance of broad-leaved agricultural plants was reduced and maintained stable. Over 95% of weed control is achieved by broadspectrum herbicides like glyphosate and glufosinate, which is a level not attained by their traditional counterparts (Westwood 1997).

Ghosheh and Al-Hajaj (2005) reported that, in general, mouldboard ploughing increased weed seedbanks when combined with frequent fallowing. Conversely, chisel ploughing combined with barley cropping generally reduced weed seedbank sizes. In one case, the plant abundance had been increased whereas the seedbank abundance decreased. Not only tillage but also herbicide use was reduced in five of these nine reviewed studies, which mitigated herbicide impacts (Schütte 2003). Problematic weed types such as grasses and perennials often increase, whereas broad leaf annual plants, which provide food for important aphid predators, decrease (Knab and Hurle 1986; Thomas and Frick 1993; Sievert 2000; Belde et al. 2000). From their study, Belde et al. (2000) concluded that wild plant abundance increases in the first years but their abundance and diversity eventually decreases. Seeds remain on the soil surface and germinate, and are then eliminated, resulting in seed bank and overall losses (Buhler et al. 1997). It shouldbe noted that larger amounts of these herbicides are used in minimum-tillage systems (Benbrook 2001; see above).

The large quantity of plant residue left on the soil surface may affect the performance of herbicides by intercepting up to 70% of the active ingredients (Sadeghi et al. 1998). Crop residues are considered partially responsible for the reduced efficacy of herbicides (Buhler 1995) and ineffective weed control limits the adoption of conservation tillage systems (Buhler et al. 1994). Froud-Williams (1988) found that crop rotation had a greater impact on weeds than tillage systems (Swanton et al. 1993). Cover crops with a highly competitive ability including, for example, legumes or mustard, can suppress weeds. The use of effective herbicides in combination with cover crops integrated into no-tillage planting systems may provide a feasible option for enhancing weed control. Small grain cover crops such as winter wheat (*Triticum aestivum* L.) or winter rye (*Secale cereale* L.) can be used to suppress weed densities in no-tillage production systems (Morse et al. 2001; Walters et al. 2008), although these crops must be killed to minimize competition with the vegetable crop (Zandstra et al. 1998). Also, cover crop residues provide many benefits toward more sustainable production practices, especially in a reduced tillage system (Russo et al. 2006).

4 Crop Yield

Intensive research has indicated that crop yields are often increased with no-tillage and reduced tillage in semi-arid areas (Table 2) but this increase depends on soil status and precipitation.

Water is the primary constraint to crop production in semi-arid regions (Fuentes et al. 2009). Higher yield with conservation tillage systems can be related to improved water retention through observing changes in pore-size distribution. Tillage and the resultant soil structure influence soil water retention and its availability to plants.

Table 2 Summary of conservation tillage on yield under semi arid environment

Conservation tillage	Crop response	Reference
Reduced tillage (RT)[a]	>Greater seed rainfed cotton (*Gossypium hirsutum*)	Blaise and Ravindran 2003
Reduced tillage (RT)[b]	>Winter wheat (*Triticum aestivum* L.)	Ozpinar and Cay 2006
Direct seeding, direct seeding with a winter crop cover	>Irrigation maize (*Zea mays* L.)	Muñoz et al. 2007
Minimum tillage (MT) with manure and mulching	>Maize, >maize emergence	Gicheru et al. 2004
Double disc tillage (DD)	>Grain common vetch (*Vicia sativa* L.) nitrogen	Ozpinar and Baytekin 2006
Shallow tillage with rototiller (ST) and double disc tillage (DD)	<Plant biomass [common vetch (*Vicia sativa* L) nitrogen content	Ozpinar and Baytekin 2006
Double disc tillage (DD)	>Grain common vetch (*Vicia sativa* L.) yield	Ozpinar and Baytekin 2006
No-tillage (NT)	>Durum wheat (*Triticum durum* Desf.), >thousand kernel weight, <protein content	De Vita et al. 2007
No-tillage (NT)[c]	No difference in barley yield	Bescansa et al. 2006
Reduced tillage (RT)		
Reduced tillage	Improved barley yield economic efficiency, <production costs	Bescansa et al. 2006
Minimum tillage with chisel plow (MT)	<Grey pea (*Pisum sativum* L.), <barley (*Hordeum vulgare* L.)	López-Fando et al. 2007
No-tillage	<Grey pea, (*Pisum sativum* L.), <barley (*Hordeum vulgare* L.)	López-Fando et al. 2007
No tillage	No difference sunflower	Aboudrare et al. 2006
Reduced tillage chiselling, paraploughing,disc harrowing,	No difference sunflower	Aboudrare et al. 2006

> – Increased; < – decreased

[a]Mouldboard ploughing, by reduction of the number of tillage operations and leaving the crop residues on the surface

[b]Rototilling followed by one discing

[c]With manure and mulching

RT systems improve soil water regimes and allow better water extraction, aeration and fertilizer use than do conventional tillage methods. Hulugalle et al. (1997) attributed yield improvements to the better water extraction in reduced till systems. In the drylands, improved soil water regimes are considered an important factor in increasing yields of crops grown under RT systems.

There is evidence that no-tillage compared to reduced tillage (and especially in comparison with conventional tillage) causes significantly increased crop yield (Table 2). Greater concentrations of organic C (Franzluebbers et al. 1996; Zibilske et al. 2002; Fernández-Ugalde et al. 2009), water availability (Fernández-Ugalde et al. 2009), microbial biomass (Bescansa et al. 2006), N mineralization (Franzluebbers et al. 1996), aggregate stability (Fernández-Ugalde et al. 2009; Muñoz et al. 2007), predator group density and diversity (López-Fando and Bello 1995), and enzymatic activities (Madejón et al. 2007), especially in the upper layer, may support higher yield in semi-arid areas. No-tillage has been shown to increase soil water content through greater infiltration and reduced evaporation (Blevins and Frye 1993; Cannell and Hawes 1994), and by increasing the proportion of smaller pores (Arshad et al. 1999; Bescansa et al. 2006).

Some studies show no significant increasing in crop yield under no tillage (NT) with high yearly precipitation (Table 2). Tillage is often justified because, without it, compaction can lead to higher bulk density and increased penetration resistance, especially in the top few centimetres of the soil. Many authors have found that semi-arid no-tillage sites have greater bulk density and penetration resistance than reduced-tillage sites (e.g. Fernández-Ugalde et al. 2009; Muñoz et al. 2007).

Few studies show any significant decrease in crop yield (Table 2). NT tillage systems may affect N availability and change the crop yield response to soil preparation. Grain yields with no-till tend to be lower than with conventional tillage at lower N rates due to the immobilization and leaching of unincorporated, surface-applied N; this decrease can exceed 40% (Kitur et al. 1984; Meisinger et al. 1985).

Precipitation is another important factor that is likely to affect the response of plant yield to tillage practices. It seems that response of crop yield to NT and conservation tillage (CT) is low during years experiencing high precipitation years. Jin et al. (2009) reported that compared to CT, winter wheat yield under NT increased significantly when there was low precipitation during fallow (291.8 mm), whereas the yield for NT and CT was not significantly different in high precipitation years. They reported that for reduced tillage (RT), averaged over 7 years, they observed a decrease of 2% of yield compared to CT. Vogeler et al. (2009) reported that winter rye, field beans, winter barley, rape seed and maize yield were not significantly affected by the tillage system. Other studies indicated that yields of winter cereals are also not affected by tillage systems (Rieger 2001; Vullioud 2000; Al-Kaisi and Kwaw-Mensah 2007). Anken et al. (2004) compared five different tillage systems and found that the plots under no tillage had a significantly lower yield in some years compared with those under various tillage systems. The other tillage systems (mouldboard plough, paraplow, chisel plough and shallow tillage) did not affect the yield. Bradford and Peterson (2000) have argued that the major benefits of conservation agriculture can be assessed only after it has been in place for a decade or more. Evidence indicated that improved soil quality positively correlated with yield. Ozpinar and Cay (2006) showed that increased organic carbon and total nitrogen in the soil boosts wheat yield; Fuentes et al. (2009) reported similar results for maize.

Soil compaction caused by agricultural machinery may adversely affect plant growth and crop yield (Miransari et al. 2009). The initial effects are on the root growth, that is, limited nutrient uptake, which eventually influences entire plant growth because of root to shoot signals (Passioura 2002). Some of the signs related to plant growth under compacted-soil field conditions include decreased plant growth and height, pale leaves, cluster growth and pancake-like growth of roots (Miransari et al. 2006). Soil compaction influences soil structure, including soil porosity, by reducing macropores, with micropores being partially affected (Tardieu 1994). This may result in the diminished movement of soil gases such as oxygen that are necessary for plant and microorganism activities, also decreased nutrient uptake and hence plant growth (Nadian et al. 1997). In this situation, the reduction of other electron receivers such as NO_3 results in the release of N_2, thus decreasing the amount of N required for plant growth (Soane and van Ouwerkerk 1995). This is one of the main reasons why N efficiency is reduced in compacted soils and more N needs to be applied (particularly in organic form), which is of environmental and economical significance. In addition, because of the reduced oxygen and slowed movement of gases, CO_2 (the product of cellular respiration) accumulates in the soil and is eventually emitted to the atmosphere as CO_2 or CH_4 (Soane and van Ouwerkerk 1995).

Under dryland farming, crop production depends on precipitation that is largely insufficient and limited to a few months of the year. As aridity increases, fewer and fewer species adapt to the conditions. The benefits of conservation tillage on soil water content are important and therefore crop yield response seems to be positive for conservation tillage. For example, it has been suggested that, agricultural production on the Chinese Loess Plateau could be increased substantially by optimizing the management of soil water and nutrients, perhaps as much as threefold (Fan and Zhang 2000). Approximately 40% (600 Mha) of the world's cropland area is affected by low and unpredictable rainfall, with 60% of these lands being located in developing countries (Johnston et al. 2002). Zero tillage (ZT) combined with crop residue retention on the soil surface can improve moisture infiltration, greatly reduce erosion and enhance water-use efficiency compared to CT (Johnston et al. 2002; Shaver et al. 2002), but lowered soil quality and reduced organic matter under insufficient precipitation combined with decades of intensive agriculture have made this opinion suspect. Reduced or conservation tillage systems may minimize nitrogen loss (Sieling et al. 1998), which is a major yield-limiting nutrient in crops. Therefore, for semi-arid dryland farming where water is the most limiting factor, the use of conventional tillage may seem to be an option for sustainable crop production; yet the symbiotic action of organisms such as mycorrhiza are believed to protect plants from the detrimental effects of drought.

5 Mycorrhiza Fungi

Drought is considered to be the single most important abiotic stress that limits crop production in arid and semi-arid areas (Kramer and Boyer 1997) and arbuscular mycorrhizal (AM) fungal symbiosis is widely believed to protect host plants from the detrimental effects of drought (Augé 2001; Ruiz-Lozano 2003).

The widely beneficial AM fungi are symbiotically associated with the roots of higher plants. An increasing number of experiments have shown that AM alters plant–water relations and prevents drought stress under certain conditions (Augé 2001). The network of mycorrhizal hyphae on the root surfaces is an important inoculum source when roots senesce. Disruption of this network is a suggested mechanism by which CT reduces root colonization and P absorption. In the same way, hyphae and colonized root fragments are transported to the upper soil layer by ploughing, decreasing and diluting their activity as viable propagules for the succeeding crop in rotation. Most semi-arid regions contain highly calcareous soils; also, rainfall and water resources for crop production are restricted. In calcareous soils, a large proportion of the phosphorus in the soil is found as insoluble precipitated calcium–phosphate minerals that are unavailable to plants in the short term (Ström et al. 2005). Therefore the use of a plant symbiont such as mycorrhiza is important in semi-arid dryland farming. Cardoso and Kuyper (2006) extensively reviewed the effects of mycorrhizas on tropical soil fertility and emphasised these effects in that area.

The symbiotic relationship between AM and the roots of higher plants contributes significantly to plant nutrition and growth (Augé 2001), and has been shown to increase the productivity of a variety of agronomic crops including maize (Sylvia et al. 1993). These positive responses in productivity to AM colonization have mainly been attributed to the enhanced uptake by AM of relatively immobile soil ions such as phosphorus (P), potassium (K), calcium (Ca), magnesium (Mg), sulfur (S), iron (Fe), zinc (Zn), copper (Cu) and manganese (Mn) (Marschner and Dell 1994; Marschner 1995; Liu et al. 2000a, b, 2007), but also involve the enhanced uptake and transport of far more mobile nitrogen (N) ions, particularly under drought conditions (Tobar et al. 1994; Liu et al. 2007). The positive effects of AM fungi on host-plant growth and development are not only noticeable in low soil fertility conditions (Jeffries 1987) but also in drought environments (Sylvia et al. 1993; Picone 2003; Liu et al. 2007). Since nutrient mobility is limited under drought conditions, AM may have a larger impact on overall plant growth and development in dry conditions compared to well-watered conditions (Sánchez-Díaz and Honrubia 1994). AM fungi have the ability to affect plant–water relations in both water-limited and well-watered conditions. Augé (2001) and Boomsma and Vyn (2008) have extensively reviewed the effects of AM symbiosis on plant–water relations in numerous host species and corn colonized by various fungal symbionts, with particular emphasis on these effects under drought conditions.

When the soil conditions are suitable, the fungal spores germinate and, by way of signal communications somewhat similar to the signal exchange process between rhizobia and legumes (Miransari and Smith 2007, 2008, 2009), begin their symbiosis with the host plant (Boglárka et al. 2005). AM also enhances soil aggregate stability through its network of hyphae, which may be very beneficial in a compacted soil in which the soil structure is considerably reduced. In addition, AM produces glomalin, a glycoprotein which binds soil particles and hence improves the soil structure and stability (Rillig and Mummey 2006). While soil compaction decreases N and P uptake at the highest level of compaction (Kristoffersen and Riley 2005; Barzegar et al. 2006), AM in a compacted sterilized soil (in which soil pathogens are absent) improves plant growth by enhancing the uptake of N and P. Since the ecological nature of the soil changes under compacted conditions as a result of altered physical, chemical and biological properties, there needs to be some property to compensate for the stress. AM possesses many of these abilities, of which the most important are the abovementioned hyphae network and production of glomalin. AM increases nitrogen uptake even at the highest level of compaction, especially in sterile conditions. This can be very advantageous to the plant under compaction, since otherwise the decreased O_2 levels may result in denitrification and reduction in nitrogen efficiency (Miransari et al. 2009). According to He et al. (2003) and Chalk et al. (2006) AM is also able to provide the plant with nitrogen by adsorbing it from the soil; the biochemical pathways for this action have recently been specified (Govindarajulu et al. 2005). In addition to all these unique abilities, AM is also able to produce enzymes, including phosphatase, that enhance the solubility and availability of immobile nutrients such as phophorus.

The practice of no-tillage has led to an increase in biodiversity, but this practice depends on higher herbicide application which may negate the advantages. The impact of herbicides on arbuscular mycorrhizae have been demonstrated in many studies. The effects of weed control practices on mycorrhizal colonization may be direct, through disturbance of hyphal networks by mechanical cultivation (McGonigle et al. 1990), or indirect, by killing weeds that host AM fungi (Schreiner et al. 2001). Physiological changes caused by herbicides in the potential host plant can create conditions where AM thrives (Nasr 1993). One study that brings this clearly into focus is the mycorrhization of a "non-host" plant, *Chenopodium quinoa* (Schwab et al. 1982), where simazine herbicide applied in sublethal doses led to an increase in root exudation, thought to be responsible for the formation of AM (Ellouze et al. 2008).

Other studies have found that herbicides have little impact on AM (Girvan et al. 2004). Preseeding field application of glyphosate was found to have no effect on AM colonization of soybean. Some herbicides have been found to be detrimental to AM formation. The introduced species *Bromus tectorum* was found to have significantly lower AM root colonization at high rates of tebuthiuron application compared to low rates, or in the control sample (Allen and West 1993); the differences between herbicide rates were not found 2 months later, although at that time AM spores were significantly less abundant in *B. tectorum* rhizosphere. Changjin and Bin (2004) found that all of the six herbicides they tested on maize decreased AM colonization, hyphal enzyme activities and hyphae in the soil, and reduced the biomass of the host plant. The decrease in mycorrhization is not always due to a direct impact on the plant or the AM under study. In a trial with conventional production practices, both sugar beet and maize were found to have reduced AM after herbicide application (Baltruschat 1987) but no reduction in AM was observed when the herbicides were tested on these crops in a greenhouse trial. The reason proposed for the reduction of AM colonization under field conditions was the elimination of weeds that may have been hosts to AM, the only inoculum for the crops being spores remaining in the soil. Weed control measures, including herbicide treatments, also affected mycorrhizal status and AM species composition in grapevines (Baumgartner et al. 2005) and cassava (Sieverding 1991).

The AM connections between target weeds and crops have been demonstrated in some studies. The herbicide bentazon applied to cocklebur has been shown to reduce the AM-colonized root length by 43%, but it had little effect on soybean (Bethlenfalvay et al. 1996). As the susceptible cocklebur succumbed to the bentazon application, an AM-mediated flux of nutrients occurred from weed to crop (Bethlenfalvay et al. 1996). A similar response was found for chlorsulfuron applied to soybean and a weed species (Mujica et al. 1998). Diclofop was found to inhibit AM root colonization in wheat; however, when grown with ryegrass (susceptible to diclofop) growth and yield of the wheat were enhanced (Rejon et al. 1997). This was attributed to interplant AM associations, with the wheat becoming a stronger sink for nutrients than the ryegrass. As found in other studies (e.g. Siquiera et al. 1991), and as illustrated above, herbicide effects on AM are complex.

Abd-Alla et al. (2000) demonstrated that in cowpea the proportion of root length colonized by AM fungi was significantly decreased with soil application of the herbicide brominal, for up to 40 days of plant growth. There are various effects of herbicides on root colonization by mycorrhizal fungi. Malty et al. (2006) found no effect of the herbicide glyphosate on mycorrhizal colonization of soybean. On the other hand, the herbicides brominal and gramoxone significantly inhibited AM root colonization and the number of spores in all legumes studied (Abd-Alla et al. 2000). Other researchers have also found lower colonization in crops due to herbicide application in soil (Paula and Zambolim 1994; Santos et al. 2006). Mycorrhizal and rhizobial performance in soybean were reduced by sulfentrazone, and both possibly contributed to reduced plant growth (Vieira et al. 2007).

As is the case for AM, earthworms may be an essential part of many agroecosystems and may be able to improve soil properties of semi-arid agroecosystems.

6 Earthworms

Earthworms (Ew) can improve plant growth, enhance soil infiltration and decrease runoff which, in turn, are affected by the soil tillage practices adopted. The impact of earthworms on physical, chemical, and biological properties of soil is well established (Edwards and Bohlen 1996; Lee 1985).

Scheu (2003) has provided an extensive review on the effects of Ew in ecosystems. Ew may be an essential part of many agroecosystems and may be useful indicators for sustainability (Lee 1995; Buckerfield et al. 1997). The role of Ew in the soil ecosystem has become of great interest as farmers, researchers, and scientists promote management practices that encourage earthworms. When present, Ew can play a major role in soil fertility and productivity (Edwards 1998; Lee 1985). They are an important component of the soil system, and can enhance plant growth by improving soil fertility and nutrient cycling (Lee 1985). Ew can stimulate microbial activity in the soil during its passage through their gut (Binet et al. 1998). Enhanced N mineralization is the best documented mechanism of Ew and is generally thought to be the most important.

Among the mechanisms by which earthworms modify plant growth at the individual or community levels (Scheu 2003; Brown et al. 2004), five have been suggested as responsible for the positive effects noted on plant production: (i) increased mineralization of soil organic matter, which increases nutrient availability (Curry and Byrne 1992; Lavelle et al. 1992; Subler et al. 1997), especially nitrogen, the major limiting nutrient in terrestrial ecosystems; (ii) the modification of soil porosity and aggregation (Blanchart et al. 1999; Shipitalo and Le Bayon 2004), which improves water and oxygen availability to plants (Doube et al. 1997; Allaire-Leung et al. 2000); (iii) the production of plant growth regulators via the stimulation of microbial activity (Nardi et al. 2002; Quaggiotti et al. 2004); (iv) the biocontrol of pests and parasites (Clapperton et al. 2001; Blouin et al. 2005); and (v) the stimulation of symbionts (Gange 1993; Furlong et al. 2002).

As previously indicated, no-tillage has been shown to increase bulk density, lower total porosity and penetration resistance of soil. Many studies have demonstrated the important role that earthworm burrows play in affecting infiltration and runoff in agricultural soils (Ehlers 1975; Lee 1985; Smettem and Collis-George 1985; Edwards et al. 1979; Friend and Chan 1995; Edwards and Bohlen 1996; Shipitalo and Butt 1999), important in semi-arid conditions.

Anecic earthworms capable of creating deep vertical burrows are particularly effective in influencing infiltration. For example, the activity of *Lumbricus terrestris* was attributed as the cause of 100 mm increase in infiltration observed under no-till compared to conventionally tilled soil (Edwards et al. 1990). The reduced organic matter in semi-arid soils results in weak infiltration so that runoff frequently occurs, and the role of earthworm burrows in infiltration and runoff is especially important (Ehlers 1975; Lee 1985; Smettem and Collis-George 1985; Edwards et al. 1979; Friend and Chan 1995; Edwards and Bohlen 1996; Shipitalo and Butt 1999).

The contribution of earthworm burrows to infiltration depends on both their geometry (length and diameter) and spatial properties (density) (Edwards et al. 1979; Smettem 1992; Wang et al. 1994), both of which tend to vary with earthworm species as well as management practices (Lee 1985). The latter includes tillage, which can affect earthworm abundance and/or diversity (Chan 2001).

Agricultural management practices such as tillage, crop rotation, or manure additions may affect earthworm populations and microbial activity positively or negatively (Li and Kremer 2000; Scullion and Malik 2001; Jordan et al. 2004). The effect of tillage on earthworm abundance is usually negative because of physical damage and adverse environmental conditions caused by the burial of residues

(Chan 2004), reducing their numbers or destroying their burrows (Edwards and Lofty 1982; Lee 1985, Buckerfield et al. 1997; Kladivko et al. 1997). Earlier studies by Jordan et al. (1997) showed that Ew are more abundant in no-tillage than conventionally tilled typical Missouri claypan soil. Chan (2001) reviews studies where conventional tillage practices have reduced the abundance of earthworms by 30–89%. Soil tillage has especially negative effects on anecic Ew, because they suffer from the destruction of their burrows and the incorporation of organic material into the soil (Edwards and Lofty 1982; Haukka 1988; Nuutinen 1992; Edwards and Shipitalo 1998). Therefore, ploughing usually shifts the earthworm community towards endogeic species (Edwards and Lofty 1982; Haukka 1988; Nuutinen 1992; Haynes et al. 1995; Pitkänen and Nuutinen 1998).

Conservation tillage, which leaves crop residues on the soil surface as a food source for soil biota, may encourage earthworm populations. Schmidt et al. (2003) and Chan (2004) demonstrated that both absence of tillage and an increased food supply were necessary for a significant increase in earthworm numbers. In many studies, absence of tillage has been found sufficient to increase the population of *L. terrestris* (Edwards and Lofty 1982; Nuutinen 1992; Edwards and Shipitalo 1998; Pitkänen and Nuutinen 1998; Chan 2001). Significant increases in earthworm numbers have occurred even 2–3 years after turning intensively cultivated field into pasture (Haynes et al. 1995). Schmidt et al. (2003), in turn, demonstrated that absence of tillage is not enough to increase earthworm numbers significantly when a lack of food is limiting their growth. Earlier, Lofs-Holmin (1983) had concluded that a yearly supply of crop residues is needed to promote earthworm activity. Therefore in semi-arid integrated management needs to support earthworm numbers. Use of conservation tillage, along with leaving more plant residues, seem necessary in this area. In an attempt to overcome the problem of low soil organic carbon and soil fertility, natural fallowing may be suitable. Another option may be the introduction of a legume crop. Cereal–legume intercropping systems combine a reduction in tillage (direct drilling) with an increase in food supply (Schmidt et al. 2003). However, the relative importance of these two factors in the regulation of earthworm populations cannot be inferred from a simple comparison between intercropping systems and conventional monocrops. Knowledge of the significance of these factors in intercropping systems could be used to select management practices which would support larger earthworm populations, and thus maximize the agronomic benefits of earthworm activity (Satchell 1958; Edwards 1983). Hendrix et al. (1992) reported that earthworm abundance was related to the quantity and quality of plant residues in different agroecosystems that they studied. Hubbard et al. (1999) also suggested that rotation and tillage affect earthworm population density and biomass. Schmidt et al. (2003) found that an absence of ploughing, alone, had only a modest effect, but combining absence of ploughing with presence of a clover understorey greatly increased earthworm populations. In some semi-arid areas, for example in Iran, crop residues are burnt or used as cattle fodder.

Convservation cropping systems require substantial herbicide input for weed control, and reduced or zero-till methods use herbicides to prepare the seed-bed for direct drilling. Herbicides in general show low toxicity toward worms, although there are some exceptions. However, herbicides have a drastic indirect effect on earthworms through their influence on the availability of organic matter (Edwards and Thompson 1973). The triazine class of herbicides have a moderate impact on earthworm numbers. Herbicides used prior to World War II, including lead arsenate and copper sulfate, are moderately toxic to earthworms. The main threat of toxicity to earthworms is from long-term buildup of these compounds in the soil (Edwards and Bohlen 1996). Earthworms directly influence the persistence of herbicides in soil by metabolizing a parent compound in their gut (Gilman and Vardanis 1974; Stenersen et al. 1974), by transporting herbicides to depth, and by increasing the soil-bound (non-extractable) fraction in soil or by absorbing herbicide residues in their tissue.

As stated, very few herbicides are directly toxic to earthworms (Edwards and Bohlen 1996), athough they may exert considerable indirect effects due to their influence on weeds as a source of supply of organic matter on which earthworms feed. There have been several reports that chlorpropham, propham, dinoseb, and triazine herbicides such as simazine have moderate effects on earthworm population (Edwards and Thompson 1973). Chio and Sanborn (1978) reportesd that *L. terrestris* could metabolize atrazine, chlorambar and dicamba.

Numerous studies support the conclusion that normal use of glyphosate formulations such as the original Roundup® will not adversely affect earthworms. A comprehensive review of the effects of agricultural chemicals on earthworms reviewed the effects of glyphosate on earthworms (Edwards and Bohlen 1996). Glyphosate was ranked as 0 on a scale of zero (relatively non-toxic) to 4 (extremely toxic). Monsanto (manufacturer of Roundup) as well as several independent researchers have conducted studies in which no adverse effects were observed when earthworms were exposed to glyphosate residues in soil at rates equal to or greater than labeled rates (Giesy et al. 2000). In field studies, it has been demonstrated that earthworms thrive under conservation-tillage cropping practices, which are facilitated by Roundup UltraMax and other glyphosate herbicides (Giesy et al. 2000).

The comparative effects of herbicide treatment, crop rotation and weed control practices on soil fauna, microflora and soil microfabric features have been measured in a multifactorial experiment. Pesticides including insecticides, herbicides, fungicides and nematicides are used extentsively on agricultural land in developed countries. It is often assumed that many pesticides are toxic to earthworms or have harmful effects on them. However most herbicides have few direct effects on earthworms, athough the triazine herbicides are slightly toxic.

7 Soil Nutrient

The efficiency of no tillage (NT) in improving soil organic matter (SOM) is universally recognized. SOM is a source of nutrient for plants. Also SOM is a source of nutrient and energy for the decomposer community. Increased losses of soil organic C have been documented where conventional tillage was employed (Rhoton 2000).

SOM retained in conservation-tillage systems may be due to reduced oxygen availability below the surface, which affects decomposition rates (Wershaw 1993).

Intensive soil cultivation worldwide has resulted in the degradation of agricultural soils, with decrease in soil organic matter and loss of soil structure, adversely affecting soil functioning and creating a long-term threat to future yields (Pagliai et al. 2004; D'Haene et al. 2008). In addition to its important effect on soil quality and crop productivity, soil organic carbon has also been identified as a possible carbon sink for sequestering atmospheric carbon dioxide (Jenkinson et al. 1991; Cox et al. 2000; Jones et al. 2005; Farage et al. 2007). Under cropping, it is generally agreed that the practice of both NT/reduced tillage (RT) and stubble retention (SR), or conservation tillage, favours soil carbon sequestration when compared to that of conventional tillage, and that the rate of soil carbon sequestration has been reported to be higher for these systems than for conventional tillage. From the limited information available, the scope for soil carbon sequestration under conservation tillage is more limited in Australia than elsewhere but the reasons are not clear (Chan et al. 2003). Because of low precipitation, soil in dryland farming areas contains small amounts of organic matter. Therefore, tillage method is important. Studies of soil organic carbon under conservation tillage and conventional tillage in 18 field trials across Australia generally found lower amounts of soil carbon than reported for USA soils (Kern and Johnson 1993). Differences in total organic carbon under different tillage/stubble/rotation treatments, and the resulting differences in soil properties, have also been reported (Chan et al. 1992, 2002; Heenan et al. 2004). Because microorganisms depend on organic matter in the soil, the lower organic matter in semi-arid regions indicates that microbial activities are very low, and agricultural practices assume more importance.

Enhancement and maintenance of soil organic matter and diminishing nutrient runoff are essential for sustainable agriculture in semi-arid regions. Degradation of soil aggregates through cultivation appears to be a primary mechanism in the loss of organic matter (Jastrow 1996; Six et al. 1999). Soil aggregates under CT are prone to the disruptive forces of drying and wetting or raindrop impact owing to the repeated mechanical disruption of macroaggregates (Tisdal and Oades 1982). NT also reduces erosion, maintains soil aggregation and greater soil organic matter levels compared with CT (Zotarelli et al. 2005; Li et al. 2007). Jacobs et al. (2009), in an extensive experiment on the effects of 40 years of MT treatment on carbon and nitrogen storage, indicated that MT led to an increased storage of organic matter (OM) measured as higher concentrations of organic carbon (C_{org}, total C in the soil) and N. they also showed that water-stable macroaggregates were more abundant in MT soils, possibly protecting OM from degradation. However, no increased storage of particulate organic matter (POM), as had been suggested for NT soils, was detected in the surface soils of MT. In several studies, MT compared to NT systems among various soil types and climatic regions accumulated increased OM and N microbes to a similar extent (Meyer et al. 1996; Salinas-Garcia et al. 1997; Ahl et al. 1998; Kandeler et al. 1999a, b; Stockfisch et al. 1999; Wright et al. 2005).

In an attempt to overcome the problem of low soil organic carbon and soil fertility, natural fallowing is a suitable approach. However, long fallow periods have become unsustainable due to population pressure (Nyamadzawo et al. 2009). It is estimated that approximately 10–20 years of natural grass fallow would be needed in order to restore soil fertility and structural stability after 1 or 2 years of cropping (Jurion and Henry 1969). Therefore, short-term improved fallowing systems, which are based on fast-growing and short-duration leguminous plants, are a possible option for soil fertility improvement and soil organic carbon (SOC) build-up. Research has demonstrated that improved legume tree fallows give significant increases in maize yields through enhanced N inputs (e.g. Kwesiga et al. 1999; Mafongoya and Dzowela 1999; Giller 2001; Mapfumo et al. 2005). Improved fallowing also adds SOC and improves soil structure (Lal 1989; Kiepe 1995; Rao and Dao 1996; Nyamadzawo et al. 2007). Also tillage practices can alter the benefits of fallow. Under fallow, NT had greater aggregate stability when compared to CT (Beare et al. 1994a; Nyamadzawo et al. 2009). In this situation, NT practices can result in greater aggregation and greater SOM levels than CT practices (Beare et al. 1994a; Nyamadzawo et al. 2009). In semi-arid areas it seems that fast-growing leguminous plants and short-term crop rotation are preferable. Karlen et al. (1994) and Fischer et al. (2002) reported that crop rotation with legumes breaks the soil pathogen cycles and restores fertility. The positive effects of crop rotation over cereal monocultures are also well documented (Halvorson et al. 2000; Galantini et al. 2000).

The clay fraction is an important soil constituent, influencing microbial activity, yields of biomass and production of microbial metabolites (Christensen and Sorensen 1985). Organic matter bound to silt particles is also important: a significant proportion of the soil organic matter can be found in the silt-sized fraction (Christensen and Sorensen 1985). Measurement of the more active fractions of soil organic matter, such as soil microbial biomass carbon and mineralizable carbon, may provide a more sensitive appraisal and indication of the effects of tillage and residue management practices on SOC contents (Franzluebbers and Arshad 1996). Microbial biomass provides an indicator of SOC degradation since it has a turnover time of less than a year and responds rapidly to changes in conditions and management practices that alter SOC levels (Nyamadzawo et al. 2009). Also minimum tillage (MT) increases microbial C and microbial N (Kandeler et al. 1999a, b; Stockfisch et al. 1999) and potential C and N mineralization (Salinas-Garcia et al. 1997; Wright et al. 2005). Kushwaha et al. (2001) studied aggregate stability in an MT system of an inceptisol of a sandy loam texture revealing a trend that tillage reduction increased macroaggregate proportion. Intensive research has shown that MT increases microbial biomass in the surface soil (Meyer et al. 1996; Salinas-Garcia et al. 1997; Ahl et al. 1998; Kandeler et al. 1999a, b; Stockfisch et al. 1999; Kushwaha et al. 2001; Spedding et al. 2004; Wright et al. 2005). Increased substrate availability to microorganisms in MT is provided through the accumulation of crop residues in the surface soil (Kandeler et al. 1999a, b; Stockfisch et al. 1999; Kushwaha et al. 2001). An increased substrate availability can be demonstrated by a high C (microbial):C (organic) ratio (Stockfisch et al. 1999).

Studies have also shown that MT enhances the stability of soil macroaggregates (Jacobs et al. 2009). Kushwaha et al. (2001) found that MT increased the proportion of macroaggregates after 1 year of diverging treatment of a tropical sandy loamy inceptisol. Jacobs et al. (2009) stated that the higher macroaggregate stability under MT systems can probably be attributed to: (1) a lower physical impact of tillage machinery on surface soils and subsoils, or (2) a higher availability of binding agents caused by higher microbial biomass which enhanced formation and cementation of aggregates mainly in the surface soil (Kandeler and Murer 1993; Kushwaha et al. 2001; Hernández-Hernández and López-Hernández 2002; Oorts et al. 2007). Soil under NT contained more macroaggregate-protected SOC than under CT, indicating that reduced soil disturbance could lead to greater SOC sequestration and improved soil quality with improvements in macroaggregation (Beare et al. 1994b; Franzluebbers and Arshad 1997; Bossuyt et al. 2002). Tisdal and Oades (1982) found greater concentrations of organic carbon in macroaggregates and suggested that the presence of decomposing roots and hyphae within macroaggregates not only increased carbon concentrations but also contributed to their stabilization.

Many research studies have shown that converting to conservation tillage changes both the physical and chemical properties of the soil (Dick 1983; Cereti and Rossini 1995). Hydraulic soil properties such as infiltration rate, bulk density, porosity, pore connectivity and water retention capacity, as well as chemical properties such as organic matter content and nutrient status, are affected (Sauer et al. 1990; Benjamin 1993; Kribaa et al. 2001). Intensive agriculture with high fertilizer input exceeding crop requirements has caused nutrient accumulation in soils. Besides nitrates, phosphorus has recently gained attention. Increased concentrations of total phosphorus in soils is known to cause increased concentrations of soluble and sediment-associated phosphorus in surface runoff (Romkens and Nelson 1974). Research has indicated that 50–80% of the phosphorus applied as fertilizer is adsorbed by the soil, but to date, the amounts of P required to attain and maintain an adequate P status in the soil are unknown. In addition to losses through surface runoff, P leaching has lately gained increased attention as an important P transport pathway, thereby contributing to eutrophication of freshwater systems. In Germany, for example, annual P losses from agriculture ranging from less than 50 to more than 200 kg/km^2 have been found (Buczko and Kuchenbuch 2007). In semi-arid areas, the soil is more prone to wind erosion and water erosion. Precipitation is restricted to several months in this regions. Rain is often irregular, and heavy falls make the soil prone to erosion, triggering preferential flows of nutrients. Soil organic carbon is important for maintaining and improving soil structure. Soil organic matter plays a crucial part in all aspects of soil quality (soil structure, soil–water relations, chemical fertility, biodiversity) and therefore is a key indicator for the integrated evaluation of soil quality (Tiessen et al. 1994; Carter 2002). Reduced tillage has been shown to decrease soil erosion (Scholz et al. 2008) and P loss associated with sediment (Baker 1985). Conservation tillage can reduce surface runoff by increasing macroporosity, but it can also trigger preferential flow of nutrients below the root zone if heavy rain falls after surface application of fertilizers. Appropriate management of fertilizer input, such as split fertilization, can minimise that risk. Grandy and Robertson (2006) observed that years of soil regeneration can be lost after a single conventional tillage event, hence NT is an option for reducing the adverse effects of CT.

8 Chemical Soil Quality

The no-tillage system based on residue retention affects the pH. Acid soils result from leaching or removal of bases from the soil with hydrogen ions becoming predominant in the colloids. The pH affects the availability of nutrient elements by affecting chemical reactions to form insoluble compounds.

The pH is alkaline in semi-arid regions. P and most micronutrients are absorbed by the plant in close to acidic conditions. Soil acidification caused by mineral fertilizers, especially ammoniac and ureic fertilizers, has a marked effect on the pH, due to the absorption of the ammonia ion by plants, or by nitrification. These processes produce hydrogen ions (Havlin et al. 1999; Johnston 1997). Fuentes et al. (2009) clearly showed that for zero-tillage (ZT) without residue retention (−r), resulted in a soil pH of 5.3 compared to its initial pH of 6.5 (Etchevers et al. 2000). This strong acidification can reduce the availability of some nutrients including Ca, K, N, Mg, Mo, P and S (Porta et al. 1999), and also indicated that, by contrast, conventional tillage with residue retention (CT + r), CT − r and ZT + r treatments showed a pH ranging from 6 to 6.5 (Fuentes et al. 2009), which is optimal for nutrient availability (Havlin et al. 1999). Fuentes et al. (2009) stated that the acidification of the soil with ZT − r was due to the addition of nitrogen fertilizers which remain in the top 5 cm of the soil profile because of the lack of moisture and increased compaction, hindering their mobility and availability to the crop (Bloom 2000).

Electrolytic conductivity (EC) has been identified as an important soil quality. Semi-arid regions have a higher EC that can change under tillage practices. Because of the pathways of conductance, EC is influenced by a complex interaction of soil properties including salinity, saturation percentage, water content, bulk density and chemical properties (SOM, cation exchange capacity) (Corwin 2003; Corwin and Lesch 2005). Fuentes et al. 2009 reported that ZT − r caused highest EC values in the soils (5–20 cm) with wheat.

9 Soil Enzyme Activities

Soil quality has been defined as 'the capacity of a soil to function within ecosystem boundaries to sustain biological productivity, maintain environmental quality, and promote plant and animal health' (Doran and Parkin 1994), Soil quality is the result of the interaction between chemical, physical and biological soil properties (Karlen et al. 2001), which therefore should all be taken into account when evaluating soil quality. However, soil organic matter changes only very slowly and therefore shorter-term changes in soil quality are often assessed using specific fractions of soil organic matter, such as particulate organic matter or biologically active components of soil organic matter, including microbial biomass and enzyme activities (Jin et al. 2009). It has been shown that microbial community composition can be altered by a change in management practice (Visser and Parkinson 1992; Schutter and Dick 2002; Liebig et al. 2006; Yao et al. 2006; Elfstrand et al. 2007), substrate availability and composition (Wardle et al. 1993; Grigera et al. 2006; Ha et al. 2008), soil type (Schutter et al. 2001), soil moisture (Chen et al. 2007; Stromberger et al. 2007), soil chemistry (Pankhurst et al. 2001), and seasonal variation in soil nutrient status (Bardgett et al. 1999; Schutter et al. 2001).

Enzyme activity varies seasonally and depend on the chemical, physical and biological characteristics of the soil (Niemi et al. 2005). Enzyme activity has been indicated as a soil property suitable for use in evaluating of the degree of alteration of soils in both natural and agroecosystems (Trasar-Cepeda et al. 2000). Some research has indicated the positive effects of conservation tillage practices on soil enzyme activity in semi-arid Mediterranean and temperate climates (Kandeler et al. 1999a, b; Riffaldi et al. 2002), and in subtropical soils (Roldán et al. 2005a, b). Microbial activity may also be affected by long-term management practices in an agroecosystem. In an earlier study by Jordan and Kremer (1994), enzyme activity as well as soil microbial biomass carbon were greatest in continuous long-term no-tillage and grass plots. Weil et al. (1993) found similar results for microbial and total carbon in undisturbed long-term grass plots.

Plant characteristics can also affect soil microbial communities by changing the soil environment (Carney and Matson 2006; Chen et al. 2007), which may be especially important for growers utilizing transgenic crops (Griffiths et al. 2007; Widmer 2007). Liu et al. (2005) has provided an extensive review on the effects of transgenic plants on soil microorganisms. They conclude that transgenic plants cause minor changes in microbal community structures that are often transient in duration. During plant litter decomposition, most transgene protein(s) appear to be rapidly degraded, but some proteins can bind to surface-active particles and reduce their availability to microbes (Liu et al. 2005).

Fewer reports have appeared in the literature concerning the temporal effects on soil biological activity under NT, compared with conventional tillage under dryland farming. They have observed that soil microbial biomass C (SMB), as a measure of the total microbial tissue of vegetative bacteria and fungi, has also often been found to respond similarly to tillage method (Doran 1987; Staley et al. 1988; Carter 1991; McCarty et al. 1995). Lynch and Panting (1980, 1982) showed that SMB in the surface layer (0–15 cm) under both direct-drilled (NT) and ploughed (CT) wheat (*Triticum aestivum* L.) was almost constant from autumn to spring, increased to a maximum during the summer, then declined to about the autumn concentration. SMB was not significantly different between tillage methods until harvest in late August, when it was 33–77% greater under NT than CT. Granatstein et al. (1987), using a wheat–barley (*Hordeum vulgare* L.)–pea (*Pisum sativum* L.) rotation site, reported little change in SMB in 5 cm depth increments to 30 cm under either tillage method from April to September, then a large increase in the 0–5 cm layer in October, but only for NT. In wheat–legume rotations, Van Gestel et al. (1992) found near-linear decreases in SMB from mid-winter to autumn in the 0–2.5 cm layer for both CT and NT. The only study that has examined tillage method effects on SMB (actually, biomass N) specifically and frequently over the growing season is that of Carter and Rennie (1984) who studied four sites on chernozemic soils planted with spring wheat. For the 0–5 cm layer, increases in biomass N during the early growing season were greater under zero tillage (NT) than shallow tillage (CT), and then declined to about the same level by the end of the growing season.

This finding is contradicted by later observations that show a strong dependence of enzyme activity on soil organic matter (SOC) (Degens et al. 2000; Kandeler et al. 2001; Jin et al. 2009). However, Gianfreda et al. (2005) investigated the activities of a range of enzymes using cultivated and non-cultivated soils from various parts of Europe and found that a high SOC does not necessarily reflect corresponding increases of enzymatic activities. Evidence indicated that returning straw into the field was also helpful to enhance enzyme production (Sun et al. 2003; Gianfreda et al. 2005).

When herbicides are applied, most of the spray solution contacts the soil and may affect soil microorganisms that are important for sustainable agriculture, such as recycling of plant nutrients and maintenance of soil structure (Vieira et al. 2007). The toxicological effects of various herbicides on legumes have been reported (Zaidi et al. 2005). However, the magnitude of these toxic effects depends on type, dose and duration of exposure as well as the species and age of the exposed plants (Abd-Alla et al. 2000). The use of herbicides may influence nodulation and biological nitrogen fixation in legumes, either by affecting rhizobia, the plant itself, or both. The herbicides trifluralin, pendimethalin (Bollich et al. 1988) and metribuzin (Mallik and Tesfai 1985) have been considered deleterious to the soybean–*Bradyrhizobium* spp. symbiosis. Recently, Arruda et al. (2001) demonstrated that nodule number and nodule dry weight in soybean decreases as a function of increased rates of the herbicide sulfentrazone (Vieira et al. 2007).

On the other hand, Gonzalez et al. (1999) have found no effects from the herbicides metribuzin, acetochlor, metolachlor, trifluralin, imazaquin, imazethapyr or chlorimuron ethyl on soybean nodulation and consequently on yield, in soil with high organic matter content. Similarly, Malty et al. (2006) observed no effect of the herbicide glyphosate, a inhibitor of 5-enolpyruvylshikimate-3-phosphate (EPSP) synthase on soybean nodulation. The presence of the herbicide may also affect the production of cell wall-degrading enzymes by AM hyphae which are essential for root infection by mycorrhizal fungi to occur (Abd-Alla et al. 2000).

10 Soil Biota

Changes in tillage and residue induce shifts in the number and composition of soil fauna, including pests and beneficial organisms. Soil organisms respond to tillage-induced changes in the soil physical/chemical environment, but they, in turn, have an impact on soil physical/chemical conditions, such as soil structure, nutrient cycling and organic matter decomposition (Govaerts et al. 2007, 2008).

Johansson et al. (2004) have provided an extensive review on the soil fauna–microbe interaction. They explained that mixing of soil and excessive soil compaction may have a negative effect on AM colonisation of plant roots, due to disruption of the extraradical mycelium, and can lead to reduced soil microbial biomass and enzyme activity. Hassall et al. (2006) explored the potential for applying broad ecological theories to interactions between soil animals and microorganisms to generate a predictive framework within which more hypothesis-led research can be undertaken. They focus on how the application of general ecological theory would increase our understanding of the consequences of these interactions. They do this with reference to some examples but mostly by posing a series of questions stemming from studies in other systems.

Tillage increases decomposition of crop residues and changes the structure of the soil food web by relocating food resources and exposing protected carbon (Hendrix et al. 1986; Moore and de Ruiter 1991; Beare et al. 1992; Wardle 1995; Six et al. 2002). Conservation tillage is used to conserve soil nutrients and structure, increase sequestration of soil carbon, and to provide habitat and substrate for biota (Hendrix et al. 1998; Lal et al. 1998; Paustian et al. 2000; Holland 2004). However, a deterrent for growers considering the transition to conservation tillage is the delay in soil response (e.g. increased soil carbon, efficient nutrient cycling, impacts on yield) associated with the equilibration of the soil food web (Phatak et al. 1999; Simmons and Coleman 2008). The activity of soil biota responsible for the mineralization of nutrients from the soil is an important component of soil function (Ingham et al. 1985; Hunt et al. 1987; Moore 1988; Beare et al. 1992). This is especially important in low input sustainable agriculture, where increased microbial diversity is expected to increase soil quality (Parr et al. 1992; Visser and Parkinson 1992). Bacteria dominate in systems where crop residue is buried or where labile substrate is abundant due to their ability to break down labile carbon sources more efficiently than saprophytic fungi (Coleman et al. 1983; Curl and Truelove 1986; Moore et al. 2003). In these systems, the rates of decomposition and nitrogen mineralization are accelerated (Moore and de Ruiter 1991; Doles et al. 2001). Microbial diversity is expected to increase with a reduction in tillage, as fungal species begin to dominate the system (Beare et al. 1992; Frey et al. 2003). In systems where crop residue is left on the surface, saprophytic fungi dominate, slowly breaking down more resistant substrates (Hendrix et al. 1986; Moore et al. 2003). The ability of an ecosystem to withstand disturbance may lie in the energy pathway, where bacteria-dominated systems are more resilient than fungus-dominated systems (Allen-Morley and Coleman 1989; Moore and de Ruiter 1991; Bardgett and Cook 1998). Moore et al. (2003) postulated that recovery times of each energy channel to disturbance may be different, resulting in an alteration of the food web. Fungi, especially arbuscular mycorrhizal (AM) fungi, may be particularly sensitive to tillage (Drijber et al. 2000). Acosta-Martinez et al. (2007) reported an increase in fungi with a combination of no-till and increased cropping intensity, potentially due to greater residue quantity. In a study involving the transition from conventional to alternative agriculture, Doran (1987) found that microbial populations and activities were regulated more by crop type and rotation than by soil physical properties. Gonzalez et al. (2003) observed an increase in humification in soils in no-till as compared to those in reduced tillage, indicating that microbial populations were probably influenced by increases in organic matter. In contrast, Spedding et al. (2004) showed no effect of tillage on the microbial community but did find a significant effect of seasonality on fungi.

Soil exposure to herbicides has a considerable side effect. Rhizobial responses to herbicides, as well as C sources utilization, discriminate the metabolically diverse isolates at the level of strains (Zabaloy and Anahí Gómez 2005). These authors also reported a remarkable rhizobial diversity with a physiological potential to use natural and xenobiotic C sources. The interaction between soil microorganisms and herbicides may influence soil quality and fertility, because these biologically active chemicals may have deleterious effects on beneficial species; otherwise, microbes utilize and degrade these compounds (Alexander 1980; Dinelli et al. 1998). During their free-living heterotrophic phase, rhizobia can degrade pesticides such as atrazine (Bouquard et al. 1997) and 2,4-dichlorophenoxyacetic acid (2,4-D) (Kamagata et al. 1997). It has been shown that catabolic pathways exist for protocatechuate in *Rhizobium* and *Bradyrhizobium*, and for catechol in *Rhizobium leguminosarum* (Sadowsky and Graham 1998). Diuron herbicides belong to an important group of pesticides which are used for pre- and post-emergence weed control in cotton, fruit and cereal crops worldwide (Tomlin 2003). Cyco and Piotrowska-Seget (2009) reported that diuron herbicide show little effect on the biodiversity and community structure of indigenous soil bacteria. They reported that the highest dosages of diuron generally increased the bacterial populations, which were able to either survive in the new conditions and/or used the pesticides as a carbon and energy source. Kara et al. (2004) showed that activity of denitrifying bacteria was stimulated by the addition of the herbicide Topogard, whereas the total number of bacteria was not influenced. They concluded that the effect of Topogard on the microbiological characteristics of coarse-textured soils is likely to be dependent on soil pH. Singh and Wright (1999, 2002) reported that terbutryn, terbuthylazine, trietazine, simazine, prometryn and bentazone, all triazine derivatives, negatively affected the growth of rhizobia. Munch et al. (1989) reported that the activity of NO_2^--oxidizing bacteria was inhibited by terbuthylazine, whereas the activity of $NH4^+$ on oxidizing and denitrifying bacteria was stimulated by the herbicide. Ferrero and Maggiore (1994) reported that alachlor + terbuthylazine application given to cattle slurry in sandy soil under maize cultivation reduced $NO3^-$ leaching, probably due to the inhibition of nitrification.

Vieira et al. (2007) tested sulfentrazone herbicide application on soil microbial biomass C and symbiotic processes associated with soybean. Only in the initial period of soybean development the microbial biomass C was lower in the presence of the herbicide. Mycorrhizal and rhizobial performance in soybean were reduced by sulfentrazone, and both possibly contributed to reduced plant growth.

Fungi are capable of degrading xenobiotics via the Fenton oxidation mechanism, using both nonenzymatic (*Gloeophyllum striatum* and *G. sepiarium*) and enzymatic (*Coniophora puteana*) mechanisms (Hyde and Wood 1997). However Vasil'chenko et al. (2002) discovered the ability of the soil fungus INBI 2-26(–), which do not produce laccase, to degrade atrazine during surface cultivation in liquid medium. Xu et al. (2006) found that the *Pseudomonas oleovorans* bacteria degraded the chloroacetamide herbicide acetochlor through pathways involving dechlorination, hydroxylation, N-dealkylation, C-dealkylation and dehydrogenation.

11 Management Strategies

The sustainability of tillage systems has received considerable attention in recent decades, with the development and increasing adoption of conservation-oriented tillage practices considered more sustainable than the soil inversion type of conventional mouldboard plough-based tillage (Rasmussen 1999; Arshad et al. 1999; Munkholm 2001; Zhang et al. 2007). Reduced tillage, a form of conservation-oriented tillage introduced as a method of reducing soil degradation and conserving soil moisture (Amezketa 1999), has increasingly been introduced in agriculture systems all over the world due to its economical and environmental effects (Birkas et al. 1989; Fowler and Rockstrom 2001; La Scala et al. 2006). Soil management involving reduced tillage has often been reported to improve soil structure (e.g. Oyedele et al. 1999). Soil aggregates were found to be more stable under reduced tillage compared with conventional mouldboard tillage (Schjonning and Rasmussen 1989; Pagliai et al. 2004). However, the success of reduced tillage depends on the local soil type and climatic conditions (Rasmussen 1999), and the effect of tillage on soil structural stability is still controversial (Amezketa 1999). Conventional tillage practices, where crop residues are incorporated into the soil by ploughing or disking, are used to aerate soils, reduce compaction, and control herbaceous pests, thereby increasing seedling germination and yield (Dickey et al. 1983; Kettler et al. 2000; Raper et al. 2000). However in many cases tillage results in soil erosion, loss of organic matter, decreased water infiltration, loss of soil structure, decreased soil fertility and reduction in overall soil quality due to the destruction of soil aggregates and structure (Parr et al. 1992; Paustian et al. 1997; Allmaras et al. 2000; Nyakatawa et al. 2000). Conservation tillage and stubble retention practices are being introduced as options to fight erosion and improve yields (and income). To benefit from crop production in dryland farming, emphasis has to be on conventional tillage practices that promote the occurrence and functioning of soil, including crop production, water use efficiency, mycorrhiza fungi, earthworms, soil nutrient, chemical soil quality and soil enzyme activities (Fig. 2).

12 Conclusion

No tillage or minimum tillage is termed conservation tillage, as it mitigates soil and nutrient losses from farmland and saves the energy taken by tillage. It is widely accepted in the USA and other developed countries, but not fully accepted by farmers in west or South-east Asian countries. Some short-term research has found that crop yield of non-tilled area is lower than that of tilled area applying chemical fertilizer. However, farming with conservation tillage may not affect the income of farmers, especially for long-term application, because of the reduced cost of tillage and chemical fertilizer. Cultural practices that increase the activity of indigenous AM fungi are: reduced tillage, crop rotations, cover crops, and phosphorus management (Douds and Reider 2003). Reduced tillage, especially no-till, leaves

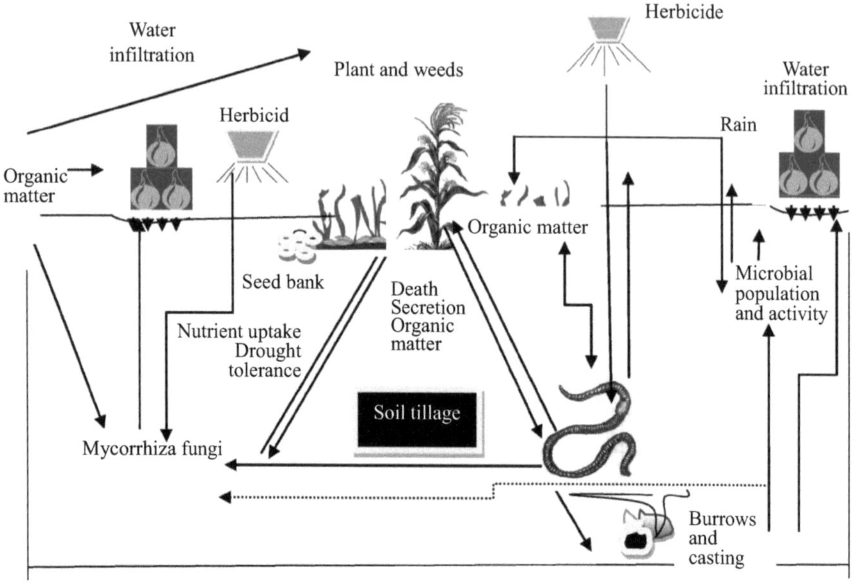

Fig. 2 Simplified conceptual model connecting the soil tillage effect on soil with its potential effects on plant, earthworm and mycorrhiza. The earthworm burrowing and casting promote soil mixing and increase infiltration. Earthworm gut and burrowing and casting increase microorganism. Earthworm extremely affect by tillage method, soil organic matter and herbicide. Earthworm can improve plant growth by improving soil fertility and nutrient cycling (Lee 1985). The effect of tillage on earthworm abundance is usually negative (Chan 2004). Herbicides in general show low toxicity toward worms, although there are some exceptions. The AMF mycelium is important in nutrient transfer in soil, but this process is affected by the activities of earthworms (Tuffen et al. 2002). Earthworms may graze preferentially on soil containing VA mycorrhizal fungal propagules and as a result, concentrate them in the casts (Gange 1993). AM colonization attribute to the enhanced uptake of relatively immobile soil ions and prevents drought stress. AM produces a glyco protein, glomalin, which is able to bind soil particles and hence, improve soil structure and stability (Rillig and Mummey 2006). Impacts of herbicides on arbuscular mycorrhizae have been shown in many studies. When herbicides are applied, most of the spray solution contacts the soil and may affect soil microorganisms that are important for recycling of plant nutrients and maintenance of soil structure. Free-living heterotrophic phase, rhizobia can degrade pesticides such as atrazine (Bouquard et al. 1997). Fungi are capable of degrading xenobiotics. Tillage practices influence vertical distribution of weed seeds in soil in addition to the rate of seedbank decline (Ball 1992; Barberi and Cascio 2000; Buhler et al. 1996, 1997; Cardina et al. 1991). Soil organic matter and diminish of nutrient runoff. SOM is retained in conservation-tillage systems. Organic matter improved soil properties such as aggregation, water-holding capacity, hydraulic conductivity, bulk density, the degree of compaction, fertility and resistance to water and wind erosion. Added organic matter increased AM hyphal length densities (Joner and Jakobsen 1995)

the extraradical mycelial network in the soil intact. This promotes rapid colonization of a new crop and enhances early season mycorrhiza-mediated P uptake (McGonigle and Miller 1993). Also mycorrhiza improves the water relationship of plants and protects them from drought stress characteristic of semi-arid dryland farming. Conventional tillage can increase soil biotas as earthworms that are important for soil water retention. Also microbial activity in these areas is affected by the kind of tillage practice adopted. Conventional tillage, but leaving more plant residue and decreasing soil erosion, enhances soil organic matter in the long term and can improve soil biological activity.

References

Abd-Alla MH, Omar SA, Karanxha S (2000) The impact of pesticides on arbuscular mycorrhizal and nitrogen-fixing symbioses in legumes. Appl Soil Ecol 14:191–200

Aboudrare A, Debaeke P, Bouaziz A, Chekli H (2006) Effects of soil tillage and fallow management on soil water storage and sunflower production in a semi-arid Mediterranean climate. Agric Water Manage 83:183–196

Acosta-Martinez V, Mikha MM, Vigil MF (2007) Microbial communities and enzyme activities in soils under alternative crop rotations compared to wheat-fallow for the Central Great Plains. Appl Soil Ecol 37:41–52

Ahl C, Joergensen RG, Kandeler E, Meyer B, Woehler V (1998) Microbial biomass and activity in silt and sand loams after long-term shallow tillage in central Germany. Soil Till Res 49:93–104

Alexander M (1980) Introducción a la microbiología del suelo. AGT Editor, México

Al-Kaisi M, Kwaw-Mensah D (2007) Effect of tillage and nitrogen rate on corn yield and nitrogen and phosphorus uptake in a corn-soybean rotation. Agron J 99:1548–1558

Allaire-Leung SE, Gupta SC, Moncrief JF (2000) Water and solute movement in soil as influenced by macropore characteristics. 1. Macropore continuity. J Contam Hydrol 41:283–301

Allen-Morley CR, Coleman DC (1989) Resilience of soil biota in various food webs to freezing perturbations. Ecology 70:1127–1141

Allen EB, West NE (1993) Nontarget effects of the herbicide tebuthiuron on mycorrhizal fungi in sagebrush semidesert. Mycorrhiza 3:75–78

Allmaras RR, Schomberg HH, Douglas CL, Dao TH (2000) Soil organic carbon sequestration potential of adopting conservation tillage in US croplands. J Soil Water Conserv 55:365–373

Amezketa E (1999) Soil aggregate stability: a review. J Sustain Agric 14:83–151

Anken T, Weisskopf P, Zihlmann U, Forrer H, Jansa J, Perhacova K (2004) Longterm tillage systems under moist cool conditions in Switzerland. Soil Till Res 78:171–183

Arruda JS, Lopes NF, Bacarin MA (2001) Nodulação e fixação do dinitrogênio em soja tratada com sulfentrazone. Pesqui Agropec Bras 36:325–330

Arshad MA, Franzluebbers AJ, Azooz RH (1999) Components of surface soil structure under conventional and no-tillage in northwestern Canada. Soil Till Res 53:41–47

Augé RM (2001) Water relations, drought and vesicular-arbuscular mycorrhizal symbiosis. Mycorrhiza 11:3–42

Bachthaler G (1974) The development of the weed flora after several years' direct drilling in cereal rotations on different soils. In: Proceedings the 12th British weed control conference, 18–21 Nov 1974, Brighton, England, pp 1063–1071

Baker DB (1985) Regional water quality impacts of intensive row-crop agriculture: a Lake Erie Basin case study. J Water Conserv 40:125–132

Ball DA (1992) Weed seedbank response to tillage, herbicides, and crop rotation sequence. Weed Sci 40:654–659

Baltruschat H (1987) Evaluation of the suitability of expanded clay as a carrier material for VA mycorrhiza spores in field inoculation of maize. Ang Bot 61:163–169

Barberi P, Cascio BL (2000) Long-term tillage and crop rotation effects on weed seedbank size and composition. Weed Res 41:325–340

Bardgett RD, Cook R (1998) Functional aspects of soil animal diversity in agricultural grasslands. Appl Soil Ecol 10:263–276

Bardgett RD, Lovell RD, Hobbs PJ, Jarvis SC (1999) Seasonal changes in soil microbial communities along a fertility gradient of temperate grasslands. Soil Biol Biochem 31:1021–1030

Barzegar AR, Nadian H, Heidari F, Herbert SJ, Hashemi AM (2006) Interaction of soil compaction, phosphorus and zinc on clover growth and accumulation of phosphorus. Soil Till Res 87:155–162

Baumgartner K, Smith RF, Bettiga L (2005) Weed control and cover crop management affect mycorrhizal colonization of grapevine roots and arbuscular mycorrhizal fungal spore populations in a California vineyard. Mycorrhiza 15:111–119

Beare MH, Parmalee RW, Hendrix PF, Cheng W, Coleman DC, Crossley DA Jr (1992) Microbial and faunal interactions and effects on litter nitrogen and decomposition in agroecosystems. Ecol Monogr 62:569–591

Beare MH, Cabrera ML, Handrix PF, Coleman DC (1994a) Aggregate protected and unprotected organic matter pools in conventional and no-tillage soils. Soil Sci Soc Am J 58:787–795

Beare MH, Cabrera ML, Handrix PF, Coleman DC (1994b) Water stable aggregates and organic matter fractions in conventional and no-tillage soils. Soil Sci Soc Am J 58:777–786

Belde M, Mattheis A, Sprenger B, Albrecht H (2000) Langfristige Entwicklung ertragsrelevanter Ackerwildpflanzen nach Umstellung von konventionellen auf integrierten und ökologischen Landbau. Z PflKrankh PflSchutz Sonderheft XVII:291–301

Benbrook C (2001) Troubled times amid commercial success for Roundup Ready soybeans; glyphosate efficacy is slipping and unstable transgene expression erodes plant defences and yields. AgBioTech InfoNet Technical Paper 4. http://www.biotech-info.net/troubledtimes.html . Accessed 11 Jan 2002

Benjamin JG (1993) Tillage effects on near-surface soil hydraulic properties. Soil Till Res 26:277–288

Bescansa P, Imaz MJ, Virto I, Enrique A, Hoogmoed WB (2006) Soil water retention as affected by tillage and residue management in semiarid. Spain Soil Till Res 87:19–27

Bethlenfalvay GJ, Mihara KL, Schreiner RP, McDaniel H (1996) Mycorrhizae, biocides, and iocontrol. 1. Herbicide-mycorrhiza interactions in soybean and cocklebur treated with bentazon. Appl Soil Ecol 3:197–204

Binet F, Fayolle L, Pussard M (1998) Significance of earthworms in stimulating soil microbial activity. Biol Fertil Soils 27:79–84

Birkas M, Antal J, Dorogi I (1989) Conventional and reduced tillage in Hungary – a review. Soil Till Res 13:233–252

Blaise D, Ravindran CD (April 2003) Influence of tillage and residue management on growth and yield of cotton grown on a vertisol over 5 years in a semi-arid region of India. Soil Till Res 70(2):163–173

Blanchart E, Albrecht A, Alegre J, Duboisset A, Gilot C, Pashanasi B, Lavelle P, Brussaard L (1999) Effects of earthworms on soil structure and physical properties. In: Lavelle P, Brussaard L, Hendrix P (eds) Earthworm management in tropical agroecosystems. CAB International, Wallingford, pp 149–172

Blevins RL, Cook D, Phillips SH, Phillips RE (1971) Influence of no-tillage on soil moisture. Agron J 63:593–596

Blevins RL, Frye WW (1993) Conservation tillage: an ecological approach to soil management. Adv Agron 51:33–78

Bloom P (2000) Soil pH and pH buffering. In: Sumner ME (ed) Handbook of soil science. CRC Press, Boca Raton, FL, pp B333–B350

Blouin M, Zuily-Fodil Y, Pham-Thi AT, Laffray D, Reversat G, Pando A, Tondoh J, Lavelle P (2005) Belowground organismactivities affect plant aboveground phenotype, inducing plant tolerance to parasites. Ecol Lett 8:202–208

Boglárka O, Brière C, Bécard G, Dénarié C, Gough G (2005) Nod factors and a diffusible factor from arbuscular mycorrhizal fungi stimulate lateral root formation in Medicago truncatula via the DMI1/DMI2 signalling pathway. Plant J 44:195–207

Bollich PK, Dunigan EP, Kitchen LM, Taylor V (1988) The influence of trifluralin and pendimethalin on nodulation, N2 (C_2H_2) fixation, and seed yield of field grown soybeans (*Glycine max*). Weed Sci 36:15–19

Boomsma CR, Vyn TJ (2008) Maize drought tolerance: potential improvements through arbuscular mycorrhizal symbiosis? Field Crops Res 108:14–31

Bossuyt H, Six J, Hendrix PF (2002) Aggregate-protected carbon in no-tillage and conventional tillage agroecosystems using carbon-14 labeled plant residue. Soil Sci Soc Am J 66:1965–1973

Bouquard C, Ouzzani J, Promé J-C, Michael-Briand Y, Plésiat P (1997) Dechlorination of atrazine by a Rhizobium sp. isolate. Appl Environ Microbiol 63:862–866

Bradford JM, Peterson GA (2000) Conservation tillage. In: Sumner ME (ed) Handbook of soil science. CRC Press, Boca Raton, FL, pp G247–G298

Brown GG, Edwards CA, Brussaard L (2004) How earthworms affect plant growth: burrowing into the mechanisms. In: Edwards CA (ed) Earthworm ecology. CRC Press, Boca Raton, FL, pp 13–49

Buckerfield JC, Lee KE, Davoren CW, Hannay JN (1997) Earthworms as indicators of sustainable production in dryland cropping in Southern Australia. Soil Biol Biochem 29:547–554

Buczko U, Kuchenbuch RO (2007) Phosphorus indices as risk-assessment tools in the U.S.A. and Europe – a review. J Plant Nutr Soil Sci 170:445–460

Buhler DD, Stoltenberg DE, Becker RL, Gunsolus JL (1994) Perennial weed populations after 14 years of variable tillage and cropping practices. Weed Sci 42:205–209

Buhler DD (1995) Influence of tillage systems on weed population dynamics and management in corn and soybean in the central USA. Crop Sci 35:1247–1258

Buhler DD, Mester TC, Kohler KA (1996) Effect of tillage and maize residue on the emergence of four annual weed species. Weed Res 40:153–165

Buhler DD, Hartzler RG, Forcella F (1997) Implications of weed seedbank dynamics to weed management. Weed Sci 45:329–336

Buschiazzo DE, Zobeckd TM, Abascal SA (2007) Wind erosion quantity and quality of an Entic Haplustoll of the semi-arid pampas of Argentina. J Arid Environ 69:29–39

Cannell RQ, Hawes JD (1994) Trends in tillage practices in relation to sustainable crop production with special reference to temperate climates. Soil Till Res 30:245–282

Cardina J, Regnier E, Harrison K (1991) Long-term tillage effects on seed banks in three Ohio soils. Weed Sci 39:186–194

Cardina J, Herms CP, Doohan DJ (2002) Crop rotation and tillage system effects on weed seedbanks. Weed Sci 50:448–460

Cardoso IM, Kuyper TW (2006) Mycorrhizas and tropical soil fertility. Agric Ecosyst Environ 116:72–84

Carney KM, Matson PA (2006) The influence of tropical plant diversity and composition on soil microbial communities. Microb Ecol 52:226–238

Carter MR, Rennie DA (1984) Dynamics of soil microbial N under zero and shallow tillage for spring wheat, using 15N urea. Plant Soil 76:157–164

Carter MR (1991) The influence of tillage on the proportion of organic carbon and nitrogen in the microbial biomass of medium-textured soils in a humid climate. Biol Fertil Soils 11:135–139

Carter MR (2002) Soil quality for sustainable land management: organic matter and aggregation interactions that maintain soil functions. Agron J 94:38–47

Cereti CF, Rossini F (1995) Effect of reduced tillage on physical properties of soils continuously cropped with wheat (Triticum aestivum L.) and maize (Zea mays L.) under dryland cultivation. Riv Agro 29:382–387

Chan KY, Roberts WP, Heenan DP (1992) Organic carbon and associated soil properties after 10 years of rotation under different stubble tillage practices. Aust J Soil Res 30:71–83

Chan KY (2001) An overview of some tillage impacts on earthworm population abundance and diversity – implications for functioning in soils. Soil Till Res 57:179–191

Chan KY, Heenan DP, Oates A (2002) Soil carbon fraction and relationship to soil quality under different tillage and stubble management. Soil Till Res 63:133–139

Chan KY, Heenan DP, So HB (2003) Sequestration of carbon and changes in soil quality under conservation tillage on light-textured soil in Australia: a review. Aust J Exp Agric 43:325–334

Chan KY (2004) Impact of tillage practices and burrows of a native Australian anecic earthworm on soil hydrology. Appl Soil Ecol 27:89–96

Chalk PM, de F Souza R, Urquiaga S, Alves BJR, Boddey RM (2006) The role of arbuscular mycorrhiza in legume symbiotic performance. Soil Biol Biochem 38:2944–2951

Changjin D, Bin Z (2004) Impact of herbicides on infection and hyphal enzyme activity on AM fungus. Acta Pedol Sin 41:750–755
Chen MM, Zhu YG, Su YH, Chen BD, Fu BJ, Marschner P (2007) Effects of soil moisture and plant interactions on the soil microbial community structure. Eur J Soil Biol 43:31–38
Chio H, Sanborn JR (1978) The metabolism of Atrazine, Chloramben, and Dicamba in earthworms (*Lumbricus terrestris*) from treated and untreated plots. Weed Sci Soc Am 26(4):331
Christensen BT, Sorensen LH (1985) The distribution of native and labeled carbon between soil particle size fractions isolated from long-term incubation experiments. J Soil Sci 36:219–229
Clapperton MJ, Lee NO, Binet F, Conner RL (2001) Earthworms indirectly reduce the effects of take-all (*Gaeumannomyces graminis* var. tritici) on soft white spring wheat (*Triticum aestivum* cv. Fielder). Soil Biol Biochem 33:1531–1538
Coleman DC, Reid CPP, Cole CV (1983) Biological strategies of nutrient cycling in soil systems. Adv Ecol Res 13:1–55
Corwin DL (2003) Soil salinity measurement. In: Stewart BA, Howell TA (eds) Encyclopedia of water science. Marcel Dekker, NewYork, NY, pp 852–860
Corwin DL, Lesch SM (2005) Apparent soil electrical conductivity measurements in agriculture. Comp Electron Agric 46:11–43
Cox PM, Betts RA, Jones CD, Spall SA, Totterdell IJ (2000) Acceleration of global warming due to carbon-cycle feedbacks in coupled climate model. Nature 408:184–187
Curl EA, Truelove B (1986) The rhizosphere. Springer-Verlag, Berlin/New York
Curry JP, Byrne D (1992) The role of earthworms in straw decomposition and nitrogen turnover in arable land in Ireland. Soil Biol Biochem 24:1409–1412
Cycoń M, Piotrowska-Seget Z (2009) Changes in bacterial diversity and community structure following pesticides addition to soil estimated by cultivation technique. Ecotoxicology 18:632–642
D'Haene K, Vermang J, Cornelis WM, Leroy BLM, Schiettecatte W, De Neve S, Gabriels D, Hofman G (2008) Reduced tillage effects on physical properties of silt loam soils growing root crops. Soil Till Res 99:279–290
De Vita P, Di Paolo E, Fecondo G, Di Fonzo N, Pisante M (2007) No-tillage and conventional tillage effects on durum wheat yield, grain quality and soil moisture content in southern Italy. Soil Till Res 92:69–78
Degens BP, Schipper LA, Sparling GP, Vojvodic-Vukovic M (2000) Decreases in organic C reserves in soils can reduce the catabolic diversity of soil microbial communities. Soil Biol Biochem 32:189–196
Derpsch R (1999) New paradigms in Agricultural Production, Tillage Research conducted by ISTRO Member Rolf Derpsch. In: ISTRO – INFO EXTRA, Volume 4, Issue 1, Springer
Dick WA (1983) Organic carbon, nitrogen, and phosphorus concentrations and pH in soil profiles as affected by tillage intensity. Soil Sci Soc Am J 47:102–107
Dickey EC, Peterson TR, Gilley JR, Mielke LN (1983) Yield comparisons between continuous no-till and tillage rotations. Trans ASAE 26:1682–1686
Dimanche PH, Hoogmoed WB (2002) Soil tillage and water infiltration in semi-arid Morocco: the role of surface and sub-surface soil conditions. Soil Till Res 66:13–21
Dinelli G, Vicari A, Acinelli C (1998) Degradation and side effects of three sulfonylurea herbicides in soil. J Environ Qual 27:1459–1464
Doles JL, Zimmerman RJ, Moore JC (2001) Soil microarthropod community structure and dynamics in organic and conventionally managed apple orchards in Western Colorado, USA. Appl Soil Ecol 18:83–96
Doran J (1987) Microbial biomass and mineralizable nitrogen distributions in no-tillage and plowed soils. Biol Fertil Soils 5:68–75
Doran JW, Parkin TB (1994) Defining and assessing soil quality. In: Doran JW, Coleman DC, Bezdicek DF, Stewart BA (eds) Defining soil quality for a sustainable environment. Soil Science Society of America, Madison, WI, pp 3–21
Doube BM, Williams PML, Willmott PJ (1997) The influence of two species of earthworm (*Aporrectodea trapezoids* and *Aporrectoedea rosea*) on the growth of wheat, barley and faba beans in three soil types in the greenhouse. Soil Biol Biochem 29:503–509

Douds DD, Reider C (2003) Inoculation with mycorrhizal fungi increases the yield of green peppers in a high P soil. Biol Agric Hortic 21:91–102
Drijber RA, Doran JW, Parkhurst AM, Lyon DJ (2000) Changes in soil microbial community structure with tillage under long-term wheat-fallow management. Soil Biol Biochem 32:1419–1430
Edwards CA, Thompson AR (1973) Pesticides and the soil fauna. Residue Rev 45:1–79
Edwards WM, vander Ploeg RR, Ehlers W (1979) A numerical study of the effects of noncapillary-sized pores upon infiltration. Soil Sci Soc Am J 43:851–856
Edwards CA, Lofty JR (1982) The effect of direct drilling and minimal cultivation on earthworm populations. J Appl Ecol 19:723–734
Edwards CA (1983) Earthworm ecology in cultivated soils. In: Satchell JE (ed) Earthworm ecology: from Darwin to vermiculture. Chapman & Hall, London, pp 123–137
Edwards WM, Shipitalo MJ, Owens LB, Norton LD (1990) Effect of *Lumbricus terrestris* L. burrows on hydrology of continuous no-till corn fields. Geoderma 46:73–84
Edwards CA, Bohlen PJ (1996) Biology and ecology of earthworms, 3rd edn. Chapman & Hall, London, 426 pp
Edwards CA (1998) Earthworm ecology. Soil and water conservation society, Ankeny, IA
Edwards WM, Shipitalo MJ (1998) Consequences of earthworms in agricultural soils: aggregation and porosity. In: Edwards CA (ed) Earthworm ecology. Soil and water conservation society, St Lucie Press, IA, pp 147–161
Ehlers W (1975) Observations on earthworm channels and infiltration in a tilled and untilled loess soil. Soil Sci 119:242–249
Ehlers W, Köpke U, Hesse F, Böhm W (1983) Penetration resistance and root growth of oats in tilled and untilled loess soil. Soil Till Res 3:261–275
Elfstrand S, Bath B, Martensson A (2007) Influence of various forms of green manure amendment on soil microbial community composition, enzyme activity and nutrient levels in leek. Appl Soil Ecol 36:70–82
Ellouze W, Hanson K, Nayyar A, Perez JC, Hamel C (2008) Intertwined existence: the life of plant symbiotic fungi in agricultural soils. In: Varma A (ed) Mycorrhiza. Springer-Verlag, Berlin, Heidelberg, pp 507–528
Etchevers J, Fisher R, Vidal I, Sayre K, Sandoval M, Oleshsko K, Román S (2000) Labranza de conservación, índices de calidad del suelo y captura de carbono. In: Memorias del Simposio Internacional de labranza de conservación. InstitutoNacional de Investigaciones Forestales y Agro Pecuarias–Produce, Mazatlán, Sinaloa
Fan XL, Zhang FS (2000) Soil water, fertility and sustainable agricultural production in arid and semiarid regions on the Loess plateau. J Plant Nutr Soil Sci 163:107–113
Farage PK, Ardo J, Olsson L, Rienzi EA, Ball AS, Pretty JN (2007) The potential for soil carbon sequestration in the tropic dryland farming systems of Africa and Latin America: a modelling approach. Soil Till Res 94:457–472
Feldman SR, Alzugaray C, Torres PS, Lewis P (1997) The effect of different tillage systems on the composition of the seedbank. Weed Res 37:71–76
Fernández-Ugalde O, Virto I, Bescansa P, Imaz MJ, Enrique A, Karlen DL (2009) No-tillage improvement of soil physical quality in calcareous, degradation-prone, semiarid soils. Soil Till Res 106:29–35
Ferrero A, Maggiore T (1994) Leaching of slurries and herbicides in subsurface water under field conditions. In: Borin M, Sattin M (eds) Proceedings of the third congress of the European society for agronomy. Padova University, Abano-Padova, Italy, pp 794–795
Fischer RA, Santiveri F, Vidal IR (2002) Crop rotation, tillage and crop residue management for wheat and maize in the subhumid tropical highland. I. Wheat and legume performance. Field Crops Res 79:107–122
Fowler R, Rockstrom J (2001) Conservation tillage for sustainable agriculture: an agrarian revolution gathers momentum in Africa. Soil Till Res 61:93–108
Frankenberger WT, Dick WA (1983) Relationships between enzyme activities and microbial growth and activity indices in soil. Soil Sci Soc Am J 47:945–951
Franzluebbers AJ, Arshad MA (1996) Soil organic matter pools with conventional and zero tillage in a cold, semiarid climate. Soil Till Res 39:1–11

Franzluebbers AJ, Haney RL, Hons FM, Zuberer DA (October–November 1996) Active fractions of organic matter in soils with different texture. Soil Biol Biochem 28(10–11):1367–1372
Franzluebbers AJ, Arshad AM (1997) Soil microbial biomass and mineralization of carbon of water stable aggregates. Soil Sci Soc Am J 61:1090–1097
Frey SD, Six J, Elliott ET (2003) Reciprocal transfer of carbon and nitrogen by decomposer fungi at the soil–litter interface. Soil Biol Biochem 35:1001–1004
Friend JJ, Chan KY (1995) Influence of cropping on the population of a native earthworm and consequent effects on hydraulic properties of Vertisols. Aust J Soil Res 33:995–1006
Froud-Williams RJ (1988) Changes in weed flora with different tillage and agronomic management systems. In: Altieri MA, Liebman M (eds) Weed management in agroecosystems: ecological approaches. CRC Press, Boca Raton, FL, pp 213–236
Fuentes M, Govaerts B, De León F, Hidalgo C, Dendooven L, Sayre KD, Etchevers J (2009) Fourteen years of applying zero and conventional tillage, crop rotation and residue management systems and its effect on physical and chemical soil quality. Eur J Agron 30:228–237
Furlong MA, Singleton DR, Coleman DC, Whitman WB (2002) Molecular and culture-based analyses of prokaryotic communities from an agricultural soil and the burrows and casts of the earthworm Lumbricus rubellus. Appl Environ Microbiol 68:1265–1279
Galantini JM, Landriscini MR, Iglesias JO, Miglierina AM, Rosell RA (2000) The effects of crop rotation and fertilization on wheat productivity in the Pampean semi-arid region of Argentina. 2. Nutrient balance, yield and grain quality. Soil Till Res 53:137–144
Gange AC (1993) Translocation of mycorrhizal fungi by earthworms during early succession. Soil Biol Biochem 25:1021–1026
Gebregziabher S, Mouazena AM, van Brussel H, Ramon H, Nyssen J, Verplancke H, Behailu M, Deckers J, de Baerdemaeker J (2006) Animal drawn tillage, the Ethiopian ard plough, maresha: a review. Soil Till Res 89:129–143
Ghaffarzadeh M, García F, Cruse RM (1994) Grain of corn, soybean, and oat grown in a strip intercropping system. Am J Altern Agric 9:171–177
Ghosheh H, Al-Hajaj N (2005) Weed seed bank response to tillage and crop rotation in a semi-arid environment. Soil Till Res 84:184–191
Gianfreda L, Antonietta RM, Piotrowska A, Palumbo G, Colombo C (2005) Soil enzyme activities as affected by anthropogenic alterations: intensive agricultural practices and organic pollution. Sci Total Environ 341:265–279
Gicheru P, Gachene C, Mbuvi J, Marea E (2004) Effects of soil management practices and tillage systems on surface soil water conservation and crust formation on a sandy loam in semi-arid Kenya. Soil Till Res 75:173–184
Giesy JP, Dobson S, Solomon KR (2000) Ecotoxicological risk assessment for Roundup herbicide. Rev Environ Contam Toxicol 167:35–120
Giller KE (2001) Nitrogen fixation in tropical cropping systems, 2nd edn. CABI Publishing, Wallingford
Gilman AP, Vardanis A (1974) Carbofuran comparative toxicity and metabolism in the worms *Lumbricus terrestris* and *Eisenia fetida*. J Food Chem 22:625–628
Girvan MS, Bullimore J, Ball AS, Pretty JN, Osborn AM (2004) Responses of active bacterial and fungal communities in soils under winter wheat to different fertilizer and pesticide regimens. Appl Environ Microbiol 70:2692–2701
Goe MR (1987) Animal traction on smallholder farms in the Ethiopian highlands. Ph.D. thesis, Cornell University, pp 127, 160
Gonzalez N, Eyherabide JJ, Barcellona MI, Gaspari A, Sanmartino S (1999) Effect of soil interacting herbicides on soybean nodulation in Balcarce, Argentina. Pesqui Agropec Bras 34:1167–1173
Gonzalez MG, Conti ME, Palma RM, Arrigo NM (2003) Dynamics of humic fractions and microbial activity under no-tillage or reduced tillage, as compared with native pasture (*Pampa argentina*). Biol Fertil Soils 39:135–138
Govaerts B, Fuentes M, Sayre KD, Mezzalama M, Nicol JM, Deckers J, Etchevers J, Figueroa-Sandoval B (2007) Infiltration, soil moisture, root rot and nematode populations after 12 years of different tillage, residue and crop rotation managements. Soil Till Res 94:209–219

Govaerts B, Mezzalama M, Sayre KD, Crossa J, Lichter K, Katrien VT, De Corte VP, Deckers J (2008) Long-term consequences of tillage, residue management, and crop rotation on selected soil micro-flora groups in the subtropical highlands. Appl Soil Ecol 38:197–210

Govindarajulu M, Pfeffer PE, Hairu J, Abubaker J, Douds DD, Allen JW, Bucking H, Lammers PJ, Shachar-Hill Y (2005) Nitrogen transfer in the arbuscular mycorrhizal symbiosis. Nature 435/9. doi:10.1038/nature03610

Granatstein DM, Bezdicek DF, Cochran VL, Elliott LF, Hammel J (1987) Long term tillage and rotation effects on soil microbial biomass, carbon and nitrogen. Biol Fertil Soils 5:265–270

Grandy AS, Robertson GP (2006) Aggregation and organic matter protection following tillage of a previously uncultivated. Soil Sci Soc Am J 70:1398–1406

Griffiths BS, Caul S, Thompson J, Birch ANE, Cortet J, Andersen MN, Krogh PH (2007) Microbial and microfaunal community structure in cropping systems with genetically modified plants. Pedobiologia 51:195–206

Grigera MS, Drijber RA, Eskridge KM, Wienhold BJ (2006) Soil microbial biomass relationships with organic matter fractions in a Nebraska corn field mapped using apparent electrical conductivity. Soil Sci Soc Am J 70:1480–1488

Ha KV, Marschner P, Bünemann EK (2008) Dynamics of C, N P and microbial community composition in particular soil organic matter during residue decomposition. Plant Soil 303:253–264

Halvorson AD, Black AL, Krupinsky JM, Merill SD, Wienhold BJ, Tanaka DL (2000) Spring wheat response to tillage system and nitrogen fertilization in rotation with sunflower and winter wheat. Agron J 92:136–144

Hassall M, Adl S, Berg M, Griffiths B, Scheu S (2006) Soil fauna–microbe interactions: towards a conceptual framework for research. Eur J Soil Biol 42:S54–S60

Haukka J (1988) Effect of various cultivation methods on earthworm biomasses and communities on different soil types. Ann Agric Fenniae 27:263–269

Havlin JL, Beaton JD, Tisdale SL, Nelson WL (1999) Soil fertility and fertilizers. An introduction to nutrient management. Prentice Hall, Upper Saddle River, NJ, p 499

Haynes RJ, Fraser PM, Williams PH (1995) Earthworm population size and composition, and microbial biomass: effect of pastoral and arable management in Canterbury, New Zealand. The significance and regulation of soil biodiversity. In: Proceedings of the international symposium on soil biodiversity, Michigan State University, East Lansing, MI, May 3–6, Kluwer, Dordrecht, pp 279–285

He X-H, Critchley CH, Bledsoe C (2003) Nitrogen transfer within and between plants through common mycorrhizal networks (CMNs). Crit Rev Plant Sci 22:531–567

Heenan DP, Chan KY, Knight PG (2004) Long-term impact of rotation, tillage and stubble management on the loss of organic carbon and nitrogen from a Chromic Luvisol. Soil Till Res 76:59–68

Hendrix PF, Parmelee RW, Crossley DA Jr, Coleman DC, Odum EP, Groffman PM (1986) Detritus food webs in conventional and no-tillage agroecosystems. BioScience 36:374–380

Hendrix PF, Mueller BR, Bruce RR, Langdale GW, Parmelee RW (1992) Abundance and distribution of earthworms in relation to landscape factors on the Georgia Piedmont, USA. Soil Biol Biochem 24:1357–1361

Hendrix PF, Franzluebbersm AJ, McCracken DV (1998) Management effects on carbon accumulation and loss in soils on the southern Appalachian Piedmont of Georgia, USA. Soil Till Res 47:245–251

Hernández-Hernández RM, López-Hernández D (2002) Microbial biomass, mineral nitrogen and carbon content in savanna soil aggregates under conventional and no-tillage. Soil Biol Biochem 34:1563–1570

Holland JM (2004) The environmental consequences of adopting conservation tillage in Europe: reviewing the evidence. Agric Ecosyst Environ 103:1–25

Hubbard VC, Jordan D, Stecker JA (1999) Earthworm response to rotation and tillage in a Missouri claypan soil. Biol Fert Soils 29:343–347

Hulugalle NR, Lobry de Bruyn LA, Entwistle P (1997) Residual effects of tillage and crop rotation on soil properties, soil invertebrate numbers and nutrient uptake in an irrigated vertisol sown to cotton. Appl Soil Ecol 7:11–30

Hunt HW, Coleman DC, Ingham ER, Ingham RE, Elliott ET, Moore JC, Rose SL, Reid CPP, Morley CR (1987) The detrital food web in a shortgrass prairie. Biol Fertil Soils 3:57–68

Hyde SM, Wood PM (1997) A mechanism for production of hydroxyl radicals by the brown-rot fungus *Coniophora puteana*: Fe(III) reduction by cellobiose dehydrogenase and Fe(II) oxidation at a distance from the hyphae. Microbiology 143:259–266

Ingham RE, Trofymow JA, Ingham ER, Coleman DC (1985) Interactions of bacteria, fungi, and their nematode grazers: effects of nutrient cycling and plant growth. Ecol Monogr 55:119–140

Jacobs A, Rauber R, Ludwig B (2009) Impact of reduced tillage on carbon and nitrogen storage of two Haplic Luvisols after 40 years. Soil Till Res 102:158–164

Jacobson AR, Dousset S, Guichard N, Baveye P, Andreux F (2005) Diuron mobility through vineyard soils contaminated with copper. Environ Pollut 138:250–259

Jastrow JD (1996) Soil aggregate formation and the accrual of particulate and mineral-associated organic matter. Soil Biol Biochem 28:665–676

Jeffries P (1987) Use of mycorrhizae in agriculture. Crit Rev Biotechnol 5:319–357

Jenkinson DS, Adams DE, Wild A (1991) Model estimates of CO2 emission from soil in response to global warming. Nature 351:304–306

Jin K, Sleutel S, Buchan D, De Neve S, Cai DX, Gabriels D, Jin JY (2009) Changes of soil enzyme activities under different tillage practices in the Chinese Loess Plateau. Soil Till Res 104:115–120

Johnston AM, Clayton GW, Wall PC, Sayre KD (2002) Sustainable cropping systems for semiarid regions. Paper presented at the international conference on environmentally sustainable agriculture for dry areas for the 2nd millennium, Shijiazhuang, Hebei Province, P.R.C, 15–19 Sept 2002

Johansson JF, Paul LR, Finlay RD (2004) Microbial interactions in the mycorrhizosphere and their significance for sustainable agriculture. FEMS Microbiol Ecol 48:1–13

Johnston AE (1997) The value of long-term field experiments in agricultural, ecological, and environmental research. Adv Agron 59:291–333

Jones C, McConnell C, Coleman K, Cox P, Falloon P, Jenkinson D, Powlson D (2005) Global climate change and soil carbon stock; predictions from two contrasting models for turnover of organic carbon in soil. Global Change Biol 11:154–166

Joner EJ, Jakobsen I (1995) Growth and extracellular phosphatase activity of arbuscular mycorrhizal hyphae as influenced by soil organic matter. Soil Biol Biochem 7(9):153–1159

Jordan D, Kremer R (1994) Potential microbial methods as indicators of soil quality in historical agricultural fields. In: Pankhurst C (ed) Management of soil biota. CSIRO, South Adelaide, pp 245–249

Jordan D, Stecker JA, Hubbard VC, Li F, Gantzer CJ, Brown JR (1997) Earthworm activity in notillage and conventional tillage systems in Missouri soils: a preliminary study. Soil Biol Biochem 29:489–491

Jordan D, Milesb RJ, Hubbardc VC, Lorenz T (2004) Effect of management practices and cropping systems on earthworm abundance and microbial activity in Sanborn Field: a 115-year-old agricultural field. Pedobiologia 48:99–110

Jurion F, Henry J (1969) Can primitive farming be modernized? Publication INEAC, Hors Série, Bruxelles, 457 pp

Kamagata Y, Fulthorpe RR, Tamura K, Takami H, Forney LJ, Tiedje JM (1997) Pristine environments harbor a new group of oligotrophic 2, 4-dichlorophenoxyacetic acid-degrading bacteria. Appl Environ Microbiol 63:2266–2272

Kandeler E, Murer E (1993) Aggregate stability and soil microbial processes in a soil with different cultivation. Geoderma 56(1–4):503–513

Kandeler E, Palli S, Stemmer M, Gerzabek MH (1999a) Tillage changes microbial biomass and enzyme activities in particle-size fractions of a Haplic Chernozem. Soil Biol Biochem 31: 1253–1264

Kandeler E, Tscherko D, Spiegel H (1999b) Long-term monitoring of microbial biomass, N mineralisation and enzyme activities of a Chernozem under different tillage management. Biol Fertil Soils 28:343–351

Kandeler E, Tscherko D, Stemmer M, Schwarz S, Gerzabek MH (2001) Organic matter and soil microorganisms – investigations from the micro- to the macroscale. Bodenkultur 52:117–131

Kara EE, Arli M, Uygur V (2004) Effects of the herbicide Topogard on soil respiration, nitrification, and denitrification in potato-cultivated soils differing in pH. Biol Fertil Soils 39:474–478

Karlen DL, Varvel GE, Bullock DG, Cruse RM (1994) Crops rotations for the 21st century. Adv Agron 53:1–45

Karlen DL, Andrews SS, Doran JW (2001) Soil quality: current concepts and applications. Adv Agron 74:1–40

Kern JS, Johnson MG (1993) Conservation tillage impacts on national soil and atmospheric carbon levels. Soil Sci Soc Am J 57:200–210

Kettler TA, Lyon DJ, Doran JW, Powers WL, Stroup WW (2000) Soil quality assessment after weed-control tillage in a no-till wheat-fallow cropping system. Soil Sci Soc Am J 64:339–346

Kiepe P (1995) Effects of Cassia siamea hedgerow barriers on soil physical properties. Geoderma 66:113–120

Kitur BK, Smith MS, Blevins RL, Frye WW (1984) Fate of 15N-depleted ammonium nitrate applied to no-tillage and conventional tillage corn. Agron J 76:240–242

Kladivko EJ, Akhouri NM, Weesies G (1997) Earthworm populations and species distributions under no-till and conventional tillage in Indiana and Illinois. Soil Biol Biochem 29:613–615

Knab W, Hurle K (1986) Enfluss der Grundbodenbearbeitung auf die Verunkrautung – ein Beitrag zur Prognose der Verunkrautung. In: Proceedings EWRS symposium economic weed control, Stuttgart, pp 309–316

Kramer PJ, Boyer JS (1997) Water relations of plants and soils. Academic, San Diego, CA

Kribaa M, Hallaire V, Curmi P, Lahmar R (2001) Effect of various cultivation methods on the structure and hydraulic properties of a soil in a semi-arid climate. Soil Till Res 60:43–53

Kristoffersen A, Riley H (2005) Effects of soil compaction and moisture regime on the root and shoot growth and phosphorus uptake of barley plants growing on soils with varying phosphorus status. Nutr Cyc Agroecosyst 72:135–146

Kushwaha CP, Tripathi SK, Singh KP (2001) Soil organic matter and water-stable aggregates under different tillage and residue conditions in a tropical dryland agroecosystem. Appl Soil Ecol 16:229–241

Kwesiga FR, Franzel S, Place F, Phiri D, Simwanza CP (1999) Sesbania sesban improved fallow in Eastern Zambia: their conception, development and farmer enthusiasm. Agrofor Syst 47:49–66

La Scala N, Bolonhezi D, Pereira GT (2006) Short-term soil CO2 emission after conventional and reduced tillage of a no-till sugar cane area in southern Brazil. Soil Till Res 91:244–248

Lal R (1989) Agroforestry systems and soil management of a tropical alfisol. IV. Effects of soil physical and soil mechanical properties. Agrofor Syst 8:197–215

Lal R, Kimble JM, Follet RF, Cole CV (1998) The potential of U.S. cropland to sequester carbon and mitigate the greenhouse effect. Ann Arbor Press, Chelsea, MI

Lavelle P, Melendez G, Pashanasi B, Schaefer R (1992) Nitrogen mineralization and reorganization in casts of the geophagous tropical earthworm Pontoscolex corethrurus (Glossoscolecidae). Biol Fertil Soils 14:49–53

Lee KE (1985) Earthworms: their ecology and relationship with soils and land use. Academic, Sydney

Lee KE (1995) Earthworms and sustainable land use. In: Hendrix PF (ed) Earthworms ecology and biogeography in North America. Lewis, Boca Raton, FL, pp 215–234

Li HW, Gao HW, Wu HD, Li WY, Wang XY, He J (2007) Effects of 15 years of conservation tillage on soil structure and productivity of wheat cultivation in northern China. Aust J Soil Res 45(5):344–350

Li J, Kremer RJ (2000) Rhizobacteria associated with weed seedlings in different cropping systems. Weed Sci 48:734–741

Liebig M, Carpenter-Boggs L, Johnson JMF, Wright S, Barbour N (2006) Cropping system effects on soil biological characteristics in the Great Plains. Renew Agric Food Syst 21:36–48

Liu B, Zeng Q, Yan F, Xu H, Xu C (2005) Effects of transgenic plants on soil microorganisms. Plant Soil 271:1–13

Liu A, Hamel C, Hamilton RI, Ma BL, Smith DL (2000a) Acquisition of Cu, Zn, Mn and Fe by mycorrhizal maize (*Zea mays* L.) grown in soil at different P and micronutrient levels. Mycorrhiza 9:331–336

Liu A, Hamel C, Hamilton RI, Smith DL (2000b) Mycorrhizae formation and nutrient uptake of new corn (Zea mays L.) hybrids with extreme canopy and leaf architecture as influenced by soil N and P levels. Plant Soil 221:157–166

Liu A, Plenchette C, Hamel C (2007) Soil nutrient and water providers: how arbuscular mycorrhizal mycelia support plant performance in a resourcelimited world. In: Hamel C, Plenchette C (eds) Mycorrhizae in crop production. Haworth Food & Agricultural Products Press, Binghamton, NY, pp 37–66

Lofs-Holmin A (1983) Influence of agricultural practices on earthworms (Lumbricidae). Acta Agric Scand 33:225–234

López-Fando C, Bello A (1995) Variability in soil nematode populations due to tillage and crop rotation in semi-arid Mediterranean agrosystems. Soil Till Res 36:59–72

López-Fando C, Dorado J, Pardo MT (2007) Pardo Effects of zone-tillage in rotation with no-tillage on soil properties and crop yields in a semi-arid soil from central Spain. Soil Till Res 95:266–276

Lynch JM, Panting LM (1980) Cultivation and the soil biomass. Soil Biol Biochem 12:29–33

Lynch JM, Panting LM (1982) Effects of season, cultivation and nitrogen fertilizer on the size of the soil microbial biomass. J Sci Food Agric 33:249–252

Madejón E, Moreno F, Murillo JM, Pelegrín F (2007) Soil biochemical response to long-term conservation tillage under semi-arid Mediterranean conditions. Soil Till Res 94:346–352

Mafongoya PL, Dzowela BH (1999) Biomass production of tree fallows and their residual effect on maize in Zimbabwe. Agrofor Syst 47:139–151

Mallik MAB, Tesfai K (1985) Pesticidal effect on soybean – rhizobia symbioses. Plant Soil 85:33–41

Malty JS, Siqueira JO, Moreira FMS (2006) Efeitos do glifosato sobre microrganismos simbiotroficos de soja, em meio de cultura e casa de vergetacao. Pessq Agropec Bras Brasilia 41:285–291

Mapfumo P, Mtambanengwe F, Giller KE, Mpepereki S (2005) Tapping indigenous herbaceous legumes for soil fertility management by resource-poor farmers in Zimbabwe Agriculture. Ecosyst Environ 109:221–233

Marschner H, Dell B (1994) Nutrient uptake in mycorrhizal symbiosis. Plant Soil 159:89–102

Marschner H (1995) Mineral nutrition of higher plants. Academic, San Diego, CA

McCarty GW, Meisinger JJ, Jenniskens FMM (1995) Relationships between total-N, biomass-N and active-N in soil under different tillage and N fertilizer treatments. Soil Biol Biochem 27:1245–1250

McGonigle TP, Evans DG, Miller MH (1990) Effect of degree of soil disturbance on mycorrhizal colonization and phosphorus absorption by maize in the growth chamber and field experiments. New Phytol 116:629–636

McGonigle TP, Miller MH (1993) Mycorrhizal development and phosphorus absorption in maize under conventional and reduced tillage. Soil Sci Soc Am J 57:1002–1006

McGonigle TP, Miller MH (1996) Development of fungi below ground in association with plants growing in disturbed and undisturbed soils. Soil Biol Biochem 28:263–269

Meisinger JJ, Bandel VA, Stanford G, Legg JO (1985) Nitrogen utilization of corn under minimal tillage and moldboard plow tillage: I. Four year results using labeled N fertilizer on an Atlantic Coastal Plain soil. Agron J 77:602–611

Meyer K, Joergensen RG, Meyer B (1996) The effects of reduced tillage on microbial biomass C and P in sandy loess soils. Appl Soil Ecol 5:71–79

Miransari M, Bahrami HA, Rejali F, Malakouti MJ (2006) Evaluating the effects of arbuscular mycorrhiza on corn nutrient uptake and yield in a compacted soil under field conditions and nutrient. Soil Water Res 20:106–121 (in Persian abstract in English)

Miransari M, Smith DL (2007) Overcoming the stressful effects of salinity and acidity on soybean [*Glycine max* (L.) Merr.] nodulation and yields using signal molecule genistein under field conditions. J. Plant Nutr 30:1967–1992

Miransari M, Smith DL (2008) Using signal molecule genistein to alleviate the stress of suboptimal root zone temperature on soybean–Bradyrhizobium symbiosis under different soil textures. J Plant Interact 3:287–295

Miransari M, Smith D (2009) Alleviating salt stress on soybean (*Glycine max* (L.) Merr.) Bradyrhizobium japonicum symbiosis, using signal molecule genistein. Eur J Soil Biol 45:146–152

Miransari M, Bahrami HA, Rejali F, Malakouti MJ (2009) Effects of arbuscular mycorrhiza, soil sterilization, and soil compaction on wheat (*Triticum aestivum* L.) nutrients uptake. Soil Till Res 104:48–55

Moore JC (1988) The influence of microarthropods on symbiotic and non-symbiotic mutualism in detrital-based belowground food webs. Agric Ecosyst Environ 24:147–159

Moore JC, de Ruiter PC (1991) Temporal and spatial heterogeneity of trophic interactions within belowground food webs. Agric Ecosyst Environ 34:371–397

Moore JC, McCann K, Setälä H, deRuiter PC (2003) Topdown is bottom up: does predation in the rhizosphere regulate aboveground dynamics? Ecology 84:846–857

Morse R, Elkner T, Groff S (2001) No-till Pumpkin production principles and practices. Pennsylvania Marketing and Research Program, College Station, TX, p 16

Moyer JR, Roman ES, Lindwall CW, Blackshaw RE (1994) Weed management in conservation tillage systems for wheat production in North and South America. Crop Prot 13:243–259

Mujica MT, Fracchia S, Menendez A, Ocampo JA, Godeas A (1998) Influence of chlorsulfuron herbicide on arbuscular mycorrhizas and plant growth of Glycine max intercropped with the weeds *Brassica campestris*. In: Proceedings of the 2nd international conference on mycorrhiza, Uppsala, Sweden, pp 5–10

Muñoz A, López-Piñeiro A, Ramírez M (2007) Soil quality attributes of conservation management regimes in a semi-arid region of south western Spain. Soil Till Res 95:255–265

Munch JC, Gloth B, Henneberger C (1989) The effect of terbuthylazine on soil microorganisms of the nitrogen cycle. Einfluss Eines Terbutylazine Praparates Bodenmikroorganizmen des Stickstoff Kreislaufs. Mitt Dtsch Bodenkd Ges 59:603–606

Munkholm LJ (2001) Non-inversion tillage effects on soil mechanical properties of a humid sandy loam. Soil Till Res 62:1–14

Nadian H, Smith SE, Alston AM, Murray RS (1997) Effects of soil compaction on plant growth, phosphorus uptake and morphological characteristics of vesicular-arbuscular mycorrhizal colonization of Trifolium subterraneum. New Phytol 135:303–311

Nardi S, Pizzeghello D, Muscolo A, Vianello A (2002) Physiological effects of humic substances on higher plants. Soil Biol Biochem 34:1527–1536

Nasr AA (1993) The effect of cytokinin and thidiazuron on tomato inoculated with endomycorrhiza. Mycorrhiza 3:179–182

Nielsen HJ, Pinnerup SP (1982) Reduceret jordbehandling og ukrudt. In: Ogr.as och ogr.asbek.ampning, 23:e svenska ogr-.askonferensen, Del 2.Rapporter, Uppsala, Sweden, 27–29 Jan 1982, pp 381–395

Niemi R, Vepsä läinen M, Wallenius M, Simpanen K, Alakukkub S, Pietola L (2005) Temporal and soil depth-related variation in soil enzyme activities and in root growth of red clover (*Trifolium pratense*) and timothy (*Phleum pratense*) in the field. Soil Biol Ecol 30:113–125

Nuutinen V (1992) Earthworm community response to tillage and residue management on different soil types in southern Finland. Soil Till Res 23:221–239

Nyakatawa EZ, Reddy KC, Mays DC (2000) Tillage, cover cropping and poultry litter effects of cotton. II. Growth and Yield parameters. Agron J 92:1000–1007

Nyamadzawo G, Nyamugafata P, Chikowo R, Giller KE (2007) Soil organic carbon dynamics of improved fallow-maize rotation systems under conventional and no-tillage in Central Zimbabwe. Nutr Cycl Agroecosyst 81:85–93

Nyamadzawo G, Nyamangara J, Nyamugafata P, Muzulu A (2009) Soil microbial biomass and mineralization of aggregate protected carbon in fallow-maize systems under conventional and no-tillage in Central Zimbabwe. Soil Till Res 102:151–157

Oorts K, Bossuyt H, Labreuche J, Merckx R, Nicolardot B (2007) Carbon and nitrogen stocks in relation to organic matter fractions, aggregation and pore size distribution in no-tillage and conventional tillage in northern France. Eur J Soil Sci 58:248–259

Osunbitan JA, Oyedele DJ, Adeklu KO (2005) Tillage effects on bulk density, hydraulic conductivity and strength of a loamy sandy soil in southwestern Nigeria. Soil Till Res 82:57–64

Ouèdraogo E, Mando A, Stroosnijder L (2006) Effects of tillage, organic resources and nitrogen fertilizer on soil carbon dynamics and crop nitrogen uptake in semi-arid West Africa. Soil Till Res 91:57–67

Oyedele DJ, Schjonning P, Sibbesen E, Debosz K (1999) Aggregation and organic matter fractions of three Nigerian soils as affected by soil disturbance and incorporation of plant material. Soil Till Res 50:105–114

Ozpinar S, Baytekin H (2006) Effects of tillage on biomass, roots, N-accumulation of vetch (*Vicia sativa* L.) on a clay loam soil in semi-arid conditions. Field Crops Res 96:235–242

Ozpinar S, Cay A (2006) Effect of different tillage systems on the quality and crop productivity of a clay–loam soil in semi-arid north-western Turkey. Soil Till Res 88:95–106

Pagliai M, Vignozzi N, Pellegrini S (2004) Soil structure and the effect of management practices. Soil Till Res 79:131–143

Pankhurst CE, Yu S, Hawke BG, Harch BD (2001) Capacity of fatty acid profiles and substrate utilization patterns to describe differences in soil microbial communities associated with increased salinity or alkalinity at three locations in South Australia. Biol Fertil Soils 33:204–217

Papendick RI (2004) Farming with the wind II: Wind erosion and air quality control on the Columbia Plateau and Columbia Basin. Special Report by the Columbia Plateau PM10 Project. Washington Agricultural Experiment Station. Report XB 1042, Pullman, WA

Parr JF, Papendick RI, Hornick SB, Meyer RE (1992) Soil quality: attributes and relationship to alternative and sustainable agriculture. Am J Altern Agric 7:5–11

Passioura JB (2002) Soil conditions and plant growth. Plant Cell Environ 25:311–318

Paula TJ, Zambolim L (1994) Efeito de fungicidas e de herbicidas sobre a micorrização de Eucalyptus grandis por Glomus etunicatum. Fitopatol Bras 19:173–177

Paustian K, Collins HP, Paul EA (1997) Management controls on soil carbon. In: Paul EA, Paustian K, Elliott ET, Cole CV (eds) Soil organic matter in temperate agroecosystems: long-term experiments in North America. CRC Press, Boca Raton, FL, pp 15–49

Paustian K, Six J, Elliott ET, Hunt HW (2000) Management options for reducing CO2 emissions from agricultural soils. Biogeochemistry 48:147–163

Phatak SC, Reed R, Fussell W, Lewis WJ, Harris GH (1999) Crimson clover cotton relay cropping with conservation tillage system. In: Hook JE (ed) Proceedings of the 22nd Annual Southern Conservation Tillage Conference for Sustainable Agriculture, Tifton, GA, pp 184–188

Picone C (2003) Managing mycorrhizae for sustainable agriculture in the tropics. In: Vandermeer JH (ed) Tropical agroecosystems. CRC Press, Boca Raton, FL, pp 95–132

Pitkänen J, Nuutinen V (1998) Earthworm contribution to infiltration and surface runoff after 15 years of different soil management. Appl Soil Ecol 9:411–415

Pleasant JM, McCollum RE, Coble HD (1990) Weed population dynamics and weed control in the Peruvian Amazon. Agron J 82:102–112

Porta J, López-Acevado M, Roquero C (1999) Edafología para la agricultura y el medio ambiente. Mundi Prensa, Espãna, p 849

Quaggiotti S, Ruperti B, Pizzeghello D, Francioso O, Tugnoli V, Nardi S (2004) Effect of low molecular size humic substances on nitrate uptake and expression of genes involved in nitrate transport in maize (*Zea mays* L.). J Exp Bot 55:803–813

Rao SC, Dao TH (1996) Nitrogen placement and tillage effect on dry matter and nitrogen accumulation and redistribution in winter wheat. Agron J 88:365–371

Raper RL, Reeves DW, Burmester CH, Schwab EB (2000) Tillage depth, tillage timing and cover crop effects on cotton yield, soil strength, and tillage energy requirements. Appl Eng Agric 16:379–385

Rapp HS, Bellinder RR, Wien HC, Vermeylen FM (2004) Reduced tillage, rye residues, and herbicides influence weed suppression and yield of pumpkins. Weed Technol 18:953–961

Rasmussen KJ (1999) Impact of ploughless soil tillage on yield and soil quality: a Scandinavian review. Soil Till Res 53:3–14

Rejon A, Garcia-Romera I, Ocampo JA, Bethlenfalvay GJ (1997) Mycorrhizal fungi influence competition in a wheat-ryegrass association treated with the herbicide diclofop. Appl Soil Ecol 7:51–57

Rhoton FE (2000) Influence of time on soil responses to no-till practices. Soil Sci Soc Am J 64:700–709

Rieger SB (2001) Impacts of tillage systems and crop rotation on crop development, yield and nitrogen efficiency. Ph.D. dissertation, ETH 14124, Swiss Federal Institute of Technology Zurich, Zurich, Switzerland, 139 pp

Riffaldi R, Saviozzi A, Levi-Minzi R, Cardelli R (2002) Biochemical properties of a Mediterranean soil as affected by long-term crop management systems. Soil Till Res 67:109–114

Rillig MC, Mummey DL (2006) Mycorrhizas and soil structure. New Phytol 171:41–53

Roldán A, Salinas-García JR, Alguacil MM, Caravaca F (2005a) Changes in soil enzyme activity, fertility, aggregation and C sequestration mediated by onservation tillage practices and water regime in a maize field. Appl Soil Ecol 30:11–20

Roldán A, Salinas-García JR, Alguacil MM, Caravaca F (2005b) Soil enzyme activities suggest advantage of conservation tillage practices in sorghum ultivation under subtropical conditions. Geodema 129:178–185

Romkens MJM, Nelson DW (1974) Phosphorus relationships in runoff from fertilized soils. J Environ Qual 3:10–13

Ruiz-Lozano JM (2003) Arbuscular mycorrhizal symbiosis and alleviation of osmotic stress. New perspectives for molecular studies. Mycorrhiza 13:309–317

Russo VM, Kindiger B, Webber CL III (2006) Pumpkin yield and weed populations following annual ryegrass. J Sustain Agric 28:85–96

Sadeghi AM, Isensee AR, Shelton DR (1998) Effect of tillage age on herbicide dissipation: a side-by-side comparison using microplots. Soil Sci 163:883–890

Sadowsky MJ, Graham PH (1998) Soil biology of the Rhizobiaceae. In: Spaink HP, Kondorosi A, Hooykaas PJJ (eds) The Rhizobiaceae. Kluwer, Dordrecht, The Netherlands, pp 155–172

Salinas-Garcia JR, Hons FM, Matocha JE, Zuberer DA (1997) Soil carbon and nitrogen dynamics as affected by long-termtillage. Biol Fertil Soils 25:182–188

Sánchez-Díaz M, Honrubia M (1994) Water relations and alleviation of drought stress in mycorrhizal plants. In: Gianinazzi S, Schü epp H (eds) Impact of arbuscular mycorrhizas on sustainable agriculture and natural ecosystems. Birkhäuser Verlag, Basel, Switzerland, pp 167–178

Santos JB, Jakelaitis A, Silva AA, Costa MD, Manabe A, Silva MCS (2006) Action of two herbicides on the microbial activity of soil cultivated with common bean (*Phaseolus vulgaris*) in conventional-till and no-till systems. Weed Res 46:284–289

Satchell JE (1958) Earthworm biology and soil fertility. Soils Fertil 21:209–219

Sauer TJ, Clothier BE, Daniel TC (1990) Surface measurements of the hydraulic character of tilled and untilled soil. Soil Till Res 15:359–369

Scheu S (2003) Effects of earthworms on plant growth: patterns and perspectives. Pedobiologia 47:846–856

Schjonning P, Rasmussen KJ (1989) Long-term reduced cultivation. I. Soil strength and stability. Soil Till Res 15:79–90

Schmidt O, Clements RO, Donaldson G (2003) Why do cereal-legume intercrops support large earthworm populations? Appl Soil Ecol 22:181–190

Scholz G, Quinton JN, Strauss P (2008) Soil erosion from sugar beet in Central Europe in response to climate change induced seasonal precipitation variations. Catena 72:91–105

Schreiner RP, Ivors KL, Pinkerton JN (2001) Soil solarization reduces arbuscular mycorrhizal fungi as a consequence of weed suppression. Mycorrhiza 11:273–277

Schütte G (2003) Herbicide resistance: promises and prospects of biodiversity for European agriculture. Agric Hum Values 20:217–230

Schutter ME, Sandeno JM, Dick RP (2001) Seasonal, soil type, and alternative agriculture management influences on microbial communities of vegetable cropping systems. Biol Fertil Soils 34:397–410

Schutter ME, Dick RP (2002) Microbial community profiles and activities among aggregates of winter fallow and covercropped soil. Soil Sci Soc Am J 66:142–153

Schwab SM, Johnson ELV, Menge JA (1982) Influence of simazine on formation of vesicularar-buscular mycorrhizae in Chenopodium quinona Willd. Plant Soil 64:283–287

Schwerdtle F (1977) Der Einfluss des Direkts.averfahrens auf die Verunkrautung. Z PflKrankh PflSchutz Sonderheft 8:155–163

Scullion J, Malik A (2001) Organic matter in restored soils as affected by earthworms and land use. In: Rees RM, Ball BC, Campbell CD, Watson CA (eds) Sustainable management of organic matter. CAB International Publishing, Wallingford, UK, pp 377–384
Selles F, McConkey BG, Campbell CA (1999) Distribution and forms of P under cultivator- and zero-tillage for continuous- and fallow-wheat cropping systems in the semi-arid Canadian prairies. Soil Till Res 51:47–59
Shaver TM, Peterson GA, Ahuja LR, Westfall DG, Sherrod LA, Dunn G (2002) Surface soil properties after twelve years of dryland no-till management. Soil Sci Soc Am J 66:1292–1303
Shipitalo MJ, Butt KR (1999) Occupancy and geometrical properties of *Lumbricus terrestris* L. burrows affecting infiltration. Pedobiologia 43:782–794
Shipitalo MJ, Le Bayon RC (2004) Quantifying the effects of earthworms on soil aggregation and porosity. In: Edwards CA (ed) Earthworm ecology. CRC Press, Boca Raton, p 441
Sieling K, Schröder H, Finck M, Hanus H (1998) Yield, N uptake, and apparent N-use efficiency of winter wheat and winter barley grown in different cropping systems. J Agric Sci 131: 375–387
Sieverding E (1991) Vesicular–arbuscular mycorrhiza management in tropical agroecosystems. Gesellschaft fur Technische Zusammenarbeit (GTZ) GmbH, Esebborn, Germany
Sievert M (2000) Aspekte des Pflanzenschutzes in Winterraps, Winterweizen und Wintergerste bei nichtwendender Bodenbearbeitung, Dissertation, Fakultät für Agrarwissenschaften, Georg-August Universität Göttingen, Germany
Sime M (1986) Field Performance of the Maresha plow and Nazret plow. AIRIC Test Report No. 14, Agricultural Implements Research and Improvement Centre, Eth/82/004. Institute of Agricultural Research, Addis Ababa, 12 pp
Simmons BL, Coleman DC (2008) Microbial community response to transition from conventional to conservation tillage in cotton fields. Appl Soil Ecol 40:518–528
Singh G, Wright D (1999) Effects of herbicides on nodulation, symbiotic nitrogen fixation, growth and yield of pea (*Pisum sativum*). J Agric Sci 133:21–30
Singh G, Wright D (2002) In vitro studies on the effects of herbicides on the growth of rhizobia. Lett Appl Microbiol 35:12–16
Siquiera JO, Safir GR, Nair MG (1991) VA-mycorrhizae and mycorrhiza stimulating isoflavanoid compounds reduce plant herbicide injury. Plant Soil 134:233–242
Six J, Elliot ET, Paustian K (1999) Aggregate and soil organic matter dynamics under conventional tillage and no-tillage systems. Soil Sci Soc Am J 63:1350–1358
Six J, Feller C, Denef K, Ogle SM, de Moraes Sa JC, Albrecht A (2002) Soil organic matter, biota and aggregation in temperate and tropical soils – effects of notillage. Agronomie 22:755–775
Smettem KRJ, Collis-George N (1985) The influence of cylindrical macropores on steady state infiltration in a soil under pasture. J Hydrol 79:107–114
Smettem KRJ (1992) The relation of earthworms to soil hydraulic properties. Soil Biol Biochem 24:1539–1543
Soane BD, van Ouwerkerk C (1995) Implications of soil compaction in crop production for the quality of the environment. Soil Till Res 35:5–22
Spedding TA, Hamel C, Mehuys GR, Madramootoo CA (2004) Soil microbial dynamics in maize-growing soil under different tillage and residue management systems. Soil Biol Biochem 36: 499–512
Staley TE, Edwards WM, Scott CL, Owens LB (1988) Soil microbial biomass and organic component alterations in a no-tillage chronosequence. Soil Sci Soc Am J 52:998–1005
Stenersen J, Gilman A, Vardanis A (1974) Carbofuran: its toxicity to and metabolism by earthworm (*Lumbricus terrestris*). J Agric Food Chem 22:342–347
Stockfisch N, Forstreuter T, Ehlers W (1999) Ploughing effects on soil organic matter after twenty years of conservation tillage in Lower Saxony Germany. Soil Till Res 52:91–101
Streit B, Rieger SB, Stamp P, Richner W (2002) The effect of tillage intensity and time of herbicide application on weed communities and populations in maize in central Europe. Agric Ecosyst Environ 92:211–224

Ström L, Owen AG, Godbold DL, Jones DL (2005) Organic acid behavior in a calcareous soil implications for rhizosphere nutrient cycling. Soil Biol Biochem 37:204–2054
Stromberger M, Shah Z, Westfall D (2007) Soil microbial communities of no-till dryland agroecosystems across an evapotranspiration gradient. Appl Soil Ecol 35:94–106
Sturz AV, Carter MR, Johnston HW (1997) A review of plant disease, pathogen interactions and microbial antagonism under conservation tillage in temperate humid agriculture. Soil Till Res 41:169–189
Subler S, Baranski CM, Edwards CA (1997) Earthworm additions increased short-term nitrogen availability and leaching in two graincrop agroecosystems. Soil Biol Biochem 29:413–421
Sun RL, Zhao BQ, Zhu LSh, Xu J, Zhang FD (2003) Effects of long-term fertilization on soil enzyme activities and its role in adjusting-controlling soil fertility. Plant Nutr Fertil 9:406–410
Swanton CJ, Clements DR, Derksen DA (1993) Weed succession under conservation tillage: a hierarchical framework for research and management. Weed Technol 7:286–297
Sylvia DE, Hammond LC, Bennet JM, Hass JH, Linda SB (1993) Field response of maize to a VAM fungus and water management. Agron J 85:193–198
Tardieu F (1994) Growth and functioning of roots and root systems subjected to soil compaction. Towards a system with multiple signaling? Soil Till Res 30:217–243
Tebrügge F, Böhrnsen A (2001) Farmers and 'experts' opinion on no-tillage in West-Europe and Nebraska (USA). In: Torres G et al (eds) Conservation agriculture, aworldwide challenge. Proceedings of the First World Congress on Conservation Agriculture of FAO-ECAF, vol. I, Madrid, Spain, pp 61–69, 1–5 Oct 2001
Temesgen M, Hoogmoed WB, Rockstrom J, Savenije HHG (2009) Conservation tillage implements and systems for smallholder farmers in semi-arid Ethiopia. Soil Till Res 104:185–191
Tessier S, Peru M, Dyck FB, Zentner FP, Campbell CA (1990) Conservation tillage for spring wheat production in semi-arid Saskatchewan. Soil Till Res 18:73–89
Thomas AG, Frick BL (1993) Influence of tillage systems on weed abundance in Southwestern Ontario. Weed Technol 7:699–705
Thomas GA, Dalal RC, Standley J (2007) No-till effects on organic matter, pH, cation exchange capacity and nutrient distribution in a Luvisol in the semi-arid subtropics. Soil Till Res 94:295–304
Tiessen H, Cuevas E, Chacon P (1994) The role of soil organic-matter in sustaining soil fertility. Nature 371:783–785
Tisdal JM, Oades JM (1982) Organic matter and water stable aggregates in soils. J Soil Sci 33:141–161
Tobar RM, Azćon R, Barea JM (1994) Improved nitrogen uptake and transport from 15N-labeled nitrate by external hyphae of arbuscular mycorrhizae under water-stressed conditions. New Phytol 126:119–122
Tomlin C (2003) The pesticide manual, 13th edn. British Crop Protection Council, UK, p 347
Tørresen KS, Skuterud R (2002) Plant protection in spring cereal production with reduced tillage. IV. Changes in the weed flora and weed seedbank. Crop Prot 21:179–193
Trasar-Cepeda C, Leiros MC, Seoane S, Gil-Sotres F (2000) Limitation of soil enzymes as indicators of soil pollution. Soil Biol Biochem 32:1867–1875
Tuffen F, Eason WR, Scullion J (2002) The effect of earthworm and arbuscular mycorrhizal fungi on growth of and ^{32}P transfer between *Allium porrum* plants. Soil Biol Biochem 34:1027–1036
Van Gestel M, Ladd JN, Amato M (1992) Microbial biomass responses to seasonal change and imposed drying regimes at in creasing depths of undisturbed topsoil profiles. Soil Biol Biochem 24:103–111
Vasil'chenko LG, Khromonygina VV, Koroleva OV, Landesman EO, Gaponenko VV, Kovaleva TA, Kozlov Yu P, Rabinovich ML (2002) Prikl Biokhim Mikrobiol 38:534–539
Vieira RF, Silva CMMS, Silveira APD (2007) Soil microbial biomass C and symbiotic processes associated with soybean after sulfentrazone herbicide application. Plant Soil 300:95–103
Visser S, Parkinson D (1992) Soil biological criteria as indicators of soil quality: soil organisms. Am J Altern Agric 7:33–37
Vogeler I, Rogasik J, Funder U, Panten K, Schnug E (2009) Effect of tillage systems and P-fertilization on soil physical and chemical properties, crop yield and nutrient uptake. Soil Till Res 103:137–143

Vullioud PA (2000) 30 years ploughless tillage experiment at Changins (Switzerland). In: Proceedings of the 15th ISTRO conference, Forth Worth, TX, 10 pp

Walters SA, Young BG, Krausz RF (2008) Influence of tillage, cover crop, and preemergence herbicides on weed control and pumpkin yield. Int J Veg Sci 14:148–161

Wang J, Hesketh JD, Woolley JT (1986) Preexisting channels and soybean rooting patterns. Soil Sci 141:432–437

Wang D, Norman JM, Lowery B, McSweeney K (1994) Non-destructive determination of hydrogeometrical characteristics of soil macropores. Soil Sci Soc Am J 58:294–303

Wardle DA, Yeates GW, Watson RN, Nicholson KS (1993) Response of soil microbial biomass and plant litter decomposition to weed management strategies in maize and asparagus cropping systems. Soil Biol Biochem 25:857–868

Wardle DA (1995) Impacts of disturbance on detritus food webs in agro-ecosystems of contrasting tillage and weed management practices. Adv Ecol Res 26:105–185

Weil RR, Lowell KA, Shade HM (1993) Effects of intensity of agronomic practices on a soil ecosystem. Am J Altern Agric 8:5–14

Wershaw RL (1993) Model for humus in soils and sediment. Environ Sci Technol 27:814–816

Westwood J (1997) Growers endorse herbicide resistant crops, recognize need for responsible use. ISB News 3:7–10

Widmer F (2007) Assessing effects of transgenic crops on soil microbial communities. Green Gene Technology. Res Area Soc Confl 107:207–234

Wright AL, Hons FM, Matocha JE Jr (2005) Tillage impacts on microbial biomass and soil carbon and nitrogen dynamics of corn and cotton rotations. Appl Soil Ecol 29(1):85–92

Xu J, Qiu X, Dai J, Cao H, Yang M, Zhang J, Xu M (2006) Isolation and characterization of a Pseudomonas oleovorans degrading the chloroacetamide herbicide acetochlor. Biodegradation 17:219–225

Yao HY, Jiao XD, Wu FZ (2006) Effects of continuous cucumber cropping and alternative rotations under protected cultivation on soil microbial community diversity. Plant Soil 284:195–203

Zabaloy C, Anahí Gómez M (2005) Diversity of rhizobia isolated from an agricultural soil in Argentina based on carbon utilization and effects of herbicides on growth. Biol Fertil Soils 42:83–88

Zaidi A, Khan MS, Rizvi PQ (2005) Effect of herbicides on growth, nodulation and nitrogen content of greengram. Agron Sustain Dev 25:497–504

Zandstra BH, Chase WR, Masabni JG (1998) Interplanted small grain cover crops in pickling cucumbers. HortTechnology 8:356–360

Zhang GS, Chan KY, Oates A, Heenan DP, Huang GB (2007) Relationship between soil structure and runoff/soil loss after 24 years of conservation tillage. Soil Till Res 92:122–128

Zibilske LM, Bradford JM, Smart JR (2002) Conservation tillage induced changes in organic carbon, total nitrogen and available phosphorus in a semi-arid alkaline subtropical soil. Soil Till Res 66:153–163

Zotarelli L, Alves BJR, Urquiaga S, Torres E, dos Santos HP, Paustian K, Boddey RM, Six J (2005) Impact of tillage and crop rotation on aggregateassociated carbon in two oxisols. Soil Sci Soc Am J 69:482–491

Synergism Among Crops to Improve Dryland Crop Production

Randy L. Anderson

Abstract Water supply is a major constraint for crop production in dryland agriculture across the world, and extensive research has been conducted to improve water use. In the grass steppe of the United States, water use has improved through a series of management advancements, such as preservation of crop residue on the soil surface, no-till, and crop diversity. We have observed an additional advancement after several years of no-till rotations; some crops synergistically improve water-use-efficiency (WUE) of following crops. For example, proso millet (*Panicum miliaceum* L.) produces 24% more grain with the same water use following corn (*Zea mays* L.) than following winter wheat (*Triticum aestivum* L.). The presence of corn and dry pea (*Pisum sativum* L.) in the rotation also improves WUE of winter wheat. Furthermore, synergism among crops increases tolerance of weed interference. The cause of synergism is not known, but identifying synergistic crop sequences and designing rotations to include these sequences can improve water conversion into grain for dryland agriculture. Because of no-till, crop diversity, and synergism, producers in the U.S. steppe have doubled land productivity with the same water supply.

Keywords Crop diversity • No-till • Water-use-efficiency • Weed tolerance

1 Introduction

Dryland farming is a major component of global crop production. Stewart et al. (2006) noted that 60% of the world's food production occurs in dryland regions, which they defined as where lack of water limits crop production during some part of the growing season. Consequently, water conservation techniques are a critical factor for successful dryland farming.

R.L. Anderson (✉)
USDA, 2923 Medary Avenue, Brookings, South Dakota, USA
e-mail: randy.anderson@ars.usda.gov

E. Lichtfouse (ed.), *Biodiversity, Biofuels, Agroforestry and Conservation Agriculture*, Sustainable Agriculture Reviews 5, DOI 10.1007/978-90-481-9513-8_8,

Extensive research has been conducted to minimize impact of limited water on crop production. One goal is to increase water-use-efficiency (WUE) of crops through breeding (Fereres 2004). A second goal is to improve effective use of water by maximizing soil water capture and transpiration by the plant (Blum 2009). For example, crop diversity can suppress root diseases, thus improving root growth to extract more soil water (Passioura 2006). Other cultural practices for effective water use include timeliness of sowing, evenness of crop establishment, and nutrient management.

In the grass steppe of the United States, we have observed an intriguing response with crop diversity that may further help alleviate water limitations in dryland farming. Some crops can synergistically improve WUE of following crops as much as 20–30% (Anderson 2005a). Synergism along with no-till and crop residue preservation on the soil surface have enabled producers to improve both WUE and effective water use with new rotations. This paper will describe the changes in farming practices in the U.S. steppe that led to this beneficial interaction. Our goal is to encourage scientists in dryland regions of the world to consider crop synergism as an aid to improving water use and crop production.

2 Evolution of Water Management Strategies

Producers have been adapting to climatic conditions in the U.S. steppe since they settled in the region. Initially, they followed cropping practices common to their place of origin. But, as producers experienced variability in precipitation, they developed numerous strategies to adjust production practices to the region's climate (Fig. 1). Strategies evolved over time as knowledge of water relations and technology improved.

The first strategy producers adopted was fallow, an interval where precipitation is stored in soil because neither crops nor weeds are allowed to grow (Black et al. 1974).

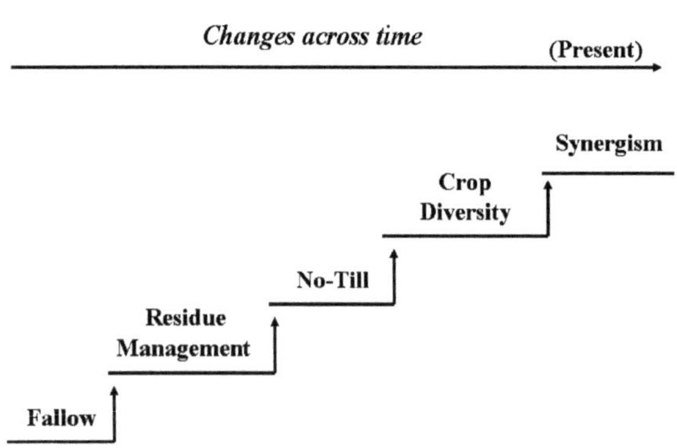

Fig. 1 Evolution of cultural practices related to water management in the semiarid steppe of the United States (Adapted from Anderson 2005b, 2009b)

Soil water gained during fallow minimized yield variability and crop loss due to drought. Weeds were controlled by tilling with the moldboard plow or tandem-disk, resulting in a condition known as 'dust mulch' fallow. This strategy led to the prevalent rotation in the region of winter wheat–fallow.

Mathews and Army (1960), assessing precipitation storage efficiency (PSE) during fallow at several locations in the steppe, found that percentage of precipitation stored during dust mulch fallow was usually less than 20%. They suggested that PSE during fallow would not improve until soil water evaporation was reduced. Duley and Russel (1939), recognizing the value of crop residues in suppressing soil water evaporation, developed the 'stubble mulch' system. Weeds are controlled with a sweep plow, which is comprised of V-shaped blades that sever weed roots with minimum soil disturbance. One operation with the sweep plow buries only 10–15% of crop residues lying on the soil surface, in contrast with tandem-disking or moldboard plowing burying 60–100% of crop residues. Stubble mulch increased PSE to more than 30%. A further benefit of preserving crop residue on the soil surface is that it reduced wind erosion that was so prevalent with the dust mulch system.

Smika and Unger (1986), examining impact of the sweep plow on residue burial and soil evaporation, found that in addition to burying some crop residue, each tillage operation also caused 0.5–0.8 cm of soil water to evaporate. They suggested that a no-till system with weeds controlled by herbicides would improve precipitation storage by eliminating these losses. Greb (1983) further supported this hypothesis by showing that PSE was related to quantity of crop residues on the soil surface; precipitation storage during fallow increased 1 cm for each 1,000 kg/ha of winter wheat residue.

Peterson et al. (1996), comparing fallow management systems across several locations in the steppe, verified that no-till fallow improved PSE compared to stubble mulch fallow. But, they also noted that PSE in no-till seldom exceeded 40%, and suggested that this may be the maximum efficiency obtainable with the wheat–fallow system. They hypothesized that including warm-season crops such as corn in the rotation may be the next step in improving PSE during fallow.

Farahani et al. (1998b) tested this hypothesis of diversifying the winter wheat–fallow rotation by examining data from cropping systems studies in the steppe. Efficiency of storing precipitation during non-crop intervals increased to almost 50% when warm-season crops were added to the winter wheat–fallow rotation. The reason for this gain is that non-crop intervals occur mainly over winter when PSE is highest (Peterson et al. 1996). Farahani et al. (1998a) also found that precipitation-use-efficiency (PUE, defined as percentage of precipitation converted to crop growth; i.e., effective water use) approached 75% with no-till, continuous cropping. In contrast, PUE of winter wheat–fallow was approximately 40%. Continuous cropping improves PUE by minimizing fallow intervals that are inefficient in PSE.

These new management systems for water have enabled producers to successfully grow other crops along with winter wheat and fallow. A further benefit of crop diversity, however, is that some crops improve WUE of following crops 20% or more (Anderson 2005a). We suggest that this change in WUE with some crop sequences, which we term synergism, may be the next step of advancement needed to improve crop production in the dryland steppe (Fig. 1).

We observed this synergism in a cropping systems study established at a site in the steppe where yearly rainfall averages 416 mm (Anderson et al. 1999). The study evaluated several crop rotations established with no-till. Rotations included crops such as corn, sunflower (*Helianthus annuus* L.), soybean (*Glycine max* Merrill), proso millet, foxtail millet [*Setaria italica* (L.) Beauv.], oat (*Avena sativa* L.), and dry pea along with winter wheat and fallow. All phases of each rotation were present in each year. Soil water extraction to a depth of 2 m was determined for each crop with a neutron probe to calculate WUE for grain yield (based on precipitation + soil water extraction during the crop growing season).

Five years after starting the study, WUE of some crops begin differing among crop sequences (Anderson 2005a). For example, proso millet WUE increased from 7.5 kg ha^{-1} mm^{-1} following winter wheat to 9.3 kg ha^{-1} mm^{-1} following corn (Table 1). Soil water level at planting time and crop water use by proso millet were similar following both crops, yet proso millet was 24% more efficient converting water into grain following corn. This trend was measured across a 4-year interval when yield of proso millet ranged from 1,600 to 3,500 kg/ha, yet the difference in WUE was consistent across years.

We also observed that WUE of winter wheat varied with crop sequence. Winter wheat WUE increased from 9.1 kg ha^{-1} mm^{-1} in winter wheat–fallow (W–F) to 11.5 kg ha^{-1} mm^{-1} in winter wheat–corn–fallow (W–C–F) (Fig. 2). We initially attributed this yield benefit to the longer interval between winter wheat crops in W–C–F suppressing root diseases of winter wheat. However, we were surprised that winter wheat WUE did not differ between W–F and winter wheat–proso millet–fallow (W–M–F). Furthermore, WUE of winter wheat did not change with W–C–M–F compared with the W–C–F rotation. But, growing dry pea (P) as a green fallow (terminated after 6–8 weeks growth) in a W–C–M–P rotation improved WUE of winter wheat to 12.8 $kg^{-1}ha^{-1}$ mm^{-1}, or 11% higher than W–C–M–F or W–C–F. Winter wheat produces 41% more grain in W–C–M–P than in W–F with the same water use.

We wondered what could cause this change in WUE. We first consider nutrient status, but N and P needs were adequate because nutrient management was based on annual soil tests and target yield goals; furthermore, P was banded with the crop

Table 1 Impact of preceding crop on proso millet yield and water use; data averaged across 4 years (Adapted from Anderson 2005b)

Agronomic data	Preceding crop	
	Wheat	Corn
Grain yield (kg/ha)	2,020	2,320[a]
Available soil water at planting (mm)	142	131
Water use (mm)	269	250
Water-use-efficiency (kg ha^{-1} mm^{-1})	7.5	9.3[a]

[a]Treatment means within an agronomic parameter were significantly different at 0.05 level of probability

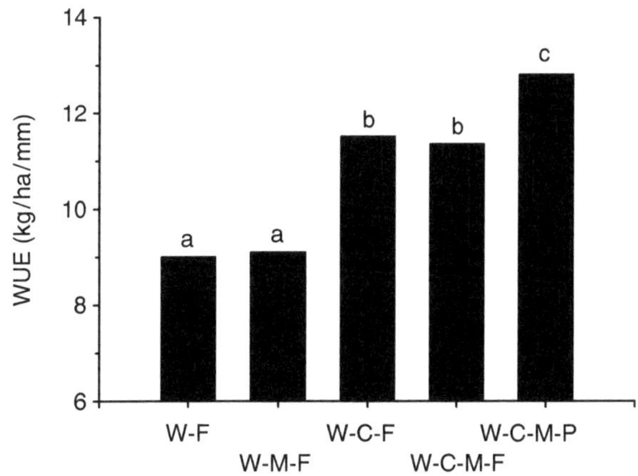

Fig. 2 Water-use-efficiency (WUE) of winter wheat in various no-till rotations in the semiarid steppe of the United States. Data collected across a 5-year interval; *bars with the same letter* are not significantly different as determined by Fischer's Protected LSD (0.05). Abbreviations: W, winter wheat; F, fallow; C, corn; M, proso millet; P, dry pea (Adapted from Anderson 2009b)

seed at planting (Anderson et al. 1999). Also, we did not observe any pest issues that would explain this difference. We were unable to explain why, but corn improved WUE of proso millet whereas winter wheat WUE increased when corn and dry pea were included in the rotation. We did not observe WUE changes with other crop sequences in the study.

Producers in the region have also noted a dramatic increase in winter wheat yields in no-till rotations (Anderson 2005b). Winter wheat yield rarely exceeds 2,650 kg/ha in producer fields with winter wheat–fallow and tillage. In contrast, winter wheat yields more than 5,400 kg/ha during favorable years with a no-till winter wheat–corn–proso millet–dry pea (as green fallow) rotation. Similarly, proso millet yields in some years exceed 4,200 kg/ha in this four-crop rotation, but with winter wheat–proso millet–fallow and tillage, proso rarely yields more than 2,000 kg/ha. The doubling of yield potential of winter wheat and proso millet reflects the interaction of residue preservation on the soil surface, no-till, crop diversity, and synergism.

3 Further Evidence for Synergism Among Crops

3.1 Crop Tolerance to Weed Interference

Because weeds compete with crops for water, we speculated that if a preceding crop increased WUE of a following crop, tolerance to weeds may also improve. To test this hypothesis, we compared winter wheat tolerance to weed interference following spring wheat, dry pea, and soybean (Anderson 2009a). This study was conducted at a site in the U.S. steppe where yearly precipitation averages 580 mm. Spring wheat, dry pea, and soybean were established in corn stubble, with winter wheat planted the fall after harvesting the above crops. Crop management was based on best management practices for the region. Nutrient management for winter wheat included N + P fertilizer applied as a starter with the seed and N broadcast during the growing season. N rates were adjusted for preceding crops to include N credits for legumes. All crops were established with no-till and the study site had been in no-till for 5 years prior to initiating the study.

To establish uniform weed interference 15 plants/m^2 of wild rye (*Secale cereale* L.) were planted by hand, 3 days after winter wheat emergence. Each plot was split into weed-free and weed-infested subplots. The study was located in fields where intensive weed management in previous years resulted in low density of the native weed community.

Dry pea improved winter wheat tolerance to wild rye interference, as winter wheat yielded 5,030 kg/ha following dry pea, but less than 3,800 kg/ha following either soybean or spring wheat when wild rye was present (Fig. 3). Compared to weed-free conditions, yield loss due to wild rye was only 11% when winter wheat followed dry pea, but 32% following soybean. Even in weed-free conditions, winter wheat yielded 10% more after dry pea than following soybean. Yield of winter wheat varied from 4,020 to 6,730 kg/ha during the 4 years of the study, yet the impact of dry pea was consistent across years.

Lower yield following spring wheat likely was due to root diseases, as legumes reduce root disease severity in wheat compared to continuous wheat (Kirkegaard

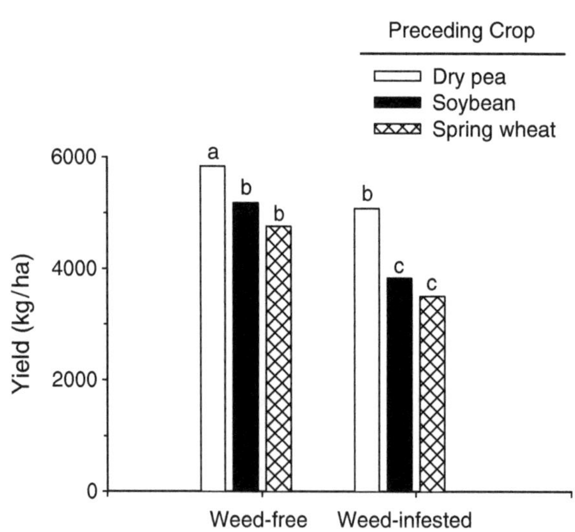

Fig. 3 Winter wheat tolerance to weed (wild rye) interference as affected by preceding crop. Dry weight of wild rye at winter wheat harvest was 350 ± 23 g/m^2. Data collected across a 4-year interval; *bars with the same letter* are not significantly different as determined by Fischer's Protected LSD (0.05) (Adapted from Anderson 2009a, 2010)

et al. 2008), but we were surprised that soybean was not favorable like dry pea. Differences in soil water levels after harvest of dry pea and soybean could be one factor, but extensive precipitation occurs during the fall and winter at this site. Consequently, the soil profile is usually at field capacity when spring growth starts, thus eliminating differences in soil water levels among preceding crops. Another factor could be N cycling in soil differed following dry pea and soybean, but wild rye biomass, measured at winter wheat harvest, did not vary among preceding crops. Yet, in some way, dry pea increased winter wheat tolerance of wild rye competing for water and nutrients.

Because winter wheat responded differently to preceding crops, we then examined these same crops for impact on corn tolerance to weeds (Anderson 2007). We also included corn as a preceding crop, and used foxtail millet to achieve uniform weed interference. Seeds of foxtail millet were broadcast on the soil surface the day of corn planting, resulting in 115 seedlings/m^2. Fresh weight of foxtail millet in corn was measured 7 weeks after emergence, and did not vary among preceding crop treatments.

Dry pea also helped corn tolerate weeds. Grain yield was reduced more than 75% by foxtail millet interference when corn followed soybean or spring wheat, but only 50% when corn followed dry pea (Fig. 4). In weed-free conditions, corn yielded 11% more following dry pea than either soybean or spring wheat. Corn yields varied from 7.5 to 10.1 Mg/ha during the 3 years of the study, yet the preceding crop effect was consistent across years.

Weed interference in corn following corn reduced yield more than 90% (Fig. 4). This drastic yield loss is due to allelopathy by corn residues, which severely stunt seedling growth during the first 5–6 weeks of corn growth (Crookston 1995). Even in weed-free conditions, corn following corn yielded 40% less than corn following dry pea.

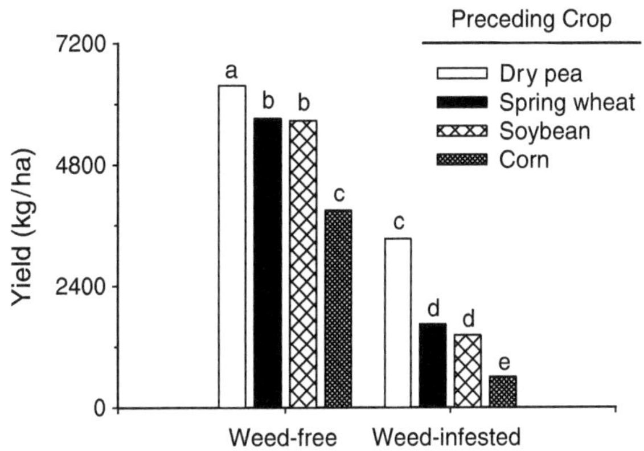

Fig. 4 Yield of corn as affected by preceding crop and foxtail millet (*Setaria italica*) interference. Fresh weight of foxtail millet 7 weeks after emergence was 1065 ± 85 g/m^2. Data averaged across 3 years; *bars with the same letter* are not significantly different as determined by Fischer's Protected LSD (0.05) (Adapted from Anderson 2007, 2009b)

We were intrigued that both winter wheat and corn responded more favorable to dry pea than soybean (Figs. 3 and 4). This trend may appear to be an anomaly, but we suggest that dry pea improves stress tolerance of winter wheat and corn to minimize weed interference. Earlier, we noted that dry pea increases WUE of winter wheat (Fig. 2), whereas Copeland et al. (1993) reported that soybean did not improve WUE of corn. Other legumes also vary in their effect on following crops. Praveen-Kumar et al. (1997) found that both WUE and nitrogen-use-efficiency in pearl millet (*Pennisetum glaucum* L.) were higher following cluster bean [*Cymopsis tetragonoloba* (L.) Tauber] than mung bean [*Vigna radiate* (L.) R. Wilczek].

3.2 Synergism Interacts with Corn Density to Affect Grain Yield

We further examined synergism by comparing corn yield at various densities as affected by the preceding crop. Seed cost is a major input of producers, and we wondered if synergism of dry pea to corn would alter the crop density–yield relationship. Therefore, we established corn at six densities from 38,000 to 73,000 plants/ha at intervals of 7,000 plants, into dry pea and soybean grown the preceding year; 73,000 plants/ha is the customary density used by producers. Corn was grown in weed-free conditions with best management practices used in the region.

Corn following soybean yielded the highest at 73,000 plants/ha, but corn following dry pea yielded similarly at 52,000 plants/ha, or with 21,000 less plants (Fig. 5). Yields ranged from 6.5 to 9.8 Mg/ha in years of the study, but yield trends among preceding crops were consistent across years. This change in optimum corn density

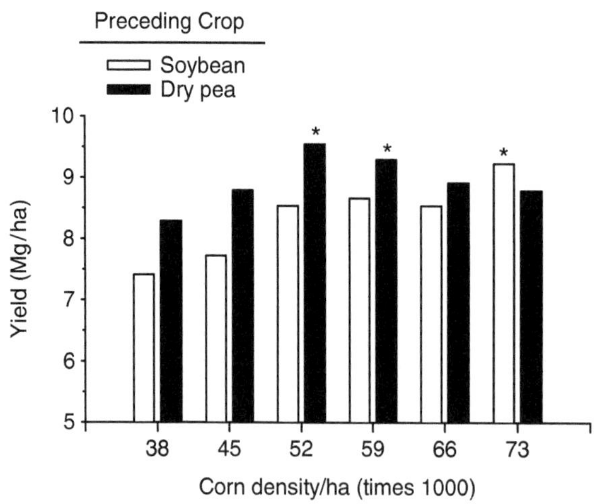

Fig. 5 Interaction between preceding crop and corn density on grain yield of corn in weed-free conditions. Data averaged across 2 years. *Asterisks* designate corn densities where yield did not differ from the highest yielding treatment, as determined by comparing treatment yields with Fisher's Protected LSD (0.05) (R.L. Anderson, research in progress)

following dry pea could reduce cost for seed more than $70/ha. Also, this effect can help dryland agriculture as lower crop densities consume less water (Debaeke and Aboudrare 2004). As found with weed tolerance, dry pea was more favorable than soybean in affecting the density–yield relationship with corn.

4 Possible Causes of Synergism Among Crops

Because of the favorable impact of dry pea on corn tolerance to weed interference, we attempted to identify the cause of this synergy in the corn density study. We monitored seedling emergence, growth, and development of corn in the 52,000 and 73,000 plants/ha treatments planted after dry pea and soybean (Anderson 2009c). We also measured concentration of N, P, Zn, and Cu in seedlings when six leaves of corn were fully emerged. The N concentration may indicate if N availability varied among preceding crops (Przednowek et al. 2004), whereas increased P, Zn, and Cu concentrations may reflect greater colonization of mycorrhizae due to preceding crop (Lambert et al. 1979; Hamel 2004). But, we found no differences in any parameter when corn followed dry pea or soybean, except higher grain yield following dry pea (Table 2). We speculate that synergism by dry pea affects corn physiology.

The cause of synergism may involve changes in the soil microbial community. Shaver et al. (2002) found that microbial biomass increases with no-till systems, whereas Mozafar et al. (2000) noted that the composition of the microbial community also changes with no-till. Lupwayi and Kennedy (2007), reviewing the

Table 2 Agronomic response of corn to dry pea or soybean as preceding crops. Data averaged across two corn densities and 2 years (Adapted from Anderson 2009c)

	Preceding crop	
Agronomic variable	Soybean	Dry pea
Corn seedling data		
Mean emergence rate (days)	16.8	17.1
V-6 leaf stage (days after June 1)	19.6	20.3
Measurements at V-6 leaf stage		
Height (cm)	39.3	39.2
Fresh weight (g/plant)	57	54
N (g/kg)	3.6	3.7
P (g/kg)	0.40	0.41
Zn (mg/kg)	29	29
Cu (mg/kg)	9.5	9.8
Height at V-9 leaf stage (cm)	107	109
Silking (days after July 1)	31	30
Yield (kg/ha)	8,490	9,520[a]

V-6; six leaves fully emerged; V-9; nine leaves fully emerged
[a]Means for a variable differed between preceding crops at 0.05 probability level

beneficial impact of legumes on small grains, suggested that rhizobacteria may be involved in the dry pea effect on following crops. Rhizobacteria improve plant growth by increasing resource-use-efficiency, crop tolerance to drought stress, and crop photosynthesis efficiency (Sturz et al. 2000; Dobbelaere et al. 2003).

Yet, attempting to identify one factor may distort understanding the underlying cause of synergy between crops. A difficulty with relating specific microbial species to a crop response is the extensive diversity of organisms present in soil and the plethora of interacting processes (Andren et al. 2008). Only a small fraction of soil organisms have been identified, which inherently limits our scope in explaining crop response. Other factors, such as growth-promoting compounds or hormones released by residue decomposition, could also be involved (Anaya 1999). A possible explanation is the crop response involves a change in the ratio or balance of multiple factors (Arshad and Frankenberger 1998).

The complexity of this interaction was noted by Kirkegaard et al. (2008) in reviewing the impact of crop diversity on root diseases. They found that inexplicable rotation effects often confounded the impact of known factors with beneficial crop sequences. Because of the complexity in soil biology, we may not be able to identify a specific cause for crop synergy. Furthermore, synergism is also related to the crops in sequence; not all sequences are synergistic. For example, corn is synergistic to proso millet (Table 1) but not to sorghum [*Sorghum bicolor* (L.) Moench] (Anderson 2005a). Similarly, dry pea is synergistic to winter wheat and corn, but not to soybean (R.L. Anderson, research in progress) in tolerating weed interference.

In an earlier program assessing the rotation effect on crop yield (Anderson 2005a), we recognized that positive responses to preceding crops could be grouped into two categories, either improving plant efficiency or increasing plant capacity for growth. Sequences that improve plant efficiency yield more with the same water use, such as when winter wheat follows dry pea or proso millet follows corn. A contrasting response occurs when the sequence improves crop capacity; yield increases only when more water is available. An example of the capacity response is canola (*Brassica napus* L.) increasing winter wheat yield in Australia. Angus and van Herwaarden (2001) found that canola did not improve WUE of winter wheat, but increased its ability to extract more water from the soil profile because of healthier roots. If extra soil water was not available, yield did not increase. Because a crop integrates multiple factors of soil biology, we suggest that measuring WUE may serve as an indicator of synergism among crops.

5 Synergism in Dryland Crop Production

Improving WUE and stress tolerance of some crops with synergism has helped producers develop sustainable production systems in the semiarid region of the U.S. steppe. For example, fallow was adopted in this region to improve success of cropping, but the 14-month fallow interval led to soil degradation and wind erosion

(Peterson et al. 1993). Bowman et al. (1990) found that more than 60% of the native organic matter has been lost in 70 years of winter wheat–fallow. No-till and residue management enabled producers to expand their rotations to include warm-season crops, yet even in rotations with three crops in 4 years, the 12- to 14-month fallow prevented restoration of organic matter levels (Sherrold et al. 2005). Initially, rotations with continuous cropping failed because of the limited water supply, but when rotations were designed to include synergistic sequences, production was successful without a 12- to 14-month fallow interval.

One rotation, winter wheat–corn–proso millet–dry pea (as green fallow), has increased land productivity and net returns more than twofold compared with winter wheat–fallow at the same precipitation level (Anderson 2009b). Organic matter levels and stability of soil aggregates are increasing; also, soil porosity has improved to further aid water management (Sherrold et al. 2005). Furthermore, these synergistic rotations are more protective of the environment because soil nitrate levels in soil have declined and less herbicides are needed for weed management (Anderson 2009b).

Even though we cannot explain why synergism among some crops occurs, we have recognized management tactics that favor this effect. The advances in water management in the U.S. steppe were shown as a series of steps (Fig. 1) because synergism has been observed only in systems that include residue preservation on the soil surface, no-till, and crop diversity.

Sustainability is critical for global agriculture. To achieve sustainability, Hobbs (2007) suggested merging crop diversity with no-till and residue preservation on the soil surface, whereas Kirschenmann (2007) encouraged redesigning the system to accentuate the biological synergies in multi-species rotations. Producers in the U.S. steppe have achieved sustainability by combining these suggestions and integrating crop synergism with residue cover, no-till, and crop diversity. We believe that crop synergism can enhance water management in other dryland regions of the world also, if synergistic sequences among crops were identified and included in the rotation.

References

Anaya AL (1999) Allelopathy as a tool in the management of biotic resources in agroecosystems. Crit Rev Plant Sci 18:697–739

Anderson RL (2005a) Are some crops synergistic to following crops? Agron J 97:7–10

Anderson RL (2005b) Improving sustainability of cropping systems in the Central Great Plains. J Sustain Agric 26:97–114

Anderson RL (2007) A changing perspective with weed management in semiarid cropping systems. Ann Arid Zone 46:1–15

Anderson RL (2009a) Impact of preceding crop and cultural practices on rye growth in winter wheat. Weed Technol 23:564–568

Anderson RL (2009b) Rotation design: a critical factor for sustainable crop production in a semiarid climate. A review. In: Lichtfouse E (ed) Organic farming, pest control, and remediation of soil pollutants. Springer, Secausus, NJ, pp 107–121

Anderson RL (2009c) Pea synergism to corn is not related to seedling growth. In: Western Society of Weed Science Research Report. Western Society of Weed Science, Las Cruces, NM, pp 102–103
Anderson RL (2010) Dry pea improves winter wheat tolerance to wild rye. 2010 West. Soc. Weed Sci. Res. Rept. Western Society of Weed Science, Las Cruces, New Mexico. pp. 115–116.
Anderson RL, Bowman RA, Nielsen DC, Vigil MF, Aiken RM, Benjamin JG (1999) Alternative crop rotations for the central Great Plains. J Prod Agric 12:95–99
Andren O, Kirchmann H, Katterer T, Magid J, Paul EA, Coleman DC (2008) Visions of a more precise soil biology. Eur J Soil Sci 59:380–390
Angus JF, Van Herwaarden AF (2001) Increasing water use and water use efficiency in dryland winter wheat. Agron J 93:290–298
Arshad M, Frankenberger JRWT (1998) Plant growth-regulating substances in the rhizosphere: microbial production and functions. Adv Agron 62:45–151
Black AL, Siddoway FH, Brown PL (1974) Summer fallow in the Northern Great Plains (winter wheat). In: Summer Fallow in the Western United States. USDA-ARS Conservation Research Report no. 17. U.S Government Printing Office, Washington, DC, pp. 36–50
Blum A (2009) Effective use of water (EUW) and not water-use-efficiency (WUE) is the target of crop yield improvement under drought stress. Field Crops Res 112:119–123
Bowman RA, Reeder JD, Lober LW (1990) Changes in soil properties after 3, 20, and 60 years of cultivation. Soil Sci 150:516–522
Copeland PJ, Allmaras RR, Crookston RK, Nelson WW (1993) Corn-soybean rotation effects on soil water depletion. Agron J 85:203–210
Crookston RK (1995) The rotation effect in corn. In: Wilkerson D (ed) Proceedings 50th annual corn sorghum research conference, 1995. American Seed Trade Association, Alexandria, VA, pp 201–215
Debaeke P, Aboudrare A (2004) Adaptation of crop management to water-limited environments. Eur J Agron 21:433–446
Dobbelaere S, Vanderleyden J, Okon Y (2003) Plant growth-promoting effects of diazotrophs in the rhizosphere. Crit Rev Plant Sci 22:107–149
Duley FL, Russel JC (1939) The use of crop residues for soil and moisture conservation. Agron J 31:703–709
Farahani HJ, Peterson GA, Westfall DG (1998a) Dryland cropping intensification: a fundamental solution to efficient use of precipitation. Adv Agron 64:197–223
Farahani HJ, Peterson GA, Westfall DG, Sherrod LA, Ahuja LR (1998b) Soil water storage in dryland cropping systems: the significance of cropping intensification. Soil Sci Soc Am J 62:984–991
Fereres E (2004) Water-limited agriculture. Eur J Agron 21:399–400
Greb BW (1983) Water conservation: Central Great Plains. In: Dregne HE, Willis WO (eds) Dryland agriculture, American Society of Agronomy Monograph 23. American Society of Agronomy, Madison, WI, pp 57–72
Hamel C (2004) Impact of arbuscular mycorrhizal fungi on N and P cycling in the root zone. Can J Soil Sci 84:383–395
Hobbs PR (2007) Conservation agriculture: what it is and why is it important for sustainable food production. J Agric Sci 145:127–137
Kirkegaard J, Christen O, Krupinsky J, Layzell D (2008) Break crop benefits in temperate wheat production. Field Crops Res 107:185–195
Kirschenmann FL (2007) Potential for a new generation of biodiversity in agroecosystems of the future. Agron J 99:373–376
Lambert DH, Baker DE, Cole H Jr (1979) The role of mycorrhizae in the interactions of phosphorus with zinc, copper, and other elements. Soil Sci Soc Am J 43:976–980
Lupwayi NZ, Kennedy AC (2007) Grain legumes in northern Great Plains: impacts on selected biological soil processes. Agron J 99:1700–1709
Mathews OR, Army TJ (1960) Moisture storage on fallowed wheatland in the Great Plains. Soil Sci Soc Am J 24:414–418

Mozafar A, Anken T, Ruh R, Frossard E (2000) Tillage intensity, mycorrhizal and nonmycorrhizal fungi, and nutrient concentrations in maize, wheat, and canola. Agron J 92:1117–1124
Passioura J (2006) Increasing crop productivity when water is scarce – from breeding to field management. Agric Water Manage 80:176–196
Peterson GA, Westfall DG, Cole CV (1993) Agroecosystem approach to soil and crop management research. Soil Sci Soc Am J 57:1354–1360
Peterson GA, Schlegel AJ, Tanaka DL, Jones OR (1996) Precipitation use efficiency as affected by cropping and tillage systems. J Prod Agric 9:180–186
Praveen-Kumar, Aggarwal RK, Power JF (1997) Cropping systems: effects on soil quality indicators and yield of pearl millet in an arid region. Am J Altern Agric 12:178–184
Przednowek DW, Entz MH, Irvine B, Flaten DN, Thiessen-Martens JR (2004) Rotational yield and apparent N benefits of grain legumes in southern Manitoba. Can J Plant Sci 84:1093–1096
Shaver TM, Peterson GA, Ahuja LR, Westfall DG, Sherrold LA, Dunn G (2002) Surface soil physical properties after twelve years of dryland no-till management. Soil Sci Soc Am J 66:1296–1303
Sherrold LA, Peterson GA, Westfall DG, Ahuja LR (2005) Soil organic pools after 12 years in no-till dryland agroecosystems. Soil Sci Soc Am J 69:1600–1608
Smika DE, Unger PW (1986) Effect of surface residues on soil water storage. Adv Soil Sci 5:111–138
Stewart BA, Koohafkan P, Ramamoorthy K (2006) Dryland agriculture defined and its importance to the world. In: Peterson GA, Unger PW, Payne WA (eds) Dryland agriculture. American Society of Agronomy Monograph 23, 2nd edn. American Society of Agronomy, Madison, WI, pp 1–26
Sturz AV, Christie BR, Nowak J (2000) Bacterial endophytes: potential role in developing sustainable systems of crop production. Crit Rev Plant Sci 19:1–30

Sustainable Irrigation to Balance Supply of Soil Water, Oxygen, Nutrients and Agro-Chemicals

Surya P. Bhattarai, David J. Midmore, and Ninghu Su

Abstract The socio-economic pressure for improvements in irrigation efficiencies is increasing due to intense competition for water between agricultural, domestic and industrial users as well as demands for compliance with environmental regulations. Precision irrigation technology involving less irrigation water and uniform application across the field is therefore important. In the context of declining water allocation for irrigation and the variations in weather and drought patterns attributed to global climate change, efficient and precise applications are necessity. Traditional irrigation methods such as furrow, flood and sprinkler are neither efficient nor environmentally benign. Precision irrigation methods such as drip and subsurface drip irrigation are advocated because they are more water use efficient and because they offer a possible approach to meet projected food demand, a doubling by the 2050. In spite of greater water use efficiency afforded by minimal soil surface evaporation and deep drainage, and ease of automation, wide scale adoption of surface and subsurface drip irrigation technology is limited. This is due to their high investment cost for installation. They often lack a significant yield benefit when compared to conventional irrigation practice. Reasons are probably linked to a sustained wetting front around emitters. These emitters impose a condition of low oxygen content in the root-dense rhizosphere surrounding emitters that impede root respiration, and negatively impact on plant uptake of water and nutrients, leading to constrained yield performance.

Here we review aspects of soil oxygen dynamics during irrigation and present evidence for sustained hypoxia in the wetting fronts associated with drip and subsurface drip irrigation. This condition of low oxygen content in the rhizosphere conditioned by the drip irrigation we term as the irrigation paradox.

S.P. Bhattarai (✉) and D.J. Midmore
Centre for Plant and Water Science, CQ University, QLD 4702, Rockhampton, Australia
e-mail: s.bhattarai@cqu.edu.au

N. Su
School of Earth and Environmental Sciences, James Cook University, QLD 4870 Cairns, Australia

E. Lichtfouse (ed.), *Biodiversity, Biofuels, Agroforestry and Conservation Agriculture*, Sustainable Agriculture Reviews 5, DOI 10.1007/978-90-481-9513-8_9,

At dissolved oxygen concentrations one half of that at saturation root respiration and function start to decline. The pros and cons of different approaches to maintain aeration of the crop root zone are reviewed and we suggest that the best approach to overcome hypoxia is through aeration of the irrigation water stream, a practice known as oxygation. This conclusion is based on the evidence derived from a number of controlled environment experiments and field trials on crops where consistent increases of 10–30% in yield and water use efficiency were reported with aerated irrigation water. Aerated drip delivery of irrigation also allows for the simultaneous application of agro-chemicals directly into the crop root zone. We define this approach of delivering multiple agro-chemicals with aerated drip irrigation as multigation. Delivering nutrients and other chemicals with irrigation water according to the crop requirements reduces the cost of application and improves input use efficiencies and becomes increasingly practical with adoption of automation using dosing equipment. The dynamics and fate of agrochemicals in the soil with aerated water irrigation differs from those of non-aerated water irrigation, e.g. oxygen provided greater salt exclusion by roots compared to non-aerated treatments. Opportunities also exists for improving the use of treated effluent water for irrigation with multigation as treated effluent water has high biological oxygen demand and often record poor dissolved oxygen concentration causing hypoxia upon crop irrigation. Review of previous research and modelling suggests that the behaviour of multiphase components in aerated irrigation water streams, and subsequently in the soil, and their interaction with plant roots influences the dynamics of salt, nutrients and pesticides and therefore provides opportunity for the sustainable management of these inputs for irrigated agriculture.

Keywords Multigation • Oxygation • Fertigation • Chemigation • Soil aeration • Water use efficiency • Vertisols • Hypoxia • Wetting fronts

Abbreviations

DI	Drip irrigation
SSDI	Subsurface drip irrigation
O_2	Oxygen
CO_2	Carbon dioxide
N	Nitrogen
Cu	Copper
Mn	Manganese
P	Phosphorus
Zn	Zinc
Fe	Iron
WUE	Water use efficiency

1 Introduction

Irrigation is an age-old practice dating back over 5,000 years (Drower 1956; Biswas 1967). Early irrigation systems used gravity as the primary means of moving water, with distribution achieved by flooding a series of canals and furrows (Nonneck 1989). Development of mass-produced piping and practical mechanical pumps in the nineteenth century underpinned a rapid development in irrigation technology. While these advances were initially used for conventional flood and furrow they also made possible drip and sprinkler irrigation.

The application of polymers to pipe production after World War II led to a practical and economical alternative to gravity water distribution and this combined with a depletion of the rural workforce following the war saw a move away from gravity flood and furrow systems (Nonneck 1989). In parallel, the development of modern lodging-resistant dwarf cereal varieties for irrigated cropping in the era of the green revolution produced a quantum jump in irrigation productivity and saw a significant increase in irrigation usage, albeit largely non-pressurised systems. The period since World War II has thus seen a rapid commercialization of water supplies and increased competition for fresh water. Currently irrigated agriculture utilizes 70% of global fresh water to irrigate 20% of the cultivated land area, producing 40% of the world's food supply (IATP 2009). However, with the projection of a world population of nine billion by 2050, world food production must be doubled, and much of this has to come from irrigated agriculture, whether through complete or supplemental irrigation. Increasing demand for water coupled with declining water allocation, increasing cost, and heightened environmental awareness has, therefore, put tremendous pressure on irrigators to adopt more efficient and environmentally benign practices.

Conventional flood and furrow irrigation are characterised by poor application uniformity and low crop water use efficiency (WUE) and associated negative environmental impacts due to runoff and percolation of dispersing pesticides, fertilisers and other pollutants (Tanji et al. 2002). These may be minimised through drip irrigation (DI). Slow and frequent application of water to the soil is possible through emitters along a surface DI or partially buried water delivery pipes. To avoid surface wetting and to minimize evaporation loss associated with DI, development of more water use efficient practices such as subsurface drip irrigation (SSDI) has taken place (Qassim 2003). Camp (1998) reviewed the development of SSDI, and highlighted the potential benefits in terms of saving irrigation water and control of runoff. First tested in Germany in the 1860s, SSDI used clay pipes to create a combination irrigation and drainage system (Lamm and Camp 2007), and was followed in the USA as early as 1913 (Lamm and Camp 2007) but its wide-scale adoption by industry did not take off until the 1970s. Interest in SSDI, with pipes buried from 5–50 cm below the soil surface, has increased over the past 20 years primarily due to ease of automation, and increased need to conserve water resources and to improve on the sustainability of irrigated crop production systems (Howell 2001).

Drip and SSDI have primarily been adopted for high-value ornamental, vegetable and fruit crops, but recently their application has been expanded to broad-acre crops such as corn, cotton, soybean, and tomato totalling about 3 million hectares (Camp et al. 2000). Further development and adoption of DI and SSDI is likely to continue in the future, primarily because of the benefits associated with savings in water when compared to other irrigation methods (Ayars et al. 1999). However, the cost associated with installation of DI and SSDI is high, and may not be quickly offset by savings in the costs of water application. To achieve more extensive adoption of SSDI additional economic returns to growers must be sought to more quickly offset the installation costs.

Conversion from surface flood or furrow irrigation to SSDI allows for more precise spatial and temporal control of soil moisture around the roots, keeping soil moisture at an optimum to avoid crop water stress. SSDI also improves disease control by keeping foliage and produce dry, is applicable to marginal soil and irrigation water, reduces evaporation, run-off and deep percolation losses and facilitates fertigation and precise water application so reducing undesirable vegetative growth and improving harvest quality. Because the soil surface remains dry, SSDI allows access to the field during irrigation, offering flexibility to concurrently conduct other management practices. It also saves labour, reduces energy requirements through low application rates and operating pressure, and thus saves on operation costs. SSDI reduces weed seed germination and growth as it keeps the soil surface dry, and the partially dry soil profile allows for capture of rainfall should there be any during the cropping cycle (McHugh et al. 2008).

Besides the high initial capital cost, root intrusion and blockage of emitters have been operational problems with SSDI. In addition, sprinkler or surface irrigations may be required on some soil types for germination and crop establishment and soil structure may deteriorate around emitters when drip systems are used over successive years. Accumulation of salt and nutrients on or close to the soil surface may be likely; but these may be flushed to roots by rainfall or leaching irrigation (Zur 1996). Concentration of root development into relatively smaller wetted volume around emitters or a strip of wet zone along the length of drip line can increase an occurrence of hypoxia. This situation can particularly occur in open hydroponics system, also called advanced fertigation systems (Falivene et al. 2005), which aim to develop an irrigation and nutrition management program where the nutrients are applied regularly to a smaller volume of soil at a low application rate and at a high frequency to meet crop demand. Open hydroponics (a recently-adopted system used in the citrus industry in Australia) crops are irrigated by DI to keep a small soil volume at field capacity for most of the time (Boland et al. 2005).The authors argue that that by achieving a small, concentrated root system, the system is able to meet all the crop's water and nutrient needs and also manipulate and control plant water and nutrient uptake through all stages of the productive cycle. However, the crop is likely to fail meeting the oxygen requirement for root respiration as the air diffusion into the root zone is slower than the demand on oxygen for mass root respiration. Hence, hypoxia can be a major limitation to productivity for crops grown under hydroponics and open hydroponics (a recently-adopted system used in the citrus industry in Australia).

Continuous cropping, soil compaction, use of marginal and saline irrigation water and heavy textured soils expose the rhizosphere of SSDI crops to a lack of oxygen, i.e. hypoxia, during and shortly after irrigation particularly in the regions of the wetting fronts, the region where most of the root activity occurs (Bhattarai et al. 2006). Hypoxia in the rhizosphere limits root respiration; consequently crop response to irrigation declines below the expected based on the potential daily ET demand. This results in decreased WUE and hence a low yield. Many benefits of SSDI are, therefore, to some extent offset by the sustained wetting fronts associated with reduced oxygen supply for root respiration in DI and SSDI cropping.

In this review we discuss opportunities of using aerated water for drip and subsurface drip irrigation to overcome hypoxia associated with these irrigation systems. We also present different approaches for aeration of irrigation water; some of the key results of controlled environment and field crop trials, and present analysis and modelling for co-application of agrochemicals with the aerated irrigation (multigation), and aeration of saline and effluent irrigation water. These not only increase yield, quality and WUE of irrigated agriculture but also for optimize drip irrigation technology so that it becomes a more attractive irrigation option to growers for adoption that contribute towards sustainable irrigation which support food and fibre need of growing world population with declining water allocation for irrigation.

2 Interdependence of Soil Constituents: The Irrigation Paradox

A soil is comprised of various proportions of solid inorganic and organic materials and pore space. The pore space is filled with either air or water plus dissolved materials. Solids supply support for roots and the aerial plant structure, water satisfies the transpiration demand and carries nutrients, and air supplies oxygen for root (and microbial) respiration. Crop health, yield and quality are largely dependent on the three constituents of soil or growing media other than the solids, i.e. air, water and nutrients. When the proportion of air to water is balanced, the crop expends a minimum amount of energy in the uptake of water and nutrients. An ideal soil mixture reputedly consists of 50% solid, 25% water and 25% air (Wolf 1999). However, this ratio (Fig. 1a) is hard to achieve when a crop is well irrigated with conventional irrigation methods. The optimum proportion of these key constituents necessary for high yields and good quality depends on the specific crop, growth stage and the environmental conditions of the rhizosphere.

An optimum balance between air, water and nutrients in the soil has been defined as that leading to a "fertile triangle" (Wolf 1999). The optimal balance of the three components of a fertile triangle in soils is rarely achieved, for traditional irrigation methods impose cycles of water-logging during and following irrigation followed by improved aeration as water is transpired or (less desirable) drained from the root zone (Meek et al. 1983) and then by periods of soil water deficit prior to subsequent irrigation. Theoretically, crop yields will continue to rise as sides of the triangle (Fig. 1b) are increased provided a desirable balance is maintained between different phases. In practice there are upper limits to the degree to which

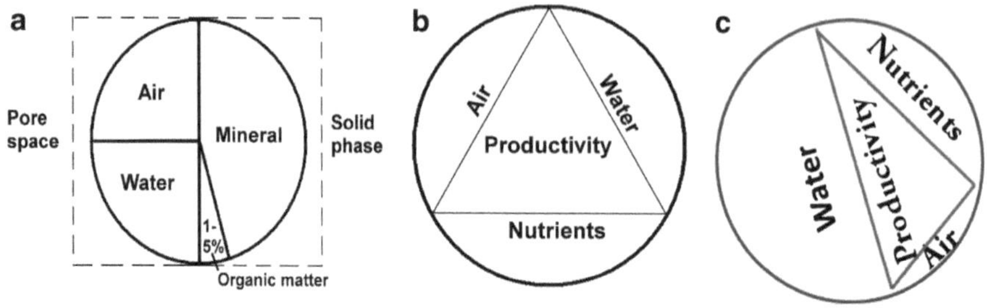

Fig. 1 (**a**) The composition of a fertile mineral soil illustrating the ratio of air and water (*Note:* in most mineral soils, air increases as water decreases and vice versa. An ideal relationship exists when they are approximately equal). (**b**) The relationship of balanced air, nutrition, and water in fertile soil. (**c**) An unbalanced relationship of air, nutrients, and water in soil. In this case one side (water) is dominating the triangle, greatly reducing the air side and limiting root access to the nutrient side, resulting in poor crop production (Modified with permission of Wolf 1999)

the sides may be expanded. These upper limits vary with different soils, being greatest in loam due to its ability to hold large amounts of air, water and nutrients without unduly affecting other parameters (Brady and Weil 2004).

Limits to crop yield are also imposed by other factors, such as availability of irradiance or carbon dioxide and sink limitations, and these may eventually restrict growth (Bugbee and Monje 1992). Some such limiting factors can be offset, making it possible to attain higher yields, but not without additional costs which at present limit treatments such as CO_2 fertigation to high value protected crops.

The interdependence of the soil inputs is most strikingly exemplified in the relationship between air, water and nutrients. Minimum quantities of water must be present to obtain the full value of added nutrients. The addition of water requires more nutrients, particularly when water leaches essential nutrients from the root zone. Irrigation increases water content at the expense of air, for water and air occupy the same soil pore space (Grieve et al. 1986). Such soil saturation and depleted oxygen slows root respiration, crucial for generating energy necessary for efficient plant uptake of nutrients and water. Alleviating the constraints on yield due to lack of soil oxygen will also minimize limits imposed by other production inputs on the crop yield (Cannell et al. 1985).

3 Soil Oxygen Dynamics during Irrigation: A Call for Delivery of Oxygen to the Rhizosphere

Soil oxygen is present both in gaseous (air-filled porosity) and dissolved states (soil solution). The gaseous state, present in soil pores, contributes to the major portion of oxygen requirements for plants and micro-organisms by acting as a reservoir of O_2. Air in the pore spaces also act as a repository for CO_2 and other gases resulting from respiration and decomposition of organic matter (OM). Some of these gases, such as carbon monoxide formed from incomplete respiration or decomposition of OM, may be toxic to plants (Grable 1966). With sufficient large pores in well-drained soils, and a porous route to the soil surface, the stored air in soil pores can readily be exchanged for fresh air from the atmosphere. The exchange of air between the atmosphere and soil takes place in both air-filled pores and pores partially filled with water. Diffusion of oxygen through water is 10,000 times slower than its diffusion through air. As a result gas exchange decreases exponentially as the water content increases in pores partially filled with water and functionally halts as saturation is reached. Hostile soil conditions, such as flooding (Fig. 1c), compaction and salinity, have negative effects on plant growth and development by impeding gas exchange between the soil and atmosphere (Powlson et al. 1997). Root development can be seriously impaired by soil oxygen deficiency and cell division and elongation slow and may stop completely depending on the severity of the deficiency (Table 1). The harmful effect of restricted air exchange on plants is exacerbated by the soil microbial populations that consume available oxygen and contribute towards the development of anaerobic conditions in the soil.

Although relatively small in volume, the exchange of air that takes place in the soil solution is crucial for it is principally in the solution phase that O_2 is taken up by roots. The small amount of O_2 present in the soil solution (generally 5–10 mg L^{-1}) is readily absorbed by roots or utilized by soil micro-organisms because the solution envelops the roots or soil particles as a film. Thus oxygen available in the water film can be quickly exhausted and the soil solution soon becomes saturated with CO_2. It has been estimated that the O_2 in a flooded clay soil can be completely exhausted in a single day whereas it may last about 3 days in a sandy soil (Sharma and Swarup 1988). The actual exhaustion rate depends on the absolute respiration by plant roots and micro-organisms and soil temperature (Table 1). Deeper in the soil the amount of O_2 declines while that of CO_2 simultaneously increases, particularly in soil subjected to flooding or compaction (Simojoki and Jaakkola 2000).

Table 1 Minimum and maximum range for different root growth resources that influence root growth, development and activities

	Requirements	
Root resources	Minimum	Maximum
Oxygen in soil atmosphere (for root survival)	3%	21%
Air pore space in soil (for root growth)	12%	60%
Soil bulk density restricting root growth (g cc^{-1})	–	1.4 clay
	–	1.8 sand
Penetration strength (water content dependent)	0.01 kPa	3 MPa
Water content in soil	12%	40%
Root initiation (O_2% in soil atmosphere)	12%	21%
Root growth (O_2% in soil atmosphere)	5%	21%
Progressive loss of element absorption in roots (O_2% in soil atmosphere)	10%	21%
Temperature limits to root growth	40°F/4°C	94°F/34°C
pH of soil (wet test)	pH 3.5	pH 8.2

Coder 1998 with permission

Oxygen concentrations in clay soil decrease sharply with increasing soil depth particularly at higher soil moisture content (Dowell et al. 2006). And this situation is more prominent with SSDI where sustained wetting fronts are prominent at depth (Machado et al. 2005).

The O_2 requirement for soil respiration is largely affected by temperature. The increased demand for O_2 at higher temperature is associated with elevated rates of root and microbial respiration. Paradoxically, the solubility of O_2 decreases with rising temperature (by about 1.6% per °C) whereas the diffusion rate of O_2 in water increases by 3–4% per °C. It is the soil solution upon which actively growing root tips depend for oxygen. Hence concentration of O_2 in the soil solution is more critical than in soil air for root processes, as limitations to root initiation and growth become significant in an hypoxic soil solution. Hypoxic and anoxic conditions exert a negative impact on root growth which results in poor plant performance. Oxygen demand for root respiration is quite high; hence O_2 in a soil solution is soon depleted leading to hypoxic/anoxic conditions, unless rapidly replenished. The uptake and mobilization of both nutrients and water are greatly reduced as soil oxygen in solution falls below a critical level of 2 ppm or an oxygen diffusion rate (ODR) $<20 \times 10^{-8}$ g cm^{-2} (Letey et al. 1964). When over-irrigated, air is depleted (purged) to the point that many microbiological soil functions (nitrification, nitrogen fixation, and organic matter decomposition) and the uptake of water and nutrients come to a halt. In most fertile porous media, the excess water quickly drains away, restoring adequate levels of air but taking some valuable nutrients with it. As water is transpired by plants or is lost to evaporation the amount of air diffusion into the soil and rhizosphere increases further. In most soils the increase in volume of air is about equal to that of water depleted except for soils which shrink upon drying and swell upon wetting (Wolf 1999). Since air in the growing media usually decreases in quantity as water increases, the addition of large quantities of water to satisfy transpiration demand needs to be carefully controlled to avoid reducing soil air below an optimum.

Susceptibility to hypoxia and adaptation to a low oxygen environment differ with species (Vartapetain and Jackson 1997). In species with tolerance to low O_2 the oxygen requirement for effective root functioning is a minimum at about 2.5 mg L^{-1}, while a concentration of 5–10 mg L^{-1} is required for effective root respiration in susceptible species (Drew and Stolzy 1996). Hewitt (1966) found that O_2 content <0.7 mg L^{-1} in solution cultures resulted in injury of avocado and citrus roots, with permanent cessation of growth by avocados but not in citrus at 0.6 mg L^{-1}. Root growth of soybean, tobacco, and tomato ceased if gaseous O_2 fell below 0.5%, and yields of tomato increased as O_2 content rose from 0.6% to 21% in soil culture (Urrestarazu and Mazuela 2005). For some species (e.g. avocado) roots loose weight at 1% O_2 in the root environment as a consequence of respiration (Stolzy 1974). The amount of O_2 required for effective root functioning, therefore, differs with plant species and growth stage. The growth of crops such as tomato, tobacco and papaya is impaired when gaseous O_2 concentrations in the medium fall below 10% in air in the soil (Khondaker and Ozawa 2004). Cereal crops or pastures on the other hand need less oxygen but can be more demanding at critical stages such as germination and early development (Blackwell and Wells 1983). In some early studies with orchards, 15% O_2 in the soil was found to give the best growth. While at least 12% was necessary for root initiation, a concentration of 5–10% was needed for growth of existing root tips, and between 0.1% and 3.0% only was required for subsistence in avocado (Coder 1998).

In terms of plant adaptation to low soil oxygen, tomato, which readily produces adventitious roots, will survive slowly induced wet conditions, while tobacco, which cannot produce adventitious roots, is usually killed under such conditions. While most flood-tolerant plants develop aerenchymatous tissue in their roots and shoots, the proportions of root tissues devoted to aerenchyma are quite variable across conditions and species, ranging from 5% to 60% of cross sectional area (Allen et al. 2004). Shortage of O_2 may lead to outbreaks of serious plant diseases caused by soil-inhabiting opportunistic pathogens. While growth of some soil pathogens, e.g. *Verticillium albo-atrum* is restricted as O_2 is reduced (Menzies 1963), the damage to plants by several organisms e.g. *Fusarium, Pythium, Phytophthora, Sclerotinia sclerotium* is greatly increased under anoxic conditions (Hiltunen and White 2002).

4 Optimization of Drip Irrigation: Fertigation and Chemigation

The application of nutrients to soil through an irrigation system is called fertigation (Wolf et al. 1985). Nitrogen is the most commonly applied plant nutrient for fertigation although many industries have started supplying other soluble macro- and micronutrients through DI and SSDI. Fertigation reduces fertilizer application costs by eliminating an extra field operation, and improves nutrient use efficiency through application closer to where and when the demand is greatest (Follett et al. 1991). Fertigation allows control of leaching, denitrification, and luxury uptake of nutrients by plants and, importantly, minimizes both amount and concentration of runoff (Camp 1998) compared to flood and furrow irrigation (McHugh et al. 2008). Chemigation is an approach for mixing agro-chemicals other than fertilizers, such as insecticides, nematicides, fungicides and surfactants with irrigation water (Werner 1990).

5 Aerated Water Irrigation: Oxygation and Oxyfertigation

Oxygation is the practice of delivering aerated water with DI (Bhattarai et al. 2005; Su and Midmore 2005) and is intended to overcome induced hypoxia in the root system associated with water saturation of the root zone. Irrigation purges the air out of the rhizosphere and reduces oxygen availability, particularly soon after irrigation (Heuberger et al. 2001). This situation prevails with DI and SSDI and point-source irrigation, particularly in the longer irrigation cycle run times needed to meet high crop evapo-transpiration demand in the tropical and subtropical environment. This leads to saturation within the major rooting zone. Improving inherent physical characteristics of soil and growth media, to promote drainage and ingress of air after irrigation may be costly (e.g., application of soil amendments, labour costs). The irrigation paradox, the antagonism between water and air in the soil pore space, can be addressed only if the irrigation technology allows a simultaneous delivery of oxygen and water. One of the approaches is through in-line mixing of atmospheric air or oxygen into the irrigation water stream (Bhattarai et al. 2006). Oxygation is achieved through the simultaneous application of water and oxygen (in air bubbles) to plant roots as an extension of traditional irrigation which supplies water only. The importance of aerating the soil in a drip irrigated crop and hydroponics media has been emphasized in early research (Bhattarai et al. 2004; Goorahoo et al. 2002; Heuberger et al. 2001) and more recent work by Marfa et al. (2005) has termed this as oxyfertigation. Maintenance of favourable levels of air, water and nutrients in the root zone throughout the crop cycle in artificial media is reasonably easy. Therefore, greater crop yields in artificial media such as used in hydroponics have been achieved than with soil-based media, but their high cost of establishment is a limitation to wider adoption. Recent analysis suggested that an oxygen limitation can occur in common artificial media such as rock wool (Schroeder and Knaack 2006). Consequently we introduced oxygation to a flood-and-drain rock wool hydroponics system using a venturi for air injection and improved yield and WUE (Bhattarai et al. 2008). The oxyfertigation technique has been developed to improve rhizosphere oxygen availability in solid media of hydroponics systems, and consists of supplying dissolved and bubbled oxygen in the irrigation water at oversaturated concentrations using pressurised water, and a sealed injection chamber linked with a small-pore diffuser. With this method Marfa et al. (2005) achieved dissolved oxygen as high as 39 mg L^{-1}. Under normal ambient conditions, i.e. at 25°C and 100 kPa the maximum solubility of oxygen in a pure oxygen atmosphere is 39 mg L^{-1}.

Large-scale root aeration of field crops was not practical until suitable delivery systems such as with drip tapes became available. Oxygation, incorporating involves incorporating air into the irrigation stream, adds air in the gaseous form as bubbles to water saturated with dissolved oxygen (~8 mg L^{-1} at 20°C). The air in the form of bubbles supplements the dissolved and subsequently consumed oxygen supply for root and microbial respiration for both soil and soil-less culture systems, and consequently shows benefits in a number of crops (Table 2).

6 Multiple Agro-Inputs Delivery with Aerated Water Irrigation: Multigation

Multigation is the application of chemigation and fertigation together with oxygation in such a way that the efficiency of application and benefits of all soil based inputs to the crop and soil are maximised and energy, cost and time are saved by a single application approach. Adoption and fine-tuning of oxygation for multigation will pave the way for major increases in productivity of drip and SSDI crops.

Table 2 Yield benefits to oxygation – aeration of irrigation water delivered by subsurface drip irrigation (SSDI) in a number of trials with different aeration methods, soil types, irrigation rates and salinity status

Crop	Treatment (f-field, g-glasshouse)	Yield response	Other factors	References
Zucchini	H_2O_2 (f)	25%	Flooded	Bhattarai et al. 2004
Cotton	Venturi, H_2O_2 (g)	14–28%	Saturated/FC	Bhattarai et al. 2004
VSB	Venturi, H_2O_2 (g)	82–96%	Saturated/FC	Bhattarai et al. 2004
Tomato	Venturi (g)	38%	Soil salinity	Bhattarai et al. 2006
Tomato	Venturi (g)	21%	FC/drier	Bhattarai et al. 2006
Chinese Cabbage	Venturi (g)	12%	Flood and drain hydroponics	Bhattarai et al. 2008[a]
VSB	Venturi (g)	35%	Depth	Bhattarai et al. 2008[b]
Pumpkin	Venturi (f)	15%	Depth	Bhattarai et al. 2008[b]
Chickpea	Venturi (g)	11%	Depth	Bhattarai et al. 2008[b]
Chickpea	Venturi (f)	0%	Irrigation rate	Pendergast 2010, unpublished
Cotton	Venturi (f)	4–15% (4 years)	Irrigation rate	Pendergast 2010, unpublished
Cotton	Venturi (g)	18%	Soil salinity	Bhattarai and Midmore 2009
VSB	Venturi (g)	13%	Soil salinity	Bhattarai and Midmore 2009
Pineapple	Venturi (f)	4–17%		Bhattarai et al. 2010
Watermelon	Venturi (f)	70%		Bhattarai et al. 2010
Lucerne	Venturi (f)	1%	Saline bore water	Bhattarai et al. 2010
Fig	Venturi (f)	133% biomass	Depth	Bhattarai et al. 2010
Capsicum	Venturi (f), Oxycrop	4%	Aeration methods	Bhattarai et al. 2010
Wheat	Venturi, Seair (g)	16–28%	Soil types	Bhattarai et al. 2010

VSB= vegetable soybean, FC= Field capacity, H_2O_2-Hydrogen peroxide

The multigation system that we are developing allows growers to incorporate an aeration system and a dosing facility for regulated and controlled injection of agro-chemicals such as insecticides, fungicides and plant growth regulators into SSDI. In this respect multigation is also an extension of the concept of oxygation. The multigation approach rests on the premise that the crop demand for water, oxygen and nutrients is not constant during a crop cycle, yet conventional irrigation methods do not allow easily tailored co-application of these multiple inputs according to crop demand. The approach of multigation fits well with precision farming where the crop supply of production inputs is matched with demand in order to improve the efficiencies and productivity and to minimize waste and contamination.

Multigation is able to improve the efficiency of these multiple plant inputs by exploiting the advantages of synergistic interactions between them. The effectiveness of inputs such as fertilisers and agro-chemicals decreases under anaerobic/hypoxic conditions caused by low oxygen availability in the plant root zone. Nutrient dynamics between the plant and soil are greatly influenced by the aeration status of the rhizosphere and this becomes important for SSDI due to the low oxygen environments that restrict root respiration (Bhattarai et al. 2006). Uptake of crop nutrients particularly N, Mn, Fe, are most susceptible to O_2 stress (Morard et al. 2004).

Plant-root interactions with nutrient elements are entirely different when the roots are exposed to an oxygen deficient environment compared to an aerated rhizosphere. Reduction in the uptake of Na and Cl by aerated roots compared to roots exposed to hypoxia is one such example (Letey 1961). Reduced uptake of Ca by hypoxic roots leading to tomato cause blossom end rot – a physiological disorder (Bhattarai et al. 2005) and rapid development of carrot cavity spot – a pathological problem in the hypoxic root environment of carrot (Hiltunen and White 2002), are examples of the issue pertinent to multigation. Such relationships clearly underpin the scope to enhance productivity when it is conditioned by plant physiological, pathological and biochemical processes which alter water and nutrient use efficiencies.

Important new applications of multigation include those for aquaculture, fluid related phenomenon, waste water treatment, phyto-remediation, water industries, vermi-aquaponics and amenity horticulture, but these are not dealt with further in this review. As water resources for agricultural irrigation become more limited and costly and ever-increasing environmental regulations force the industry towards more efficient water usage, the water saving properties of multigation need to be embraced for improved crop production. Multigation has the potential to increase WUE, defined as crop yield per unit applied irrigation water, during multigation compared with other irrigation methods, and, by minimizing leakage and waste, multigation can aid irrigators in complying with environmental regulations.

6.1 *Approach and Significance of Multigation*

The fertile triangle deals with an optimum balance between the air, water and nutrients in the root zone (Fig. 1). The interaction of plant roots with soil-water and nutrients is

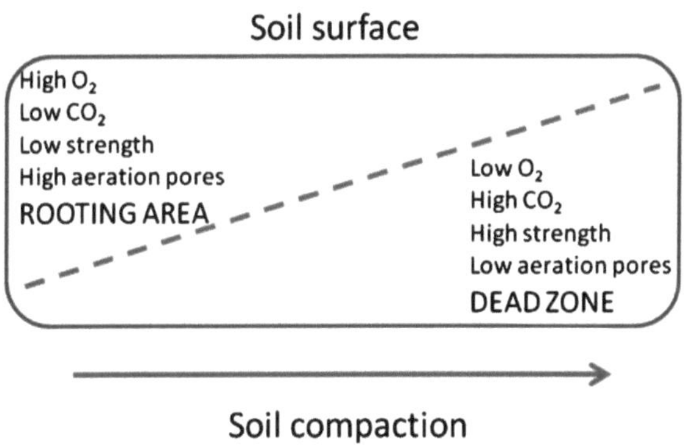

Fig. 2 Soil compaction and its associated effects on soil physical properties, gas diffusion and plant rooting abilities (Redrawn with the permission of Kim D. Coder 1998)

very different with and without imposed aeration of the rhizosphere. A positive response to oxygation in terms of transpiration and uptake of nutrients (as reported by Bhattarai et al. 2004) opens an avenue to change the crop nutrition and other soil and water management practices such as application of water-soluble pesticides, micro-nutrients, herbicides and growth regulators in aerated irrigation streams.

Soil compaction around the emitters of drip and SSDI crops is profound especially in the clayey and loam soil that diminishes the oxygen diffusion to rhizosphere when saturated during irrigation (Zhao et al. 2007). Soil compaction and soil aeration are inversely related in most soil types (Fig. 2); compaction is increasingly becoming a constraint for crop production.

6.2 Crop Response to Multigation

Plants, through their root systems, respond dynamically to the soil environment. Plant performance in any soil is not only a response to the nutrient content of the soil but it is also strongly influenced by the rhizosphere environment. In the soil, the interaction between different applied inputs plays a major role in plant productivity. Plant response to one input changes in relation to others in the rhizosphere. Justus von Liebig's theory of limiting factors attempted to explain such relationships between different plant nutrients in terms of plant production. Sprengel (1828) developed the principle, later popularized by von Liebig, and it states that growth is controlled not by the total of resources available, but by the scarcest resource.

Liebig's Law can be safely extended to a population of a field crop and is also used in crop ecosystem models (Loomis et al. 1979). Applied to multigation, soil oxygen may be limiting when soil moisture and nutrients are not limiting factors in an SSDI crop on a fertile soil. For example, the growth of a crop is dependent on a number of edaphic factors, such as soil oxygen (A), soil moisture (M) and mineral nutrient availability (N). The availability of these may vary, such that at any given time one is more limiting than the others. As such, growth is limited in situations where there are steady state conditions, and factor interactions are tightly controlled. Liebig's Law states that growth only occurs at the rate permitted by the most limiting factors. For instance, in Eq. 1, the growth of crop O is a function of the minimum of three Michaelis-Menten terms representing limitation by factors A (air), M (Moisture), N (Nitrogen) and k (constant).

$$\frac{dO}{\partial t} = \min\left(\frac{A}{kA + A'} \frac{M}{kM + M'} \frac{N}{kN + N'}\right) \tag{1}$$

A significant interaction exists between the components of a plant rhizosphere environment and inputs that are applied to the soil. The crop demand and response to the nutrient input is a dynamic phenomenon so that nutrients availabilities need to be adjusted to different growth stages and environments. For a given crop and environment, the interactions between the inputs determine optimum levels for each input supplied to the crop. Optimization of production inputs is therefore essential to achieve the highest yield. SDI offers opportunities to tailor the concentration of input mixtures to optimum levels and to effect direct delivery to the root zone. Therefore, the concept of multigation, with SSDI as the delivery mechanism, regulates and optimises the intricate relationships between the production inputs that are supplied to crops.

7 Modelling and Measurements of Multigation

Different production inputs such as oxygen, other gases, nutrients, plant growth regulators, pesticides, heat (for raising soil temperature), osmotica, metabolites and other compounds may be directly injected into the root zone through SSDI. Optimising multigation involves around the investigation of concurrent multiphase flow and transport of water, oxygen and solutes delivered to the soil via a single irrigation stream. The flow dynamics of the fluid and individual components change with respect to shape, density, and behaviour of the multiphase flow components (Choi et al. 2002). Therefore, multigation must be seen as a highly interactive fluid movement that does not necessarily deliver a uniform application of all those inputs to the soil during irrigation. Therefore, a critical evaluation and detailed assessment is required on the nature of multiphase flow components in an aerated irrigation water stream in order to maximise the application uniformity of the production inputs during multigation. Critical amounts of oxygen in the root zone mineral nutrients and water vary in spatio-temporal dimensions. Multigation as a dynamic system can accommodate such variations in the crop demand for these factors. It represents a special form of multiphase flow in heterogonous soil conditions, hence, the methods for analysing multiphase flow in porous media can be applied.

Distribution of applied inputs in the soil follows the principle of multiphase flow. The term multiphase flow is used to refer to any fluid flow consisting of more than one phase or component. The flows in multigation have some levels of phase or component separation at a scale well above the molecular level. One could classify them according to the state of the different phases or components and therefore refer to gas/solid flows, or liquid/solid flows or gas/particle flows or bubble flows. Some treatises define multiphase flow in terms of a specific type of fluid flow and deal with low Reynolds number (Re) suspension flows and dusty gas dynamics. The Reynolds Number is a nondimensional parameter defined by the ratio of intertia forces (ρu^2) and viscous forces ($\mu u/L$) and can be expressed as: $R_e = Lu\rho/\mu$ for a pipe of length L. Other specific types of multiphase flows include slurry flows, cavitating flows, aerosols, debris flows and fluidized flow. In this article we discuss the basic fluid mechanical phenomena and illustrate those phenomena with examples from a broad range of applications and types of flow pertinent to multigation.

Previous and current research worldwide reflects on the diverse and ubiquitous challenges of multiphase flow. The ability to predict the fluid flow behaviour of these processes is central to the efficiency and effectiveness of multigation. Two general typologies of multiphase flow can be categorised, as disperse flows and separated flows. Disperse flows refer to those consisting of finite particles, drops or bubbles (the disperse phase) distributed in a connected volume of the continuous phase, e.g. water. Separated flows consist of two or more continuous streams of different fluids such as agricultural wettable powder (WP) which is a dispersible pesticide separated by interfaces in the irrigation stream and used to control crop root-related diseases and pests.

7.1 Multiphase Flow Concept and Applications

The effective analysis of multiphase flows in irrigated soils requires understanding of the processes and patterns of those flows and the phenomena that they manifest. Such models can be explored experimentally, theoretically or computationally. There are some applications in which full-scale laboratory models are not possible. The predictive capabilities of the models rely on the physical understanding of the processes. Our incomplete understanding of the processes imposes a major hurdle in the formulation of theories and computational methods of multiphase flow. Even though the current computer power and speed allow for modelling of most of the flows that are commonly experienced, of when one or both of the phases becomes turbulent (as often happens) the magnitude of the challenge becomes even more complex. Therefore, simplifications are essential in realistic models of most multiphase flows.

Multigation is an application of multiphase flow to agricultural soils. An understanding of multiphase flow and its interactions with the porous media, such as the soil is essential in order to utilise the benefits of multigation. The topic of multiphase flow was originally studied in petroleum engineering and has later been extended to address environmental problems (Drew and Passman 1998; Brennen 2005). Multiphase models are often used to study the flow of liquid contaminants in aquifers and of air injected below the groundwater table for remediation purposes, so-called air-sparging (Philip 1998). However, the majority of multiphase flow studies in porous media to date are of two-liquid phases, such as oil and water (Fukushima 1999). Two-phase flow of water and air, as a special case of multiphase flow, has been an issue in soil physics since Green and Ampt (1911) proposed the first infiltration model. The role of oxygen in soils during irrigation is generally ignored and the mechanisms of multiphase flow in soils in the agricultural environment are poorly understood. Because of the differences in the hydraulic and physico-chemical properties of the multiphase components, their interactions with the porous media (soils), and the influence of temperature gradients, the multiphase components separate out in the same porous soil (Fig. 3). These differences for example in soil moisture and oxygen gradients lead to multiple concentration fronts and spatiatemporal distributions of air and nutrients. At variable spatial and temporal scales the distributions of each component in the multiphase flow will vary. Information on the mechanics of this unique fluid flow facilitated by air enrichment in soils provides vital knowledge that guides the application of this technique (Su and Midmore 2005).

Miller et al. (1998) present a comprehensive review of multiphase flow in porous media. They discuss the status of the approaches to modelling multiphase flow, including the formulations of balance equations, constitutive relations for both pressure-saturation-conductivity and inter-phase mass transfer, and stochastic as well as computational issues. As pointed out by Miller et al. (1998), most multiphase environmental models published to that date did not consider an energy equation. Under natural conditions, the differences in soil temperatures on the surface and within the soil create thermal gradients which drive heat transfer in the soil as a concurrent process to multiphase flow. Furthermore, porous media can also undergo spallation (i.e. the process in which fragments of

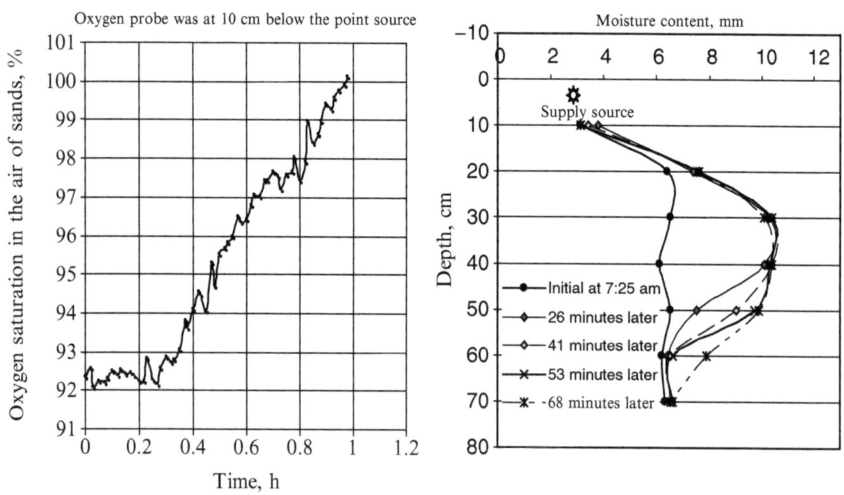

Fig. 3 Measured oxygen (*left*) and moisture contents (*right*) in an irrigated sandy soil with oxygation without plants (Su and Bhattarai, 2006, Unpublished data)

materials are ejected from a body due to impact or stress) under larger thermal gradients (Dimitrienko 1999), a phenomenon of phase transformation in the porous particles under a large thermal gradient known as pyrolysis. The spallation, pyrolysis, and mass interphase transfer complicate the physical processes and therefore, the capacity of soil to hold oxygen in the root zone. The nature of multiphase flow in irrigation has been recognised by many such as Gish et al. (2004). The equations governing the coupled flow of water and gas (vapour) and heat transfer are given by de Vries (1958, p. 912, Eq. 4–6, p. 912, Eq. 19). With measured oxygen and moisture contents, both U and m can be modelled using Eq. 8 and is presented as Fig. 4 (detailed definition of the symbols and derivation of the equations are presented in pp. 909–910 in de Vries, 1957) and are as follows:

$$\frac{\partial \theta}{\partial t} = \nabla\left(\frac{q_m}{\rho_l}\right) + \nabla \bullet \left(D_\theta \nabla \theta\right) + \nabla \bullet \left(D_T \nabla T\right) + \frac{\partial K}{\partial z} \tag{2}$$

$$\frac{\partial \theta_l}{\partial t} = \nabla\left(\frac{q_l}{\rho_l}\right) + \nabla \bullet \left(D_{\theta l} \nabla \theta_l\right) + \nabla \bullet \left(D_{Tl} \nabla T\right) + \frac{\partial K}{\partial z} \tag{3}$$

$$\frac{\partial \theta_v}{\partial t} = \nabla\left(\frac{q_v}{\rho_l}\right) + \nabla \bullet \left(D_{\theta v} \nabla \theta_v\right) + \nabla \bullet \left(D_{Tv} \nabla T\right) + \frac{\partial K}{\partial z} \tag{4}$$

$$\begin{aligned} &\left[C + L\left(S - \theta_l\right)h\beta\right]\frac{\partial T}{\partial t} + \left[\frac{L\rho_l D_{\theta v}}{\alpha \upsilon D_m} - L\rho_v + \frac{\rho_l g}{j}\left(\psi - T\right)\frac{\partial \psi}{\partial T}\right]\frac{\partial \theta_l}{\partial t} \\ &= \nabla \bullet \left(\lambda \nabla T\right) + L\rho_l \nabla \bullet \left(D_{\theta v} \nabla \theta\right) + \rho_l c_l\left[\left(D_{\theta l} \nabla \theta_l + D_{Tl} \nabla T - Kk\right)\nabla T\right] \end{aligned} \tag{5}$$

Under isothermal conditions, and for solute transport due to diffusion and convection in the liquid phase only, the usual equation of solute transport in unsaturated flow is written as Warrick et al. (1971), and Smiles et al. (1978):

$$\frac{\partial\left(\theta_l c\right)}{\partial t} = \frac{\partial}{\partial x}\left(\theta_l D(u) \frac{\partial c}{\partial x}\right) - \frac{\partial}{\partial x}\left(vc\right) \tag{6}$$

for solute transport in a one-dimensional flow, and

$$\frac{\partial\left(\theta_l c\right)}{\partial t} = \nabla \bullet \left(\theta_l D_i(u) \nabla c\right) + \nabla \bullet \left(v_i \nabla c\right) \tag{7}$$

for solute transport in a three-dimensional flow.

Under non-isothermal conditions, the flow of liquid is coupled to the energy (expressed as temperature) through Eq. 2. In such a case, the fraction (or concentration) of the liquid phase is dependent on the temperature, then Eqs. 4–7 are also dependent on temperature through. Sources/sinks for liquid, gas and energy

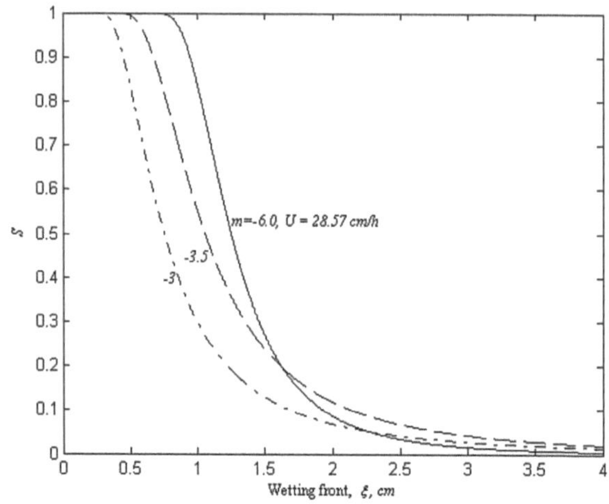

Fig. 4 Computed saturation profile, *S*, versus the wetting front depth, ξ; where m= Mass flow exponent, and U = Velocity components) (Su and Bhattarai, 2006, Unpublished data)

can be added to the above formulations to take into account interphase mass transfer (i.e. bubble phase change to dissolved oxygen phase), internal reactions (i.e. oxygen reactions ions in the solution) and external source/sink dynamics.

In a simple three phase flow of water, air and heat, the saturation position relationship is derived as follows:

$$\xi = \left[\frac{(2m-1)In(1-s)}{3(1-U)}\right]3/(2m-1) \tag{8}$$

The computed front velocities (Su and Bhattarai, unpublished data) are presented in Fig. 4. When nutrients and agro-chemicals are not present, multigation reduces to a two-phase problem. With data on soil temperature, soil air relative humidity and soil moisture contents, we demonstrated that the periodic changes in soil temperature bring about negative periodic changes in relative humidity of the soil. This information will be useful for management of liquids and gases in porous media during aerated irrigation. For example, to maintain a higher level of soil air and hence soil oxygen, a lower soil temperature is preferred. This implies that night or early morning irrigation is most appropriate from the viewpoint of storing soil oxygen that can be used later in the day as the demand by root respiration increases to its post-midday peaks (Huck et al. 1962).

7.2 *Flow of Water in Heavy Clay Soils*

Soil volume change associated with water flow characterizes very large areas of low-lying, often organic rich soils as well as soils with high contents of clay, especially montmorillonite clay. These soils are generally fertile for agriculture, but physically clay soils are very difficult to deal with because they are sticky when they are wet and hard when they are dry. Understanding of water flow and accompanying volume change aids the management of these soils.

According to Smiles and Raats (2005), the earliest measurement of volume change as made in the early nineteenth century by Schubler as reported by Tempany (1917). The concept of flow in swelling soils was conceived by Terzaghi (1923), and the shrinkage curve concept was developed by Haines (1923) to characterise swelling agricultural soils. The equations governing the flow of liquids in swelling media formulated in a material coordinate were formulated more than 4 decades ago (Raats 1965; Raats and Klute 1968) as were the more convenient form of the equations (Smiles and Rosenthal 1968; Smiles and Philip 1978; Philip 1969). It is important to note that these developments are focused on one-dimensional flow and material characteristics were defined per unit cross sectional area of the system. Since the development in 1960s on flow in swelling media, limited progress was made with respect to multi-dimensional flow for the next 40 years (Raats 2002). Although a preferential flow of water and transport of solutes is evident in a drained dual porosity model that occurs due to presence of macro-pores in well-structured clay soil, added air in water could influence the multidimensional flow and transport of solutes and water during irrigation events (van den Eertwegh et al. 2001).

Infiltration and water movement in swelling soils such as clays are investigated by Smiles (1974) and Philip (1969), however, new concepts such as those of fractal porous media were not investigated. Recently, it has been shown (Su 2009) that adsorption onto swelling soils is subject to different mechanisms known as super-diffusion (Tsallis and Bukman 1996). Su (2009) shows that for absorption onto swelling soils, the cumulative absorption or horizontal infiltration is given as

$$I(t) = St^{\beta/2} \tag{9}$$

and the adsorption rate is given as

$$i(t) = St^{(\beta/2)-1} \tag{10}$$

For $\beta = 1$, the above equations become their classic counterparts given by Philip (1969).

7.3 Transport of Solutes in Unsaturated Swelling Soils

Most of the innovation on solute transport in unsaturated swelling soil is based on Smiles and Philip (1978), and suggests that during absorption of water and a non-reactive solutc the water content and the solute concentration both preserve similarity in terms of the Boltzmann variable, i.e., distance divided by the square root of time (Smiles et al. 1978). The use of material coordinates of the water to describe transport and dispersion of solutes in unsaturated soils are given by Raats (2002). This approach has been extended by Smiles et al. (1978) to include hydrodynamic dispersion and chemical reaction in swelling systems where the solid, the water and the solute are all in motion relative to an external observer.

White (1995) and van Genuchten et al. (1999) considered the influence of the chemical composition of soils upon their hydraulic properties. Lessoff and Konikow (1997) analysed solute transport and infiltration-redistribution cycles in heterogeneous soils on the basis of three mechanisms: (i) advection by gravitational flow of water, (ii) linear-equilibrium sorption, and (iii) linear decay. Broadbridge et al. (2000) discussed analytical solutions for two-dimensional solute transport with velocity-dependent dispersion. Using symmetry analysis, they found new solutions for non-radial solute transport against a background of radial water flow. This Lie Group approach is akin to earlier studies for the Richards equation leading to solutions for the class of versatile nonlinear soils, including the subclass of Knight soils (Sposito 1990).

7.4 Spatio-Temporal Dimension of Multiphase Flow and Transport Processes

Water movement and solute transport during multigation for agricultural soils must be studied at the plot, farm and catchments scales and appropriate models are needed for each level. The plot scale studies comprise of micro-scale of solid particles and pores, the farm scale implies field scale studies of soil structural elements and of individual plants and associated volumes of soil, and the catchment scale implies regional studies of the soil profile and of plant communities.

At the micro-scale water retention in soil pores is acted upon by surface tension at the air-water interface and by the diffusive double layer of solid-air or the solid-water interface, or by both. The study by Tuller (1986) suggested that not only are water retention characteristics but also soil hydraulic conductivity characteristics conditioned by soil physico-chemical influence of the solid phase upon the water. For the mesoscale geometries Phillips (1968) considered the characteristic times associated with local equilibration which turned out to be small, and concluded that the lack of local equilibrium was not a threat to the validity of analyses based on the Richards equation. However, a lack of internal equilibrium was implied regularly by laboratory experiments in the 1960s and early 1970s (Smiles et al. 1971) and this received considerable attention in the following decades. The general impression is that layering, aggregates, cracks, and bio-channels left behind by penetrating roots and burrowing animals often have a large influence on movement of air-water and transport of solutes in the soil. Many models for flow and transport in such soils distinguish a mobile and a stagnant phase, roughly corresponding to networks of large and small pores. Important further ingredients in these models are mechanisms of transport in mobile phase and nature of storage capacities of phases and associated exchanges between phases. Such models are often referred to as dual-porosity models.

A review by van Genuchten et al. (1999) showed that the root water uptake, multi-component transport and preferential flow can be reasonably handled by multiphase flow models. More complete overviews of flow and transport in structured soils and related image analysis are given in van Genuchten et al. (1991) Mermut and Norton (1992) and Selim and Ma (1998). Unfortunately, effects arising from genuine lack of local equilibrium are difficult to distinguish from effects arising from unstable flows described as shown by Parlange et al. (1982). For the micro- and mesoscales non-destructive techniques such as dose rate measurements, typically adapted from medical technology, are now used to map the pore space and the fluids filling it (Císlerova 1999).

The field-scale implementation of local scale models is suitable for the design and interpretation of detailed field experiments and extrapolation from the limited number of experiments that can be afforded.

The catchment scale is of greatest interest to policy makers. It is the scale at which pollution of ground and surface water and damage to biotopes becomes most evident. Developing and implementing models at this scale remains an enormous challenge. It appears that at the catchment scale the concept of a single model with effective parameters generally becomes untenable and that catchment models by necessity are aggregates of representative elements, i.e. fields.

Multiphase flow and transport phenomena in the unsaturated and saturated zones of the subsurface environment are the current focus of research (Darnault et al. 2002). Transient flow of nutrients and pollutants in soil after irrigation and between two irrigation events with respect to multigation is less well understood. Consequently, there is a need for fast, non-destructive, and accurate measurements of transient multiphase-fluid flow in porous media. Transient visualization utilizing radiation-based probes allows determination of fluid content in three-phases. Synchrotron x-ray allows accurate and fast measurements of fluid content in transient flow fields in any soil type in a small section of a flow field (Tomomasa et al. 2005) whereas the radiation transmission is a non-destructive method that allows visualization and measurement of fluid content in transient three phase flow occurring in sandy porous media over the whole flow field with a small (millimeters per hour) time resolution (Darnault et al. 2002).

8 Multigation Dynamics with Other Inputs

8.1 Nutrient Dynamics Under Multigation

Soil oxygen is one of the important determinants of soil fertility and therefore must be conserved during irrigation and post-irrigation events. It is the energy produced by aerobic respiration that drives root growth for uptake of nutrients and water (Barrett-Lennard, 2003). When the O_2 concentration is reduced below an optimum, growth of plant roots is slowed, with a corresponding loss in yield (Buwalda et al. 1988). Prolonged deficiency of O_2 leads to anoxia and eventual death of roots. The decomposition of dead roots in turn can lead to the formation of toxic substances, such as phenyl acetic, 4-phenyl-butyric, salicylic and 0-coumaric acids and salicylaldehyde (Chou and Patrick 1976). These toxins and those formed from chemical reduction e.g. nitrites, Fe^{2+}, Mn^{2+}, H_2S in the soil can lead to plant death within a relatively short period. Oxygen is also essential for most soil micro-organisms (Subba Rao 1999) and it has been estimated that in a fertile soil more O_2 is consumed by micro-organisms than by crop plants (Grable 1966). Insufficient oxygen can slow down or halt important soil microbial functions that (a) govern the breakdown of OM, the formation of humus that increases porosity and resistance to wind and water erosion, and (b) promote symbiotic and non-symbiotic fixation of nitrogen (e.g. by *Azotobacter*), and the nitrification of urea, ammonium and nitrite nitrogen (Powlson 2000).

Oxygen limitation for root functioning can occur in solution culture as much as it occurs in irrigated field crops. The state of many nutrient elements and their dynamics in the soil depends on the aeration status of the rhizosphere. Nitrate-nitrogen (NO_3-N) is particularly susceptible to leaching as the root uptake and transpiration slow down under anaerobic conditions. Nitrate, manganese (Mn), copper (Cu) and iron (Fe) are some of the important nutrients susceptible to leaching with anoxia. The amount of soil O_2 required for adequate growth is greater if ammonium rather than nitrate-nitrogen is present or if Mn or Cu are deficient (Garcia-Novo and Crawford 1973). In solution culture, tomato and apple, which have higher O_2 requirements, did best with 16 mg L^{-1} dissolved O_2 while avocado, barley, citrus, soybean, with moderate demand, did very well with only 6–8 mg L^{-1} (Huang et al. 1994). In gravel culture, a concentration of about 5–6 mg L^{-1} of O_2 in the nutrient solution appears to be ample for carnation if continuous circulation of the solution is used. However, ideal amounts are related to number of plants per unit soil volume and time of day, being 0.4–0.55 mg min^{-1} $plant^{-1}$ during the day and 0.2–0.4 mg $min^{-1}plant^{-1}$ during the night (Huang et al. 1994). Ironically, solution culture plants can also be potentially injured with an excess of O_2. While soybean showed Fe deficiency in the presence of higher O_2 when nitrates were present in the solution, oats and wheat did not (Bourget et al. 1966).

Irrigation methods such as drip are excellent tools to optimise fertilizer use efficiency, if managed effectively. However, our review reveals that wetting fronts associated with drip can often exhibit hypoxia and anoxia (e.g. Silberbush et al. 1979). In irrigated agriculture water and fertilizer management are interlinked, especially with fertigation. Flushing of the subsurface sodic zone can be achieved with introduction of a smaller volume of gypsum to displace sodium by calcium and an additional supply of sulphur (Reyhan and Amiraslani 2006). On occasions SSDI may be used in conjugation with rainfall events exceeding 12 mm to leach accumulated salts around emitters (Roberts et al. 2008). Multigation allows such controlled application of water, air, nutrients and gypsum together in the irrigation stream to maximise the leaching of salt beyond the rooting zone (Bhattarai and Midmore 2009). Aerated roots exclude salt; therefore salt accumulation occurs in the periphery of the wetting fronts.

Multigation allows for reduction of salt loading primarily through three different processes. Firstly the crop WUE with multigation is greater, and timing irrigation to crop evapo-transpiration demand can be effectively achieved through multigation. This way multigation reduces the amount of water required per unit crop production, and hence reduces salt input into the soil. Secondly, it promotes salt exclusion, enabling the productive use of saline soils or saline irrigation water over the crop period. Thirdly an appropriate shandying can be made of different proportions of saline and non-saline water which permit irrigation with a regulated mix from saline irrigation water sources.

8.2 *Watertable Depletion and Pollution Management with Multigation*

In many regions irrigation is sourced from groundwater, associated with fragmented, shallow aquifers the recharge for which is mainly through rainfall. Excessive pumping can deplete underground water resources resulting in increased salinity by successively lowering the water table that concentrates salts in the underground (Abu-zeid 1993; Tanji and Keyes 2002). Management of watertable is more plausible with drip irrigation compared to that of the furrow. Lowering the watertable provides more water storage capacity in the field to accommodate excess water from heavy rainstorms.

Increasing competition among industries for fresh water has forced agricultural industries to use marginal quality water for irrigation. When irrigation water is allowed to infiltrate into soils with a ground watertable at a rate faster than that of extraction, the watertable starts to rise and may, in time, bring the zone of saturation close to the surface. This leads to water-logging and often secondary salinization, causing an accumulation of salts in the root zone. In areas where the climate is hot and dry, irrigated land is subject to substantial water losses through evaporation. Salts contained in precipitation and irrigation water remain in the soil and increase in concentration when the water evaporates from the soil or when the plants take up water for transpiration. If the salt is not leached its concentration increases constantly and subsequently constrains crop yield. If the salinization process is allowed to continue, the land eventually has to be abandoned. Flood and furrow irrigation produce significant runoff and deep drainage, contributing to pollution and rising watertables. Matching irrigation to crop demand with SSDI leads to less likelihood for drainage and rises in groundwater level.

Multigation can also play an increasing role in many agricultural areas in future. The goal of an irrigation system should be towards optimizing crop production through prudent management of limited ground water resources that result in economic gain for producers. By controlling the rooting depth SSDI allows uptake of water from the soil when groundwater is fully recharged and minimizes depletion. Bhattarai et al. (2006) reported that across a number of crop species a reduction in supply of irrigation water resulted in a less than proportional reduction in yield. Deficit irrigation combined with conservation tillage could potentially extend groundwater resources and minimize nitrate movement within the soil profile. Additional research is needed to verify early results on deficit irrigation. Deficit irrigation and nutrient management are more amenable to multigation compared to conventional irrigation methods, and oxygation has been shown to result in benefits even at deficit irrigation rates (Bhattarai et al. 2006). Multigation can also reduce the pressure on groundwater where groundwater depletion is rapid by improving irrigation efficiency (IE).

8.3 Reduction in Use of Pesticides

Precise application of fertilizers and agro-chemicals through aerated irrigation water to plant roots can considerably increase the efficient use of these inputs compared to their use under conventional irrigation. Benefits of fertigation have been reported by many researchers, for example Grimes et al. (1990), and Lyannoi and Shevchenko (1985) on grape vines, Bar-Yoseph et al. (1989) on sweet corn and Phene et al. (1986) on tomatoes. Less weed germination takes place in multigation crops due to the dry soil surface in between the rows; this in turn leads to a reduced need for herbicides. As an example, Bhattarai et al. (2006) observed very low weed infestation for deficit SSDI rates in cotton compared to furrow irrigation. The higher yields in tomatoes grown with SSDI compared to furrow irrigation reported by Grattan et al. (1988) were largely attributed to the lessened competition from weed with SSDI. Ferrell (1990) suggested that one of the biggest cost savings with SSDI crops comes from reduced labour and herbicide requirements. Reduced weed infestation also makes other cultural practices easier and enhances quality of some crops, for example cotton (McHugh et al. 2008).

A dry growing environment reduces the incidence of fungal disease such as *Botrytis* (Hiaring 1987) and gomosis thereby reducing the need for application of fungicides. Irrigating raisin grapes solely with SSDI reduced the use of nematicides by 83% (Ferrell 1990). Chemigation through SSDI systems has been found to be easier, safer and more effective than spraying and drenching (Ptacek 1986). Some herbicides, nematicides and systemic insecticides are very effectively when applied through SSDI systems. Application of Vapam (*a.i.* 42% metam sodium) through the drip system has been found to be effective against root knot nematodes (Phene et al. 1986; Tollefson 1988). Application of many of the more toxic chemicals through SSDI may make them safer to use. For example, SO_2 applied through SSDI to control *Phylloxera* in vines in Sonoma County, California, USA was found by Granett et al. (2001) to be safe and effective. Soil aeration has a bearing on the epidemiology and epidemics of many soil pathogens. As an example, poor soil aeration caused by poor soil structure, soil type or water logging is associated with the development of cavity spot caused by *Pythium* species in carrot (Hiltunen and White 2002). Multigation can effectively deliver oxygen in the rooting zone and directed application of safe fungicides can effectively control root diseases caused by *Pythium*. Nursery and horticultural industries experience losses due to crop root rot diseases. Multigation will pave the way toward more effective management of soil borne and root rot related diseases in many horticultural crops, including cut flowers (Sullivan et al. 2000).

Pesticide runoff has been a major concern for irrigated agriculture. Although the total quantity of aerially-sprayed pollutant runoff under furrow irrigation is greater than from managed SSDI, for individual run-off events, the concentration of canopy or soil applied pesticides can be higher in SSDI than conventional irrigation methods. Concentrations in SSDI runoff above a critical threshold level can be toxic and harmful to other aquatic and non-aquatic biota (McHugh et al. 2008). As multigation aims to apply pesticides more directly to root systems, it can significantly lower the off-farm movement of chemicals. Aerating the irrigation water stream that contains pesticides will reduce runoff and deep drainage of these chemicals as an aerated rhizosphere increases transpiration and plant water use (McHugh et al. 2008).

8.4 Multigation and Improving the Use of Reclaimed Water

Demand for fresh water is increasing, and water of better quality is preferentially used for domestic purposes, hence water of lower quality is increasingly used for irrigation. One challenge for the future will be to maintain, or better increase, crop production with less water. Water that may be of poor quality such as saline waters is successfully used for agriculture in Israel, Italy, and the USA (Malash et al. 2002). Reclaimed water often has a high biological oxygen demand (BOD); once delivered to the soil microbial consumption of O_2 in irrigation water reduces the O_2 available for root respiration. To overcome this, oxygation has been successfully employed to enrich the O_2 concentration in reclaimed water to satisfy both microbial and root respiration (Kele et al. 2005).

The potential for salinity hazards in the root zone increases with SSDI when using wastewater. Generally pre-treatment procedures do not ameliorate wastewater salinity or sodicity. Beneficial and safe use of reclaimed wastewater for SSDI will depend on management strategies that focus on irrigation pre-treatment, virus monitoring, field and crop selection, and periodic leaching of salts (Assadian et al. 2005). Choi and Suarez Rey (2004) highlight rises in surface EC and clogging of emitters in a multi-year experiment using SSDI and reclaimed water for pasture production. Multigation can deter crop root intrusion into emitters by the action of impregnated herbicides in the emitters.

SSDI shows promise to safely deliver reclaimed wastewater and to minimize the exposure of soil surfaces, above-ground plant parts, and groundwater to reclaimed wastewater (Newham 1992). However, the persistence and movement of waterborne viruses are of growing scientific and public concern with the recent increase in wastewater reclamation efforts. The wastewater blend can be pre-treated to reduce faecal *coliforms* by utilizing in-line UV treatments during multigation to meet wastewater reuse guidelines for edible crops. Because human viruses can persist in the soil for extended periods if using an SSDI delivery system (Assadian et al. 2005; Yates 1992), virus inactivation strategies may need to be integral to treating reclaimed wastewater, regardless of irrigation delivery system.

Secondary treated sewage effluent is frequently disposed of by flood irrigation on "sewage farms". These farms often apply water at considerably higher levels than is used by plants, and excess water, salts and nitrogen are carried down to the aquifer. In the long term 70–80% recoveries of N are possible when producing healthy crops with SSDI (Pettygrove 1992). Pollution by nitrates from reclaimed water may, therefore, be reduced by multigation. Such application of multigation to minimize nitrate pollution also equally applies to schemes where the prime purpose is to safely dispose of the effluent and to schemes where the main purpose is to extend the regional water resources. Multigation improves the reclaimed water utilization by high value crops of the horticultural industries such as flowers and fruits.

The pollution of ponds, streams, lakes and aquifers by domestic septic systems is well documented (Harter et al. 2003). On-site treatment plants combined with air injection system may address this problem (Ruskin et al. 1990). SSDI is particularly suited to disposal of wastewater when watertable are high, soils have low permeability, and slope of the ground or the available area does not allow for conventional soakage drains. Excreta from animal feed lots and dairies are a major source of nitrate pollution of aquifers surrounding these industries. Potential exists to treat excreta in an on-site packaged plant and then dispose of the liquid effluent with SSDI. Feasibility studies are valuable to develop alternatives for pollution control of aquifers in a sustainable way, and multigation will be one of the appropriate approaches to the same.

9 Future Directions

The first step when reviewing issues of irrigation is to find ways to improve WUE with associated reduction in drainage volumes. This can be accomplished by improving management of existing drip irrigation systems through incorporation of multigation, and promoting grower change to a more efficient irrigation system that captures the benefits of multigation. Optimised multigation systems improve WUE, minimize the application of fertilizers and pollutants, reduce potential pollution of aquifers and can become an effective tool for the use of reclaimed water for irrigation. Disposal of saline and reclaimed water will present a new opportunity but will also impose significant problems for irrigated agriculture. The value of multigation for pollution abatement lies in its precision application of aerated water with other farm chemicals, matching demand by the crop for these inputs. To enable the irrigation industry to use this tool efficiently, and to promote incentives for its adoption, more research is needed towards a greater understanding of multiphase phenomena and a quantification of costs and benefits.

Multigation will potentially become commonplace: decreasing allocation and increasing cost of irrigation water to agricultural industries will provide incentives to growers to adopt SSDI. Increased adoption of SSDI will make multigation more attractive to the industry, for incorporating the features for multigation once the industry already has drip or SSDI has negligible cost implications. Plants need a dynamic yet balanced supply of water, nutrients and air during their entire life cycle. Therefore, it is crucial that the delivery system is designed in such a way that the input needs of the crop are matched with the supply. Multigation allows objective delivery of all the required soil-based inputs in a coordinated manner directly into the root system. This includes the often-forgotten supply of oxygen that is in short supply during and soon after irrigation events. Wide-scale adoption of multigation can potentially bring a quantum increase in yield, a decrease in the use of agro-chemicals such as fertilizers and pesticides, and a halt to the escape of pesticides to other ecosystems. Streamlining the current research base and expanding developments of multigation technology will make it a viable irrigation option for the industry in the changing scenario of water allocation, environmental regulation and water pricing.

Acknowledgements We thank Laurence Tait and Lance Pendergast for their comments and feedback on the manuscript, and Dr. Eric Lichtfouse for suggested improvements. Financial contribution of CQUniversity Australia and the National Program for Sustainable Irrigation (NPSI) Australia is gratefully acknowledged.

References

Abboud A (1990) Phosphorus nutrition of crops: a sustainable approach on sustainable farming. http://eap.mcgill.ca/magrack/sf/spring%2090%20k.htm

Abu-zeid M (1993) Review of water table depth planning for a multi objective water management system. Irrig Drain Syst 6(4):265–274

Allen S, Nehl D, Kochman J, Salmond G (2004) Irrigation and cotton disease interactions. In "WATERpak – a guide for irrigation management in cotton", Australian Cotton Cooperative Research Centre, Cotton Research and Development Corporation, Narrabri, Australia, pp 152–156

Assadian NW, Di Giovanni GD, Enciso J, Iglesias J, Lindemann W (2005) The transport of waterborne solutes and bacteriophage in soil sub-irrigation with a wastewater blend. Agric Ecosyst Environ 111:279–171

Ayars JE, Phene CJ, Hutmacher RB, Davis KR, Schoneman RA, Vali SS, Mead RM (1999) Subsurface drip irrigation of row crops: a review of 15 years of research at the water management research laboratory. Agric Water Manage 42:1–27

Barrett-Lennard EG (2003) The interaction between waterlogging and salinity in higher plants: causes, consequences and implications. Plant Soil 253:35–54

Bar-Yoseph B, Sagiv B, Markovitch T (1989) Sweet corn responses to surface and subsurface trickle phosphorous fertigation. Agron J 81(3):443–447

Bhattarai S, Huber S, Midmore DJ (2004) Aerating subsurface irrigation gives growth and yield benefits to zucchini, vegetable soybean and crops in heavy clay soils. Ann Appl Biol 144:285–298

Bhattarai S, Pendergast L, Midmore DJ (2006) Root aeration improves yields and water use efficiency of tomato in heavy clay and saline soils. Sci Hortic 108:278–288
Bhattarai SP, Salvaudon C, Midmore DJ (2008a) Oxygation of the rockwool substrate for hydroponics. Aquaponics J 49 (2): 29–33
Bhattarai SP, Midmore DJ, Pendergast L (2008b) Yield, water-use efficiencies and root distribution of soybean, chickpea and pumpkin under different subsurface drip irrigation depths and oxygation treatment in vertisols. Irrig Sci 26: 439–450
Bhattarai S, Su N, Midmore D (2005) Oxygation unlocks yield potentials of crops in oxygen limited soil environments. Adv Agron 88:314–365
Bhattarai SP, Midmore DJ (2009) Oxygation effects on physiology and water use efficiency and salt tolerance of cotton and vegetable soybean in a vertisol. J Integ Plant Biol 51(7):675–688
Bhattarai SP, Midmore DJ, Dhungel J, Pendergast L (2010) Diversifying the application of oxygation to lawn, landscape, and large scale agricultural irrigation. Paper presented in the Australian Irrigation Conference & Exhibition, 7-9 June 2010, Sydney, Australia
Biswas AK, HY5 (1967) Hydrologic engineering prior to 600 BC. J Hydraul Div Am Soc Civil Eng 93:115–135
Blackwell PS, Wells EA (1983) Limiting oxygen flux densities for oat root extension. Plant Soil 73:129–139
Boland A, Falivene S, Goodwin I, Williams D (2005) Open hydroponics: opportunities and risks stage. Final report LWA DAN 22. National Program for Sustainable Irrigation, Land & Water Australia, Canberra
Bourget SJ, Finn BJ, Dow RK (1966) Effects of different soil moisture tensions on flax and cereals. Can J Soil Sci 46:213–216
Brady NC, Weil RR (2004) Soil aeration and temperature. In: Brady NC, Weil RR (eds) The nature and properties of soils. Prentice Hall, New York, pp 265–306
Brennen CE (2005) Fundamentals of multiphase flow. Cambridge University Press, Cambridge, p 257
Broadbridge P, Hill JM, Goard JM (2000) Symmetry reductions of equations for solute transport in soil. Nonlin Dyn 22(1):15–27
Bugbee B, Monje O (1992) The optimization of crop productivity: theory and validation. Bioscience 42:494–502
Buwalda F, Barrett-Lennard EG, Greenway H, Davies BA (1988) Effects of growing wheat in hypoxic nutrient solutions and of subsequent transfer to aerated solutions. II. Concentrations and uptake of nutrients and sodium in shoots and roots. Aus J Plant Physiol 15:599–612
Camp CR (1998) Subsurface drip irrigation: a review. Trans Am Soc Agric Eng 41:1353–1367
Camp CR, Lamm FR, Evans RG, Phene CJ (2000) Subsurface drip irrigation: past, present and future. Proceedings of the fourth decennial irrigation symposium, 14–16 November, Phoenix, Arizona. American Society of Agricultural Engineers, St Joseph, Michigan, USA, pp 363–372
Cannell MGR, Murray MB, Sheppard LJ (1985) Frost avoidance by selection for late budburst in *Picea sitchensis*. J Appl Ecol 22:931–941
Choi CY, Suarez Rey EM (2004) Subsurface drip irrigation for Bermuda grass with reclaimed water. Trans Am Soc Agric Eng 47(6):1943–1951
Choi HM, Terauchi T, Monji H, Matsui G (2002) Visualization of bubble-fluid interaction by a moving object flow image analyzer system. Ann New York Acad Sci 972(Oct):235–241
Chou CC, Patrick ZA (1976) Identification and phytotoxic activity of compounds produced during decomposition of corn and rye residues in soil. J Chem Ecol 2:369–387
Císlerova M (1999) Characterization of pore geometry. In: Feyen J, Wiuyo K (eds) Modelling of transport processes in soils, vol 1. Wageningen Press, The Netherlands, pp 103–117
Coder KD (1998) Root growth requirements and limitations. University of Georgia cooperative extension service forest resources publication for March 1998–1999, pp 7
Cook SE, Bramley RGV (1998) Precision agriculture – opportunities, benefits and pitfalls of site-specific crop management in Australia. Aus J Exp Agric 38:753–763
Dangler JM, Locascio SJ (1990a) Yield of trickle-irrigated tomatoes as affected by time of N and K application. J Am Soc Hortic Sci 115:585–589

Dangler JM, Locascio SJ (1990b) External and internal blotchy ripening and fruit elemental content of trickle-irrigated tomatoes as affected by N and K application time. J Am Soc Hortic Sci 115:547–549

Darnault CJG, DiCarlo DA, Bauters TWJ, Steenhuis TS J, Parlange Y, Montemagno CD, Baveye P (2002) Visualization and measurements of multiphase flow in porous media using light transmission and synchrotron X-rays. Ann New York Acad Sci 972:103–110

Davidson JM (1992) Water and nutrient requirements for drip-irrigated vegetables in humid regions. University of Florida. http://ww5.5wfwmd.state.fl.us/

de Vries DA (1958) Simultaneous transfer of heat and moisture in porous media. Trans. Amer. Geophys. Union 39(5): 909–916

Dimitrienko YI (1999) Thermo-mechanics of composites under high temperatures. Kluwer, Dordrecht

Dowell RJ, Crees R, Burford JR, Cannell RQ (2006) Oxygen concentrations in a clay soil after ploughing or direct drilling. Euro J Soil Sci 3(2):239–245

Drew DA, Passman SL (1998) Theory of multicomponent fluids. Applied Mathematical Science, 135. Springer, New York

Drew MC, Stolzy LH (1996) Growth under oxygen stress. In: Waisel Y, Eshel A, Kafkafi U (eds) Plant roots: the hidden half. Marcel Decker, New York, pp 397–414

Drower MS (1956) Water-supply, irrigation and agriculture. In: A History of Technology, vol. 1: From Early Times to Fall of Ancient Empires, Singer CS, Holmyard EJ, Hall AR (eds). Clarendon Press: Oxford, UK; 520–557

Eertwegh GAPH van den, Nieber JL, Feddes RA (2001) Multidimensional flow and transport in a drained, dual-porosity soil. In: Bosch DD, King KW (eds) Preferential flow, water movement and chemical transport in the environment. Proceedings of 2nd International Symposium, 3–5 January 2001, Honolulu, Hawaii, USA, St. Joseph, Michigan, ASAE 701P0006, pp 241–244

Falivene S, Goodwin I, Williams D, Boland A (2005) Introduction to open hydroponics, NPSI fact sheet. National Program for Sustainable Irrigation, Australia

Howell TA (2001) Enhancing water use efficiency in irrigated agriculture. Agron J 93:281–289

IATP (2009) Integrated solutions to the water, agriculture and climate crisis. Fact Sheet. Trade and global governance, Institute for agriculture and trade policy, Minnesota, USA, pp 1–4

Ferrell JE (1990) Less water, more crops. This World Magazine, The San Francisco Chronicle, 11 March 1990, pp 17–18

Follett RF, Porter LK, Halvorson AD (1991) Border effects on nitrogen-15 fertilized winter wheat microplots grown in the Great Plains. Agron J 83:608–612

Fukushima E (1999) Nuclear magnetic resonance as a tool to study flow. Ann Rev Fluid Mech 31:95–107

Garcia-Novo F, Crawford RMM (1973) Soil aeration, nitrate reduction, and flooding tolerance in higher plants. New Phytol 72:1031–1039

Gish TJ, Kung KJS, Perry DC, Posner J, Bubenzer G, Helling CS, Kladivko EJ, Steenhuis TS (2004) Impact of preferential flow at varying irrigation rates by quantifying mass fluxes. J Environ Qual 33:1033–1040

Goorahoo D, Carstensen G, Zoldoske DF, Norum E, Mazzei A (2002) Using air in sub-surface drip irrigation (SSDI) to increase yields in bell peppers. Int Water Irrig Tel-Aviv Israel 22:39–42

Grable AR (1966) Soil aeration and plant growth. Adv Agron 18:57–106

Graetz DA, Fiskell JGA, Locascio SJ, Zur B, Meyers JM (1978) Chloride and bromide movement with trickle irrigation of bell peppers. P Fl St Hortic Soc 91:319–322

Granett J, Walker MA, Kocsis L, Omer AD (2001) Biology and management of grape phylloxera. Ann Rev Entomol 46:387–412

Grattan SR, Schwankl LJ, Lanini WT (1988) Weed control by subsurface drip irrigation. Calif Agric 42(3):22–24

Green W, Ampt G (1911) Studies of soil physics: 1.The flow of air and water through soils. J Agric Sci 4:1–24

Grieve AM, Dunford E, Marston D, Martin RE, Slavich P (1986) Effects of waterlogging and soil salinity on irrigated agriculture in the Murray Valley: a review. Aus J Exp Agric 26:761–77

Grimes DW, Munk DS, Goldhammer DA (1990) Drip irrigation emitter depth placement in a slowly permeable soil. Visions of the future. Proceedings of 3rd national irrigation symposium, American society of agricultural engineers, St. Joseph, MI, pp 248–254

Hall DW, Risser DW (1993) Effect of agricultural nutrient management on nitrogen fate and transport in Lancaster County, Pennsylvania. J Am Water Res Assoc 29(1):55–76

Harris GA (2005) Subsurface drip irrigation: crop management. DPI & F Note, Brisbane, Australia. http://www.nrw.qld.gov/au/rwue/factsheets/sdi_crop_management.pdf

Harter T, Mayer RD, Mathews MC (2003) Nonpoint source problem from animal farming in semi-arid regions: spatio-temporal variability and ground water monitoring strategies. In: Riberio L (ed) Future groundwater resources at risk (2002). Proceedings of the 3rd international conference, June 2001, pp 363–372

Hartz TK (2003) Deficit drip irrigation for soluble solids improvement. Annual report submitted to the California Tomato Research Institute

Hartz TK, Hochmuth GJ (2008) Fertility management of drip-irrigated vegetables. UC Davis vegetable research and information centre. http://vric.uvdavis.edu/veginfo/topics/fertilizer/fertilitymanagement.pdf

Heuberger H, Livet J, Schnitzler W (2001) Effect of soil aeration on nitrogen availability and growth of selected vegetables - preliminary results. Acta Hortic 563:147–154

Hewitt EJ (1966) Sand and water culture methods used in the study of plant nutrition. Technical communication no 22, 2nd edn rev. Commonwealth Agricultural Bureaux, Farnham Royal, Bucks, England, pp 212–214

Hiaring PE (1987, Jan) Murphy-Goode vineyard in Alexander Valley: going underground with drip. Wines and Vines

Hiltunen LH, White JG (2002) Cavity spot of carrot (*Daucus carota*). Ann Appl Biol 141:201–223

Huang BR, Johnson JW, Nesmith DS, Bridges DC (1994) Root and shoot growth of wheat genotypes in response to hypoxia and subsequent resumption of aeration. Crop Sci 34:1538–1544

Huber S (2000). New uses for drip irrigation: partial root zone drying and forced aeration. Masters' Thesis, CQU Australia & TUM Germany, pp192

Huck MG, Hageman RH, Hanson JB (1962) Diurnal variation in root respiration. Plant Physiol 37(3):371–375

Kele B, Midmore DJ, Harrower K, McKennariey BJ, Hood B (2005) An overview of the Central Queensland University self-contained evapotranspiration beds. Water Sci Technol 51(10):273–281

Khondaker NA, Ozawa K (2004) Papaya plant growth as affected by soil air oxygen deficiency. Acta Hortic 740:137–143

Lamm FR, Camp CR (2007) Subsurface drip irrigation. In: Lamm FR, Ayars JE, Nakayama FS (eds) Microirrigation for crop production – design, operation and management. Elsevier, Amsterdam, pp 473–551

Lamm FR, Trooien TP (2003) Subsurface drip irrigation for corn productivity: a review of 10 years of research in Kansas. Irrig Sci 22:195–200

Lessoff SC, Konikow LF (1997) Ambiguity in measuring matrix diffusion with single-well injection/recovery tracer tests. Ground Water 35(1):166–176

Letey J (1961) Aeration, compaction and drainage. Calif Turfgrass Cult 11:17–21

Letey J, Stolzy LH, Lunt OR, Yoongner VB (1964) Growth and nutrient uptake of Newport bluegrass as affected by soil oxygen. Plant Soil 20(2):143–148

Lewis DC, Grant IL, Maier NA (1993) Factors affecting the interpretation and adoption of plant analysis services. Aus J Exp Agric 33:1053–66

Locascio SJ, Olson SM, Rhoads FM (1989) Water quantity and time of N and K application for trickle-irrigated tomatoes. J Am Soc Hortic Sci 114:265–268

Locascio SJ, Smajstrla AG (1989) Drip-irrigated tomato as affected by water quantity and N and K application timing. P Fl St Hortic Soc 102:307–309

Loomis RS, Rabbinge R, Ng E (1979) Explanatory models in crop physiology. Ann Rev Plant Physiol 30:339–67

Lyannoi AD, Shevchenko IV (1985) Effectiveness of fertilizer application with drip irrigation in vineyards. Vinodelie i Vinogadarstvo 1:35–37

Machado RMA, do Rosario M, Oliveira G, Oliveira G (2005) Tomato root distribution, yield and fruit quality under different subsurface drip irrigation regimes and depths. Irrig Sci 24:15–24

Malash N, Ghaibeh A, Abdelkarim G, Yeo A, Flowers T, Ragab R, Cuartero J (2002) Effect of irrigation water on yield and fruit quality of tomato. Acta Hortic 573:423–427

Marfa O, Cáceres R, Guri S (2005) Oxyfertigation: a new technique for soilless culture under Mediterranean conditions. Acta Hortic 697:65–72

McHugh AD, Bhattarai SP, Lotz G, Midmore DJ (2008) Effects of subsurface drip irrigation rates and furrow irrigation for cotton grown on a vertisol on off-site movements of sediments, nutrients and pesticides. Agron Sustain Develop 28:507–519

Meek BD, Ehlig CF, Stolzy LH, Graham LE (1983) Furrow and trickle irrigation: effects on soil oxygen and ethylene and tomato yield. Soil Sci Soc Am J 47:631–635

Menzies JD (1963) Survival of microbial plant pathogens in soil. Bot Rev 29(1):79–122

Mermut AR, Norton LD (1992) Digitization, processing and quantitative interpretation of image analyses in soil science and related areas. Geoderma 53:3–4

Mikkelsen RL (1989) Phosphorus fertilization through drip irrigation. J Prod Agric 2:279–286

Miller CT, Christakos G, Imhoff P, McBride J, Pedit J, Trangenstein J (1998) Multiphase flow and transport modelling in heterogeneous porous media: challenges and approaches. Adv Water Res 21(2):77–120

Morard P, Lacoste L, Silvestre J (2004) Effect of oxygen deficiency on mineral nutrition of excised tomato roots. J Plant Nutr 27:651–661

Newham D (1992) Reclaimed water in the united States with a worldwide overview. Proceedings of the Recycled Water Seminar, Wagga Wagga, NSW, Australian Water and Wastewater Association, Australia, 19–20 May 1992

Nonneck IL (1989) Vegetable production. Van Nostrand, New York, pp 229–235

Oster JD (1992) Water quality impacts: issues, information sources and mitigation strategies. Proceedings of reclaimed wastewater: practical approaches to developing an alternative water supply, University of California Davis, CA 95616, 18 May 1992

Parlange JY, Lisle I, Braddock RD, Smith RE (1982) The three-parameter infiltration equation. Soil Sci 133:337–341

Pendergast L (2010) Oxygation: enhanced root functions, yields and water use efficiencies through oxygated subsurface drip, with a focus on cotton. PhD Thesis submitted to CQUniversity Australia, pp387

Persaud N, Locascio SJ, Geraldson CM (1976) Effect of rate and placement of nitrogen and potassium on yield of mulched tomato using different irrigation methods. P Fl St Hortic Soc 89:135–138

Persaud N, Locascio SJ, Geraldson CM (1977) Influence of fertilizer rate and placement and irrigation method on plant nutrient status, soil soluble salt and root distribution of mulched tomatoes. Proc Soil Crop Sci Soc Fl 36:122–125

Pettygrove S (1992) Nitrogen in reclaimed wastewater: asset or liability? Proceedings of reclaimed wastewater: practical approaches to developing an alternative water supply, University of California Davis, CA 95616, 18 May 1992

Phene CJ, Bar-Yoseph B, Hutmacher RB, Patton SH, Davis KR, McCormick RL (1986) Fertilization of high yielding subsurface trickle irrigated tomatoes. The proceedings of the 34th annual California fertilizer conference, Fresno, CA, pp 33–43

Philip JR (1969) Theory of infiltration. Adv Hydrol Sci 5:215–296

Philip JR (1998) Full and boundary-layer solutions of the steady air sparging problem. J Contam Hydrol 33(3–4):337–345

Phillips E (1968) Washington climate for these counties, King, Kitsap, Mason, Pierce. Cooperative Extension Service, Washington State University, Pullman Wa, p 66

Powlson DS (2000) Tackling nitrate from agriculture. Soil Use Manage 16:141–141

Powlson DS, Goulding KWT, Willison TW, Webster CP, Hutsch BW (1997) The effect of agriculture on methane oxidation in soil. Nutr Cycl Agroecosys 49:50–70

Ptacek LR (1986) Subsurface irrigation and the use of chemicals. Proceedings of the irrigation association annual conference, pp 225–234

Qassim A (2003) Subsurface irrigation: a situation analysis. Water Conservation and Use in Agriculture. http://www.wca – infonet.org/cds_upload/1058151636725_SUBSURFACE_IRRIGATION.pdf. Accessed 10 Feb 2006

Raats PAC (1965) Development of equations describing transport of mass and momentum in porous media, with special reference to soils. Ph.D. Dissertation, University of Illinois, Urbana-Champaign

Raats PAC (2002) Flow of water in rigid and non-rigid, saturated and unsaturated soils. In: Capriz G, Ghionna VN, Giovine P (eds) The modelling and mechanics of granular and porous materials. Birkhäuser, Boston, MA, pp 181–211

Raats PAC, Klute A (1968) Transport in soils: the balance of mass. Proc Soil Sci Soc Am J 32(2):161–166

Reyhan MK, Amiraslani F (2006) Studying the relationship between vegetation and physio-chemical properties of soil, Case Study: Region Ian Tabus. Pakistan J. of Nutrition 5 (2):169 – 171

Roberts TL, White SA, Warrick AW, Thompson TL (2008) Tape depth and germination method influence patterns of salt accumulation with subsurface drip irrigation. Agric Water Manage 95(6):669–677

Rogers DH, Lamm FR, Mahbub A (2008) SSDI water quality assessment guidelines. http://www.oznet.K-state.edu/sdi_waterquality/2008

Rolston DE, Rauschkolb RS, Phene CJ, Miller RJ, Urier K, Carlson RM, Henderson DW (1981) Applying nutrients and other chemicals to trickle-irrigated crops. California division of agricultural science bulletin **1893**

Ruskin R, Van Voris P, Cataldo DA (1990) Root intrusion protection of buried drip irrigation devices with slow release herbicides. Proceedings of the third international irrigation symposium, Albury-Wadonga, Australia, IA-ASAE, pp 211–216

Schroeder FG, Knaack H (2006) Gas concentration in the root zone of cucumber grown in different substrates. Acta Hortic 761:493–500

Selim HM, Ma L (1998) Physical nonequilibrium in soils: modeling and application. Ann Arbor Press, Chelsea, MI, p 520

Sharma DB, Swarup A (1988) Effects of short-term flooding on growth, yield and mineral composition of wheat on sodic soil under field conditions. Plant Soil 107(1):137–143

Silberbush M, Gornat B, Goldberg D (1979) Effect of irrigation from a point source (trickling) on oxygen flux and on root extension in the soil. Plant Soil 52(4):507–514

Simojoki A, Jaakkola A (2000) Effect of nitrogen fertilization, cropping and irrigation on soil air composition and nitrous oxide emission in a loamy clay. Euro J Soil Sci 51:413–424

Smiles D, Vachaud G, Vauclin M (1971) A test of the uniqueness of the soil moisture characteristic during transient non hysteretic flow of water in a rigid soil. Soil Sci Soc Am Proc 35:534–539

Smiles DE (1974) Infiltration into a swelling material. Soil Sci 117(3):140–147

Smiles DE (2000) Hydrology of swelling soils: a review. Aus J Soil Sci 38:510–521

Smiles DE, Philip JR (1978) Solute transport during absorption of water by soil: laboratory studies and their practical implications. Soil Sci Soc Am J 42:537–44

Smiles DE, Philip JR, Knight JH, Elrick DE (1978) Hydrodynamic dispersion during absorption of water by soil. Soil Sci Soc Am J 42:229–234

Smiles DE, Raats PAC (2005) Hydrology of swelling clay soils. In: Anderson MG (ed) Encyclopedia of hydrological sciences. Wiley, Chichester, England, pp 1011–1026

Smiles DE, Rosenthal MJ (1968) The movement of water in swelling materials. Aus J Soil Res 6:237–248

Smith RE (2002) Infiltration theory for hydrologic applications. Water Res Monogr 15, p 13. AGU, Washington DC

Sposito G (1990) Lie group invariance of the Richards equation. In: Cushman JH (ed) Dynamics of fluids in hierarchical porous media. Academic, London, pp 327–347

Sprengel C (1828) Von den Substanzen der Ackerkrume und des Untergrundes (About the substances in the plow layer and the subsoil). J für Technische und Ökonomische Chemie 2:423–474/3/42–99/313–352/397–421
Stolzy LH (1974) Soil atmosphere. In: Carson EW (ed) The plant root and its environment. University Press of Virginia, Charlottesville, pp 335–362
Su N (2009) Equations of anomalous absorption onto swelling porous media. Materials Letters 63: 2483–2485
Su N, Midmore DJ (2005) Two-phase flow of water and air during aerated subsurface drip irrigation. J Hydrol 313:158–163
Subba Rao NS (1999) Soil microbiology, 4th edn of soil microorganism and plant growth. Science Publishers Inc., USA
Sullivan K, Martin DJ, Carwell RD, Toll JE, Duke S (2000) An analysis of the effects of temperature on salmonids of the Pacific Northwest with implications for selecting temperature criteria. Sustainable Ecosystems Institute, Portland, Ore. pp192
Sviklas A, Shlinkshene R (2003) Liquid fertilizer based on dolomite, nitric acid and ammonia. Russ J Appl Chem 76(12):1885–1890
Sweeney DW, Graetz DA, Bottcher AB, Locascio SJ, Campbell KL (1987) Tomato yield and nitrogen recovery as influenced by irrigation method, nitrogen source, and mulch. Hortic Sci 22:27–29
Tanji KK, Keyes CG (2002) Irrigation and drainage division with a focus on water quality aspects. J Irrig Drain Eng 128:332–340
Tazuke A (1997) Effects of adding NaCl and reducing aeration to nutrient culture solution on the growth of cucumber fruit. J Japan Soc Hortic Sci 66(3–4):563–568
Tempany HA (1917) The shrinkage of soils. J Agric Sci 8:312–330
Terzaghi K (1923). Die berechnung der durchlassigkeitsziffer des tones aus dem verlauf der hydrodynamischen spannungserscheinungen. Akademie der Wissenschaften in Wein, Sitzungberichte, Mathematisch-Naturewissenschaftliche Klasse, Part IIa, pp 132/3–4/125–138
Tollefson S (1988) Commercial production of field and vegetable crops with subsurface drip irrigation. Irrigation association 1988 technical conference proceedings, pp 144–153
Tomomasa U, Yasufumi Y, Shigeru N, Shigem M, Kentaro U (2005) PIV imaging for multiphase flow measurement utilizing synchrotron radiation x-ray. J Visual Soc Japan 25(2):205–208
Tsallis C, Bukman DJ (1996) Anomalous diffusion in the presence of external forces: exact time-dependent solutions and their thermostatistical basis. Phys Rev E 54(3):2197–2198
Tuller S (1986) The spatial distribution of heavy precipitation in the greater Victoria Region. Climatol Bull 20:2–19
Urrestarazu M, Mazuela PC (2005) Effect of slow release oxygen supply by fertigaiton on horticultural crops under soil less culture. Sci Hortic 106(4):484–490
van Genuchten MT, Leij FJ, Wu L (1999) Characterisation and measurement of hydraulic properties of unsaturated porous media. University of California, Riverside, CA
van Genuchten MT, Leij FJ, Yates SR (1991) The RETC code for quantifying the hydraulic functions of unsaturated soils, EPA/600/2-91/065. Environmental Protection Agency, Ada, Okla
Vartapetian BB, Jackson MB (1997) Plant adaptations to anaerobic stress. Ann Bot 79 (suppl A): 3–20
Warrick AW, Biggar JW, Nielsen DR (1971) Simultaneous solute and water transfer for an unsaturated soil. Water Resour Res 7:1216–1225
Werner H (1990) Chemigation safety. South Dakota state university cooperative extension service. Fact Sheet 860
White SP (1995) Multiphase non-isothermal transport of systems of reacting chemicals. Water Resour Res 31:1761–1772
Wolf B (1999) The fertile triangle: the interrelationship of air, water and nutrients in maximizing soil productivity. Haworth Press, pp 211
Wolf B, Fleming J, Batchelor J (1985) Fluid fertilizer manual, vol 1. National Fertilizer Solutions Association, Peoria, IL

Yasuhito H, Masabika K, Arata K, Shozo K (2006) Optimization of nutrient supply drip-application (fertigation) culture: 1. Phosphorus application adjusted with the soil phosphorus status. Jpn J Soil Sci Plant Nutr 70(5):555–561

Yates MV (1992) Human health effects – pathogens. Proceedings of reclaimed wastewater: practical approaches to developing an alternative water supply, University of California Davis, CA 95616, 18 May 1992

Zhao FJ, Javier LF, Gray CW, Whalley WR, Clark LJ, McGrath SP (2007) Effects of soil compaction and irrigation on the concentrations of selenium and arsenic in wheat grains. Sci Total Environ 372:433–439

Zur B (1996) Wetted soil volume as a design objective in trickle irrigation. Irrig Sci 16:101–105

Agricultural Practices in Northeast India and Options for Sustainable Management

Mritunjay Majumder, Awadhesh Kumar Shukla, and Ayyanadar Arunachalam

Abstract The north east part of India has seven states comprising an area of 255,083 km^2 with hills, valley and plateau. This region is inhabited by 100 major tribes and immigrant communities. Due to topographical and environmental conditions this region is rich in biodiversity and is one of the hot spots of the world. Altitude ranges from 150 to 7,300 m a.s.l and temperature varies from freezing point to 37°C. Mostly tribal people and immigrant communities depends on farming and forest products for their food and livelihood. Local people have been maintaining traditional agricultural practices, agro-biodiversity and knowledge. Generally farmers practice jhum or shifting agricultural system with other sedentary agricultural practices. About 400,000 families practice jhum cultivation covering land area approximately 386,300 ha annually. Other agricultural system are wet rice cultivation which is practiced in valley land and Aji system where rice and millet are cultivated with fish in deep water. In valley land mono cropping as well as mixed cropping is practiced by farmers. Terrace land cultivation system introduced by government could not get wide acceptability by farmers due to high input of labour and fertilizers. Farmers also have cultivation systems such as homegardens and agroforestry that link their families to the forest ecosystem. Recently government and non governmental organization have introduced agri-horti-silvipastoral system for good harvest and yield. The population density of the region is 324 person per km^2 that is lower than the whole country. However, the growth rate during 1991–2001 has been recorded 31.2 person/km^2, which is higher than the national rate of 21.4 person/km^2. If population growth continues at this rate then a serious threat may occur to the sustainability of agroecosystem and rich biodiversity of the region.

M. Majumder
Vivekananda Kendra Vidyalaya, Jairampur 792121, India

A.K. Shukla (✉)
Department of Botany, Rajiv Gandhi University, Rono Hills, Itanagar 791112, India
e-mail: ashukla21@rediffmail.com

A. Arunachalam
Department of Forestry, North Eastern Regional Institute of Science and Technology, Nirjuli 791109, India

E. Lichtfouse (ed.), *Biodiversity, Biofuels, Agroforestry and Conservation Agriculture*, Sustainable Agriculture Reviews 5, DOI 10.1007/978-90-481-9513-8_10,

An attempt is made here to focus on agricultural practices, their productive capability and viable sustainable land use strategies for people of the region.

The northeastern area is rich in diversity of wild relatives of cultivated crops and out of 355 reported from all over India, 132 are found in this region. This area is also considered as the native origin of more than 20 major agricultural and horticultural crops and native home of about 160 domesticated species of cultivated crops. The utilization of bioresources by tribes and other communities is based on indigenous and traditional knowledge that help in sustainable use and conservation of natural resources. The tribal farmers have been using hundred of locally adapted major and minor crops in their various agricultural systems that helped them to survive under risk and hard prone conditions. The yield and energy efficiency of different agricultural systems depends on the type of crops cultivated. The more efficient were found where rice is cultivated with maize or millet or any other crop. Maximum yield has been reported in homegardens and Aji agricultural system practice by Apatani tribes. The efficiency of different agricultural practices varied between 1.7 and 75.2 and 0.7 and 8.8 respectively from ecological and economical view point. The maximum energy efficiency was recorded for the Aji system. As far as efficiency of jhum agricultural system is concerned optimum efficiency was reported with jhum cycle of 10 years period otherwise on shortening or increasing the cycle period efficiency declines. In general terrace land has the lowest efficiency among the different existing agricultural systems. In jhum system farmers grow several crops under mixed cultivation, therefore known as one of the rich agro-biodiversity system. This system, despite being rich in agro-biodiversity, does not harbor good yield and energy as the Aji system. The jhum system is generally practiced on hill slopes and the major causes of nutrients loss are due to blown off, run-off and through percolation of mineral nutrients that lead to poor yield and efficiency. Perhaps because of this reason farmers cultivate mixed crops comprising variety of cultivers in jhum system so that they can get maximum yield and output. In this context a number of studies have been carried out and workers have suggested many alternatives and modified practices for overall improvement of agricultural systems and socio-economic status of the people of this region. Popularization of agroforestry and horticultural practices, improved fallow management by introduction of native nitrogen fixing plants, recycling of agricultural waste in the form of composting are important among them.

Keywords Agro-biodiversity • Natural resource • Agricultural practices • Efficiency • Soil nutrients • Management practices • Sustainable land use

1 Introduction

The northeastern part of India, covering an area of 255,083 km^2 of hills, valleys and plateau is ethnically and culturally very distinct from the rest of country. The region comprises seven states of which Arunachal Pradesh is the largest with an area of 83,743 km^2 and Tripura is the smallest with an area of 10,486 km^2 (Table 1). The northeastern region occupies 7.8% of the total geographical area and is the

Table 1 A comparative analysis of demographic pattern of states of northeast India and the country as a whole

Parameters	Arunachal Pradesh	Assam	Manipur	Meghalaya	Mizoram	Nagaland	Tripura	NE	India
Area (km^2)	83,743	78,438	22,327	22,427	22,081	16,579	10,486	255,083	3268,090
Population (persons)	1091,117	266,38,407	2388,634	2306,069	891,058	19,88,636	31,91,168	3,75,13,089	1,027,015,247
Density (person km^{-2})	13	340	107	103	40	120	304	151	324
Sex ratio (female per 1,000 male)	901	932	978	975	938	909	950	940	933
Literacy (%)	54.74	64.28	68.87	63.31	88.49	67.11	73.66	65.71	65.38
Forest cover (km^2)	51,540	30,710	15,150	8,510	15,930	8,620	6,310	1,36,770	–
Cultivable land	293	3,387	164	1,074	445	626	310	6,299	1,94,680
(000 ha as per 1995–1996)	(3.5)	(43.18)	(7.35)	(47.88)	(21.11)	(37.76)	(29.56)	(24.69)	(59.22)
Annual area under Jhum cultivation (ha)	70,000	69,600	90,000	53,000	63,000	19,000	22,300	3,86,900	–
Families involved in jhumming	54,000	58,000	70,000	52,290	50,000	1,16,046	43,000	4,43,336	–
Per capita availability of food									
Grains (kg $year^{-1}$ as per 1995–1996)	254.00	157.58	212.67	85.59	193.98	175.19	201.69	164.71	235.52
Annual compound growth rate (%)									
Food grains	3.91	2.01	3.54	1.29	9.35	5.51	1.57	2.11	2.54
Population	3.11	2.15	2.74	2.85	3.72	3.45	2.90	2.41	2.19

(continued)

Table 1 (continued)

Parameters	Arunachal Pradesh	Assam	Manipur	Meghalaya	Mizoram	Nagaland	Tripura	NE	India
Average land holding size									
(ha) (1991)	3.62	1.31	1.24	1.81	1.34	6.92	0.97	1.60	1.57
Percentage of population below									
Poverty (1993)	45	45	45	45	45	45	45	45	37.3
Livestock population (X1000) (1992)									
Cattle	180	7,308	885	579	53	160	696	9,861	2,62,236
Sheep and Goats	157	3,603	52	217	21	152	434	4,636	1,66,062
Livestock growth rate (1982–1992)	5.82	4.92	–0.61	1.13	0.59	8.8	3.0		1.14
Density of livestock (per 100 person)	97	72	70	67	30	89	58		56

homeland of 3.7% total population of the country. The region is inhabited by 100 major tribes, many sub-tribes, indigenous and many immigrant communities. Altitude ranges between 150 m a.s.l and 7,300 m a.s.l (m a.s.l: meters above sea level), whereas, temperature varies from below freezing point to 37°C and monthly rain fall very low during winter to 512.6 mm in rainy season. Topographical and environmental conditions have contributed to the rich and unique biodiversity of the region. The remoteness and inaccessibility has contributed to the thinner population density of 151 person km^{-2}, which facilitate the local people to maintain their traditional agricultural practices, agro-biodiversity and knowledge. Shifting cultivation also known as jhum or slash and burn is the traditional agricultural practice of almost all the major tribes of the region, with other sedentary agricultural practices where environmental conditions permit. In jhum and other agro-ecosystems the farmers maintain high species diversity, which contributes to the agro-ecosystem stability (Ramakrishnan 1992). With high crop diversity it would be possible to achieve increased harvestable food production with the need for maintaining high organic biomass content in the system as a whole. Without this high organic matter production it would become necessary to constantly input costly inorganic fertilizers, which are hard to come by and whose effectiveness in the face of high temperature and heavy rainfall is questionable. As reported above, such climatic conditions are present only in some parts of the region.

Though population density of the region is much lower as compared to the country as a whole i.e. 324 person km^{-2}, but the decadal growth rate of the northeastern region is much higher i.e. 31.18 during 1991–2001 then the nation rate of 21.35 person km^{-2} as reported between 1991 and 2001 (Anonymous 2001). If this continues then serious threats may occur in regard the rich agro-biodiversity and sustainability of the different agro-ecosystems as they are vulnerable to soil erosion and land slide due to hilly terrain. In this review an attempt has been taken to focus on the diverse agricultural practices, their productive capability, strength of agro-biodiversity, traditional management practices and the viable sustainable land use strategies for the region.

2 Natural Resources: Utilization and Their Traditional Management Practices in North East India

The different communities of northeast region of India are mainly dependent upon biodiversity linked land use activities for their livelihood concerns. The forest is an indispensable component of mountain inhabited people of northeast India, which provide supplementary food, fodder, medicine, fuel wood and other livelihood resources. For example each household of Chakma community living in the adjoining areas of Namdapha national park collected in total, about 25 quintals of bamboo and timber, 52 quintals fuel-wood, one quintal of *Zalacca secunda* leaves for roofing and one quintal of wild vegetables and medicinal plants during the year 2002 (Arunachalam et al. 2004) (Fig. 1).

Natural resource utilizations of the traditional societies are based on their traditional ecological knowledge (TEK). The mountain areas of northeast have a variety of natural

Fig. 1 Map of north east India

and human managed ecosystems, and even sacred groves or sacred landscapes protected through cultural and religious reasons. The sacred values of *Mesua ferrera*, *Ficus religiosa*, *Alstonia scholaris* have been recorded among the Buddhists of Srilanka (Withanage 1998) and similarly Buddhists residing in eastern and western part of Arunachal Pradesh have honour to above mentioned plant species. Local people believe that God will punish them if the natural habitat of those particular areas and trees in the vicinity of the temple areas are disturbed. Concept of the sacred values and sacred groves is common in Meghalaya and Manipur states and are widely studied (Tiwari et al. 1998; Khumbongmayum 2004). The Chakma community harvests the forest products mostly during the winter and no or a little harvesting is made during the summer season. This process of harvesting shows the traditional conservation method. Most of the species naturally regenerates during the summer season and the young shoots would be destroyed if the harvesting is made during summer. Thus the Chakmas' play unknowingly the conservation and natural regeneration practices traditionally (Fig. 2).

The complexities of a variety of agro-ecosystems maintained by traditional societies are due to TEK-based biodiversity management, both in space and time. This forms the basis for their ability to cope up with uncertainties in the environment and maintain a sustained production level. The TEK of mountain societies also plays vital role in

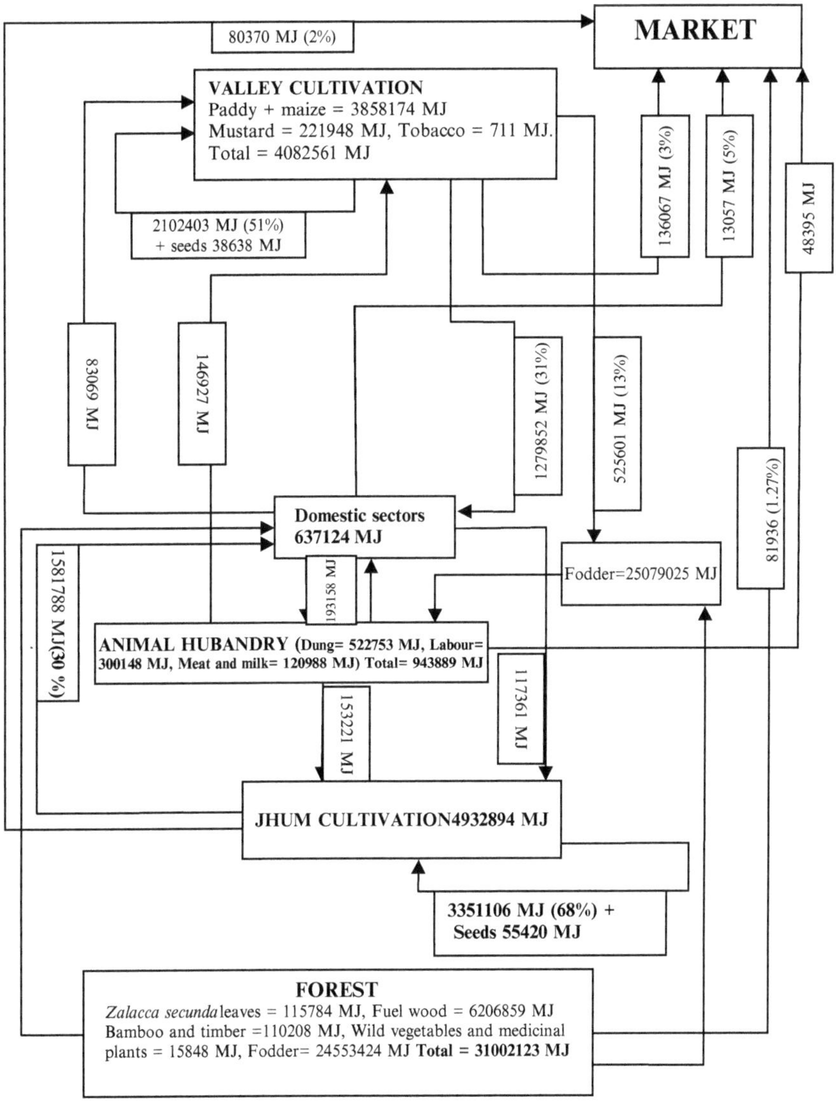

Fig. 2 Energy flow through a Chakma village ecosystem

conservation of crop diversity as well as soil fertility (Box 1). The spiritual beliefs, cosmologies are playing vital role in cultural aspect of natural resource management. Indigenous knowledge of natural resource management about their environment and manipulation for better needs were studied by Richards (1985). These studies demonstrate that, this knowledge is not only a function of utility but it is also an intellectual process for the better management of environment. For example the village council of Changki village in Mokokchung district of Nagaland state has certain conservation-oriented measures to be followed by the community (Choudhury 1998). Some of the

Box 1 Description of the Terms Used in the Article

Multiple cropping (intercropping; mixed cropping; home gardens) – Maintains biodiversity as interdependent crop variability. Often reduces damage from pests and diseases favouring maintenance of agro-biodiversity.

Varietals mixtures – Maintains interspecific crop variability.

Crop selection – Cultivation of cereal crops under long jhum cycle of 30-years or more whereas tuber and vegetable crops under shorter cycle of 5-years or less is to emphasize upon the nutrient use efficient species under shorter cycle, which in turn also conserve agro-biodiversity.

Crop rotation – Decreases insect pests and pathogen damage and increases agro-biodiversity. Help in soil nutrient amendment.

Fallowing and rotation – Maintains soil biodiversity and fertility. Manages soil pathogens and pests (through interruption of life cycles) and soil nutrient amendment favouring crop health and maintenance of agro-biodiversity.

Organic amendments – Soil enrichment favours soil biodiversity. Development of suppressive soils. Manages soil pathogens and pests favouring crop health and maintenance of agro-biodiversity.

Flooding – Nutrient enrichment favours soil biodiversity. Reduces damage from weeds, pests and diseases favouring crop health (especially in paddy field) and maintenance of agro-biodiversity.

Burning – Slash and burn systems maintain considerable agrobiodivesity. Contributes to pest and disease management, crop health and maintenance of agro-biodiversity.

Mulching – Lowers soil temperature, protects against erosion, improves soil texture, provides nutrients and organic matter, reduces weed problems and suppresses soil borne pathogens contributing to crop health and maintenance of agro-biodiversity.

Raised beds – Improve drainage, fertilization, frost control and irrigation, support management of soilborne pathogens and pests contributing to crop health and maintenance of agro-biodiversity.

Site selection – Avoids diseases, pests and weeds associated with previous crops, matches soil fertility and drainage to crop and variety contributing to crop health and maintenance of agro-biodiversity.

Traditional ecological knowledge (TEK) – A particular form of knowledge which includes empirical observations about the local environment. The observations may be in different forms like management of agriculture, use of medicinal plants, interaction among plants and animals and other traits of biophysical environment. The knowledge acquire is transmitted through oral traditions to coming generations.

(continued)

Box 1 (continued)

Manipulating shade – Maintains biodiversity as interdependent multiple crop variability, e.g. in coffee, cocoa and tea cultivation systems. Manages pathogens and pests favouring crop health and maintenance of agro-biodiversity.

Selective logging in agricultural fields (agroforestry) – The Jhumias of northeast India of northeast India conserves various tree species in their jhum field, which protect the field from soil erosion and wind. The tree species like *Alnus nepalensis*, species of *Albizia* spp., *Flemingia vestita* etc. are specially conserved if present in the field, as they are natural nitrogen fixer. Some of the bamboo species (e.g. *Dendroclamus hamiltonii*) are also conserved which can concentrate and conserve N, P and K.

most important measures taken up by the people include e.g. conservation of land surrounding the village as reserve forest; ban on the use of fish poison both chemical and indigenous herbal form; not to allow the trapping of nesting birds; prohibition of hunting during the breeding seasons of animals, or of female animals; and strict prohibition on the cutting of edible and wild fruit trees.

In general inhabitants of this region are dependent on farming and forest products for their food and livelihood. The utilization of bio-resources by tribes and other communities is based on indigenous and traditional knowledge that helps in sustainable use and conservation of natural resources.

3 Agricultural Practices

The northeastern region of India falls under the agro-climatic zone-II that is Eastern Himalayan Agro-climatic region, as identified by the Planning commission, Government of India, in year 1989 which consist five different agro-climatic zones viz., alpine >3,500 m a.s.l, temperate sub-alpine 1,500–3,500 m a.s.l, sub-temperate 1,000–1,500 m a.s.l, mid tropical hill 200–1,000 m a.s.l and mid tropical plain <200 m a.s.l. The agro-climatic, ethnic, cultural, socio-economic and environmental variations have diversified the agricultural systems of the region. Jhum is the predominant form of agricultural practice among the farmers of upland communities of northeast India and over 400,000 families comprising almost all the major communities' practice jhuming. Average land holding under jhum cultivation system varies from 0.16 to 1.29 ha per family and almost covers 386,900 ha land annually under this practice through out the region (Choudhury and Sundriyal 2003). Mixed cropping with fallow period of 3–10 years is the main characteristic features of the system. However, fallow length of up to 60-years has been reported in remote areas and less then 3-years in the densely populated areas of the region. As a result of increasing population pressure there has been a shift from more extensive to more intensive land use systems. Under shorter fallow agriculture period is 1–2 years where the weed biomass is slashed in January and organized in parallel rows covered by thin layer of soil and allowed to decompose. The crops are sown on these ridges in March. Double cropping is done in a year, one between March and June and another between August and November (Mishra and Ramakrishnan 1981). The composition of crop mixture varies from communities to communities and from place to place (Photo 1).

Photo 1 Different agricultural systems

Among different settled cultivation systems wet rice cultivation is predominantly practiced throughout the hill terrain, both at low and high elevations. Apatani tribe of Arunachal Pradesh is well known practitioners of modified version of wet rice cultivation locally known as ‘Aji’. The ‘Aji’ system is combination of rice and fish together with millet on the bunds separating each plot has been referred as one of the most productive and efficient agricultural systems of the region (Ramakrishnan 1994). The Apatanis also practiced upland dry farming, growing millets, maize and vegetables and some extant of jhuming (Maikhuri and Ramakrishanan 1990). The Monpas and Sherdukpens of Arunachal Pradesh, plain tribes and communities of Assam, Tripura and Imphal valley also exclusively depends on wet rice cultivation. In the valley areas lacking water storage facilities mono-cropping as well as mixed cropping of seasonal crops are practiced, where seeds are directly broadcasted instead of plantation as in wet rice cultivation system. The Chakmas’ residing in the adjoining villages of Namdapha national park in Arunachal Pradesh practices double cropping in this system. The first cropping is from May to August using paddy and maize. The second cropping is from October to January with mono cropping of mustard or any of the winter vegetables.

Terrace cultivation was introduced in the region by Government agencies in order to discourage farmers from jhuming. However, due to the higher input in the form of labour i.e. 2,478 MJ ha^{-1} in the first year and 984 MJ ha^{-1} during subsequent year and application of inorganic fertilizers such as 60 kg ha^{-1} $year^{-1}$ N, 30 kg ha^{-1} $year^{-1}$ P and 30 kg ha^{-1} $year^{-1}$ K this system could not get wide acceptability among the farmers (Ramakrishnan 1992). In Arunachal Pradesh this system was practiced in small scale in Lower Subansiri district (Gangwar and Ramakrishnan 1987) and in some parts like Burnihat, Shillong and Nayabuglow of Meghalaya state.

Along with jhum and valley cultivation systems some secondary form such as home gardens, and plantation crop cultivation are also practiced by the farmers of the region. Through these the farmer has linked his family to forest ecosystem and also effectively incorporated animal husbandry. Thus agriculture, animal husbandry and domestic sub-systems of the village are all closely linked with the forest ecosystem, providing food, fodder, fuel-wood, timber, medicine and other day to day requirements. Home gardens are complex and highly diversified systems, an interesting agro-ecosystem from the point of view of resource management for sustainable agriculture (Gliessman 1989). The home gardens have rich plant species diversity, dominated by woody perennials and are stratified forming a multistoried structure and resembled natural forest. At some places in the region farmers have started cultivation of cash crops due to low productivity and decrease in period of jhum fallows. Broom grass (*Thysanolaena maxima*) and bamboo (*Dendrocalamus hamiltonii*) have been harvested from the wild or from cultivated areas. Such as *Citrus* species and *Cinnamomum obtusifolium* have been established frequently with understorey of ginger (*Zingiber officinale*), banana (*Musa* sp.) or pineapple (*Ananas comosus*). Other crops recently introduced by the government agencies on an experimental basis include tea (*Camellia sinensis*), coffee (*Coffea* sp), rubber (*Hevea* sp) and cashew-nuts (*Anacardium occidentale*) (Table 2).

Table 2 Edible plants, weeds and other tree species grown in different systems

Botanical name	Family	Habit	Uses	Parts used	Valley	Jhum
Crop species						
Cereals						
Oryza sativa L.	Poaceae	Herb	E	Endosperm	+	+
Zea mays L.	Poaceae	Herb	E	Endosperm	+	+
Millets						
Eleusine coracana (L.) Gaertn.	Poaceae	Herb	E	Endosperm	–	+
Pennisetum typhoides (Burm.f.) Staf and Hubb	Poaceae	Herb	E	Endosperm	–	+
Vegetables and pulses						
Abelmoschus esculentus (L.) Moench.	Malvaceae	Herb	E	Fruits	–	+
Benincasa hispida (Thunb.) Cogn.	Cucurbitaceae	Climber	E	Fruits	–	+
Cajanus cajan (L.) Huth.	Papilionaceae	Shrub	E	Seeds	–	+
Cucumis sativus L.	Cucurbitaceae	Climber	E	Fruits	–	+
Cucurbita maxima Duch ex Lamk.	Cucurbitaceae	Climber	E	Fruits	–	+
Dioscorea hispida Dennst.	Dioscoreaceae	Climber	E	Tubers	–	+
Dolichos lablab L.	Papilionaceae	Climber	E	Fruits	–	+
Garcinia cowa Roxb ex DC.	Cluciaceae	Climber	E	Fruits	–	+
Ipomaea batatas (L.) Lamk.	Convolvulaceae	Climber	E	Tubers	–	+
L. cylindrica (L.) Roem.	Cucurbitaceae	Climber	E	Fruits	–	+
Lagenaria ciceraria (Molina.) Standl.	Cucurbitaceae	Climber	E	Fruits	–	+
Luffa acutangula (L.) Roxb.	Cucurbitaceae	Climber	E	Fruits	–	+
Lycopersicum esculentum Miller.	Solanaceae	Herb	E	Fruits	–	+
Manihot utilissima Pohl.	Euphorbiaceae	Shrub	E	Root	–	+
Momordica charantia L.	Cucurbitaceae	Climber	E	Fruits	–	+
Momordica cochinchinensis (L.) Spreng.	Cucurbitaceae	Climber	E	Fruits	–	+
Phaseolus vulgaris L.	Papilionaceae	Climber	E	Fruits	–	+
Raphanus sativus L.	Brassicaceae	Herb	E	Stem	–	+
Sechium edule Sw.	Convolvulaceae	Climber	E		–	+

Solanum melongena L.	Solanaceae	Shrub	E	Fruits	–	+
Vicea faba L.	Papilionaceae	Climber	E	Fruits	–	+
Oil yielding						
Brassica campestris L.	Brassicaceae	Herb	C	Seeds	+	+
Pogostemon plectranthoides Desf.	Lamiaceae	Shrub	E	Seeds	–	+
Sesamum indicum DC.	Pedalineaceae	Shrub	E	Seeds	–	+
Narcotics						
Cannabis sativa L.	Cannabiaceae	Shrub	E	Leaves	–	+
Nicotiana tabaccum L.	Solanaceae	Shrub	E	Leaves	+	–
Condiments and spices						
Capsicum frutescens L.	Solanaceae	Shrub	E + C	Fruits	–	+
Coriandrum sativum L.	Umbelliferae	Herb	E	Whole plant	+	–
Curcuma longa L.	Zingiberaceae	Herb	E	Rhizome	–	+
Zingiber officinale Rosc.	Zingiberaceae	Herb	E + C	Rhizome	–	+
Fibre yielding						
Gossypium hirsutum L.	Malvaceae	Shrub	F + C	Seed fibres	–	+
Fruit plants						
Dillenia indica L.	Dillaniaceae	Tree	E	Fruit	–	+
Carica papaya L.	Caricaceae	Tree	E	Fruit	–	+
Tamarindus indica L.	Caesalpinaceae	Tree	E	Fruit	–	+
Tree species						
Albizia odoratissima Benth.	Mimosaceae	Tree	T + Fw	Wood	+	+
Albizia procera (Roxb.) Benth.	Mimosaceae	Tree	T + Fw	Wood	+	+
Altingia excelsa Noronha.	Hamamelidaceae	Tree	T + Fw	Wood	–	+
Artocarpus lacucha Wall.	Moraceae	Tree	T + Fw	Wood	–	+
Bombax ceiba L.	Bombaceae	Tree	Hb	Whole plant	+	+

(continued)

Table 2 (continued)

Botanical name	Family	Habit	Uses	Parts used	Valley	Jhum
Cinnamomum glanduliferum Meissn.	Lauraceae	Tree	T	Wood	–	+
Dalbergia latifolia	Papilionaceae	Tree	T + Fw	Wood	–	+
Dillenia indica L.	Dillaniaceae	Tree	E	Fruits		+
Duabanga grandiflora (Roxb. Ex DC.) Walp.	Sonneratiaceae	Tree	T + Fw	Wood	+	+
Dysoxylum sp.	Meliaceae	Tree	Fw	Wood	–	+
Erythrina indica (Lamk.) O. Ktze	Papilionaceae	Tree	–	–	–	+
Ficus bengalensis L.	Moraceae	Tree	Hb	Whole plant	–	+
Ficus hispida L.	Moraceae	Tree	V	Fruits	–	+
Musa paradisiaca L.	Musaceae	Herb	E	Fruits	–	+
Sapindus mukorossi Gaertn.	Sapindaceae	Tree	Fw	Wood	–	+
Sterculia villosa Roxb. Ex Smith.	Sterculariaceae	Tree	R	Bark	–	+
Stereospermum chelonoides (L.) DC.	Bigniniaceae	Tree	Hb	Whole plant	–	+
Tamarindus indicus L.	Caesalpinaceae	Tree	E	Fruits	–	+
Terminalia myriocarpa Heurck and Muell.	Combretaceae	Tree	T + Fw	Wood	–	+

E – edible, C – Cash crop, M – medicinal, V – vegetables, R – rope, T – timber, Fw – fuel wood, Hb – harbor of honeybees, (–) Indicated no data

Majumder 2007

In recent past government agencies and other non governmental organizations have suggested to farmers for practicing of agri-horti-silvipastoral system which is more productive. This system is practiced on land with 80–90% slope with soil depth greater than 1 m. The system comprises agricultural land use towards the foot-hills, horticultural in the mid portion and silvipastoral crops in the top portion of hill slopes. Counter bunds, bench terraces, half moon terraces, grassed ways are the major conservation measures. Agri-horti-silvipastoral system is slightly labour intensive and needs input of about 190 man days ha^{-1} (Verma et al. 2001).

Generally farmers practice jhum or shifting cultivation agricultural system. About 0.4 million families practice jhum agricultural system covering a land area of approximately 386,300 ha annually. Another agricultural system is wet rice cultivation which is practiced by tribes inhabiting in valley land area. Apatani tribe along with rice cultivates millet and fish that is locally known as Aji system. In valley land mono cropping as well as mixed cropping is practiced by farmers. Terrace land cultivation system introduced by government could not get wide acceptability by farmers due to high input in the form of labour and fertilizers. Besides these system farmers also have cultivation systems such as home gardens and agro-forestry that link their families to forest ecosystem. Recently government and non governmental organization have introduced agri-horti-silvipastoral system for good harvest and yield.

4 Status of Agro-Biodiversity and Management Practices

Agro-biodiversity is a fundamental basis for agricultural production and food security, as well as a valuable ingredient of environmental conservation. The wide variation in topographical and geographical position, climatic conditions and variations of agricultural and management practices within northeastern region attributed to rich diversity of agricultural crops. This region is also rich in diversity of wild relatives of cultivated crops. Out of 355 wild relatives of cultivated crops reported from all over India 132 (37.2%) occur in this region (Table 3), which indicates the richness of the area in terms of agro-biodiversity. The northeastern Indian

Table 3 Distribution of wild relatives of cultivated crops in north east and India as a whole

Crop	Number of species	
	NE Himalaya	India
Cereals	16	60
Legumes	6	33
Fruits	51	109
Vegetables	27	64
Oil seeds	1	12
Fibre crops	5	24
Spices and condiments	13	27
Miscellaneous	13	26
Total	132 (37.18%)	355

Upadhyay and Sundriyal 1998

Himalayas is the centre of origin of more than 20 major agricultural and horticultural crops (Vavilov 1950). This region is the native home of about 160 domesticated and 355 wild relatives species of cultivated crops (Upadhyay and Sundriyal 1998). The wild relatives of rice e.g. *Oryza granulata*, *O. rufipogon*, *O. jeyporensis*, *O. malampuzhaensis*, and *O. sativa var spontanea*, *Digitaria*, *Coix*, *Panicum*, *Setaria*, *Elusine*, *Zingiber*, *Circuma*, *Cinnamomum*, *Elettaria*, *Gossypium* and legumes such as pigeon pea, rice beans, green gram, winged beans, broad beans, Dolichos, and sword beans are available in this region (Borthakur 1992). According to an estimate of National Bureau of Plant Gene Resources (NBPGR) about 50,000 land races of paddy are expected to exist in India and of that 5,000 alone from the northeastern Indian Himalayas. The genetic diversity of rice in this region can also be assumed for our survey among a small group of Chakma population inhabiting in the far remote area of Arunachal Pradesh, where they are cultivating 41 different varieties in their diverse agricultural systems of which 22 were upland varieties and 18 were wetland varieties (Majumder 2007). One variety of rice locally known as "Begun bichi" can be grown in both condition. There is wide variability in the rice germplasm collected from different parts of northeastern region, but glutinous and japonica forms dominate the endemic types. Out of 37 reported citrus species of India 17 species with 52 varieties are from Assam (Bhattacharya and Dutta 1951). This region is also native home to many sub-tropical fruits such as *Garcinia*, *Artocarpus*, *Phylanthus*, *Anona*, *Averrhoea*, *Persia*, *Aegale*, *Flacourtica*, *Passiflora*, *Avocado*, *Actinidia*, *Dillenia laeocarpus*, *Eugenia*, *Ficus*, *Juglans*, *Vitis*, *Spondias*, and *Syzygium*. A number of species belonging to the genera *Malus, Prunus, Pyrus, Sorbus, Docynia, Rubus, Cotoneaster, Ribes, Fragaria*, and *Actinidia* grow in the wild. The wild relatives of A*belmoschus, Alocasia, Alpinia, Amomum, Brassica, Camellia, Canavalia, Citrus, Colocasia, Corchorus, Cucumis, Curcuma, Digitaria, Dioscorea, Docynia, Erianthus, Eurya, Hedychium, Hibiscus, Mangifera, Momordica, Morus, Mucuna, Musa, Oryza, Prunus, Rubus, Setaria, Sorbus, Trichosanthes* and *Vitis* are native to the northeastern India (Upadhyay and Sundriyal 1998).

Until recently, the immense agro-biodiversity of the region were safe as people used to practice mixed cropping of wide indigenous crop cultivars in their traditional shifting cultivation system and in home gardens. In jhum cultivation system five or up to 45 species of traditional crops are mixed together, where the number of species decline drastically with the shortening of the jhum cycle. A total of 59 edible plant species have been documented from different agricultural systems of Chakmas in Arunachal Pradesh. About fifty one plant species were found grown in home gardens, thirty-three species in different jhum fields and fifteen in valley cultivation system. The selection of crop species under different jhum cycle is based on the traditional ecological knowledge of the farmers (Ramakrishnan 2001). Different traditional practices also play important role in the conservation of agro-biodiversity (Thurston et al. 1999) and soil nutrients.

North east region of India, harbour rich agro-biodiversity due to variability in geographical and climatic conditions. The region is one of the hot spots of mega diversity of the world. This area consist 37.2% wild relatives of crops of India and is also considered centre for origin of more than 20 major agricultural and horticultural crops. Large number of crop varieties and trees are grown in different agricultural systems.

5 Yield Pattern of Different Agro-Ecosystems

The crop yield differs markedly in different agricultural practices depending upon the crop components, number of crops mixed and mode of practices. The crop yield from different jhum cultivation systems has been estimated between 986 and 3,745 kg ha^{-1} year^{-1} of jhum cycle length between 5 and 60 years (Maikhuri and Ramakrishanan 1990; Ramakrishnan 1992). In case of 3-years jhum cycle of Chakmas the yield was 2,915 kg ha^{-1} year^{-1} (Table 4) which is higher then 5-years cycle of other communities may be due to emphasizing more on seed and fruit

Table 4 Yield and efficiencies under different jhum cultivation systems of northeast India

		Input		Output		Efficiencies	
		Energy	Monetary	Energy	Monetary	Energy	Monetary
Practices	Yield	(MJ ha^{-1})	(Rs ha^{-1})	(MJ ha^{-1})	(Rs ha^{-1})		
Arunachal Pradesh							
[1]60-years cycle	3,745	2,855	4,568	67,171	7,430	24	1.6
[1]30-years cycle	3,125	2,294	3,888	60,839	6,657	27	1.7
[1]20-years cycle	3,215	1,599	3,016	51,660	5,634	32	1.9
[2]15-years cycle	2,069	1,062	2,435	33,649	7,176	32	3.0
[1]10-years cycle	3,225	1,194	2,766	51,774	6,464	43	2.3
[1]5-years cycle	2,450	853	2,215	35,474	4,336	41	2.0
[3]3-years cycle	2,915	8,579	10,671	41,626	16,585	5.5	1.6
Meghalaya							
[5]30-years	3,460	1,665	2,616	56,766	5,586	34.1	2.1
[5]20-years	3,430	16,88	NA	60,277	NA	35.7	NA
[6]-do-	2,662	1,352	NA	48,985	NA	36.2	NA
[7]-do-	1,786	1,043	NA	32,978	NA	31.6	NA
[6]15-years	2,443	3,675		19,790		5.4	
[5]10-years	3,366	1,191	1,830	56,601	3,354	47.5	1.8
[6]-do-	2,267	1,200	10,548	52,142	18,370	22.6	1.7
[7]-do-	1,359	794	NA	23,158	NA	29.0	NA
[5]5-years	1,584	810	896	44,758	1,524	55.2	1.9
[6]-do-	1,590	1,470	7,431	26,686	7,520	18.1	1.01
[7]-do-	986	546	NA	15,829	NA	29.0	NA
[6]3-years	NA	9,054	8,986	41,030	12,096	4.5	1.4

Systems practiced by [1]Nishis; [2]Sulungs; [3]Chakmas; [4]Apatanis, [5]Garos, [6]Khasis, [7]Mikirs, [8]Nepali
NA – data not available
Maikhuri and Ramakrishanan 1990; Gangwar and Ramakrishnan 1989; Toky and Ramakrishnan 1981; Patnaik and Ramakrishnan 1989; Gangwar and Ramakrishnan 1987; Mishra and Ramakrishnan 1981, 1982; Kumar and Ramakrishnan 1990

crops like paddy, maize and chillies and out of that about 62% of total yield was obtained from paddy, 28% from maize, 6% from chillies and the remaining 4% from other crops. From different studies (Maikhuri and Ramakrishanan 1990; Ramakrishnan 1992; Majumder 2007) it has been observed that the yield per year from jhum system gradually declines with the shortening of the cycle length, however, exceptional cases were also recorded. The terrace cultivation of Shulungs' in Arunachal Pradesh is the lowest productive system with an annual yield of only 1,172 kg ha^{-1}(Gangwar and Ramakrishnan 1987).

The Aji system, in which paddy, millet and fish are cultivated together seems to be highly productive in comparison with jhum and other valley system of the region, where the production was recorded as 3,456 kg ha^{-1} $year^{-1}$ with early varieties of rice and 4,046 kg ha^{-1} $year^{-1}$ with late varieties of rice (Table 5). The yield from wet rice cultivation system of Chakmas was estimated 2,932 kg ha^{-1}. The Agro-ecological Research Centre at Jorhat, Assam (northeast India) reported an average paddy yield in the hill region of northeastern India as 800–900 kg ha^{-1}. Mishra and Ramakrishnan (1981) also reported 900 kg ha^{-1} yield of paddy from valley cultivation system in Meghayala. However, Aurora et al. (1977) have reported about 1,200 kg ha^{-1} paddy yield from Tripura state. The paddy production in different agricultural systems of Chakmas were higher than that of other communities in the region, however, the overall production of the systems were lower in comparison with the jhum system under longer fallow period (10–60 years).

From paddy and maize mixed cropping system of Chakmas in the valley, the estimated yield was 2,468 kg ha^{-1} and whereas during second cropping i.e. monocropping of mustard only 597 kg ha^{-1} yield were obtained.

The yield in home gardens of Apatanis in Arunachal Pradesh was estimated 5,811 kg ha^{-1} (Kumar and Ramakrishnan 1990), which seem to be probably the most productive then any other reported systems in the northeast. However, the yield from home gardens of Mikirs in Meghalaya was estimated only 2,590 kg ha^{-1} (Maikhuri and Ramakrishanan 1990). The total yield from the home gardens of Chakmas was 3,454 kg ha^{-1} $year^{-1}$ that includes 1,116 kg ginger, 1,550 kg vegetables, 480 kg fruits and 25 kg tobacco

The yield from different cash crop systems varies greatly depending upon the crop components. The yield from pure cultivation of broom grass of Khasis from Meghalaya were reported 620, 1,095 and 1,500 kg ha^{-1} respectively from first, second and third year cultivation, whereas in mixed cropping system the yield of broom grass was estimated 390, 660 and 660 kg ha^{-1} respectively for first, second and third year cultivation. In case of thatch grass and bamboo of the same community the estimated yield was 4,133 and 3,695 kg ha^{-1} respectively.

As such mixed cultivation of paddy with millet and fish practice by Apatani tribe was found to have high productivity in comparison to jhum, terrace and wet rice cultivation system. Also maximum yield of some crops was reported in the home gardens of Apatani tribes. Terrace land cultivation seems to be less productive.

Table 5 Yield and efficiencies under other land use systems of northeast India

Practices	Yield	Input		Output		Efficiencies	
		Energy (MJ ha^{-1})	Monetary (Rs ha^{-1})	Energy (MJ ha^{-1})	Monetary (Rs ha^{-1})	Energy	Monetary
Arunachal Pradesh							
Valley cultivation							
[3]Wet rice	2,932	7,518	6,474	42,221	13,194	5.6	2.0
[3]Mixed cropping	2,468	7,416	7,061	36,634	10,012	4.9	1.4
[3]Mustard	597	3,219	3,259	14,806	5,373	4.6	1.7
[4]Aji							
Early variety (paddy + millet)	3,456	946	2,798	58,480	7,817	61.8	2.8
Late variety (paddy + millet + fish)	4,046	907	2,753	68,182	10,062	75.2	3.7
[4]Home gardens	5,811	1,774	3,162	58,873	10,524	33.2	3.3
[2]Terrace cultivation	1,172	847	1,894	19,058	2,637	22.5	1.4
		(4,475)	(6,460)				
Meghalaya							
Valley cultivation							
[5]Double cropping	NA	2,843	4,843	50,596	5,565	17.8	1.1
[7]Single cropping	NA	11,601	2,388	41,938	3,876	3.6	1.6
[8]-do-	NA	3,479	1,316	43,172	4,460	12.4	3.4
Home gardens							
[6]	NA	NA	13,093	NA	23,155	NA	1.8
[7]	2,590	1,140	1,650	26,794	14,667	23.5	8.8
Terrace cultivation							
[5]	NA	6,509	2,542	43,602	3,658	6.7	1.4
	NA	(8,003)	(4,544)			(5.4)	(0.8)

(continued)

Table 5 (continued)

Practices	Yield	Input Energy (MJ ha^{-1})	Input Monetary (Rs ha^{-1})	Output Energy (MJ ha^{-1})	Output Monetary (Rs ha^{-1})	Efficiencies Energy	Efficiencies Monetary
[6]	NA	12,878	6,004	21,889	12,561	1.7	2.1
	NA	(13,968)	(12,243)			(1.6)	(1.0)
[8]		4,846	6,217	1,01,525	11,791	21	1.9
Cash crops (Broom grass)							
[6]Pure							
First-year	620	NA	4,670	NA	3,348	NA	0.7
Second-year	1,095	NA	3,930	NA	5,913	NA	1.5
Third-year	1,500	NA	3,870	NA	8,100	NA	2.1
[6]Mixed							
First-year	390	NA	4,995	NA	6,474	NA	1.3
Second-year	660	NA	3,886	NA	7,953	NA	2.0
Third-year	904	NA	3,303	NA	9,250	NA	2.8
Coffee	NA	8,855	2,754	8,450	4,560	1.0	1.7
Tea	NA	19,425	14,314	1,81,310	37,125	9.3	2.6
[6]Thatch grass	4,133	NA	1,716	NA	3,827	NA	2.2
[6]Bamboo	14,385	NA	3,695	NA	6,850	NA	1.9
Pineapple (mixed cropping)	NA	973	3,096	17,085	12,090	17.6	3.9
Ginger	NA	1,302	1,830	56,655	3,354	43.5	1.8

Systems practiced by [1]Nishis; [2]Sulungs; [3]Chakmas; [4]Apatanis, [5]Garos, [6]Khasis, [7]Mikirs, [8]Nepali

Values in parenthesis are for the first year; NA – data not available

Maikhuri and Ramakrishanan 1990; Gangwar and Ramakrishnan 1989; Toky and Ramakrishnan 1981; Patnaik and Ramakrishnan 1989; Gangwar and Ramakrishnan 1987; Mishra and Ramakrishnan 1981, 1982; Kumar and Ramakrishnan 1990

6 Energy and Economic Efficiencies of Different Agricultural Practices

The efficiencies of the jhum cultivation gradually increased with the reduction of the jhum cycle period up to 10 years and start declining with further shortening of the cycle (Maikhuri and Ramakrishanan 1990). The net return under a 10-year cycle was higher than all other jhum cycle, because of reduced labour costs involved in slash-and-burn operations under this cycle than under longer cycles on one hand, and the poor crop yield due to reduced soil fertility under a too short 5-year cycle. The energy and economic efficiencies of 3-year jhum cycle system of Chakma were 5.5 and 1.6 respectively (Photo 2). The efficiencies of jhum system practiced by different tribes with varied fallow length in Arunachal and Meghalaya states have been summarized in Table 4.

The Aji system of Apatanis was the most efficient with energy efficiency values of 75.2 and 61.8 respectively for late and early varieties of rice and the respective monetary efficiencies were 3.7 and 2.8 (Table **5**). In case of wet rice cultivation system of Chakmas the respective energy and economic efficiencies were 5.6 and 2.0. From several studies the energy efficiency of valley cultivation among different communities like Garos, Mikirs, Nepalis, Khasis, Nishis and Apatanis of Meghalaya and Arunachal Pradesh was reported in between 3.6 and 17.8, whereas for the same system of these communities the monetary efficiency was reported in between 1.1 and 3.4 (Gangwar and Ramakrishnan 1989; Patnaik and Ramakrishnan 1989; Maikhuri and Ramakrishanan 1990, 1991; Kumar and Ramakrishnan 1990). In case of rice and maize mixed cropping system of Chakmas the total energy invested was 7,416 MJ ha^{-1} which is less then that of both wet rice cultivation as well as the jhum cultivation systems. The total energy output in rice and maize mixed cropping system practiced by Chakma community was 36,634 MJ ha^{-1}. The energy output and input ratio of the system was 4.9 and the monetary efficiency for this system was 1.42. During second cropping (with mono-cropping of mustard) total energy investment was around 3,219 MJ ha^{-1}. The energy output was about 14,806 MJ ha^{-1} and the energy output-input ratio was 4.6, which is higher than any other cropping system practiced by them. Here the

Photo 2 Chakma house made of bamboo, wood and roofing with *Zalacca Secunda* leaves

economic output was found to be Rs. 5,376 ha^{-1} in return of the total input of Rs. 3,259 ha^{-1} i.e. the monetary efficiency is 1.7 (Mishra and Ramakrishnan 1982).

The energy and monetary efficiencies of home gardens of Apatanis were recorded 33.2 and 3.3 respectively (Kumar and Ramakrishnan 1990). Though the home gardens are less efficient in comparison of Aji system and 10-years jhum cycle systems of different communities of the state however, they provide with a variety of food items and their day to day requirements. From the energy view point the home garden system of Apatanis are more efficient then Mikirs (23.5), whereas from monetary point of view Mikirs (8.8) system is much more efficient. The economic efficiency of home gardens of Khasis was estimated 8.8.

The terrace cultivation system has the lowest efficiency among the different existing systems of the region. In terrace cultivation practiced by Sulungs, the energy and monetary efficiency was recorded 22.5 and 1.4 respectively (Ramakrishnan 1992). The energy efficiencies for terrace cultivation systems from Meghalaya were estimated between 1.7 and 21.0, whereas the monetary efficiencies were found in between 1.4 and 2.1.

The economic efficiency of broom grass cash crop system as estimated by Karki (2001) showed a gradual increase up to fourth year of plantation and then started declining as the plants grow older (Table **6**). The total monetary return up to sixth year was estimated Rs. 35,600 ha^{-1} in return of an investment of Rs. 9,450 ha^{-1} with an efficiency value of 3.8. Gangwar and Ramakrishnan (1989) estimated the economic efficiencies for pure cultivation of broom grass were 0.7, 1.5 and 2.1 respectively for first, second and third to seventh year and from mixed cropping with *Cinnamomum obtusifolium*, the respective values were 1.3, 2.0 and 2.8. For tea, coffee, pineapple mixed cropping and ginger based cash crop systems the economic efficiencies were estimated between 1.7 (coffee) and 3.9 (pineapple) and energy values between 1for coffee and 43.5 for ginger.

Among the all systems Aji system practice by Apatani was found more efficient in terms of energy input and output. The energy efficiency of different agricultural systems depends on the type of crops cultivated and more efficient were found

Table 6 Cost and Return (Rs ha^{-1}) Analysis for *Thysanolaena maxima*

	Year						
Item	First	Second	Third	Fourth	Fifth	Sixth	Total
Revenue	3.0	5.2	9.6	12.4	4.5	0.9	35.6
Production cost	3.7	1.4	1.55	1.55	0.85	0.4	9.45
Labour							
– Site Clearance	1.0						1.0
– Weeding (2x per year)	1.2	1.2	1.2	1.2	0.65	0.25	5.7
– Pit digging and rhizome planting	0.8						0.8
– Transportation to godowns	0.2	0.2	0.35	0.35	0.2	0.15	1.45
Materials							
Small tools and implements	0.5	–	–	–	–	–	0.5
Efficiency	0.8	3.7	6.2	8.0	5.3	2.3	3.8

Karki 2001

where rice is cultivated with millet or maize or any other crop. As far as efficiency of jhum agricultural system is concerned optimum efficiency was reported with jhum cycle of 10 years period otherwise on shortening or increasing the cycle period efficiency declines. In general terrace land has the lowest efficiency among the different existing agricultural systems.

7 Soil Nutrients in Different Agricultural Practices

It is evident that nutrient losses are greater in the agricultural fields than in the forest ecosystem and that their replenishment is very low, which deteriorates soil fertility (Sharma et al. 2001). The major cause of depletion of soil fertility in agricultural system is the removal of plant cover. The removal through run-off water and leaching processes could be substantial under situations of uneven topography and poor soil physical qualities of northeastern India. In slash and burn agriculture (jhum) the burning of slashed plant materials is done in order to release the plant nutrients in a single flush after fire (Table 7) and to capitalize on the nutrient released by growing

Table 7 Nutrients accumulated (kg $ha^{-1}year^{-1}$) through burning and their loss form different jhum fields and different cash crop plantation

	Jhum cycle			Cash crop plantation			
	30-years	10-years	5-years	Coffee	Tea	Pineapple	Ginger
Ash							
Released	17.4	13.8	6.9	–	–	–	–
Blown off	8.2	8.2	1.9	–	–	–	–
P							
Released	313.0	262.2	150.7	–	–	–	–
Blown off	147.1	155.6	42.7	–	–	–	–
Run-off	1.1	1.3	0.9	0.59	2.56	0.61	1.68
Percolated	0.1	0.1	0.1	0.22	0.30	0.19	1.0
K							
Released	1,739.0	2,070.0	685.0	–	–	–	–
Blown off	817.0	1,228.5	194.0	–	–	–	–
Run-off	64.7	91.2	51.0	22.01	54.67	15.15	41.03
Percolated	15.1	21.2	13.7	9.45	5.90	3.80	15.86
Ca							
Released	956.5	193.2	116.5	–	–	–	–
Blown off	449.4	114.7	33.0	–	–	–	
Run-off	15.1	15.9	13.8	10.83	26.0	8.01	12.94
Percolated	5.3	4.9	4.6	3.84	3.0	2.33	6.56
Mg							
Released	208.7	151.8	113.7	–	–	–	–
Blown off	98.0	90.1	32.2	–	–	–	–
Run-off	6.3	5.4	9.5	8.97	36.94	6.37	10.86
Percolated	2.5	2.1	2.3	2.33	4.55	1.62	5.58

Toky and Ramakrishnan 1981, 1982

mixture of crop species for a year or two after which the land is reverted back to its natural vegetation so as to restore soil chemical fertility and to improve its physical properties. However, during the process of cultivation a number of perturbations take place due to slash, fire, hoeing and ploughing, introduction of crop species, weeding and crop harvest, which causes rapid depletion of nutrients and this process, continues through the early secondary successional phases. The major physical causes of loss of nutrients from jhum fields of northeastern region are blown off, run-off and deep percolation (Table 8). The burning also causes loss of carbon and nitrogen due to volatilization (Ramakrishnan 1992). The decrease of organic carbon content on burning is more pronounced particularly when the temperature exceeds 150°C during burning (Ramakrishnan 1992). The pH, potassium and exchangeable calcium and magnesium content of the soil increased after burning (Table 8). However, the available phosphorus content did not change appreciably (Chauhan 2000). The data on the loss of organic carbon, phosphate and potash in jhum cultivation showed that the loss of these nutrients in the first cropping year was 84.7, 0.1 and 1.6 kg ha^{-1}, respectively (Table 9). During second year cropping the loss of these nutrients were found to be 1,321.0, 0.2 and 12.5 kg ha^{-1}, respectively (Chauhan 2000). The loss of these nutrients from jhum cultivation suggests that the practice is detrimental to soil fertility particularly in case of shorter jhum cycle. During the cropping phase the nutrients are taken up by crops and weeds, some of which are recycled back into the system as plant residues whereas substantial quantities are removed through crop harvest and weed removal from the plots. The net consequence of these input/output events is often a net loss from the system and a decline in soil fertility at the end of the cropping period. The recovery of the loss would take place during the fallow phase and the extent of recovery depends on the length of the fallow phases.

Table 8 Changes in surface soil before and after burning in different jhum cycle

Properties	15-year		10-year		5-year	
	Before	After	Before	After	Before	After
pH	5.1	7.5	5.3	7.6	5.5	7.5
Carbon (%)	1.9	1.6	1.8	1.7	1.6	1.6
Nitrogen (%)	0.26	0.25	0.26	0.25	0.21	0.20
Phosphorus (%)	3.5	3.6	3.4	3.6	3.3	3.5
K (mg 100 gm^{-1} soil)	13.0	61.0	11.0	56.0	12.0	51.0
Ca (mg 100 gm^{-1} soil)	10.0	32.0	12.0	28.0	9.0	21.0
Mg (mg 100 gm^{-1} soil)	8.0	23.0	10.0	21.0	9.0	20.0

Mishra and Ramakrishnan 1982

Table 9 Loss of organic carbon and plant nutrients in jhum cultivation

Year	Organic C	P_2O_5	K_2O
	(kg ha^{-1})		
First year	84.70	0.08	1.60
Second year	1,321	0.21	12.50
Average	702.90	0.15	7.10

Chauhan 2000

Our studies on different agricultural practices of Chakmas in the adjoining villages of Namdapha national park indicated that the field soil under wet rice cultivation had rich nutrient as compared to other systems including the jhum field. The C/N ratio varied between a narrow range of 12.9–15.7 (Table 10). The highest C/N ratio during mustard cultivation could be due to greater rates of microbial immobilization of soil nitrogen due to increasing surface area for microbial colonization by the incorporation of residues of the previously harvested crops like paddy and maize.

Overall occurrence of mineral nutrients in different agricultural systems shows that the jhum cultivation system is detrimental to soil fertility due to blown off, run off and percolation of elements. Among the systems it is observed that wet rice cultivation have high nutrient concentration in soil than other systems which might be due to accumulation through run off from adjoining areas by rain water.

8 Option for Sustainable Land-Use Development

Current agricultural practices for sustainable management of natural resources needs minor alterations and some of the strategies in this context are as follows:

- With wide variations in cropping and yield patterns under jhum practiced in diverse ecological situations, the transfer of technology from one area to another alone could improve jhum, valley land and home-garden ecosystems. Thus, for example emphasis on potato at higher elevations has led to a manifold increase in monetary efficiency (Ramakrishnan 1992).
- When the jhum cycle length cannot be increased beyond 5-year period, redesign and strengthen the agro-forestry system by incorporating ecological insights on tree architecture. During fallow period the regeneration could be accelerated by

Table 10 Physico-chemical properties of soil during cropping period in different agricultural systems

	Valley cultivation			Jhum cultivation
Properties	Wet rice	Paddy + maize	Mustard	
Textural class	Loamy sand	Sandy loam	Sandy loam	Loamy sand
Clay (%)	12.24	14.52	14.52	10.30
Silt (%)	14.02	21.04	21.04	10.25
Sand (%)	69.73	64.44	64.44	69.46
Moisture (%)	24.77 ± 1.30	16.68 ± 1.46	14.40 ± 1.58	22.6 ± 0.60
pH (1:2.5 w/v H_2O)	5.17 ± 0.03	5.67 ± 0.25	5.78 ± 0.13	5.88 ± 0.51
Organic C(%)	2.43 ± 0.03	2.190 ± 0.06	2.03 ± 0.06	1.96 ± 0.02
Total N(%)	0.18 ± 0.02	0.17 ± 0.02	0.13 ± 0.01	0.15 ± 0.01
C/N	13.21	12.88	15.27	13.22
Ammonium-N (μg g^{-1})	36.23 ± 2.14	33.18 ± 1.33	30.62 ± 2.79	30.17 ± 1.45
Nitrate-N (μg g^{-1})	23.37 ± 0.17	11.04 ± 0.14	10.93 ± 0.10	11.47 ± 0.10
Available P (μg g^{-1})	35.03 ± 0.04	22.78 ± 0.40	20.88 ± 0.12	24.86 ± 0.15

introducing fast growing native plants such as *Alnus nepalensis*, *Flemingia vestita, Clarodendrum collebrookenum, Albizia lebbeck, Cassia stipulate*, etc. and suitable fodder grasses having social and ecological values.

- The Sloping Agricultural Land Technology (SALT) developed in Philippines can also be a viable alternative to jhuming in the northeastern India, where cropping, livestock, horticulture and forestry can be incorporated by farmers in different forms like SALT-1 (Sloping Agricultural Land Technology), SALT-2 (Simple Agro-livestock Technology), SALT-3 (Simple Agro-Forest Land Technology) and SALT-4 (Small Agro-fruit Livestock Technology). The SALT can be established on farmland with slopes between 5% and 25% or more (Caleda and Esteban 1981). The salient features in terms of design of different SALT are summarized in Table 11.
- Since citrus, pineapple and banana are the major fruit crops of the region, pure horticultural land use can be developed with plantation of mandarin variety of orange at a distance of 5 m and pineapple (semi-shady species) may be planted in between the orange plants in the same row and the space between the rows can be used for vegetable cultivation. Here various tree species can be grown as wind breaks, shelterbelts or fillers in this system to protect the orange plants from the high speed of winds (Verma et al. 2001). Plant species *Salix* sp., *Populus* sp. and *Alnus nepalensis* have been proved successfully around the fruit farms without any adverse effect on the fruit production.
- Improve the nitrogen economy of jhum in the cropping and fallow phase by the introduction of nitrogen fixing leguminous and non-leguminous plants. Farmers have already adopted the *Alnus nepalensis* and *Albizia* species in the agricultural systems based on their traditional knowledge to meet modern needs. Another example is the less known food crop legume *Flemingia vastita* (Ramakrishnan 1992).
- Important bamboo species (e.g. *Dendrocalamus hamiltonii*), is highly valued by the tribals, can concentrate and conserve important nutritive elements such as N, P, and K (Rao and Ramakrishnan 1989). They could also be used as windbreaks against the loss of ash and nutrient losses in water.

Table 11 Land use characteristics of different SALT systems

Production system	SALT-1	SALT-2	SALT-3	SALT-4
Base	Staple crops	Fodder	Trees	Horticulture
Major product	Food grains	Meat/milk/ manure	Fuelwood/timber	Plantation crops
Planting area (%)				
• Staple crops	75	20	20	40
• Food/cash crops	25	20	20	60
Perennials/trees				
• Forage/fodder	–	40	–	–
• Private forestry	–	20	60	–

Conceived, tested and recommended by Mindanao Baptist Rural Life Center (MBRLC)/Asian Rural Life Development Center (ARLDF)

- In case of shorter jhum cycle burning should be avoided to prevent volatilization of nutrients (Ramakrishnan and Toky 1981). The crops and herbaceous weeds residues can be recycled back inside the systems in a scientific and well managed manner. In this context vermicompost may represent an efficient strategy to improve agriculture and residue management. It has been observed that for 99% (t_{99}) decomposition of foliage materials required less then 1 year and if these residues could be recycled, then 53–105 kg ha^{-1} nitrogen, 7–14 kg ha^{-1} phosphorus and 18–36 kg ha^{-1} potassium could be recycled into the system (Majumder et al. 2005).
- Redevelop village ecosystems through the introduction of appropriate technology to reduce hard work and improve energy efficiency through cooking stoves, agricultural implements, biogas generation, small hydroelectric projects, etc.
- Strengthen conservation measures based upon the traditional knowledge and value system with which the tribal communities can identify, e.g. the revival of the sacred grove concept based on cultural tradition, which enabled each village to have a protected forest.
- Encourage the cooperative efforts for carrying out forest based activities, i.e. basket making, rope making, cane furniture products processing of minor forest produce, honey collection etc. have to be made commercially viable by providing proper marketing facilities. This will help not only in decreasing dependence of farmers on shifting cultivation but will also help them monetarily.

Among the various agricultural practices carried out by tribal and other communities of north eastern India maximum energy efficiency and yield was recorded for Aji system practiced by Apatani tribe. In general terrace land cultivation has the lower efficiency among the different existing agricultural systems. In jhum system farmers grow number of crops under mixed cultivation and therefore known as one of the rich agro-biodiversity system. This system despite being rich in agro-biodiversity does not harbor good yield and energy like Aji system. The jhum system is generally practiced on hill slopes and the major causes of nutrients loss are due to blown off, run-off and through percolation of mineral nutrients that lead to poor yield and efficiency. Perhaps because of this reason farmers cultivate mixed crops with more variety in jhum system so that they can get maximum output. Different studies suggested many alternatives and modified practices for overall improvement of agricultural systems and socio-economic status of the people of this region. In jhum cultivation system during fallow period native nitrogen fixing plants may be grown to enhance soil fertility. Sloping Agricultural Land Technology may a viable alternative to jhum cultivation system. Popularization of agro-forestry and horticultural practices, improved fallow management by introduction of native nitrogen fixing plants, recycling of agricultural waste in the form of composting may be helpful for sustainability of traditional agricultural practices.

Acknowledgements The authors are thankful to University Grant Commission and UNESCO-MacArthur Foundation for financial support. Academic support given by Professors P.S. Ramakrishnan, K.G. Saxena and Uma Melkania is acknowledged.

References

Anonymous (2001) Census report. Government of India

Arunachalam A, Khan ML, Arunachalam K (2002) Balancing traditional jhum cultivation with modern agroforestry in eastern Himalaya- A biodiversity hot spot. Curr Sci 83(2):117–118

Arunachalam A, Sarmah R, Adhikari D, Majumder M, Khan ML (2004) Anthropogenic threats and biodiversity conservation in Namdapha nature reserve in Indian Eastern Himalayas. Curr Sci 87(4):447–454

Aurora GS, Billorey RK, Patton A, Myrchiang P (1977) Socio-economic impact of shifting cultivation control schemes in north-eastern region. Department of Sociology and Anthropology, North Eastern Hill University, Shillong (mimeographed)

Bhattacharya SC, Dutta S (1951) Citrus varieties of Assam. Ind J Genet Pl Br 11(1):57–62

Borthakur DN (1992) Agriculture of the North Eastern region with special reference to hill agriculture. Beecee Prakashan, Guwahati, India, p 265

Caleda A, Esteban ID (1981) Agroforestry in the Phillipines. Proceedings of Environmentally sustainable agroforestry and fuel wood production with first growing, nitrogen fixing, multipurpose legumes. Environment and policy institution, East-West center, Honolulu, USA (Mimeograph)

Chauhan BS (2000) Economics of ecosystem degradation due to shifting cultivatin in North Eastern region, India. J Assam Sci Soc 141(3):145–162

Choudhury D, Sundriyal RC (2003) Factors contributing to the marginalization of shifting cultivation in north-east India: micro-scale issues. Outl Agric 32(1):17–28

Choudhury D (1998) Conservation by local communities in north-east India. In: Kothari A (ed) Communities and conservation of natural resource management in South and Central Asia. UNESCO/Sage, New Delhi, India

Gangwar AK, Ramakrishnan PS (1987) Agriculture and animal husbandry among the Sulungs and the Nishis of Arunachal Pradesh. Soc Act 37:345–372

Gangwar AK, Ramakrishnan PS (1989) Ecosystem function in a Khasi village of the desertified Cherrapunji area in north-east India. Proceedings of Indian academic of sciences. Plant Sci 99:199–210

Gliessman SR (1989) Integrating trees into agriculture: the home garden agro-ecosystems as an example of agro-forestry in the tropics. In: Gliessman SR (ed) Agro-ecology: researching the ecological basis for sustainable agriculture. Springer, New York, pp 160–168

Karki M (2001) Institutional and socioeconomic factors and enabling policies for non-timber forest products-based development in northeast India. In IFAD report no. 1145-IN, Rome, p 23

Khumbongmayum AD (2004) Studies on plant diversity and regeneration of few tree species in sacred groves of Manipur. Ph.D. thesis North Eastern Hill University, Shillong, p 247

Kumar Y, Ramakrishnan PS (1990) Energy flow through an Apatani village ecosystem of Arunachal Pradesh in north-east India. Human Ecol 18:315–336

Maikhuri RK, Ramakrishanan PS (1990) Ecological analysis of cluster of villages emphasizing land use of different tribes in Meghalaya in north-east India. Agric Ecosys Environ 31:17–37

Maikhuri RK, Ramakrishanan PS (1991) Comparative analysis of the village ecosystem function of different tribes living in the same area in Arunachal Pradesh in north eastern India. Agric Sys 35:377–399

Majumder M (2007) Crop diversity, microbial biomass and soil nutrient dynamics of agro-ecosystems in 'Chakma' villages adjoining Namdhapa national park, Arunachal Pradesh. Ph.D. thesis, Rajiv Gandhi University, Itanagar, India

Majumder M, Arunachalam A, Melkania U, Adhikari D, Sharma R (2005) Agriculture as a component of village ecosystem function: Chakmas living around Namdapha national park, Arunachal Pradesh. In: Ramakrishnan PS, Saxena KG, Rao KS (eds) Shifting agriculture and sustainable development of North-Eastern India: tradition in transition. Oxford/IBH Publishing, New Delhi, India, pp 207–276

Mishra BK, Ramakrishnan PS (1981) The economic yield and energy efficiency of hill agro-ecosystems at higher elevations of Meghalaya in north-eastern India. Acta Oecol-Oecol Appl 4:237–245

Mishra BK, Ramakrishnan PS (1982) Energy flow through a village ecosystem with slash and burn agriculture in North-Eastern India. Agric Sys 9:57–72

Patnaik S, Ramakrishnan PS (1989) Comparative study of energy flow through village ecosystems of two co-existing communities (the Khasis and the Nepalis) of Meghalaya in north-east India. Agric Sys 30:245–267

Ramakrishnan PS, Toky OP (1981) Soil nutrient status of hill agroecosystems and recovery pattern after slash-and-burn agriculture (*jhum*) in north-eastern India. Plant Soil 60:41–64

Ramakrishnan PS (1992) Shifting agriculture and sustainable development (Man and Biosphere series: V.10). The Parthenon Publishing Group, Paris, p 424

Ramakrishnan PS (1994) The Jhum agro-ecosystem in north-eastern India: a case study of biological management of soils in a shifting agricultural system. In: Woomer PL, Swift MJ (eds) The biological management of tropical soil fertility. Wiley-Sayce, Chichester, pp 189–207

Ramakrishnan PS (2001) Increasing population and declining biological resources in the context of global change and globalization. J Biosci 26(4):465–479

Ramakrishnan PS, Toky OP, Mishra BK, Saxena KG (1978) Slash and burn agriculture in North-Eastern India, pp 570–586. In: Mooney H, Bonnicksen JM, Christensen NL, Lotan JE, Relners WA (eds) Fire regumes and ecosystem properties. USDA Forest Service General Technical Report, Washington DC

Rao KS, Ramakrishnan PS (1989) Role of Bamboo in nutrient conservation during secondary succession following slash and burn agriculture (JHUM) in North East India. J Appl Ecol 26:625–633

Richards P (1985) Indeginous agricultural revolution. Hutichinson & Company, London, p 192

Sharma E, Rai SC, Sharma R (2001) Soil, water and nutrient conservation in mountain farming systems: case study from the Sikkim Himalaya. J Environ Manage 61:123–135

Thurston HD, Salick J, Smith ME, Trutmann P, Pham JL, McDowell R (1999) Traditional management of agro-biodiversity. In: Wood D, Lenne JM (eds) Agro-biodiversity: characterization, utilization and management. CABI, Wallingford, pp 211–243

Tiwari BK, Barik SK, Tripathi RS (1998) Sacred groves of Meghalaya in conserving the sacred groves. In: Ramakrishanan PS (ed) Biodiversity management. UNESCO/Oxford/IBH Publications, New Delhi, India, pp 253–263

Toky OP, Ramakrishnan PS (1981) Cropping and yields in agricultural systems of the north-eastern hill region of India. Agro-ecosys 2:127–132

Toky OP, Ramakrishnan PS (1982) Run-off and infiltration losses related to shifting agriculture (Jhum) in north-east India. Environ Conserv 8:313–321

Upadhyay RC, Sundriyal RC (1998) Crop gene pools in the Northeast Indian Himalayas and threats. In: Partap T, Sthapit B (eds) Managing agro-biodiversity-farmers' changing perspectives and institutional responses in the Hindu Kush-Himalayan region. ICIMOD & IPGRI, Kathmandu, Nepal, pp 167–173

Vavilov NI (1950) The origin, variation, immunity and breeding of cultivated plant. Chron Bot 13:364

Verma ND, Satapathy KK, Singh RK, Singh JL, Dutta KK (2001) Shifting agriculture and alternative framing systems. In: Verma ND, Bhatt BP (eds) Steps towards modernization of agriculture in NEH region. ICAR, Meghalaya, India, pp 345–364

Withanage H (1998) Role of sacred groves in conservation and management of biodiversity in Sri Lanka. In: Ramakrishanan PS, Saxena KG, Chandrasheekara UM (eds) Conserving the sacred for biodiversity management. Oxford/IBH Publishing, New Delhi, India, pp 169–186

Microbial Community Structure and Diversity as Indicators for Evaluating Soil Quality

Sushil K. Sharma, Aketi Ramesh, Mahaveer P. Sharma, Om Prakash Joshi, Bram Govaerts, Kerri L. Steenwerth, and Douglas L. Karlen

Abstract The living soil system is of primary importance in sustainable agricultural production. Soil quality is considered as an integrative indicator of environmental quality, food security and economic viability. Therefore, soil itself serves as a potential indicator for monitoring sustainable land management. As part of the soil quality concept, a healthy soil supports high levels of biological diversity, activity, internal nutrient cycling and resilience to disturbance. The use of microbial community structure and diversity as an indicator to monitor soil quality is challenging due to little understanding of the relationship between community structure and soil function. This review addresses two critical questions regarding soil quality: (1) which soil microbial properties, particularly diversity and community structure, most effectively characterize soil quality and can be used as indicators, and (2) how can soil quality assessed by such indicators be improved or maintained?

We provide an overview of available techniques to characterize microbial community structure and diversity, and furnish information pertaining to strategies that can improve microbial diversity, including mycorrhizae, in relation to soil quality by adopting suitable agricultural practices to sustain soil and crop productivity. These techniques include those for structural profiling, i.e. fatty acid methyl ester analysis, genetic profiling, i.e. PCR-DGGE, SSCP, T-RFLP, functional profiling,

S.K. Sharma (✉), A. Ramesh, M.P. Sharma, and O.P. Joshi
Directorate of Soybean Research (ICAR), Khandwa Road, Indore 452001, Madhya Pradesh, India
e-mail: sks_micro@rediffmail.com

B. Govaerts
International Maize and Wheat Improvement Center (CIMMYT), Apdo. Postal 6-641, 06600, Mexico, DF, Mexico

K.L. Steenwerth
USDA-ARS, Crops Pathology and Genetics Research Unit, 95616 Davis, CA, USA

D.L. Karlen
USDA-ARS, National Soil Tilth Laboratory, 2110 University Blvd, Ames, Iowa 50011, USA

E. Lichtfouse (ed.), *Biodiversity, Biofuels, Agroforestry and Conservation Agriculture*, Sustainable Agriculture Reviews 5, DOI 10.1007/978-90-481-9513-8_11,

i.e. catabolic profiling, diversity of enzyme activity, and to profile both structural and functional communities comprehensively, i.e. gene chip. We identify the importance of minimum data sets (MDS) of microbial indicators, such that they must be (i) compatible with basic ecosystem processes in soil as well as physical or chemical indicators of soil health, (ii) sensitive to management in acceptable time frames, (iii) easy to assess or measure, (iv) composed of robust methodology with standardized sampling techniques, (v) cost-effective, and (vi) relevant to human goals, food security, agricultural production, sustainability and economic efficiency. We focus on specific agricultural strategies such as tillage, crop rotations, organic amendments and microbial inoculation to improve soil quality by managing microbial communities and diversity. Overall, we provide techniques to assess microbial communities and diversity, and their management through agricultural practices to improve quality of soil.

Keywords Soil quality • Microbial community • Diversity • Gene chip • AMF • MDS • Tillage • Crop rotation • Inoculation

1 Introduction

Agriculture today is often characterized by a high degree of intensity, particularly in developed countries. Heavy machines for tillage, planting and harvesting are repeatedly used during the growing season and crops are often given high amounts of fertilizers and pesticides to maximize yields. One outcome of this intensification during the last century was the Green Revolution, which increased food production and reduced hunger for millions of people by increasing both biological input such as high yielding cultivars as well as non-biological inputs like agrochemicals, fertilizers and irrigation. This approach has encouraged many developing countries of Asian and African continents to grow crops using monoculture and irrigation to ensure a maximum economic status. However, many rural communities in the tropics and sub-tropics are still persistently affected by insufficient household food production (Dalgaard et al. 2003).

The Food and Agricultural Organisation (FAO) defines food security as "when all people, at all times, have physical and economic access to sufficient, safe nutritious food to meet their dietary needs and food preference for an active and healthy life". Technologies such as irrigation, mechanization and improved crop varieties have changed the socio-economic status of some people, but food insecurity still persists amongst the poorest and most vulnerable people. Therefore, food security is a major concern around the globe, because more than a billion people are still undernourished and have no access to food (Stocking 2003; Reynolds and Borlaug 2006).

Sustainable food security is ultimately dependent on the availability and condition of natural resources including soils, which are gradually deteriorating and increasing the pressure on food availability to human beings. Some agricultural soils can endure intensive cultivation practices, but many gradually show a lower ability to support high productivity due to impaired soil quality. In recent years, there has been an increasing awareness of soil quality to ensure a greater sustainability of agricultural soils. This review addresses two critical questions regarding soil quality: (1) which soil microbial properties, in particular diversity and community structure, most effectively characterize soil quality and should be used as indicators, and (2) how can soil quality assessed by such indicators be improved or maintained? In particular, we provide an overview of techniques available to characterize microbial community structure and diversity for evaluating soil quality, and furnish information pertaining to strategies that can improve microbial diversity in relation to soil quality by adopting suitable agricultural practices to sustain soil and crop productivity.

2 The Concept of Soil Quality

In 1971, Alexander proposed for the first time development of soil quality criteria in the context of agriculture's role in environmental improvement. The soil quality concept per se was introduced by Warkentin and Fletcher (1997) as an approach to facilitate better land use planning for multiple functions. Their concept of soil quality was based on four criteria, upon which future concepts of soil quality were developed. These criteria were that (1) soil resources were constantly being evaluated for an ever-increasing range of uses, (2) several different stakeholder groups were concerned about the state of soil resources, (3) priorities and demand of society were changing, and (4) soil-resource and land-use decisions were made in a human and institutional context. In a broad sense, the concept of soil quality was not introduced until the mid-1980s, wherein emphasis was mainly given to soil resource management, particularly in controlling soil erosion and minimizing its effects on crop productivity (Pierce et al. 1984). Later, soil management gradually shifted from minimizing soil erosion to broader issues like sustainable agriculture, environmental health and prevention of soil degradation (Karlen et al. 2003a). In the 1990s, the pace in soil quality research was further accelerated by the recommendation of the U.S. National Research Council's (NRC) Board on Agriculture that "we conserve and enhance soil quality as a fundamental step toward environmental improvement" and that the concept of soil quality be in principle a guide to agricultural policies and practices (NRC 1993). Thereafter, many researchers contributed to developing a soil quality concept in the publications entitled, "Defining Soil Quality for Sustainable Environment" (Doran et al. 1994) and "Methods for Assessing Soil Quality" (Doran and Jones 1996).

Soil quality has been defined in several ways including 'fitness for use' and dependent upon the extent to which a soil fulfills its destined role (Larson and Pierce 1994; Singer and Edwig 2000). In a broad ecological sense, soil quality has been defined as the capacity of a soil to function within ecosystem boundaries to sustain plant-animal productivity, maintain or enhance water and air quality, and support human health and habitation (Karlen et al. 1997). Doran and Safely (1997) further defined soil quality by considering the continuous and dynamic nature of the soil as "the continued capacity of soil to function as a vital living system, within ecosystem and land-use-boundaries, to sustain biological productivity, promote the quality of air and water and maintain plant, animal and human health". More recently, a healthy soil as part of the soil quality concept is defined as a stable soil system with high levels of biological diversity and activity, internal nutrient cycling and resilience to disturbance (van Bruggen et al. 2006). Overall, soil quality is considered as an integrative indicator of environmental quality, food security and economic viability (Herrick 2000) and therefore, it would serve as a good indicator for monitoring sustainable land management.

The concept developed in this review differs from traditional technical approaches that focus solely on productivity. Instead, soil quality is examined as a holistic concept, recognizing soil as a part of a dynamic and diverse production system with biological, chemical and physical attributes that relate to the demands of human society (Swift 1999; Sanchez et al. 2003). Society, in turn, actively adapts soil to its needs, mining it of its nutrients on demand and replenishing these nutrients in times of excess.

3 Indicators of Soil Quality

Assessment of soil quality is a major challenge because it is highly dependent on management of soil through resources available in a given agroecosystem and the agroclimatic conditions (Karlen et al. 2003b). Common approaches used for assessing the soil quality are either qualitative or quantitative. Qualitative indicators are often sensory descriptors e.g. appearance, smell, feel and taste recorded through direct observations usually made by the growers' (Garlynd et al. 1994; Dang 2007). Other observations include soil colour, yield response, frequency of ploughing or hoeing, and visual documentation of plant growth, selected weed species, and earthworm casts. The use of indigenous local knowledge and experience of growers provides a simple approach to characterize the status of and to diagnose any change in soil quality (Roming et al. 1995; Barrios et al. 2006).

Quantitative assessments of soil quality involve more sophisticated analytical approaches (Harris and Bezdicek 1994). Generally, soil quality is assessed by the combination of the physical, chemical and biological properties acting as indicators (He et al. 2003), and a large number of different physical, chemical and biological properties of soil are being employed as quantitative indicators to define soil quality (Roming et al. 1995; Dang 2007). Typical soil physical indicators include texture, bulk density and infiltration, water holding capacity and retention characteristics, porosity, aggregate stability and soil depth. Organic carbon, pH, electrical conductivity, cation exchange capacity, extractable N, P, K, S are important chemical indicators, and biological indicators include quantity, activity, and diversity of soil fauna and flora and soil enzymes. Several bio-indicators of soil quality have been developed (Trasar-Cepeda et al. 2000; Nielsen and Winding 2002; Anderson 2003). A number of soil biological properties respond to changes in agricultural practices, showing potential use as indicators of soil quality. Other biological indicators include organic matter content; soil macrofauna like earthworms, springtails, collembulas and nematodes; and the overall litter decomposition ability of living organisms (Pfiffner and Mäder 1997; Wardle et al. 1999). Among biological parameters, soil microorganisms and their functions (i.e. enzyme activities such as FDA, phosphatase, amidohydrolase, nitrogen mineralization, nitrification, etc.) are also widely recognized as integral component of soil quality because of their crucial involvement in ecosystem functioning and their capability to respond quickly to environmental changes (Aseri and Tarafdar 2006; Sharma et al. 2005).

In comparison to the rapid shifts in biochemical and biological properties that occur after soil disturbance (Le Roux et al. 2008), changes in physical properties may occur relatively less quickly. Among the biological properties, soil microorganisms are very sensitive to external perturbations and can act as a sensor for monitoring soil response, and more generally soil quality. Soil microbial biomass, soil enzymes and basal soil respiration are among the most important biological parameters and have proven to be powerful tools in monitoring soil quality (Karlen et al. 2006; Nogueira et al. 2006), although some authors have reported that soils experiencing different treatments can have similar microbial biomass whereas their functioning can markedly differ (Patra et al. 2005). Other microbial indicators of soil status encompass the diversity and structure of microbial communities. Many methods for analyzing microbial diversity have been developed in recent years and utilized as indicators for assessing soil quality in congruence with established indicators. Numerous studies have reported the beneficial impacts of conservation tillage management, organic amendments, crop rotation and application of microbial inoculants on enzyme activities (Naseby and Lynch 1997; Acosta-Martinez et al. 2003; Melero et al. 2006), microbial biomass (Liebeg et al. 2004; Monokrousos et al. 2006; Franchini et al. 2007; Saini et al. 2004) and microbial community structure and diversity (Sun et al. 2004; Roesti et al. 2006; Mathimaran et al. 2007; Acosta-Martinez et al. 2007; Govaerts et al. 2008).

4 Rationales for Using Microorganisms as Soil Quality Indicators

Microorganisms are a component of the 'biological engine of the earth' and provide an integrated measure of soil quality, an aspect that cannot always be obtained with physical and chemical measures and/or analysis of higher organisms. Microorganisms are driving many fundamental nutrient cycling processes, soil structural dynamics, degradation of pollutants, various other services (Bloem et al. 1994) and respond quickly to natural perturbations and environmental stress due to their short generation time and their intimate relation with their surroundings, attributed to their higher surface to volume ratio. This allows microbial analyses to discriminate soil quality status, and shifts in microbial population and activity could be used as an indicator of changes in soil quality (Kennedy and Smith 1995; Pankhurst et al. 1995).

Microbial indicators have been defined as “properties of the environment or impacts that can be interpreted beyond the information that the measured or observed [indicator] represents itself” (Nielsen and Winding 2002). Stenberg (1999) listed five different levels at which microorganisms can be studied. These are: (1) as individuals; (2) at population levels (Hill et al. 2000); (3) at the functional group level, including autotrophic nitrification (Stenberg et al. 1998), arbuscular mycorrhiza (Kahiluoto et al. 2001) and specific soil enzymes; (4) as the whole microbial community studied using genetic or physiological diversity or quantitative methods to enumerate the total community including microbial biomass, basal respiration rate, nitrogen mineralization, denitrification and general soil enzymes (Griffiths et al. 2001) and (5) at the ecosystem level which can describe data from all the other levels. It is not possible to use all ecosystems or soil attributes as indicators of soil quality (Karlen and Andrews 2000) and thus, there is a need to select specific indicators having high discriminating potential and high value to account for actual soil quality status of agricultural systems: an indicator would not be so useful if it is very sensitive to disturbances. In particular, the search for indicator organisms associated with healthy or deteriorated soil requires a unified concept of soil quality. In this context, microbial indicators can be divided into general, or universal, and specific indicators (Nielsen and Winding 2002). Universal indicators may include biodiversity, stability and self-recovery from stress (Parr et al. 2003). *Rhizobium*, mycorrhizae and nitrifying bacteria could be used as specific indicators because of their high sensitivity to agrochemicals (Domsch et al. 1983) or management regimes (Le Roux et al. 2008), and clearly defined roles among soil functions. Specific indicators are dependent on the geographic zone, climate, soil type and land use history.

Although the relationship between soil quality and microbial diversity is not completely understood, a medium to high diversity in agricultural soil is generally considered to indicate a ‘good’ soil quality (Winding 2004). This statement is based on the assumption that there is a functional redundancy in a healthy soil, so that soil ecosystem will recover from a stress factor that eliminates part of the microbial community (Yin et al. 2000) (Fig. 1) In addition, the active microbial pool is a reserve pool of quiescent microorganisms, which can respond to foreign substances in the soil (Zvyaginstsev et al. 1984). This diverse microbial pool maintains soil homeostasis. The larger the microbial diversity and functional redundancy, the quicker the ecosystem can return to stable initial conditions after exposure to stress or disturbance. This concept is highly debated. Indeed, several removal experiments (in which microbial taxa are successively removed from an innate community through a stressing agent or dilution of the original community) have shown that the functioning and stability of soil microbial communities can be maintained following strong erosion of microbial diversity (Griffiths 2000; Wertz et al. 2006; 2007). Furthermore, although some observational studies show some links between soil microbial community structure and functioning (Patra et al. 2006), the shifts in functioning often appear to be linked to key species rather than due to richness.

Besides these controversies, many authors argue that measurements of the structure and activities of specific microbial communities contributing to soil processes has the potential to provide rapid and sensitive means of characterizing changes to

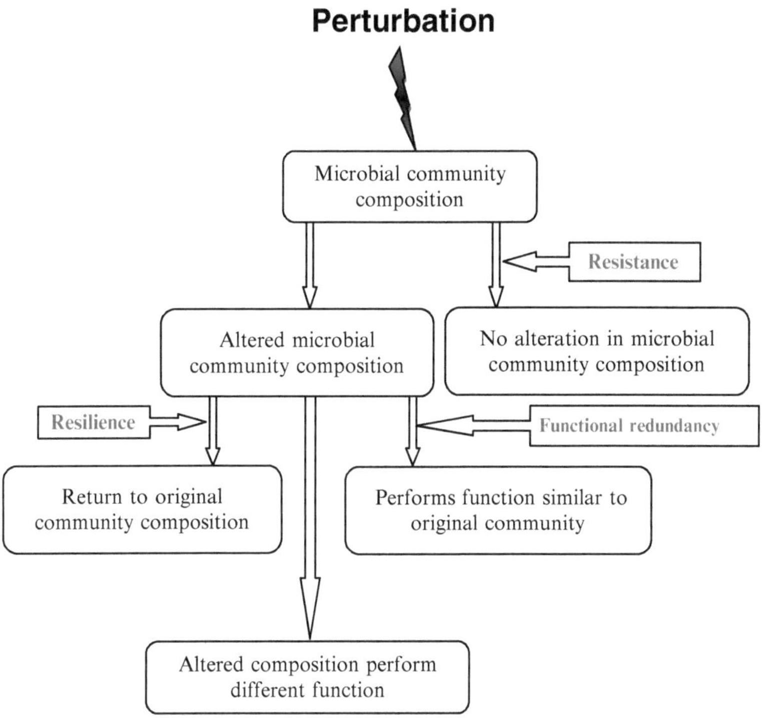

Fig. 1 Schematic representation of impact of perturbation on changes in microbial community composition and function (Modified from Allison and Martiny 2008)

soil quality (Waldrop et al. 2000; Bending et al. 2004; Enwall et al. 2005; Bressan et al. 2008). In particular, the size and diversity of specific functional microbial groups such as AM fungi and nitrifying bacterial communities have the potential to characterize the effects of management on the sustainability of soil (Chang et al. 2001). Additionally, a number of features viz. fast growth rate, high degree of physiological flexibility and rapid evolution (mutation) of microorganisms could make microbial communities more resilient to the new environment (Fig. 1) (Allison and Martiny 2008).

5 Evaluation of Microbial Community Structure and Diversity: Tools, Their Use and Misuse

Microbial diversity viz. structural and functional diversity in soil is increasingly assessed for measurement of soil health (Visser and Parkinson 1992). In the following sections, different methods for evaluating microbial community structure and diversity will be described in detail (Fig. 2).

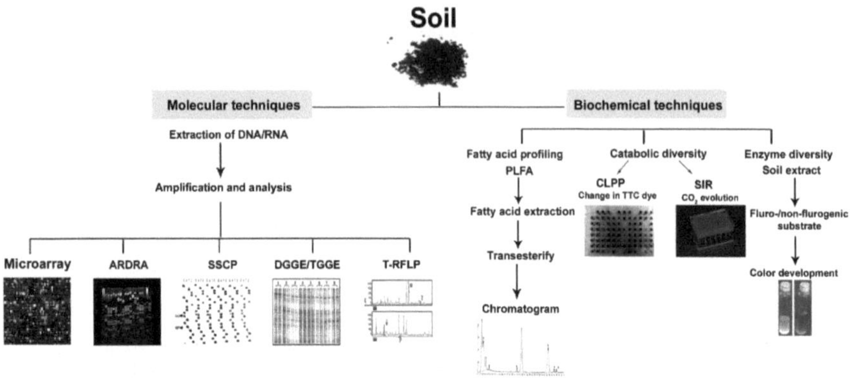

Fig. 2 An overview of techniques used for soil microbial community structure and diversity

5.1 *Structural Profiling Technique*

Structural diversity is defined as the number of parts or elements within a system, indicated by such measures as the number of species, genes, communities or ecosystems (Avidano et al. 2005). Several indices such as species richness and evenness are used to describe the structural diversity of a community (Ovreas 2000). However, these indices cannot be used for soil microbes as easily as for macroorganisms. Indeed, with the rise of molecular tools in microbial ecology, it became evident that we have described only a very small portion of the diversity in the microbial world. Most of this unexplored microbial diversity seems to be hiding in the high amount of yet uncultured bacteria. New direct methods independent of culturing and based on the genotype and phenotype of microbes allow a deeper understanding of the composition of microbial communities in a soil ecosystem (Amann et al. 1995). Based on molecular studies, it could be estimated that 1 g of soil consists of more than 10^9 bacteria belonging to about 10,000 different microbial species (Ovreas and Torsvik 1998) or even much more (Gans et al. 2005). This huge level of diversity makes it difficult to employ the microbial community structure as an indicator of soil quality. A widely observed result is that the structural diversity of a bacterial community is often sensitive to environmental changes and exhibits a shift in its composition (Kandeler et al. 1999; Saison et al. 2006). Ovreas and Torsvik (1998) compared the influence of crop rotation and organic farming on microbial diversity and community structure and found higher values for proxies of diversity in soils under organic farming management as compared to conventional practices. In addition to shifts in community structure, there have been reports that indices of bacterial diversity suggested a reduced diversity in soils contaminated with phenyl-urea herbicides, fumigants etc. (El Fantroussi et al. 1999; Yang et al. 2000; Ibekwe et al. 2001) Although, these management practices certainly induce change in microbial community, the extent of soil function loss in relation to reduction in microbial diversity is not known. With regard to soil quality assessment, it is also important to note that in addition to examining microbiological effects of various management practices (e.g. herbicides, fungicides, tillage) these changes must also be weighted against chemical- and physical-indicators changes that may also occur in response to these practices.

5.1.1 Fatty Acid Methyl Ester (FAME) Analysis

Phospholipid fatty acids (PLFAs) are a potentially useful biomarker molecule that is being used to elucidate structure of microbial community in soil because of their presence in all living cells and rapid degradation upon cell death (White et al. 1979; Pinkart et al. 2002). In microorganisms, PLFAs are found exclusively in cell membranes and not in other parts of the cell such as storage products. Fatty acid methyl ester (FAME) analysis, which directly extracts PLFAs from soil, is a biochemical method that does not rely on culturing of microorganisms and provides information on the microbial community composition based on groupings of the fatty acids (Ibekwe and Kennedy 1998; Drenovsky et al. 2004; Drenovsky et al. 2008). PLFAs compose a relatively constant proportion of the cell biomass and signature fatty acids exist that can differentiate major taxonomic groups within a community. Individual PLFAs or signature fatty acids are specific for subgroups of microorganisms, e.g. gram-negative or gram-positive bacteria, methanotrophic bacteria, fungi, mycorrhiza, and actinomycetes (Zelles 1999). It is possible to quantify different groups of microorganisms by this method, and PLFA profiles can be related to microbial community structure using multivariate analysis (e.g., canonical correspondence analysis, principal components analysis). Therefore, a change in the fatty acid profile would represent a change in the microbial populations. It has been used in the study of microbial community composition and population changes due to chemical contaminants (Siciliano and Germida 1998; Kelly et al. 1999), land use history (Myers et al. 2001; Steenwerth et al. 2003), agricultural practices (Bossio et al. 1998), and rhizosphere effects (Ibekwe and Kennedy 1998). Based on phospholipid fatty acid profiles, Bossio et al. (1998) detected changes in microbial communities consistent with different farming practices. When these researchers calculated the Shannon diversity index based on PLFA relative abundance, no difference could be detected. This could be because of a difference in the community structure but not in diversity (Bossio et al. 1998). These studies clearly demonstrated the utility of this method in determining gross community changes associated with soil management practices. This method has been recommended for soil quality monitoring programme in Scotland and Northern Ireland (Chapman et al. 2000).

5.2 Genetic Profiling Techniques

The genetic diversity of soil microorganisms is an indicator that provides the basis for all actual and potential functions. Techniques for determining genetic diversity include several molecular methods, a few of which have been suggested to be implemented for Dutch soil monitoring programme (Bloem and Breure 2003). Taxonomic diversity of microorganisms at the genetic level is most commonly studied by determining the DNA gene coding for ribosomal RNA. The 16S rRNA genes are used for phylogenetic affiliation of Eubacteria and Archaea, while 18S rRNA genes are used for fungi. The conserved regions within the rRNA genes have facilitated the design of primers targeting the majority of members of defined groups of bacteria or fungi. Several comparable molecular methods based on DNA analyses using polymerase chain reaction (PCR) followed by an analysis of the diversity of PCR products through denaturing gradient gel electrophoresis (PCR-DGGE), temperature gradient gel electrophoresis (PCR-TGGE), terminal restriction fragment length polymorphism (T-RFLP), single strand conformation polymorphism (SSCP), restriction fragment length polymorphism (RFPL) and amplified ribosomal DNA restriction analysis (ARDRA) targeting 16S rDNA gene have been employed for community analysis.

5.2.1 Polymerase Chain Reaction-Denaturing Gradient Gel Electrophoresis (PCR-DGGE) and PCR-Temperature Gradient Gel Electrophoresis (PCR-TGGE)

The PCR-DGGE (Muyzer et al. 1993) and PCR-TGGE (Heuer and Smalla 1997) are widely used methods for estimation of microbial community fingerprining. They are based on variation in base composition and secondary structure of fragments of the 16S rDNA molecule. A fragment of 16S rDNA gene of known size can be amplified by PCR, with primers mainly targeting all eubacteria or selected subgroups (Table 1). Following PCR, denaturing gradient gel electrophoresis separates the products. In DGGE, the gel itself contains a chemical-denaturing gradient, making

Table 1 Gene specific primers used in DGGE for the amplification of 16S rRNA gene of bacteria or archaea

Primer name	Sequence (5′–3′)	16S rDNA target (base number)[a]
PRBA338F	59[b]AC TCC TAC GGG AGG CAG CAG 39	Bacteria V3 region (338–358)
PRUN518R	59ATT ACC GCG GCT GCT GG 39	Universal V3 region (534–518)
PRBA968F	59[b]AA CGC GAA GAA CCT TAC 39	Bacteria V6 region (968–983)
PRBA1406R	59ACG GGC GGT GTG TAC 39	Bacteria V9 region (1406–1392)
PRA46F	59C/TTA AGC CAT GCG/A AGT 39	Archaea (46–60)
PREA1100R	59T/CGG GTC TCG CTC GTT G/ACC 39	Archaea (1117–1100)
PARCH340F	59[b]CC TAC GGG GC/TG CAG/C CAG 39	Archaea V3 region (340–358)
PARCH519R	59TTA CCG CGG CG/TG CTG 39	Archaea V3 region (534–519)

[a]Bases numbered relative to *E. coli* 16S rRNA sequence

[b]GC clamp added to the 59 end of the primer, 59CGC CCG CCG CGC GCG GCG GGC GGG GCG GGG GCA CGG GGG G 39

Nakatsu et al. (2000)

the fragments denature along the gradient according to their base composition. In PCR-TGGE, a temperature gradient is created across the gel, resulting in the same type of denaturation. The number and position of the fragments reflect the dominant genus in the community. Similar to other profiling methods, PCR-DGGE/TGGE detects only a limited part of the microbial diversity in a community, due to generally high diversity. Soil communities may easily contain more than 10,000 different species per 100 g of soil (Torsvik et al. 1998), while the resolution of more than 20–50 bands on a gel is difficult. To show up as a visible band on the gel, a species has to constitute approximately 1% of the entire population (Casamayor et al. 2000). Sequencing and identification of visible bands on the gel following PCR-DGGE may further improve the resolution (Casamayor et al. 2000). DGGE/TGGE has been used to assess the diversity of bacteria and fungi communities in rhizosphere (Smalla et al. 2001) caused by changes in nutrient applications (Iwamoto et al. 2000). It has also been used for forest soils (Marschner and Timonen 2005), grasslands (Ritz et al. 2004), and to evaluate agricultural management effects of manure and fertilizers (Sun et al. 2004), and anthropogenic chemicals (MacNaughton et al. 1999; Whiteley and Bailey 2000). Continuous cereal crops had similar rhizoplane communities while communities from cereal-legumes rotation showed greater variability in West African soils (Alvey et al. 2003). PCR-DGGE of bacterial 16S rRNA genes has recently been implemented in the Dutch Soil Monitoring Programme (Bloem and Breure 2003). Results from the first round visit of 60 farms showed that the number of DNA bands was dependent on soil type and also, to a lesser extent, land use. Such changes in PCR and all other indicators of microbial diversity confirm the responsiveness of the soil microbial community to soil and crop management practices, but a critical unknown is what constitutes a "good community" and specifically how does or doesn't this affect soil quality. This is a critical question which can only be answered with investment in basic soil science research not only in India but throughout the world.

5.2.2 Single Strand Conformation Polymorphism (SSCP)

Like DGGE/TGGE, the SSCP technique was originally developed to detect point mutations in DNA (Orita et al. 1989). When DNA is denatured into single strands, each strand folds up into a configuration based not only on size but also on sequence. This feature can separate single stranded DNA on an agarose gel according to folding and secondary structure (Lee et al. 1996). Reannealing of the DNA during electrophoresis remains a potential problem of the method. Schwieger and Tebbe (1998) further refined the method by removing one of the two DNA strands before electrophoresis. Each band in the agarose gel should then represent a single species; however, multiple sequences within a single band have been reported (Schmalenberger and Tebbe 2003). SSCP has been used to measure succession of bacterial communities (Peters et al. 2000), rhizosphere communities (Schwieger and Tebbe 1998; Schmalenberger et al. 2001), bacterial population changes in an anaerobic bioreactor (Zumstein et al. 2000) and AMF species in roots (Simon et al. 1993; Kjoller and

Rosendahl 2000). To date, SSCP alone has been applied in soil quality assessment but it may be optimized and then integrated with other well established tools and techniques of soil quality assessment.

5.2.3 Restriction Fragment Length Polymorphism (RFLP)

RFLP, also known as amplified ribosomal DNA restriction analysis (ARDRA) is another tool used to study microbial diversity that relies on DNA polymorphism. In a study by Liu et al. (1997), PCR amplified rDNA is digested with a 4-base pair cutting restriction enzyme. Different fragment lengths are detected using agarose or non-denaturing polyacrylamide gel electrophoresis in the case of community analysis (Liu et al. 1997; Tiedje et al. 1999). RFLP banding patterns can be used to screen clones (Pace 1996) or to measure bacterial community structure (Massol-Deya et al. 1995). This method is useful for detecting structural changes in bacterial communities in soil inoculated with biocontrol agents (Bakker et al. 2002).

5.2.4 Terminal Restriction Fragment Length Polymorphism (T-RFLP)

The T-RFLP, a polymerase-chain reaction-fingerprinting method, is an improved alternative method for examining comparative microbial community analysis (Liu et al. 1997; Marsh 1999). T-RFLP is a technique that addresses some of the limitations of RFLP (Tiedje et al. 1999). The method can be used to analyse communities of bacteria, archaea, fungi, other phylogenetic groups or subgroups, as well as functional genes (Thies 2007). For 16S rRNA genes, a number of primer sequences have been published that are complementary to highly conserved sequences of the bacteria or the archaea or are conserved among specific subgroups within these domains such as alpha- or beta-proteobacteria (Table 2). It follows the same principle

Table 2 Primers used commonly to amplify short-subunit rRNA genes from microbial community DNA extracts

Primer name	Sequence (5′–3′)	Specificity
1511R	YGCAGGTTCACCTAC	Universal
1492R	ACCTTGTTACGACTT	Universal
27F	AGAGTTTGATCMTGGCTCAG	Bacteria 16S
63F	CAGGCCTAAYACATGCAAGTC	Bacteria 16S
1387R	GGGCGGWGTGTACAAGGC	Bacteria 16S
21F	TTCCGGTTGATCCYGCCGGA	Archaea 16S
25F	CTGGTTGATCCTGCCAG	Eukarya 18S
BLS342F	CAGCAGTAGGGAATCTTC	Bacilli
BETA680F	CRCGTGTAGCAGTGA	Beta-proteobacteria
Pln930R	CTCCACCGCTTGTGTGA	Planctomycetes
Act1159R	TCCGAGTTRACCCCGGC	Actinomycetes
Bas1105F	CCGTTGTAGTCTTAACAG	Basidiomycota

Theis (2007)

as RFLP except that one PCR primer is labeled with a fluorescent dye, such as TET (4,7,2′,7′-tetrachloro-6-carboxyfluorescein) or 6-FAM (phosphoramidite fluorochrome 5-carboxyfluorescein). This allows detection of only the labeled terminal restriction fragment (Liu et al. 1997). The primers are labeled with a fluorescent tag at the terminus resulting in labeled PCR-products. The products are cut with several restriction enzymes, one at a time, which result in labeled fragments that can be separated according to their size on agarose gels. As the PCR products are labeled at the terminus, only restriction enzyme fragments containing either of the terminal ends of the PCR product will be detected. The digested PCR products are subsequently loaded on a sequencer. The output includes fragment size and quantity. Marsh (2005) provided detailed protocols for performing T-RFLP analysis.

T-RFLP has been used for bacterial community analysis in response to spatial and temporal changes (Acinas et al. 1997; Lukow et al. 2000; Mummey and Stahl 2003), organic amendments (Wang et al. 2006), microbial inoculants (Conn and Franco 2004), tillage (Buckley and Schmidt 2001), inorganic fertilization (Mohanty et al. 2006), changes in farming systems (Hartmann et al. 2006), in different soil types (Singh et al. 2006) and cultivation practices (Buckley and Schmidt 2001). Lasting changes in the composition of soil bacterial population due to soil solarization (Culman et al. 2006), herbicides (Moran et al. 2006), pesticide use (Rousseaux et al. 2003) and soil pollutants (Jung et al. 2005) were readily detected by T-RFLP analysis. Recently in Switzerland, Hartmann and Widmer (2006) emphasized that changes in microbial community structure but not soil bacterial diversity analyzed through T-RFLP offer a better understanding of the impact on soil quality in agriculturally managed systems (biodynamic, bioorganic, conventional and conventional with inorganic fertilizers) thus making it highly useful tool for soil monitoring after its optimization.

5.3 *Functional Profiling Techniques*

The functional diversity of microbial communities includes the range and relative expression of activities involved in functions namely decomposition of organic carbon, nutrient transformation, plant growth promotion/suppression and soil physical processes as influenced by microorganisms (Giller et al. 1997). The functional diversity of microbial communities has been found to be very sensitive to environmental changes (Kandeler et al. 1999). Among the functional diversity indicators, the carbon utilization pattern and the measurement of enzymatic activity profiles expressed by the whole bacterial community have been suggested as useful tools to evaluate the soils (Nielsen and Winding 2002). The metabolic profile, obtained by a Biolog assay and MicroResp, provides a physiological fingerprinting of the potential functions of the microbial community (Garland and Mills 1991; Campbell et al. 2003). Since enzymatic activities in the soil are mainly of bacterial and fungal origin, the characterization of soil enzyme patterns can improve our knowledge on microflora activity, soil productivity and the impact of pollutants (Pankhurst et al. 1995).

5.3.1 Catabolic Profiling-Based on Substrate Utilization

The diversity in decomposition functions performed by heterotrophic microorganisms represents one of the important components of microbial functional diversity. A simple approach to measure functional diversity is to examine the number of different C-substrates utilized by the microbial community. The two most common methods of measuring substrate utilization patterns are the community-level physiological profiling by Biolog plates methods (Garland and Mills 1991) and in situ substrate-induced respiration (SIR) (Degens and Harris 1997; Campbell et al. 2003).

Community-Level Physiological Profiling (CLPP)

Garland and Mills (1991) developed a technique using a 96-well Biolog microtitre plate utilizing sole carbon source to assess physiological profiles of bacterial communities to reflect their functional diversity. This culture-dependent technique is widely used for analyzing soil microbial communities. Gram-negative (GN) and gram-positive (GP) plates containing 95 different carbon sources and one control well per plate without a substrate, growth medium and redox dye tetrazolium salt are available from the Biolog (Hayward, CA, USA, www.biolog.com). Subsequently, Biolog introduced an Eco-plate (Insam 1997; Choi and Dobbs 1999) containing three replicates of 31 different environmentally relevant carbon sources and one control well per replicate. The tetrazolium salt changes colour as bacteria metabolize the substrate. Since many fungal species are not capable of reducing the tetrazolium salt (Praveen-Kumar and Tarafdar 2003), Biolog developed fungal specific plates SFN2 and SFP2, having the same substrates as GN and GP plates but without tetrazolium salt (Classen et al. 2003). The important considerations in the use of this method for community analysis are: (1) density of initial inoculum must be standardized, (2) functional diversity is based on the assumption that color development in each well is solely a function of the proportion of organisms present in the sample able to utilize a particular substrate, and (3) substrates found in commercially available Biolog plates are not necessarily ecologically relevant and most likely do not reflect the diversity of substrates found in the environment (Hill et al. 2000).

The CLPP is used extensively in analyzing microbial communities because it is sensitive, reproducible and has the power to distinguish between tillage systems (Govaerts et al. 2007a), contaminated soil sites (Boivin et al. 2002), rhizosphere (Grayston et al. 1998, 2004; Soderberg et al. 2004), inoculation of microorganisms (Bej et al. 1991) and, soil management practices (Schouten et al. 2000; Mäder et al. 2002). This method is currently implemented in the Dutch Soil Monitoring Programme where the CLPP of microbial communities is determined in the Biolog ECO plate to discriminate between different types of soil and management practices (Schouten et al. 2000, 2002). The assay is also recommended in the soil-monitoring programme of Scotland and Northern Ireland (Chapman et al. 2000). For the monitoring of soils, commercial plates must contain a uniform composition and concentration, and they should be available in the market.

Substrate-Induced Respiration

Catabolic conversion can be used to assess catabolic diversity in soil microbial communities utilizing many simple organic substances through the Substrate-Induced Respiration (SIR) test. Degens and Harris (1997) developed this concept using multiple SIR tests for the measurement of patterns of in situ catabolic potential of microbial communities and does not require extraction and culturing of microorganism while in Biolog system this problem exists. They tested 83 simple sugars, carboxylic acids, amino acids, polymers, amines and amides, and subsequently identified 36 substrates providing the greatest difference in SIR responses among five different soils of ecological importance. It is expected that only a limited number of species will contribute to the SIR response in a specific test. Using this catabolic response profile to estimate heterotrophic evenness and diversity, Degens and co-workers found that a decrease in microbial diversity does not consistently result in declines in soil functions (Degens 1998), consistent with Wertz et al. (2006). Furthermore, reduction in the catabolic diversity due to changed land use could reduce the resistance of microbial communities to stress or disturbance (Degens et al. 2001). This resistance might be coupled to organic matter content in soil, as depletion of organic C may cause a decline in catabolic diversity (Degens et al. 2000).

Since Degens's method is more laborious and time consuming, a comprehensive micro-respiratory system (MicroResp) containing 96- deep- wells-microtitre plate has been developed by Campbell et al. (2003) based on CO_2 evolution within a short period of time (4–6 h). Incubation of a substrate for a short period allows microorganisms to grow to some extent and allows active microorganism groups to act directly on the substrate applied. This whole-soil method discriminates vegetation types and soil treated with wastewater sludge (Campbell et al. 2003). Hence, both methods, although not yet applied in any routine soil assessment programs, do offer promising opportunities for community profiling and subsequent application in soil quality monitoring programmes.

5.3.2 Diversity of Enzyme Activity

Kandeler and Böhm (1996) suggested that the enzyme diversity of a soil provides an effective approach to examine its functional diversity. The responsiveness of enzymes to environmental disturbance makes them a potential indicator of the soil biological quality (Dick 1994). Only extracellular enzymatic activity is used to determine the diversity of enzyme patterns in soil extracts. Activity of ecto- and free- enzymes can be quantified by incubation of the soil extract with commercial fluorogenic enzyme substrates like 4-methylumbelliferin (MUF or MUB) and 4-methylcoumarinyl-7- amide or 7-amino-4-methyl coumarin (MC or AMC) (Kemp et al. 1993; Marx et al. 2001). The use of a microplate fluorometric assay using MUB and AMC to study the enzyme diversity in soils has recently been reported (Marx et al. 2001). Colorimetric substrates like remazol brilliant blue, p-nitrophenol, or tetrazolium salt-coupled specific compounds of interest (e.g., cellulose or phosphate) also can be used to assay functional diversity of microbial communities (Wirth and Wolf 1992). For example, Verchot and Borelli (2005) have reported application of para-nitrophenol (pNP) conjugated with β-glucopyranoside, N-acetyl-β-D-glucosaminide, β-D-cellobioside and phosphate for measuring respective enzyme activity in degraded tropical soils. Measurement of only released pNP derived from all the pNP-linked substrates is the advantage of this method because only one method is employed for all enzyme analysis. The advantages of these assays include the independence of cellular growth and new enzyme synthesis due to a shorter incubation time and thus closer approximation of the in situ function. However, a few dominating organisms expressing high enzyme activity may give a biased result while measuring a diverse set of enzyme activities (Miller et al. 1998).

5.4 The Gene-Chip for Profiling of Structural and Functional Communities of Microorganisms

The development and application of a microarray-based genomic technology for microbial detection and community analysis has received a great deal of attention. Because of its increasingly high-density and high-throughput capacity, it is expected that microarray-based genomic technologies will revolutionize the analysis of microbial community structure, function and dynamics. The basic principle of DNA microarray technology is the identification of an unknown nucleic acid mixture (targets) by hybridization to numerous known diagnostic nucleic acids (probes), which are immobilized in an arrayed order on a miniaturized solid surface (Loy et al. 2006). They are originally developed for the analysis of gene expression in a variety of model organisms, but have great potential in community analysis and in detection of different functional characteristic (Sessitsch et al. 2006). For the first time in 1997, the microarray approach was introduced to environmental microbiology for microbial community composition analysis using a prototype array consisting of nine 16S rRNA-targeted probes for the identification of selected nitrifying bacteria (Guschin et al. 1997). Since then, this field has grown rapidly and today many different microarray systems consisting of more than 30,000 probes (Wilson et al. 2002) are available for detection of target nucleic acids (Taylor et al. 2007; Zhou 2003). Microarrays for microbial community analysis have been classified into two main categories: Phylochips and Functional Gene Arrays (FGAs).

Phylochips contain short nucleotide probes, targeting a phylogenetic marker gene (rRNA genes), and they are usually applied in order to detect specific bacteria such as pathogens (Franke-Whittle et al. 2005) in an ecosystem or to study diversity and structure of microbial communities (Loy et al. 2004; Günther et al. 2005). A functional group of microorganisms with considerable ecological and economic importance in the terrestrial ecosystem is the nitrifying bacteria. Nitrifying bacteria convert ammonium to nitrate by the process of nitrification.

Nitrification measurements are included in the soil-monitoring programme of Austria, Czeck Republic and Germany. A microarray for the nitrifying bacteria (Nitrifier-Phylochip) group (Loy et al. 2006) containing about 200 probes of 18-mer oligonucleotides has enabled extensive monitoring of this functional group. The majority of DNA microarray applications in microbial ecology have focused on determination of community structure based upon phylogenetic markers such as the 16S rRNA gene (Gentry et al. 2006). Although this approach provides powerful and detailed pictures of microbial community structure in complex environmental samples, it generally provides little insight into microbial function.

Functional gene arrays contain DNA probes targeting genes that encode key enzymes conferring a specific functional capability to the respective microorganisms. Some examples of functional enzymes catalyzing different steps in the global nitrogen, sulphur and carbon cycles are nitrite reductase (*nir*S) for denitrification, ammonium monooxygenase (*amo*A) for ammonia oxidation, nitrogenase (*nif*A) for nitrogen fixation (Taroncher-Oldenburg et al. 2003; Wu et al. 2001), dissimilatory bisulphate reductase (*dsr*AB) for sulphate reduction (Wagner et al. 2005) and methane monooxygenase (*pmo*A) for methane oxidation (Bodrossy et al. 2003). The FGAs composed of the formerly cited genes have been mainly developed to understand microbial ecology and biogeochemical cycle of aquatic systems. Functions linked to these identified genes are highly important for soils. These arrays used in aquatic systems may lay the foundation for developing further specific arrays that target soil microorganisms. However, additional probes with increased specificity for soil microorganisms may be required. For example, a 70-mer long oligonucleotide FGA containing *nir*S, *nir*K, *nif*H and *amo*A probes has been used to study change in the community involved in nitrogen cycle in aquatic environment (Taroncher-Oldenburg et al. 2003).

The greatest advantage of FGAs is that microorganism identification is directly linked to potential physiological traits. One of the greatest challenges in using FGAs for detecting functional genes and/or microorganisms in the environment is to design oligonucleotide probes specific to the target genes/microorganisms of interest because sequences of a particular functional gene are highly homologous and/or incomplete, especially in sequences derived from laboratory cloning of environmental samples. Another challenge for using FGAs to study the microbial communities in natural systems is the lack of arrays containing comprehensive probe sets. To tackle these challenges, recently, He et al. (2007a) developed a comprehensive FGA, termed *GeoChip 2.0* version, which contains more than 24,000 oligonucleotide (50-mer) probes covering more than 150 functional groups of 10,000 gene sequences involved in biogeochemical cycling of carbon, nitrogen, phosphorus and sulphur along with metal resistance, metal reduction and organic contaminant degradation (Table 3). Almost all (approximately 98.2%) of the gene sequences were from bacteria whereas the rest (approximately 1.8%) were from fungi. Two major types of applications of the developed *GeoChip* can be visualized. One is to track microbial community dynamics under different environmental/treatment conditions. The developed *GeoChip* has been successfully used to track the changes of the responsible microbial populations during the bioremediation processes of groundwater

Table 3 *Geochip* 2.0 containing number of probes of the functional genes

Gene category	Total gene probes	Percentage (%) of probe target the genes
Nitrogen fixation	1,225	5.0
Denitrification	2,306	9.5
Nitrification	347	1.4
Nitrogen mineralization	1,432	5.9
Carbon fixation	1,018	4.2
Cellulose, lignin and chitin degradation	2,542	11.6
Sulphate reduction	1,615	6.7
Metal reduction and resistance	4,546	18.8
Contaminant degradation	8,028	33.1

Modified from He et al. (2007a)

contaminated with uranium (He et al. 2007a). The other is to use it as a genetic tool for profiling the differences between microbial communities. For this purpose, the *GeoChips* have been used to analyze microbial communities from a variety of habitats, including bioreactors, soils, marine sediments and animal guts. *GeoChip 2.0* also has been employed in ecological applications to detect carbon- and nitrogen-cycle genes that were significantly different across different sample locations and vegetation types of an Antarctic latitudinal transect (Yergeau et al. 2007).

A new generation of the *GeoChip* (v. 3.0) is being developed with several new features compared to *GeoChip* 2.0. *GeoChip* 3.0 is expected to cover >37,000 gene sequences of 290 gene families, allowing access to more information about microbial communities across more diverse environmental samples. It also includes phylogenic markers, such as *gyrB* (the structural gene for the DNA gyrase *b* subunit) and verifies the homology of automatically retrieved sequences by key words using seed sequences so that unrelated sequences are removed. Additionally, a software package (including databases) has been developed for sequence retrieval, probe and array design, probe verification, array construction, array data analysis, information storage, and automatic updates, which greatly facilitate the management of such a complicated array, especially for future updates. Finally, *GeoChip 3.0* also includes *GeoChip 2.0* probes, and those *GeoChip 2.0* probes are checked against new databases for changes in ecosystem management, and environmental cleanup and restoration (He et al. 2007b). In particular, it provides direct linkages of microbial genes/populations to ecosystem processes and functions. All of these results suggest that the developed *GeoChip* is useful for studying various biogeochemical, ecological and environmental processes and associated microbial communities in natural settings in a rapid, high throughput and potentially quantitative fashion. With the developed *GeoChips*, it is possible to address many fundamental and applied research questions in microbial ecology important to human health, agriculture, energy, global climate changes, ecosystem management and environmental cleanup and restoration. Hence, FGAs contain probes from the genes with known biological functions, and they will be useful in linking microbial diversity to ecosystem processes and functions. Due to their ability to connect microbial community analysis to the structural and functional levels, these chips are expected to be a future tool in soil quality monitoring programme for agricultural systems.

5.5 Arbuscular Mycorrhizae Fungi (AMF) Community Structure and Diversity

Arbuscular mycorrhizal (AM) fungi are ubiquitous in nature and constitute an integral component of terrestrial ecosystems, forming symbiotic associations with plant root systems of over 80% of all terrestrial plant species, including many agronomically important species (Smith and Read 1997; Harrier and Watson 2003). Mycorrhizae exist alone or in association with helper rhizospheric bacteria that maintain soil health, and hence, AMF can serve as key species for monitoring soil quality (Jeffries et al. 2003). AMF efficiently deliver soil minerals particularly phosphorus (P) (Sharma and Adholeya 2004) and nitrogen (N) to the plant (Govindarajulu et al. 2005), and in turn the fungi are energized by sugar from the plant (Pennisi 2004). Sharma and Adholeya (2004) reported that mycorrhization of strawberry can save 35 kg^{-1} ha^{-1} of phosphorus fertilizer when compared to non-mycorrhizal strawberry plants grown at a particular P applied level. They play an important role in P uptake and growth of many cereals, legumes and other crop plants (George et al. 1995; Sharma and Sharma 2006). This process of enhancing P absorption by plants appears to be particularly important in highly weathered, fine textured, and acid tropical soils, where great proportions of applied P fertilizer are not available to plants due to strong fixation of P on iron and aluminum oxides (Jama et al. 1997; Bunemann et al. 2004; Sharma et al. 1996). Mycorrhizal associations can also exert a positive influence on plant diversity, stress, disease tolerance, and soil aggregation (Gosling et al. 2006).

Colonization by AMF has been shown to be highly dependent on the presence of host plants, land use and management practices (Kling and Jakobson 1998). Spore abundance and diversity can be distinct between extensively and intensively managed soils (Oehl et al. 2003). AMF diversity has been reported to be sensitive to heavy metal contamination, organic pollutant and atmospheric deposition. (Egeston-Warburton and Allen 2000; Egli and Mozafar 2001). Furthermore, nitrogen enrichment induces a shift in AM community composition. In particular, an increasing input of nitrogen was associated with the displacement of the larger-spore species of *Scutellospora* and *Gigaspora* (due to a failure to sporulate) with a concomitant proliferation of small-spore *Glomus* species (e.g., *G. aggregatum, G. leptotichum*). Such changes also indicated that AMF species are sensitive indicators of nitrogen enrichment (Egeston-Warburton and Allen 2000). Abundance and diversity of AMF is determined by extraction of spores from soil samples and subsequent counting in a microscope (Oehl et al. 2003). Thus, spores build up in soil and plant root colonization by AMF has been proposed as an important indicator of plant and soil ecosystem health (van der Heijden et al. 1998; Stenberg 1999; Oehl et al. 2003). More than 150 species have been described within the phylum Glomeromycota on the basis of their spore development and morphology, although recent molecular analyses indicate that the number of AMF taxa may be much higher (Daniell et al. 2001; Vandenkoornhuyse et al. 2002). Finally, methods for direct detection and quantification of AMF in soil samples or roots include 18S rRNA gene PCR (Chelius and Triplett 1999), and nested PCR at the species level (Jacquot et al. 2000). Quantitative analysis of the density of a particular AMF based on spore morphology has been implemented as a microbial indicator in the Swiss soil quality-monitoring network for ascertaining the heavy metal contamination (Egli and Mozafar 2001).

6 Minimum Data Set (MDS) of Microbial Indicators

Given the large number of soil microbial characteristics that can be measured as indices of the soil quality, the question is: what is the 'minimum data set', i.e. a set of specific soil measurements considered as the basic requirement for assessing the soil quality (Doran and Parkins 1996). Microbial indicators of MDS must be (i) compatible with basic ecosystem processes in soil as well as physical or chemical indicators of soil health, (ii) sensitive to management in acceptable time frames, (iii) easy to assess or measure, (iv) composed of robust methodology with standardized sampling techniques, (v) cost-effective, and (vi) relevant to human goals, food security, agricultural production, sustainability and economic efficiency (Bunning and Jimenez 2003)

Many countries have developed their own MDS of microbiological indicators for monitoring of soil quality (Table 4) where microbial biomass and soil respiration are the most commonly used indicators. However, some of the recent tools such as Biolog, PLFA, DGGE/TGGE etc have been recommended for microbial diversity assessment to monitor soil quality in certain countries (Chapman et al. 2000, Winding et al. 2005). In India, no 'minimum data set' for monitoring soil quality has been recommended but soil enzyme activities, respiration and microbial biomass are being used widely (Ramesh et al. 2004a, b; Rao et al. 1995; Sharma et al. 2005). Therefore, to help improve soil management and retain or remediate soil quality throughout India, a MDS to assess quality of agricultural soils within the country should be developed. This would provide a structured approach that could be followed to determine if the soils have deteriorated or are deteriorating in terms of soil productivity or other critical soil functions. Then, based on the magnitude of soil deterioration, stakeholders can develop better practices to manage their soils and increase the sustainability of their agricultural practices. Assuming that soil microbial diversity and function are linked, the progress of soil rehabilitation could later be evaluated by the resultant microbial diversity and various functional attributes.

Table 4 Minimum data set (MDS) of microbial indicators for soil quality monitoring as defined in different countries

	Country	Microbial indicators
1.	United Kingdom[a]	Microbial biomass; soil respiration; microbial diversity by Biolog; *Rhizobium* population; biosensor bacteria
2.	United States of America[a]	Microbial biomass; potential N-mineralization; soil respiration; soil enzymes
3.	Germany[a]	Soil respiration; microbial biomass; potential N-mineralization; soil enzymes; metabolic quotients
4.	The Netherlands[a]	Microbial biomass; potential C-mineralization; potential N-mineralization; microbial diversity by Biolog and DGGE
5.	Switzerland[b]	Microbial biomass; soil respiration; potential N-mineralization; arbuscular mycorrhizae
6.	Czech Republic[a]	Microbial biomass; soil respiration; nitrification; N-mineralization; soil enzymes
7.	Russia, Sweden[a], Finland	Soil respiration, soil enzymes; potential N-mineralization
8.	Austria[a]	Microbial biomass; soil enzyme; nitrification; mycorrhizae
9.	India[c]	Soil enzymes; soil respiration; microbial biomass

Modified from [a]Winding et al. 2005; [b]Mader et al. (2002); [c]Ramesh et al. (2004a, b); [c]Rao et al. (1995); [c]Sharma et al. (2005)

7 Agricultural Strategies to Improve Soil Quality by Managing Microbial Communities and Diversity

In recent years, agricultural practices that improve soil quality and agricultural sustainability have received increased attention from researchers and growers. An understanding of soil processes is key to estimate the influence of farming practices on the fertility and quality status of the soil, and thus, on the environment. Species diversity can give rise to ecosystem stability through the ability of the species or functional groups it contains to respond differentially and in compensatory fashion to perturbations in the soil environment (Sturz and Christie 2003). Shifts in bacterial community structure or diversity and associated physiological responses can be used as indicators of these perturbations or disturbances in agroecosystems (Calderón et al. 2001), although some results cast some doubts on the strength of biodiversity-ecosytem functioning relationships as observed for macroorganisms when extrapolated to bacteria (Griffiths 2000; Wertz et al. 2006; 2007). Changes in the community structure have been caused by changes in agronomic practices such as types of amendment (Kennedy et al. 2004; Marschner et al. 2004), reduced or no-tillage (Drijber et al. 2000), crop rotations (Lupwayi et al. 1998) and microbial inoculation (Roesti et al 2006; Srivastava et al. 2007). Some even suggest that it is appropriate to adopt agricultural practices that preserve and restore microbial diversity than practices that destroy it (Lupwayi et al. 1998). In addition, the soil microbial

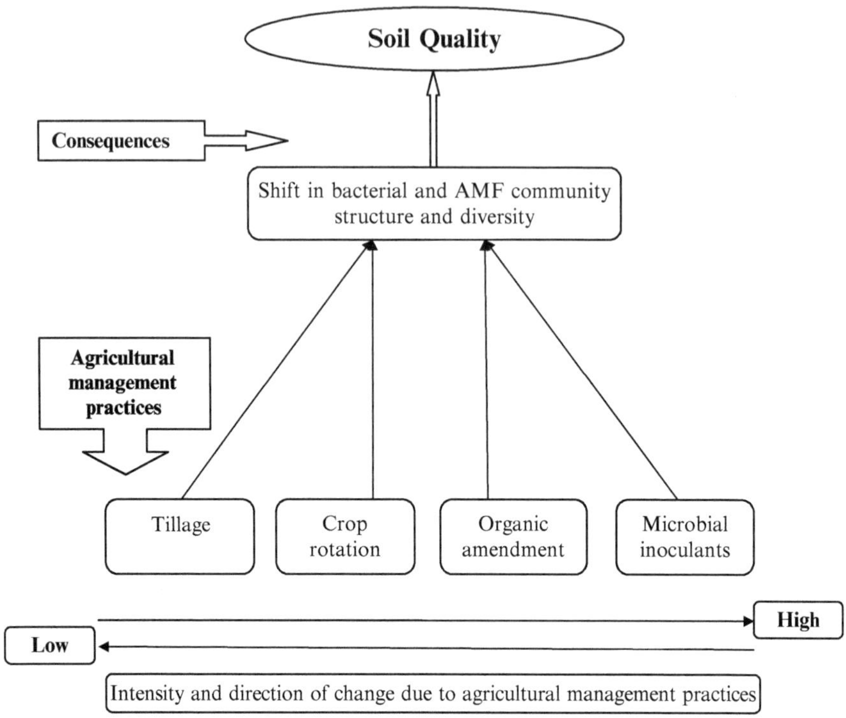

Fig. 3 An overview of agricultural management practices on soil microbial community structure and diversity and their possible influence on soil quality

community may serve as a fingerprint associated with certain land use practices, soil conditions, and associated function, suggesting that achievement of such a fingerprint and its associated soil characteristics may be gained through adoption of a suitable site-specific suite of agricultural practices (Steenwerth et al. 2003, 2005). In this section, we address impacts of agricultural practices on soil microbial communities, with a primary focus on arbuscular mycorrhizal fungi due to their readily defined functions in agricultural systems (Fig. 3).

7.1 Tillage

Studies on tillage indicate that many critical soil quality indicators and functions can be improved by decreasing tillage intensity (Jackson et al. 2003; Karlen 2004; Govaerts et al. 2006; Govaerts et al. 2007b). Recent approaches aim to reduce excessive cultivation in favor of limited or more strategic tillage practices. Such practices are grouped under the term conservation or reduced tillage as opposed to conventional tillage (Carter 1994). Compared to conventional tillage, reduced tillage practices offer not only long term benefits to soil stability, reduce erosion, but also enhance soil microbial diversity (Davis et al. 2002; Phatak et al. 2002; Welbaum et al. 2004; Govaerts et al. 2008). Conventional tillage promotes a vicious cycle of soil aggregate disruption and reconsolidation that result in denser soils with the loss of organic matter (Six et al. 1999). Soils in humid thermic regimes are more sensitive to degradation from repeated tillage due to loss of soil organic matter and erosion (Lowrance and Williams 1988; Langdale et al. 1992). Thus, they are better candidates for adoption of no-tillage systems. It is also well-known that no-till practices combined with crop residue retention increase soil organic matter content in the surface layer, improve soil aggregation, and preserve the soil resources better than conventional till practices (Govaerts et al. 2006, 2007c). Increased soil organic matter content associated with no-till practices not only improves soil structure and water retention, but also serves as a nutrient reservoir for plant growth and a substrate for soil microorganisms.

In terms of functional diversity, tillage influences microbial communities in many different ways. It causes a physical disruption of AMF fungal mycelia and may change physico-chemical properties of the soil (Evans and Miller 1990, Kabir, 2005). Soil disturbances reduce the density of AMF spores, species richness and the length of extraradical mycelium of AMF relative to undisturbed soil (Kabir 2005; Boddington and Dodd 2000). Conversely, no-tillage often stimulates mycorrhizal activity in soil, thereby influencing nutrient uptake by plants (Dodd 2000). Glomalin, an exudation product of AMF hyphae having a role in soil aggregation, was 1.5 times higher in no-till than tilled soils (Wright et al. 1999). Based on spore morphology and sequencing of ITS rDNA, at least 17 AMF species were identified including five genera (*Glomus*, *Gigaspora*, *Scutellospora*, *Aculospora* and *Entrophospora*) from soils exposed to different tillage practices (Jansa et al. 2002). Under reduced tillage, the incidence of certain AMF in agricultural soils increased, excluding *Glomus* spp. In contrast, *Glomus* species (G. *mosseae, G. claroideum, G. caledonium, G. constrictum, G. clarum*-like) were predominant under conventional tillage.

Giller (1996) and Lupwayi et al. (1998) suggested that the diversity in microbial communities provoked by tillage resulted from a reduction in substrate richness and microbial uniformity under conventional tillage (CT). In terms of functional diversity, as measured with the Biolog method, soils under reduced tillage (RT) had a higher average well color development (AWCD) and a higher Shannon's Diversity index (H) compared to those under CT. This confirmed the adverse effect of intense tillage on microbial diversity (Diosma et al. 2006). Giller (1996) suggested that soil microbial diversity could be reduced by such disturbances as desiccation, mechanical destruction, soil compaction, reduced pore volume and food resources and/or access to them. For example, in semi-arid highland of Mexico, soil microbes under zero tillage and crop residue addition resulted in higher AWCD than those under zero tillage without residue addition. This suggested that zero tillage, in the absence of crop residue retention, is an unsustainable practice that may lead to poor soil health (Govaerts et al. 2007a). Different tillage intensities can also select for specific dominant microbial populations within the soil bacterial community, as depicted by 16S rRNA and *rpoB* genes using DGGE (Peixoto et al. 2006).

7.2 Crop Rotation

Crop rotation is a very ancient cultural practice (Howard 1996) that has a strong influence on soil structure, organic matter, and microbial communities (Janvier et al. 2007). Traditionally, it has been used primarily to disrupt disease cycles (Curl 1963) and fix atmospheric nitrogen by legumes for subsequent non-leguminous crops (Pierce and Rice 1998). Crop rotation can cause changes in substrate utilization patterns, suggesting that soil bacterial communities under crop rotation have greater species diversity than under continuous wheat (*Triticum aestivum* L.) or summer fallow (Lupwayi et al. 1998). For instance, functional diversity measured by the Biolog method increased in soil under wheat/maize rotation with crop residue addition as compared to that exposed to a monoculture of maize alone (Govaerts et al. 2007a).

Relatively limited work has been executed to characterize mycorrhizal community composition and diversity, which is crucial in furthering our understanding of mycorrhizal functions in agro-ecosystems (Johnson and Pfleger 1992). Traditionally, AMF communities in field soil employ spore surveying, which is sometimes complemented by trap culturing (Douds et al. 1993; Jansa et al. 2002; Oehl et al. 2004). These spore-based surveys are considered as a baseline to assess the impact of agricultural practices on AMF communities (Douds and Millner1999). However, it has become clear that morphological characterization of the AMF spore community and its diversity might not reflect the actual functional symbiosis that refers to active fungal structures within and outside roots (Clapp et al. 1995; Jansa et al. 2003).

It has also been shown that introduction of leguminous crops for a season into a conventional system of continuous cultivation of maize (*Zea mays* L.) increased microbial diversity (Bunemann et al. 2004; Bossio et al. 2005). In a Kenyan ferrasol, the species diversity of AMF spores was neither affected by crop rotation nor by P fertilization. However, the composition of AMF spore communities was significantly affected by crop rotation (Mathimaran et al. 2007). Johnson et al. (1992) found that maize had higher yield and nutrient uptake on soils that had previously cultivated continuously with soybean (*Glycine max* (L.) Merrill) for 5 years than on soil that had grown continuously with maize for the 5 years. Conversely, soybean had both lower yields and nutrient uptake on soil supporting 5 years of continuous soybean as compared to its increased growth on soils exposed to 5 years of continuous maize. The most abundant AMF species in the continuous maize soil was negatively correlated with maize yield, but positively correlated with soybean yield; there was a similar effect with soybean soil. This yield decline after continuous cropping of soybean and maize is attributed to selection of AMF species which grow and sporulate most rapidly and these AMF species offer the least benefit to the respective monocrop because they divert more resources to their own growth and reproduction and the mycorrhizal group acts as 'resource cheater'. The non-specific association of mycorrhizae with monocropping was further confirmed by Bever (2002) who demonstrated a negative feedback between AMF and plants. A substantial part of soil microbial communities belongs to the AMF (Leake et al. 2004). Agricultural management practices affect AMF communities both qualitatively and quantitatively (Sieverding 1990; Miller et al. 1995). This has been documented in many studies showing that crop rotation, fertilization, and tillage affect the composition and diversity of AMF communities as well as spore and mycelium densities in temperate and tropical agro-ecosystems (Sieverding 1990; Jansa et al. 2002; Oehl et al. 2003). However, active structures such as fungal hyphae and arbuscules in the roots and the soil can only be properly identified by means of molecular or immunological approaches (Treseder and Allen 2002; Redecker et al. 2003; Sanders 2004), which may require calibration for each specific field site (Jansa et al. 2003).

7.3 Organic Amendments

Organic amendments cover a wide range of inputs, including animal manure, solid waste, and various composts, and often improve soil quality and productivity. Girvan et al. (2004) and Melero et al. (2006) showed that these amendments, as well as crop residues, resulted in significant increases in total organic carbon (TOC), Kjeldahl-N, available-P, soil respiration, microbial biomass, and enzyme activities (e.g., protease, urease, and alkaline phosphatase). Microbial diversity and crop yields also increased as compared to conventional management. Applying cattle manure increased the amount of readily available organic C and mineral nutrients. This improved soil structure and promoted growth of both r-strategists (fast growing microorganisms having high reproductive capacity and successful only in resource-rich environment) and K-strategists (slow growing microorganisms having slow reproductive capacity and successful in resource-limited situations) while chemical fertilizers enriched the K-strategists bacterial community. As a result, the richness, evenness and diversity of the microbial community in manure-treated soil were enhanced and were positively correlated with soil productivity (Parham et al. 2003). In the Netherlands, organically managed soils had also shown higher biological diversity in both nematodes and eubacteria (van Diepeningen et al. 2006). In another long-term experiment comparing organic and synthetic soil fertility amendments, cotton (*Gossypium hiristium* L.) gin trash application was found to maintain significantly higher bacterial community diversity as assessed by CLPP and DGGE analyses compared to synthetic fertilizer (Liu et al. 2007).

Organic amendments do not always elicit a shift in bacterial diversity. Srivastava et al. (2007) reported that the incorporation of okra (*Hibiscus esculetus* L.), pea (*Pisum sativum* L.) and cowpea (*Vigna unguiculata (L.) Walp.*) residues significantly increased fungal activity but bacterial community composition as revealed by DGGE analysis remained the same. Furthermore, in semiarid conditions of southern Italy, the composition of diversity of total bacteria as well as ammonium oxidizers exhibited no significant change after incorporation of crop residues in soil under monoculture of durum wheat. However, a change was detected after applying nitrogenous fertilizer (Crecchio et al. 2004, 2007). In both cases, despite a lack of change in microbial community, soil fertility was found to be high. In contrast, Saison et al. (2006) detected clear changes in the bacterial community structure after amendment of rape compost in sandy soil.

Oehl et al. (2004), using a soil from Switzerland, found that *Glomus* spp. were similarly abundant whether fertilized with mineral or organic fertilizers, but spores of *Acaulospora* and *Scutellospora* spp. were more abundant in soil that only received organic fertilizers. Spore dominance of two genera viz., *Gigaspora* sp. and *Glomus* sp. was recorded in a rehabilitated site where *Gigaspora* genera showed a strong positive correlation with organic carbon content (Gaur et al. 1998). Different forms of organic fertilizers also differentially affected AMF communities in other studies. For example, addition of leaf compost combined with either chicken litter (poultry) or cow (*Bovine* spp.) manure, enhanced spore populations of some AMF species (*Glomus etunicatum* and *G. mosseae*) relative to those found in soils fertilized with raw dairy-cow manure or with mineral fertilizer (Douds et al. 1997). Supplementation of organic amendments continuously for 10 years in corn, soybean and citrus did not show any AMF diversity and richness (Franke-Snyder et al. 2001). Nevertheless, the organic apple orchard had the highest AMF richness, even though sporulation and the Shannon diversity index were higher for the conventional orchard (Purin et al. 2006). Broadly, organic amendments application in soil either increased or do not affect microbial diversity but improve soil quality and crop productivity.

7.4 Microbial Inoculation

Inoculation of microbial inoculants is generally being done to improve soil fertility and crop productivity through various microbe-mediated mechnaisms. The introduction of microbial inoculant to soils either through seed bacterization or direct application results in a disturbance of the rhizosphere's biological equilibrium. Significant increases in soil enzymes such as α-galactosidase, β-galactosidase, α-glucosidase and β-glucosidase, chitibiosidase and urease and a decrease in alkaline phospahatese activity has been observed upon soil perturbation through separate inoculation of *Flavobacter* spp and *Pseudomonas fluorescens* (Mawdsley and Burns 1994; Naseby and Lynch 1997). De Leij et al. (1995) reported only transient perturbations in the indigenous microbiota in the rhizosphere with the introduction of wild and genetically modified- *P. fluorescens* in rhizosphere of wheat. A minor impact on the composition of the microbial rhizosphere community of *Medicago sativa* and *Chenopodium album* was reported after inoculation with *Sinorhizobium meliloti* L33 (Meithling et al. 2000; Schwieger and Tebbe 2000). The inoculation of plant growth promoting rhizobacteria (PGPR) alone through seed bacterization or soil application near seeds may cause negligible to extreme shifts in microbial community composition (Nacamulli et al. 1997; Lottmann et al. 2000; Marschner et al. 2001a,b; Kozdroj et al. 2004). Furthermore, inoculation of PGPR and AMF alone or in combination in soils supporting wheat and vegetable crops modified the bacterial community (Roesti et al. 2006; Srivastava et al. 2007).They also showed that an increase in crop yield occurred, further suggesting that inoculations with selected beneficial microorganisms can enhance crop yield.

AM symbiosis influences the structure and function of surrounding bacterial communities (Marschner and Timonen 2005). In canola, which is considered to be a non-mycorrhizal species, inoculation with *Glomus intraradices* increased the shoot dry weight compared to *G. versiforme* and the non-mycorrhizal control plants and also induced changes in the bacterial community composition in the rhizosphere, as analyzed by DGGE. Surprisingly, less than 8% of the canola's root length was colonized. In contrast, although 50% of the clover's root length was colonized and inoculation with *G. versiforme* resulted in a higher shoot dry weight compared to *G. intraradices* or the control plants, no change in the rhizosphere bacterial community composition was recorded (Marschner and Timonen 2005). In another study, inoculation of *G. intraradices* induced greater plant growth in autoclaved soil than in non-autoclaved soil, an effect that was positively related to inoculum density (Dabire et al. 2007). Catabolic evenness and richness were positively correlated with the number of inoculated AM propagules in the autoclaved soil, but negatively correlated in the non-autoclaved soil. In non-autoclaved soil, application of G. *intraradices* inoculum induced disequilibria in microbial functionalities. Hence, it was suggested that AM inoculation of the non-autoclaved soil increased the susceptibility of soil microflora to stress and disturbance (Degens et al. 2001). Although *G. intraradices* inoculation had stimulated plant growth, this fungal inoculant had not improved soil microbial diversity. In addition, after soil autoclaving and AM inoculation, catabolic evenness and richness were significantly higher than in the control (non-inoculated soil) and in the non-disinfected (non-autoclaved) soil without AM inoculation (Dabire et al. 2007).

It has also been reported that rhizosphere soils contain a higher proportion of culturable bacteria that were r-strategists and were, therefore able to respond and multiply quickly in presence of available nutrients (Sarathchandra et al. 1997). An increase in the number of AM propagules in autoclaved soil enhanced catabolic diversity by stimulating the growth of contaminant r-strategist microorganisms (*e.g. Bacillus* spp., *Pseudomonas)* that were probably received from water used during irrigation. However, *G. intraradices* inoculation in non-autoclaved soil resulted in decreased soil microbial catabolic diversity by inhibiting the growth of r-strategist microorganisms. It has also been reported that populations of total culturable bacteria decreased in mycorrhizal-inoculated rhizospheres (Vazquez et al. 2000). Hence, it seems that AMF inoculation not only favors plant growth but also microbiological activities contributing to soil quality.

8 Future Perspectives

The importance of microorganisms in soil functions by mediating various processes for nutrient cycling has long been acknowledged, underscoring the importance of understanding microbial diversity and associated functions for sustainable agriculture. The indices accounting for microbial diversity or community structure in soil are considered to be of immense significance to manage soil quality in order to sustain productivity of crops. Agricultural systems that sustain or enhance soil quality through creation of higher biodiversity can provide sustainability to production of crops and often partially substitute for nutrition that is currently being managed through external chemical inputs.

So far consensus exists only to a limited extent on the importance of utilizing microbial biodiversity indices as a measure of soil quality. Currently, several different indices for microbial diversity have been established, but it is very difficult to know which index is most suitable for a given situation. To address this question, more in depth research is needed to determine the most suitable microbial indicators and how they should be interpreted. Moreover, while we have documented shifts in soil microbial communities in response to various conditions, we have yet to determine if depressions in microbial diversity begets a shift in soil quality or if the extent of microbial functional redundancy is so great that the link between microbial diversity and function is weak, and whether these relationships shift in response to different disturbance intensities. The related challenge is to develop quantitative relationships between any apparent functional redundancy and genetic diversity. Despite the debate that functional redundancy is not commonly existent, exceptions may occur. For example, it was once thought that AMF were functionally redundant given a lack of host specificity, but it has been demonstrated that AMF provide different benefits to the different plant hosts.

Another important concern regarding assessment of microbial diversity for soil quality monitoring are the issues of sampling, sample preparation, and handling for analyses. Most microbial analyses are very sensitive to water content and temperature, thus sample collection and storage become major barriers for many analyses. As new protocols are developed to include microbial diversity in the suite of soil quality indicators, specific guidelines for sample collection and preservation must be standardized to facilitate comparisons among independent studies. Should samples be stored moist at a low temperature, processed immediately, or air dried before analysis? Will it be more efficient and cost effective to pursue enzymatic measures or genetic profiles considering that commercial soil-testing laboratories will ultimately be called upon for soil quality assessments? These are difficult questions to be answered because they involve the human element and bias, but must be addressed before microbial indicators will be as easily incorporated into soil quality assessment as current physical and chemical indicators.

9 Conclusion

The diversity of microorganisms in agro-ecosystems is immense but critical to maintaining 'good' soil quality because they are involved in so many important soil processes. With an increased number of monitoring programs utilizing microbial diversity, composition and function as an indicator for evaluating soil quality, comprehensive comparisons among geographic zones and cropping systems will both strengthen our understanding of links between microbial diversity and function, and develop regional standards for microbial fingerprints associated with 'good' soil quality. The judicious use of biological inputs including inoculums, manures, cover crops, and plant residues is recognized as a practical way to promote a healthy soil and support sustainable crop production. To guide the use of these materials and to achieve the desired change in microbial communities, the first step will be to confirm critical linkages between specific soil microbial groups and critical soil functions. Then, baseline parameters such as soil respiration, organic carbon pool, and soil enzymes that are routinely utilized worldwide for monitoring soil quality can be incorporated into an overall soil health assessment programme. However, we advise that factors that highly influence soil microbial community composition e.g. soil pH, texture, and water content be incorporated into the monitoring programs to avoid false conclusions regarding association between microbial diversity, function and soil quality. Although recent advances in molecular techniques for analysis of soil microorganisms have occurred, we emphasize that exclusive use of a single technique in a monitoring program may provide biased and distorted interpretations of microbial diversity, emphasizing the importance of establishing common standards among soil monitoring programs. Likewise, coordination among independent research programs to develop a minimum common database (MCD) and methods standardization, such as has been demonstrated by the National Ecological Observatory Network in the United States, would facilitate greater understanding of microbial diversity, function, and soil quality, and increase its accessibility to both growers and policy makers across a broader geographic scale.

Acknowledgements We thank Dr. Xavier Le Roux, INRA, CNRS, Université de Lyon, Université Lyon 1, UMR 5557 Ecologie Microbienne, Villeurbanne, France, for his critical comments and corrections of earlier draft of this paper

References

Acinas S, Rodriguez-Valera R, Pedros-Alio C (1997) Spatial and temporal variation in marine bacterioplankton diversity as shown by RFLP fingerprinting of PCR amplified 16S rDNA. FEMS Microbiol Ecol 24:27–40

Acosta-Martinez V, Zobeck TM, Gill TE, Kennedy AC (2003) Enzyme activities and microbial community structure in semiarid agricultural soils. Biol Fert Soils 38:216–227

Acosta-Martinez V, Mikh MM, Vigil MF (2007) Microbial communities and enzyme activities in soils under alternative crop rotations compared to wheat-fallow for the Central Great Plains. Appl Soil Ecol 37:41–52

Allison SD, Martiny BH (2008) Resistance, resilience, and redundancy in microbial communities. Proc Nat Acad Sci USA 150:11512–11519

Alvey S, Yang CH, Buerkert A, Crowley DE (2003) Cereal/legume rotation effects on rhizosphere bacterial community structure in West African soils. Biol Fert Soil 37:73–82

Amann R, Ludwig W, Schleifer KH (1995) Phylogenetic identification and in situ detection of individual microbial cells without cultivation. Microbiol Rev 59:143–149

Anderson TH (2003) Microbial eco-physiological indicators to assess soil quality. Agric Ecosys Environ 98:285–293

Aseri GK, Tarafdar JC (2006) Fluorescien diacetate: a potential biological indicator for arid soils. Arid Land Res Manage 20:87–99

Avidano L, Gamalero E, Cossa GP, Carraro E (2005) Characterization of soil health in an Italian polluted site by using microorganisms as bio-indicators. Appl Soil Ecol 30:21–33
Bakker PAHM, Glandorf DCM, Viebahn M, Ouwens TWM, Smit E, Leeflang P, Wernars K, Thomashow LS, Thomas-Oates JE, van Loon LC (2002) Effects of *Pseudomonas putida* modified to produce phenazine-1-carboxylic acid and 2, 4-diacetylphloroglucinol on the microflora of field grown wheat. Anton van Leeuwenhoek 81:617–624
Barrios E, Delve RJ, Bekunda M, Mowo J, Agunda J, Ramisch J, Trejo MT, Thomas RJ (2006) Indicator of soil quality: a south-south development of a methodological guide for liking local and technical knowledge. Geoderma 135:248–259
Bej AK, Perlin M, Atlas RM (1991) Effect of introducing genetically engineered microorganisms on soil microbial community diversity. FEMS Microbiol Ecol 86:169–176
Bending GD, Turner MK, Rayns F, Marx MC, Wood M (2004) Microbial and biochemical soil quality indicators and their potential for differentiating areas under contrasting agricultural management regimes. Soil Biol Biochem 36:1785–1792
Bever JD (2002) Negative feedback within a mutualism: host-specific growth of mycorrhizal fungi reduces plant benefit. Proc Royal Soc London 269:2595–2601
Bloem J, de Ruiter P, Bouwman LA (1994) Soil food webs and nutrient cycling in agroecosystems. In: van Elsas JD, Trevors JT, Wellington HME (eds) Modern Soil Microbiology. Marcel Dekker, New York, pp 245–278
Bloem J, Breure AM (2003) Microbial indicators. In: Markert BA, Breure AM, Zechmeister HG (eds) Bioindicators and iomonitors. Elsevier, Oxford, pp 259–282
Boddington CL, Dodd JC (2000) The effect of agricultural practices on the development of indigenous arbuscular mycorrhizal fungi. I. Field studies in an Indonesian ultisol. Plant Soil 218:137–144
Bodrossy L, Stralis-Pavese N, Murrell JC, Radajewski S, Weilharter A, Sessitsch A (2003) Development and validation of a diagnostic microbial microarray for methanotrophs. Environ Microbiol 5:566–582
Boivin MEY, Breure AM, Posthuma L, Rutgers M (2002) Determination of field effects of contaminants: significance of pollution-induced community tolerance. Human Ecol Risk Assess 8:1035–1055
Bossio DA, Girvan MS, Verchot L, Bullimore J, Borelli T, Albrecht A, Scow KM, Ball AS, Pretty JN, Osborn AM (2005) Soil microbial community response to land use change in an agricultural landscape of western Kenya. Microb Ecol 49:50–62
Bossio DA, Scow KM, Gunapala N, Graham KJ (1998) Determinants of soil microbial communities: effects of agricultural management, season and soil type on phospholipid fatty acid profiles. Microb Ecol 36:1–12
Bressan M, Mougel C, Dequiedt S (2008) Response of soil bacterial community structure to successive perturbations of different types and intensities. Environ Microbiol 10:2184–2187
Buckley DH, Schmidt TM (2001) The structure of microbial communities in soil and the lasting impact of cultivation. Microb Ecol 42:11–21
Bunemann EK, Bossio DA, Smithson PC, Frossard E, Oberson A (2004) Microbial community composition and substrate use in a highly weathered soil as affected by crop rotation and P fertilization. Soil Biol Biochem 36:889–901
Bunning S, Jimenez JJ (2003) Indicators and assessment of soil biodiversity/soil ecosystem functioning for farmers and governments. Presented in OECD expert meeting on soil erosion and biodiversity, Rome, Italy, 25–28 March
Calderón FJ, Jackson LE, Scow KM, Rolston DE (2001) Short-term dynamics of nitrogen, microbial activity, and phospholipid fatty acids after tillage. Soil Sci Soc Am J 65:118–126
Campbell CD, Chapman SJ, Cameron CM, Davidson MS, Potts JM (2003) A rapid microtitre plate method to measure carbon dioxide evolved from carbon substrate amendments so as to determine the physiological profiles of soil microbial communities by using whole soil. Appl Environ Microbiol 69:3593–3599
Carter MR (1994) Conservation tillage in temperate agro-ecosystems. Lewis Publishers/CRC Press, Boca Raton, FL, 390

Casamayor EO, Schafer H, Baneras L, Pedros-Alio C, Muyzer G (2000) Identification of and spatio-temporal differences between microbial assemblages from two neighboring sulfurous lakes: comparison by microscopy and denaturing gradient gel electrophoresis. Appl Environ Microbiol 66:499–508

Chang YJ, Hussain AKMA, Stephen JR, Mullen MD, White DC, Peacock A (2001) Impact of herbicides on the abundance and structure of beta-subgroup ammonia oxidizer communities in soil microcosms. Environ Toxicol Chem 20:2462–2468

Chapman SJ, Campbell CD, Edwards AC, McHenery JG (2000) Assessment of the potential of new biotechnology environmental monitoring techniques. Report no. SR (99) 10F, Macaulay Research and Consultancy Services, Aberdeen

Chelius MK, Triplett EW (1999) Rapid detection of arbuscular mycorrhizae in roots and soil of an intensively managed turfgrass system by PCR amplification of small subunit rDNA. Mycorrhiza 9:61–64

Choi KH, Dobbs FC (1999) Comparison of two kinds of Biolog microplates (GN and ECO) in their ability to distinguish among aquatic microbial communities. J Microbiol Method 36:203–213

Clapp JP, Young JPW, Merryweather J, Fitter AH (1995) Diversity of fungal symbionts in arbuscular mycorrhizas from a natural communitiy. New Phytol 130:259–265

Classen AT, Boyle SI, Haskins KE, Overby ST, Hart SC (2003) Community-level physiological profiles of bacteria and fungi: plate type and incubation temperature influences on contrasting soils. FEMS Microbiol Ecol 44:319–328

Crecchio C, Gelsomino A, Ambrosoli R, Minati JL, Ruggiero P (2004) Functional and molecular responses of soil microbial communities under differing soil management practices. Soil Biol Biochem 36:1873–1883

Crecchio C, Curci M, Pellegrino A, Ricciuti P, Tursi N, Ruggiero P (2007) Soil microbial dynamics and genetic diversity in soil under monoculture wheat grown in different long-term management systems. Soil Biol Biochem 39:1391–1400

Conn VM, Franco CMM (2004) Effects of microbial inoculants on the indigenous actinobacterial endophyte population in the roots of wheat as determined by terminal restriction fragment length polymorphism. Appl Environ Microbiol 70:6407–6413

Culman SW, Duxbury JM, Lauren JG, Theis JE (2006) Microbial community response to soil solarization in Nepal's rice-wheat cropping system. Soil Biol Biochem 38:3359–3371

Curl E (1963) Control of plant diseases by crop rotation. Botan Rev 29:413–479

Dabire AP, Hien V, Kisa M, Bilgo A, Sangare KS, Plenchette C, Galiana A, Prin Y, Duponnois R (2007) Responses of soil microbial catabolic diversity to arbuscular mycorrhizal inoculation and soil disinfection. Mycorrhiza 17:537–545

Dalgaard T, Hutchings NJ, Porter JR (2003) Agroecology, scaling and interdisciplinarity. Agric Ecosyst Environ 100:39–51

Dang MV (2007) Qualitative and quantitative soil quality assessments of tea enterprises in Northern Vietnam. Afr J Agric Res 2:455–462

Daniell TJ, Husband R, Fitter AH, Young JPW (2001) Molecular diversity of arbuscular mycorrhizal fungi colonising arable crops. FEMS Microbiol Ecol 36:203–209

Davis LC, Castro-Diaz S, Zhang O, Erickson LE (2002) Benefits of vegetation for soils with organic contaminants. Crit Rev Plant Sci 21:457–491

Degens BP (1998) Decreases in microbial functional diversity do not result in corresponding changes in decomposition under different moisture conditions. Soil Biol Biochem 30:1989–2000

Degens BP, Harris JA (1997) Development of a physiological approach to measuring the catabolic diversity of soil microbial communities. Soil Biol Biochem 29:1309–1320

Degens BP, Schipper LA, Sparling GP, Duncan LC (2001) Is the microbial community in a soil with reduced catabolic diversity less resistant to stress or disturbance. Soil Biol Biochem 33:1143–1153

Degens BP, Schipper LA, Sparling GP, Vojvodic-Vukovic M (2000) Decreases in organic C reserves in soil can reduce the catabolic diversity of soil microbial communities. Soil Biol Biochem 32:189–196

De Leij FAAM, Sutton EJ, Whipps JM, Fenlon JS, Lynch JM (1995) Impact of field release of genetically modified *Pseudomonas fluorescens* on indigenous microbial populations of wheat. Appl Environ Microbiol 61:3443–3453

Dick RP (1994) Soil enzyme activities as indicators of soil quality. In: Doran JW, Leman DC, Bezdicek DF, Stewart BA, SSSA (eds) Defining soil quality for a sustainable environment. Proceedings of a symposium, Minneapolis, MN, 4–5 November 1992, pp 107–124

Diosma G, Aulicino M, Chidichimo H, Balatti PA (2006) Effect of tillage and N fertilization on microbial physiological profile of soils cultivated with wheat. Soil Till Res 91:236–243

Dodd JC (2000) The role of arbuscular mycorrhizal fungi in agro-and natural ecosystems. Out Agric 29:55–62

Doran JW, Safely M (1997) Defining and assessing soil health and sustainable productivity. In: Pinehurst C, Doube BM, Gupta VVSR (eds) Biological indicators of soil health. CABI, Wallingford, Oxon, pp 1–28

Doran JW, Coleman DF, Bezdick DF, Stewart BA (eds) (1994) Defining soil quality for sustainable environment. SSSA Special publication number 35. Soil Science Society of America. Inc. and America Society of Agronomy, Inc. Madison, Wisconsin

Doran JW, Jones AJ (1996) Methods for assessing soil quality. SSSA Special publication number 49. Soil Science Society of America, Madison, Wisconsin, p 410

Doran JW, Parkins TB (1996) Quantitative indicators of soil quality: a minimum data set. In: Doran JW, Jones AJ (eds) Methods for assessing soil quality. Soil Science Society of America, Madison, WI

Douds DD, Galvez L, Franke-Snyder M, Reider C, Drinkwar LE (1997) Effect of compost addition and crop rotation point upon VAM fungi. Agric Ecosyst Environ 65:257–266

Douds DD, Janke RR, Peters SE (1993) VAM fungus spore populations and colonization of roots of maize and soybean under conventional and low-input sustainable agriculture. Agric Ecosyst Environ 43:325–335

Douds DD, Millner P (1999) Biodiversity of arbuscular mycorrhizal fungi in agroecosystems. Agric Ecosyst Environ 74:77–93

Domsch KH, Jagnow G, Anderson TH (1983) An ecological concept for the assessment of side effects of agrochemical on soil microorganisms. Residue Rev 86:65–105

Drenovsky RE, Elliot GN, Graham KJ, Scow KM (2004) Comparison of phospholipid fatty acid (PLFA) and total soil fatty acid methyl esters (TSFAME) for characterizing soil microbial communities. Soil Biol Biochem 36:1793–1800

Drenovsky RE, Feris KP, Batten KM, Hristova K (2008) New and current microbiological tools for ecosystem ecologists: towards a goal of linking structure and function. Am Midl Nat 160: 140–159

Drijber RA, Doran JW, Pankhurst AM, Lyon DJ (2000) Changes in soil microbial community structure with tillage under long-term wheat-fallow management. Soil Biol Biochem 32:1419–1430

Egeston-Warburton LM, Allen EB (2000) Shifts in arbuscular mycorrhizal communities along an anthropogenic nitrogen deposition gradient. Ecol Appl 10:484–496

Egli S, Mozafar A (2001) Eine Standardmethod zur Erfassung des Mykkorrhiza-Infektionspotenzials in Landwirtschaftsböden. VBB Bull 5:6–7

El Fantroussi S, Verschuere L, Verstraete W, Top EM (1999) Effect of phenylurea herbicides on soil microbial communities estimated by analysis of 16S rRNA gene fingerprints and community-level physiological profiles. Appl Environ Microbiol 65:982–988

Evans DG, Miller MH (1990) The role of the external mycelial network in the effect of soil disturbance upon vesicular arbuscular mycorrhizas colonization of maize. New Phytol 114:65–71

Enwall K, Philippot L, Hallin S (2005) Activity and composition of the denitrifying bacterial community respond differently to long-term fertilization. Appl Environ Microbiol 71: 8335–8343

Franchini JC, Crispino CC, Souza RA, Torres E, Hungria M (2007) Microbiological parameters as indicators of soil quality under various soil management and crop rotation systems in southern Brazil. Soil Till Res 92:18–29

Franke-Snyder M, Douds DD, Galvez L, Philip JG, Wagoner P, Drinkwater L, Morton J (2001) Diversity of communities of arbuscular mycorrhizal (AM) fungi present in conventional versus low-input agricultural sites in eastern Pennsylvania, USA. Appl Soil Ecol 16:35–48
Franke-Whittle I, Klammer S, Insam H (2005) Design and application of an oligonucleotide microarray for the investigation of compost microbial communities. J Microbiol Method 62:37–56
Gans J, Murray W, Dunbar J (2005) Computational improvements reveal great bacterial diversity and high metal toxicity in soil. Science 309:1387–1389
Garland JL, Mills AL (1991) Classification and characterisation of heterotrophic microbial communities on the basis of patterns of community level sole-carbon-source utilization. Appl Environ Microbiol 57:2351–2359
Garlynd MJ, Roming DE, Harris RF, Kukakov AV (1994) Descriptive and analytical characterization of soil quality/health. In: Doran JW, Coleman DC, Bezdicek DF, Stewart BA (eds) Defining soil quality for a sustainable environment. SSSA Special Publication Number 35, Wisconsin, pp 159–168
Gaur A, Sharma MP, Adholeya A, Chauhan SP (1998) Variation in the spore density and percentage of root length colonized by arbuscular mycorrhizal fungi at rehabilitated waterlogged sites. J Trop Forest Sci 10:542–551
Gentry TJ, Wickham GS, Scadt CW, He Z, Zhou J (2006) Microarray applications in microbial ecology research. Microb Ecol 52:159–175
George E, Marschner H, Jakobsen I (1995) Role of arbuscular mycorrizal fungi in uptake of phosphorus and nitrogen from soil. Crit Rev Biotechnol 15:257–270
Giller PS (1996) The diversity of soil communities, the poor man's tropical forest. Biodivers Conserv 5:135–168
Giller KE, Beare MH, Lavelle P, Izac AMN, Swift MJ (1997) Agricultural intensification, soil biodiversity and agro-ecosystem function. Appl Soil Ecol 6:3–16
Girvan MS, Bullimore J, Ball AS, Pretty JN, Osborne AM (2004) Responses of active bacterial and fungal communities in soils under winter wheat to different fertilizer and pesticides regimes. Appl Environ Microbiol 70:2692–2701
Gosling P, Hodge A, Goodlass G, Bending GD (2006) Arbuscular mycorrhizal fungi and organic farming. Agric Ecosyst Environ 113:17–25
Govaerts B, Sayre KD, Deckers J (2006) A minimum data set for soil quality assessment of wheat and maize cropping in the highlands of Mexico. Soil Till Res 87:163–174
Govaerts B, Mezzalama M, Unno Y, Sayer KD, Luna-Guido M, Vanherck K, Dendoovan L, Deckers J (2007a) Influence of tillage, residue management, and crop rotation on microbial biomass and catabolic diversity. Appl Soil Ecol 37:18–30
Govaerts B, Sayre KD, Lichter K, Dendooven L, Deckers J (2007b) Influence of permanent raised bed planting and residue management on physical and chemical soil quality in rain fed maize/wheat systems. Plant Soil 291:39–54
Govaerts B, Fuentes M, Sayre KD, Mezzalama M, Nicol JM, Deckers J, Etchevers J, Figueroa-Sandoval B (2007c) Infiltration, soil moisture, root rot and nematode populations after 12 years of different tillage, residue and crop rotation managements. Soil Till Res 94:209–219
Govaerts B, Mezzalama M, Sayre KD, Crossa J, Lichter K, Troch V, Vanherck K, De Corte P, Deckers J (2008) Long-term consequences of tillage, residue management, and crop rotation on selected soil micro-flora groups in the subtropical highlands. Appl Soil Ecol 38:197–210
Govidarajulu M, Pfeffer PE, Jin H, Abubaker J, Douds DD, Allen JW, Bucking H, Lammers PJ, Schachar-Hill Y (2005) Nitrogen transfer in the arbuscular mycorrhizal sysmbiosis. Nature 435:819–823
Grayston SJ, Wang S, Campbell CD, Edward AC (1998) Selective influence of plant species on microbial diversity in the rhizosphere. Soil Biol Biochem 30:369–378
Grayston SJ, Campbell CD, Bardgett RD, Mawdsley JL, Clegg CD, Ritz K, Griffiths BS, Rodwell JS, Edwards SJ, Davies WJ, Elston DJ, Millard P (2004) Assessing shift in microbial community structure across a range of grasslands of differing management intensity using CLPP, PLFA and community DNA techniques. Appl Soil Ecol 25:63–84

Griffiths BS (2000) Ecosystem response of pasture soil communities to fumigation-induced microbial diversity reductions: an examination of the biodiversity-ecosystem function relationship. Oikos 90:279–294
Griffiths BS, Bonkowski M, Roy J, Ritz K (2001) Functional stability, substrate utilisation and biological indicators of soils following environmental impacts. Appl Soil Ecol 16:49–61
Günther S, Groth I, Grabley S, Munder T (2005) Design and evaluation of oligonucleotide microarrays for the detection of different species of the genus *Kitasatospora.* J Microbiol Method 65:226–236
Guschin DY, Mobarry BK, Proudnikov D, Stahl DA, Rittmann BE, Mirzabekov AD (1997) Oligonucleotide microchips as genosensors for determinative and environmental studies in microbiology. Appl Environ Microbiol 63:2397–2402
Harrier LA, Watson CA (2003) The role of arbucular mycorrhizal fungi in sustainable cropping systems. Adv Agron 79:185–225
Harris RF, Bezdicek DF (1994) Descriptive aspects of soil quality/health. In: Doran JW, Coleman DC, Bezdicek DF, Stewart BA (eds) Defining soil quality for a sustainable environment. SSSA Special publication number 35, Wisconsin, pp 23–36
Hartmann M, Widmer F (2006) Community structure analyses are more sensitive differences in soil bacterial communities than anonymous diversity indices. Appl Environ Microbiol 72:7804–7812
Hartmann M, Fliiessbach A, Oberholzer HR, Widmer F (2006) Ranking the magnitude of crop and farming system effects on soil microbial biomass and genetic structure of bacterial communities. FEMS Microbiol Ecol 57:378–388
He ZL, Yang XE, Baligar VC, Calvert DV (2003) Microbiological and biochemical indexing systems for assessing quality of acid soils. Adv Agron 78:89–138
He Z, Gentry TJ, Schadt CW, Wu L, Liebich J, Chong SC, Huang Z, Wu W, Gu B, Jardine P, Criddle C, Zhou J (2007a) GeoChip: a comprehensive microarray for investigating biogeochemical, ecological and environmental processes. Int Soc Microb Ecol (ISME) J 1:67–77
He Z, Deng Y, Nostrand JV, Wu L, Hemme C, Liebich J, Gentry TJ, Zhou J (2007b) Geochip 3.0: further development and applications for microbial community analysis. ASA-CSSA-SSSA/2007 international annual meetings, 4–8 Nov, New Orleans, Louisianas
Herrick JE (2000) Soil quality: an indicator of sustainable management? Appl Soil Ecol 15:75–83
Heuer H, Smalla K (1997) Application of denaturing gradient gel electrophoresis (DGGE) and temperature gradient gel electrophoresis (TGGE) for studying soil microbial communities. In: van Elsas JD, Trevors JT, Welligton EMH (eds) Modern soil microbiology. Mercel Dekker, New York, pp 353–373
Hill GT, Mitkowski NA, Aldrich-Wolfe L, Emele LR, Jurkonie DD, Ficke A, Maldonado-Ramirez S, Lynch ST, Nelson EB (2000) Methods for assessing the composition and diversity of soil microbial communities. Appl Soil Ecol 15:25–36
Howard RJ (1996) Cultural control of plant diseases: a historical perspective. Can J Plant Pathol 18:145–150
Ibekwe AM, Kennedy AC (1998) Phospholipid fatty acid profiles and carbon utilization pattern for analysis of microbial community structure under field and green house conditions. FEMS Microbiol Ecol 26:151–163
Ibekwe AM, Papiernik SK, Gan J, Yates SR, Yang CH, Crowley DE (2001) Impact of fumigants on soil microbial communities. Appl Environ Microbiol 67:3245–3257
Insam H (1997) A new set of substrates proposed for community characterization in environmental samples. In: Insam H, Rangger A (eds) Microbial communities: functional versus structural approaches. Springer, Berlin, pp 259–260
Iwamoto T, Tani K, Nakamura K, Suzuki Y, Kitagawa M, Eguchi M, Nasu M (2000) Monitoring impact of *in situ* bio-stimulation treatment on groundwater bacterial community by DGGE. FEMS Microbiol Ecol 32:129–141
Jackson LE, Calderón FJ, Steenwerth KL, Scow KM, Rolston DE (2003) Responses of soil microbial processes and community structure to tillage events and implications for soil quality. Geoderma 114:305–317

Jacquot E, van Tuinen D, Gianinazzi S, Gianinazzi-Pearson V (2000) Monitoring species of arbuscular mycorrhizal fungi in planta and in soil by nested PCR: application to the study of the impact of sewage sludge. Plant Soil 226:179–188

Jama B, Swinkels RA, Buresh RJ (1997) Agronomic and economic evaluation of organic and inorganic sources of phosphorus in western Kenya. Agron J 89:597–604

Jansa J, Mozafar A, Anken T, Ruh R, Sanders IR, Frossard E (2002) Diversity and structure of AMF communities as affected by tillage in a temperate soil. Mycorrhiza 12:225–234

Jansa J, Mozafar A, Kuhn G, Anken T, Ruh R, Sanders IR, Frossard E (2003) Soil tillage affects the community structure of mycorrhizal fungi in maize roots. Ecol Appl 13:1164–1176

Janvier C, Villeneuve F, Alabouvette C, Edel-Hermann V, Mateille T, Steinberg C (2007) Soil health through soil disease suppression: which strategy from descriptors to indicators? Soil Biol Biochem 39:1–23

Jeffries P, Gianinazi S, Perotto S, Turnau K, Barea JM (2003) The contribution of arbuscular mycorrhizal fungi in sustainable maintenance of plant health and soil fertility. Biol Fertil Soils 37:1–16

Johnson NC, Pfleger FL (1992) VA mycorrhizae and cultural stresses. In: Bethlenfalvay GJ, Linderman RG (eds) Mycorrhizae in sustainable agriculture. American Society of Agronomy, Madison, WI, pp 71–100

Johnson NC, Copeland PJ, Crookston RK, Pflegger FL (1992) Mycorrhizae: possible explanation for yield decline with continuous corn and soybean. Agron J 84:387–390

Jung SY, Lee JH, Chai YG, Kim SJ (2005) Monitoring of microorganisms added into oil-contaminated microenvironments by terminal-restriction fragment length polymorphism analysis. J Microbiol Biotechnol 15:1170–1177

Kahiluoto H, Ketoja E, Vestberg M, Saarela I (2001) Promotion of AM utilization through reduced P fertilization II Field studies. Plant Soil 231:65–79

Kabir Z (2005) Tillage or no-tillage: impact on mycorrhizae. Can J Plant Sci 85:23–29

Kandeler E, Böhm KE (1996) Temporal dynamics of microbial biomass, xylanase activity, N-mineralization and potential nitrification in different tillage systems. Appl Soil Ecol 5:221–230

Kandeler E, Tscherko D, Spiegel H (1999) Long-term monitoring of microbial biomass, N mineralization and enzyme activities of a Chernozem under different tillage management. Biol Fertil Soil 28:343–351

Karlen DL (2004) Soil quality as an indicator of sustainable tillage practices. Soil Till Res 78: 129–130

Karlen DL, Andrews SS (2000) The soil quality concept: a tool for evaluating sustainability. In: Elmholt S, Stenberg B, Grontund A, Nuutinen V (eds) Soil stress, quality and care. DIAS Report no 38, Danish Institute of Agricultural Sciences, Tjele, Denmark

Karlen DL, Andrew SS, Wienhold BJ, Doran JW (2003a) Soil quality: Humankind's foundation for survival. J Soil Water Conserv 58:171–179

Karlen DL, Andrew SS, Wienhold BJ (2003b) Soil quality, fertility and health- historical context, status and perspective. In: Schjonning P, Christensen BT, Elmholt S (eds) Managing soil quality-challenges in modern agriculture. CABI, Oxon, UK

Karlen DL, Hurrley EG, Andrews SS, Cambardella CA, Meek DW, Duffy MD, Mallarino AP (2006) Crop rotation effects on soil quality at three Northern corn/soybean-belt locations. Agron J 98:484–495

Karlen DL, Mausbach MJ, Doran JW, Cline RG, Harris RF, Schuuman GE (1997) Soil quality: a concept, definition, and framework for evaluation. Soil Sci Soc Am J 61:4–10

Kelly JJ, Hoggblom M, Tate RL III (1999) Changes in soil microbial communities over time resulting from one time application of zinc: a laboratory microcosm study. Soil Biol Biochem 31:1455–1465

Kemp PF, Sherr BF, Sherr EB, Cole JJ (1993) Handbook of methods in aquatic microbial ecology. Lewis, Boca Raton, FL

Kennedy P, Smith KL (1995) Soil microbial diversity and sustainability of agricultural soils. Plant Soil 170:75–86

Kennedy N, Brodie E, Conolly J, Clipson N (2004) Impact of lime, nitrogen and plant species on bacterial community structure in grassland microcosms. Environ Microbiol 6:1070–1080
Kjoller R, Rosendahl S (2000) Detection of arbuscular mycorrhizal fungi (Glomales) in roots by nested PCR and SSCP (single stranded conformation polymorphism). Plant Soil 226:189–196
Kling M, Jakobsen I (1998) Arbuscular myccorhiza in soil quality assessment. Ambio 27:29–34
Kozdroj J, Trevor JT, van Elsas JD (2004) Influence of introduced biocontrol agents in maize seedling growth and bacterial community structure in the rhizosphere. Soil Biol Biochem 36:1775–1784
Langdale GW, West LT, Bruce RR, Miller WP, Thomas AW (1992) Restoration of eroded soil with conservation tillage. Soil Technol 5:81–90
Larson WE, Pierce FJ (1994) The dynamics of soil quality as a measure of sustainable management. In: Coran JW, Coleman DC, Bezdicek DF, Stewart BA (eds) Defining soil quality for sustainable environment. SSSA Special publication number 35, Madison, WI, pp 37–51
Leake JR, Johnson D, Donnelly DP, Muckle GE, Boddy L, Read DJ (2004) Networks of power and influence: the role of mycorrhizal mycelium in controlling plant communities and agroecosystem functioning. Can J Bot 82:1016–1045
Lee DH, Zo YG, Kim SJ (1996) Non-radioactive method to study genetic profiles of natural bacterial communities by PCR-single stranded conformational polymorphism. Appl Environ Microbiol 62:3112–3120
Le Roux X, Poly F, Currey P, Commeaux C, Hai B, Nicol GW, Prosser JI, Schloter M, Attard E, Klumpp K (2008) Effects of aboveground grazing on coupling among nitrifier activity, abundance and community structure. ISME J 2:221–232
Liebeg MA, Tanaka DL, Weinhold BJ (2004) Tillage and cropping effects on soil quality indicators in the northern Great Plains. Soil Till Res 78:131–141
Liu WT, Marsh TL, Cheng H, Forney LJ (1997) Characterization of microbial diversity by determining terminal restriction fragment length polymorphisms of genes encoding 16S rRNA. Appl Environ Microbiol 63:4516–4522
Liu B, Gumpertz ML, Hu S, Ristaino JN (2007) Long-term effects or organic and synthetic soil fertility amendments on soil microbial communities and the development of southern blight. Soil Biol Biochem 39:2302–2316
Lottmann J, Heuer H, de Vries J, Mahn A, Düering K, Wackernagel W, Smalla K, Berg G (2000) Establishment of introduced antagonistic bacteria in the rhizosphere of transgenic potatoes and their effect on the bacterial community. FEMS Microbiol Ecol 33:41–49
Lowrance R, Williams RG (1988) Carbon movement in runoff and erosion under simulated rainfall conditions. Soil Sci Soc Am J 52:1445–1448
Loy A, Küsel K, Lehner A, Drake HL, Wagner M (2004) Microarray and functional gene analyses of sulphate-reducing prokaryotes in low-sulfate, acidic fens reveal co-occurrence of recognized genera and novel lineages. Appl Environ Microbiol 70:6998–7009
Loy A, Taylor MW, Bodrossy L, Wagner M (2006) Applications of nucleic acid microarrays in soil microbial ecology. In: Cooper JE, Rao JR(eds) Molecular approaches for soil, rhizosphere and plant microorganism analysis. CAB International, Wellingford, Oxforshire, U.K. pp 18–24
Lukow T, Dunfield PF, Liesack W (2000) Use of the T-RFLP technique to assess spatial and temporal changes in the bacterial community structure within an agricultural soil planted with transgenic and non-transgenic potato plants. FEMS Microbiol Ecol 32:241–247
Lupwayi NZ, Rice WA, Clayton GW (1998) Soil microbial diversity and community structure under wheat as influenced by tillage and crop rotation. Soil Biol Biochem 30:1733–1741
MacNaughton SJ, Stephen JR, Venosa AD, Davis GA, Chang YJ, White DC (1999) Microbial population changes during bioremediation of an experimental oil spill. Appl Environ Microbiol 65:3566–3574
Mäder P, Fließbach A, Dubois D, Gunst L, Fried P, Niggli U (2002) Soil fertility and biodiversity in organic farming. Science 296:1694–1697
Marsh TL (1999) Terminal restriction fragment length polymorphisms (T-RFLP): An emerging method for characterizing diversity among homologus population of amplification products. Curr Opin Microbiol 2:323–327
Marsh TL (2005) Culture-independent microbial community analysis with terminal restriction fragment length polymorphisms. Method Enzymol 397:308–329

Marschner P, Timonen S (2005) Interactions between plant species and mycorrhizal colonization on the bacterial community composition in the rhizosphere. Appl Soil Ecol 28:23–36

Marschner P, Crowley DE, Lierberei R (2001a) Arbuscular mycorrhizal infection changes in the bacterial 16S rRNA gene community composition in rhizosphere of maize. Mycorrhiza 11:297–302

Marschner P, Yang CH, Lierberei R, Crowley DE (2001b) Soil and plant specific effects on bacterial communities composition in the rhizosphere. Soil Biol Biochem 33:1437–1445

Marschner P, Crowley DE, Yang CH (2004) Development of specific rhizosphere bacterial communities in plant species, nutrition and soil type. Plant Soil 261:199–208

Marx M-C, Wood M, Jarvis SC (2001) A microplate fluorimetric assay for the study of enzyme diversity in soils. Soil Biol Biochem 33:1633–1640

Massol-Deya AA, Odelson DA, Hickey RF, Tiedje JM (1995) Bacterial community fingerprinting of amplified 16S and 16S-23S ribosomal DNA gene sequences and restriction endonuclease analysis (ARDRA). In: Akkerman ADL, van Elsas JD, De-Briijn FJ (eds) Molecular Microbial Ecology Manual. Kluwer, Boston, pp 3.221–3.328

Mathimaran N, Ruh R, Jama B, Verchot L, Frossard E, Jansa J (2007) Impact of agricultural management on arbuscular mycorrhizal fungal communities in Kenyana ferralsol. Agric Ecosyst Environ 119:22–32

Mawdsley JL, Burns RG (1994) Inoculation of plants with *Flavobacterium* P25 results in altered rhizosphere enzyme activities. Soil Biol Biochem 26:871–882

Melero S, Porras JCR, Herencia JF, Madejon E (2006) Chemical and biochemical properties in a silty loam soil under conventional and organic management. Soil Till Res 90:162–170

Meithling R, Weiland G, Backhaus H, Tebbe CC (2000) Variation of microbial rhizosphere communities in response to crop species, soil origin and inoculation with *Sinorhizobium meliloti* L 33. Microb Ecol 41:43–56

Miller MH, McGonigle TP, Addy HD (1995) Functional ecology of vesicular-arbuscular mycorrhizas as influenced by phosphate fertilization and tillage in an agricultural ecosystem. Crit Rev Biotechnol 15:241–255

Miller M, Palojarvi A, Rangger A, Reeslev M, Kjoler A (1998) The use of fluorogenic substrates to measure fungal presence and activity in soil. Appl Environ Microbiol 64:613–617

Mohanty SR, Bodelier PLE, Floris V, Conard R (2006) Differential effects of nitrogenous fertilizers on methane-consuming microbes in rice field and forest soils. Appl Environ Microbiol 72:1346–1354

Monokrousos N, Papatheodorou EM, Diamantopoulos JD, Stamou GP (2006) Soil quality variables in organically and conventionally cultivated field sites. Soil Biol Biochem 38:1282–1289

Moran AC, Muller A, Manzano M, Gonzalez B (2006) Simazine treatment history determines a significant herbicide degradation potential in soils that is not improved by bio-augmentation with *Pseudomonas* sp ADP. J Appl Microbiol 101:26–35

Mummey DL, Stahl PD (2003) Spatial and temporal variability of bacterial 16S rDNA-based T-RFLP patterns derived from soil of two Wyoming grassland ecosystems. FEMS Microbiol Ecol 46:113–120

Muyzer G, Waal ECD, Uitterlinden AG (1993) Profiling of complex microbial populations by denaturing gradient gel electrophoresis analysis of polymerase chain reaction-amplified gene coding for 16S rRNA. Appl Environ Microbiol 59:695–700

Myers RT, Zak DR, White DC, Peacock A (2001) Landscape-level patterns of microbial community composition and substrate use in upland forest ecosystems. Soil Sci Soc Am J 65:359–367

Nacamulli C, Bevivino A, Dalmastri C, Tabacchioni S, Chiarini L (1997) Perturbation of maize rhizosphere microflora following seed bacterization with *Burkholderia cepacia* MCI 7. FEMS Microbiol Ecol 23:183–193

Nakatsu CH, Torsvik V, Ovreds L (2000) Soil community analysis using DGGE of 16S rDNA polymerase chain reaction. Soil Sci Soc Am J 64:1382–1388

Naseby DC, Lynch JM (1997) Rhizosphere soil enzymes as indicators of perturbations caused by enzyme substrate addition and inoculation of a genetically modified strain of *Pseudomonas fluorescens* on wheat seed. Soil Biol Biochem 29:1353–1362

National Research Council (1993) Soil and water quality: an agenda for agriculture. National Academy, Washington, DC, p 516

Nielsen MN, Winding A (2002). Microorganisms as indicators of soil health. National Environmental Research Institute, Technical Report no. 388. National Environmental Research Institute, Denmark. (http://www.dmu.dk/1_viden/2_Publicakationer/_3_fagrapporter/rapporter?FR388.pdf)
Nogueira MA, Albino UB, Brandao-Junior O, Braun G, Cruz MF, Dias BA, Duarte RTD, Gioppo NMR, Menna P, Orlandi JM, Raiman MP, Rampazo LGL, Santos JMD, Hungria M, Andrade G (2006) Promising indicators for assessment of agroecosystems alternation among natural, reforestation and agricultural land use in Southern Brazil. Agric Ecosys Environ 115:237–247
Oehl F, Sieverding E, Mader P, Dubois D, Ineichen K, Boller T, Wiemken A (2004) Impact of long-term conventional and organic farming on the diversity of arbuscular mycorrhizal fungi. Oecologia 138:574–583
Oehl F, Sieverding E, Ineichen K, Mader P, Boller T, Wiemken A (2003) Impact of land use intensity on the species diversity of arbuscular mycorrhizal fungi in agro-ecosystems of central Europe. Appl Environ Microbiol 69:2816–2824
Orita M, Suzuki Y, Sekiya T, Hayashi K (1989) A rapid and sensitive detection of point mutations and genetic polymorphisms using polymerase chain reaction. Genomics 5:874–879
Ovreas L (2000) Population and community level approaches for analyzing microbial diversity in natural environments. Ecol Lett 3:236–251
Ovreas L, Torsvik VV (1998) Microbial diversity and community structure in two different agricultural soil communities. Microbiol Ecol 36:303–315
Pace NR (1996) New perspective on the natural microbial world: molecular microbial ecology. ASM News 62:463–470
Pankhurst CE, Hawke BG, MacDonald HJ, Kirkby CA, Buckerfield JC, Michelsen P, O'brien KA, Gupta VVSR, Doube BM (1995) Evaluation of soil biological properties as potential bioindicators of soil health. Aus J Exp Agric 35:1015–1028
Parham JA, Deng SP, Da HN, Sun YH, Raun WR (2003) Long-term cattle manure application in soil. II. Effect on soil microbial populations and community structure. Biol Fertil Soil 38:209–215
Parr JF, Papendick RI, Hornick SB, Meyer RE (2003) Soil quality: attributes and relationship to alternative and sustainable agriculture. Am J Altern Agric 7:5–7
Patra AK, Abbadie L, Clays A, Degrange V, Grayston S, Loiseau P, Louault F, Mahmood S, Nazaret S, Philippot L, Poly F, Prosser JI, Richaume A, Le Roux X (2005) Effect of grazing on microbial functional groups involved in soil N dynamics. Ecol Monogr 75:65–80
Patra AK, Abbadie L, Clays A, Degrange V, Grayston S, Guillaumaud N, Loiseau P, Louault F, Mahmood S, Nazaret S, Philippot L, Poly F, Prosser JI, Le Roux X (2006) Effects of management regime and plant species on the enzyme activity and genetic structure of N-fixing, denitrifying and nitrifying bacterial communities in grassland soils. Environ Microbiol 8:1005–1016
Peixoto RS, Coutinho HLC, Madari B, Machado PLOA, Rumjanek NG, Van Elsas JD, Seldin L, Rosado AS (2006) Soil aggregation and bacterial community structure as affected by tillage and cover cropping in the Brazilian Cerrados. Soil Till Res 90:16–28
Pennisi E (2004) The secret life of fungi. Science 304:1620–1621
Peters S, Koschinsky S, Schwieger F, Tebbe CC (2000) Succession of microbial communities during hot composting as detected by PCR—single stranded-conformation-polymorphism-based genetic profiles of small-subunit rRNA genes. Appl Environ Microbiol 66:930–936
Pfiffner L, Mäder P (1997) Effects of biodynamic, organic and conventional production systems on earthworm populations. Entomol Res Org Agric (special edition of Biological Agriculture and Horticulture) 15:3–10
Phatak SC, Dozier JR, Bateman AG, Brunson KE, Martini NL (2002) Cover crops and conservation tillage in sustainable vegetable production. In: van Santen E (eds) Making conservation tillage conventional building a future on 25 years research. Proceeding 25th annual southern conservation tillage conference for sustainable agriculture. Auburn, AL, pp 401–403
Pierce FJ, Larson WE, Dowdy RH (1984) Soil loss tolerance: maintenance of long-term soil productivity. J Soil Water Conserv 39:136–138
Pierce FJ, Rice CW (1998) Crop rotation and its impact on efficiency of water and nitrogen use. In: Hargrove WL (ed) Cropping strategies for efficient use of water and nitrogen, special publication no 51. American Society of Agronomy, Madison, WI, pp 21–36

Pinkart HC, Ringelberg DB, Piceno YM, Macnaughton SJ, White DC (2002) Biochemical approaches to biomass measurements and community structure analysis. In: Hurst CJ, Crawford RL, Knudsen GR, McInerney MJ, Stetzenbach LD (eds) Manual of environmental microbiology, 2nd edn. American Society for Microbiology Press, Washington, DC, pp 101–113

Praveen-Kumar, Tarafdar JC (2003) 2, 3, 5-triphenyltetrazolium chloride (TTC) as electron acceptor of culturable soil bacteria, fungi and actinomycetes. Biol Fertil Soil 38:186–189

Purin S, Filho OK, Sturmer SL (2006) Mycorrhizae activity and diversity in conventional and organic apple orchards from Brazil. Soil Biol Biochem 38:1831–1839

Ramesh A, Billore SD, Singh A, Joshi OP, Bhatia VS, Bundela VPS (2004a) Arylsulphatase activity and its relationship with soil properties under soybean (*Glycine max*) -based cropping systems. Indian J Agric Sci 74:9–13

Ramesh A, Billore SD, Joshi OP, Bhatia VS, Bundela VPS (2004b) Phosphatase activity, phosphorous utilization efficiency and depletion of phosphate fractions in rhizosphere of soybean (*Glycine max*) genotypes. Indian J Agric Sci 74:150–152

Rao AV, Tarafdar JC, Sharma SK, Praveen-Kumar, Aggarwal RK (1995) Influence of cropping systems on soil biochemical properties in an arid rainfed environment. J Arid Environ 31:237–244

Redecker D, Hijri I, Wiemken A (2003) Molecular identification of arbuscular mycorrhizal fungi in roots: perspectives and problems. Folia Geobotan 38:113–124

Reynolds MP, Borlaug NE (2006) Applying innovations and new technologies for international collaborative wheat improvement. J Agric Sci 144:95–115

Ritz K, Mac Nicol JW, Nunan N, Grayston S, Millard P, Atkinson A, Gollotte A, Habeshaw D, Boag B, Clegg CD (2004) Spatial structure in soil chemical and microbiological properties in upland grassland. FEMS Microbiol Ecol 49:191–205

Roesti D, Gaur R, Johri BN, Imfeld G, Sharma S, Kawaljeet K, Aragno M (2006) Plant growth stage, fertilizer management and bioinoculant of arbuscular mycorrhizal fungi and plant growth promoting rhizobacteria affect the rhizobacterial community structure in rain-fed wheat fields. Soil Biol Biochem 38:1111–1120

Roming DL, Garlynd MJ, Harris RF, Mc Sweeney K (1995) How farmers assess soil health and soil quality? J Soil Water Conserv 50:229–235

Rousseaux SA, Hartmann N, Rouard N, Soulas G (2003) A simplified procedure for terminal restriction fragment length polymorphism analysis of the soil bacterial community to study the effects of pesticides on soil microflora using 4, 6-dinitroorthocresol as a test case. Biol Fertil Soil 37:250–254

Saini VK, Bhandare SC, Tarafdar JC (2004) Comparison of crop yield, soil microbial C, N, and P, N-fixation, nodulation and mycorrhizal infection in inoculated and non-inoculated sorghum and chickpea crops. Field Crop Res 89:39–47

Sanchez PA, Palm CA, Buol SW (2003) Fertility capability soil classification: a tool to help assess soil quality in the tropics. Geoderma 114:157–185

Saison C, Degrange V, Oliver R, Millard P, Commeaux C, Montange D, Le Roux X (2006) Alteration and resilience of the soil microbial community following compost amendment: effects of compost level and compost-borne microbial community. Environ Microbiol 8:247–257

Sanders IR (2004) Plant and arbuscular mycorrhizal fungal diversity – are we looking at the relevant levels of diversity and are we using the right techniques? New Phytol 164:415–418

Sarathchandra SU, Burch G, Cox NR (1997) Growth patterns of bacterial communities in the rhizoplane and rhizosphere of white clover (*Trifolium repense* L.) and perennial rye grass (*Lolium perenne* L.) in long-term pasture. Appl Soil Ecol 6:239–299

Schmalenberger A, Tebbe CC (2003) Bacterial diversity in maize rhizospheres: Conclusions on the use of genetic profiles based on PCR-amplified partial small subunit rRNA genes in ecological studies. Mol Ecol 12:251–262

Schmalenberger A, Schwieger F, Tebbe CC (2001) Effect of primers hybridizing to different evolutionarily conserved regions of the small-subunit rRNA gene in PCR-based microbial community analyses and genetic profiling. Appl Environ Microbiol 67:3557–3563

Schouten T, Bloem J, Didden WAM, Rutgers M, Siepel H, Posthuma L, Breure AM (2000) Development of a biological indicator for soil quality. Soc Environ Toxicol Chem Globe 1:30–33

Schouten AJ, Bloem J, Didden W, Jagers OP, Akkerhuis G, Keidel H, Rutgers M (2002) Biological indicator for soil quality. Ecological quality of Dutch pastures on sandy soil in relation to grazing intensity. Report 607604003.RIVM,Bilthoven (in Dutch with English summary)
Schwieger F, Tebbe CC (1998) A new approach to utilize PCR-single-strand-conformation polymorphism for 16S rRNA gene-based microbial community analysis. Appl Environ Microbiol 64:4870–4876
Schwieger F, Tebbe CC (2000) Effects of field inoculation with *Sinorhizobium meliloti* L33 on the composition of bacterial communities in rhizospheres of a target plant (*Medicago sativa*) and a non-target plant (*Chenopodium album*)-linking of 16S rRNA gene-based single-strand conformation polymorphism community profiles to the diversity of cultivated bacteria. Appl Environ Microbiol 66:3556–3565
Sessitsch A, Hackl E, Wenzl P, Kilian A, Kostic T, Stralis-Pavese N, Tankouo Sandjong B, Bodrossy L (2006) Diagnostic microbial microarrays in soil ecology. New Phytol 171:719–736
Sharma KL, Mandal UK, Srinivas K, Vittal KPR, Mandal B, Grace JK, Ramesh V (2005) Long-term soil management effects on crop yields and soil quality in dryland Alfisol. Soil Till Res 83:246–259
Sharma MP, Adholeya A (2004) Influence of arbuscular mycorrhizal fungi and phosphorus fertilization on the *post-vitro* growth and yield of micropropagated strawberry in an alfisol. Can J Bot 82:322–328
Sharma MP, Sharma SK (2006) Arbuscular mycorrhizal fungi: an emerging bioinoculant for production of soybean. SOPA Digest III:10–16
Sharma MP, Gaur A, Bhatia NP, Adholeya A (1996) Mycorrhizal dependency of *Acacia nilotica* var. *cupriciformis* to indigenous vesicular arbuscular mycorrhizal consortium in a wasteland soil. Mycorrhiza 6:441–446
Siciliano SD, Germida JJ (1998) Biolog analysis and fatty acid methyl ester profiles indicate that pseudomonad inoculants that promote phytoremediation alter the root-associated microbial community of *Bromus biebersteinii*. Soil Biol Biochem 30:1717–1723
Sieverding E (1990) Ecology of VAM fungi in tropical agrosystems. Agric Ecosyst Environ 29:369–390
Simon L, Levesque RC, Lalonde M (1993) Identification of endomycorrhizal fungi colonizing roots by fluorescent single strand conformation polymorphism-polymerase chain reaction. Appl Environ Microbiol 59:4211–4215
Singh BK, Munro S, Reid E, Reid B, Ord JM, Potts JM, Paterson E, Millard P (2006) Investigating microbial community structure in soils by physiological, biochemical and molecular fingerprinting methods. Euro J Soil Sci 57:72–82
Singer MJ, Edwig S (2000) Soil quality. In: Summer ME (ed) Handbook of soil science. CRC Press, Boca Raton, pp G271–G298
Six J, Elliot ET, Paustian K (1999) Aggregate and soil organic matter dynamics under conventional and no-tillage systems. Soil Sci Soc Am J 63:1350–1358
Smalla K, Wieland G, Buchner A, Zock A, Parzy J, Kaiser S, Roskot N, Heur H, Berg G (2001) Bulk and rhizosphere soil bacterial communities studied by denaturing gradient gel electrophoresis: plant-dependent enrichment and seasonal shifts revealed. Appl Environ Microbiol 64:1220–1225
Smith SE, Read DJ (1997) Mycorrhizal symbiosis instead of symbios, 2nd edn. Academic, San Diego, CA
Soderberg KH, Probanza A, Jumpponen A, Baath E (2004) The microbial community in the rhizosphere determined by community-level physiological profile (CLPP) and direct soil- and cfu-PLFA techniques. Appl Soil Ecol 25:135–145
Srivastava R, Roesti D, Sharma AK (2007) The evaluation of microbial diversity in a vegetable based cropping system under organic farming practices. Appl Soil Ecol 36:116–123
Steenwerth KL, Jackson LE, Calderón FJ, Stromberg MR, Scow KM (2003) Soil microbial community composition and land use history in cultivated and grassland ecosystems in coastal California. Soil Biol Biochem 35:489–500
Steenwerth KL, Jackson LE, Calderón FJ, Scow KM, Rolston DE (2005) Response of microbial community composition and activity in agricultural and grassland soils after a simulated rainfall. Soil Biol Biochem 37:2249–2262

Stenberg B, Pell M, Torstensson L (1998) Integrated evaluation of variation in biological, chemical and physical soil properties. Ambio [Royal Swedish Academy of Sciences] 27:9–15
Stenberg B (1999) Monitoring soil quality of arable land: microbiological indicators. Acta Agric Scandinavia 49:1–24
Stocking MA (2003) Tropical soil and food security: the next 50 years. Science 21:1356–1359
Sturz AV, Christie BR (2003) Beneficial microbial allelopathies in the root zone: the management of soil quality and plant disease with rhizobacteria. Soil Till Res 72:107–123
Sun HY, Deng SP, Raun WR (2004) Bacterial community structure and diversity in a century-old manure-treated agro-ecosystem. Appl Environ Microbiol 70:5868–5874
Swift MJ (1999) Integrating soils, systems and society. Nat Resour 35:12–20
Taroncher-Oldenburg G, Griner EM, Francis CA, Ward BB (2003) Oligonucleotide microarray for the study of functional gene diversity in the nitrogen cycle in the environment. Appl Environ Microbiol 69:1159–1171
Taylor MW, Loy A, Wagner M (2007) Microarrays for studying the composition and function of microbial communities. In: Seviour RJ, Blackall LL (eds) The microbiology of activated sludge. IWA Publishing, London
Thies JE (2007) Soil microbial community analysis using terminal restriction fragment length polymorphisms. Soil Sci Soc Am J 71:579–591
Tiedje JM, Asuming-Brempong S, Nusslein K, Marsh TL, Flynn SJ (1999) Opening the black box of soil microbial diversity. Appl Soil Ecol 13:109–122
Torsvik V, Daae FL, Sandaa RA, Ovrea OSL (1998) Novel techniques for analysing microbial diversity in natural and perturbed environments. J Biotechnol 64:53–62
Trasar-Cepeda C, Leiro's MC, Seoane S, Gil-Sotres F (2000) Limitations of soil enzymes as indicators of soil pollution. Soil Biol Biochem 32:1867–1875
Treseder KK, Allen MF (2002) Direct nitrogen and phosphorus limitation of arbuscular mycorrhizal fungi: a model and field test. New Phytol 155:507–515
van Bruggen AHC, Semenov AM, van Diepeningen AD, de Vos OJ, Blok WJ (2006) Relation between soil health, wave-like fluctuations in microbial populations, and soil-borne plant disease management. Euro J Plant Pathol 115:105–122
Vandenkoornhuyse P, Husband R, Daniell TJ, Watson IJ, Duck JM, Fitter AH, Young JPW (2002) Arbuscular mycorrhizal community composition associated with two plant species in a grassland ecosystem. Mol Ecol 11:1555–1564
van Diepeningen AD, de Vos OJ, Korthals GW, van Bruggen AHC (2006) Effects of organic versus conventional management on chemical and biological parameters in agricultural soils. Appl Soil Ecol 31:120–135
van der Heijden MGA, Klironomos JN, Ursic M, Moutoglis P, Streitwolf-Engel R, Boller T, Wiemken A, Sanders IR (1998) Mycorrhizal fungal diversity determines plant biodiversity, ecosystem variability and productivity. Nature 396:69–72
Vazquez MM, César S, Azc n R, Barea JM (2000) Interactions between arbuscular mycorrhizal fungi and other microbial inoculants (*Azospirillum*, *Pseudomonas*, *Trichoderma*) and their effects on microbial populations and enzyme activities in the rhizosphere of maize plants. Appl Soil Ecol 15:261–272
Verchot LV, Borelli RB (2005) Application of *para*-nitrophenol (pNP) enzyme assays in degraded tropical soils. Soil Biol Biochem 37:625–633
Visser S, Parkinson D (1992) Soil biological criteria as indicators of soil quality: soil microorganisms. Am J Altern Agric 7:33–37
Wagner M, Loy A, Klein M, Lee N, Ramsing NB, Stahl DA, Friedrich MW (2005) Functional marker genes for identification of sulphate-reducing prokaryotes. Method Enzymol 397:469–489
Wardle DA, Bonner KI, Baker GM, Yeatas GW, Nicholson KS, Bardgett RD, Watson RN, Ghani A (1999) The plant removals in perennial grassland: vegetation dynamics, decomposers, soil biodiversity, and ecosystem properties. Ecolog Monogr 69:535–568
Waldrop MP, Balser TC, Firestone MK (2000) Linking microbial community composition to function in a tropical soil. Soil Biol Biochem 26:1837–1846
Wang P, Durkalski JT, Yu WT, Hoitink HAJ, Dick WA (2006) Agronomic and soil responses to compost and manure amendments under different tillage systems. Soil Sci 171:456–467

Warkentin BP, Fletcher HF (1997). Soil quality for intensive agriculture. Proceedings of the international seminar on soil environment and fertlizer management in intensive agriculture. National Institute of Agricultural Science, Tokyo, Japan, pp 594–598
Welbaum GE, Sturz AV, Dong Z, Nowak J (2004) Managing soil microorganisms to improve productivity of agro-ecosystems. Crit Rev Plant Sci 23:175–193
Wertz S, Degrange V, Prosser JI, Poly F, Commeaux C, Freitag T, Guillaumaud N, Le Roux X (2006) Maintenance of soil functioning following erosion of microbial diversity. Environ Microbiol 8:2162–2169
Wertz S, Degrange V, Prosser JI, Poly F, Commeaux C, Guillaumaud N, Le Roux X (2007) Decline of soil microbial diversity does not influence the resistance and resilience of key soil microbial functional groups following a model disturbance. Environ Microbiol 9:2211–2219
Whiteley AS, Bailey MJ (2000) Bacterial community structure and physiological state within an industrial phenol bioremediation system. Appl Environ Microbiol 66:2400–2407
White DC, Davis WM, Nickels JS, King JD, Bobbie RJ (1979) Determination of the sedimentary microbial biomass by extractable lipid phosphate. Oecologia 40:51–62
Wilson KH, Wilson WJ, Radosevich JL, DeSantis TZ, Viswanathan VS, Kuczmarski TA (2002) High density microarray of small-subunit ribosomal DNA probes. Appl Environ Microbiol 68:2535–2541
Winding A (2004) Indicators of soil bacterial diversity. In: Agricultural impacts on soil erosion and soil biodiversity: developing indicators for policy analysis. Proceedings of an OECD expert meeting on soil erosion and soil biodiversity indicators, 25–28 March 2003, Rome, Italy, OECD, Paris, pp 495–504
Winding A, Hund_Rinke K, Rutgers M (2005) The use of microorganisms in ecological soil classification and assessment concepts. Ecotoxicol Environ Safety 62:230–248
Wirth SJ, Wolf GA (1992) Microplate colorimetric assay for endo-acting cellulase, xylanase, chitinase, 1, 3-beta-glucanase and amylase extracted from forest soil horizons. Soil Biol Biochem 24:511–519
Wright SF, Starr JL, Paltineanu IC (1999) Changes in aggreagate stability and concentration of glomalin during tillage management transistion. Soil Sci Soc Am J 63:1825–1829
Wu L, Thompson DK, Li R, Hurt RA, Tiedje JM, Zhou J (2001) Development and evaluation of functional gene arrays for detection of selected genes in the environment. Appl Environ Microbiol 67:5780–5790
Yang Y, Yao J, Hu S, Qi Y (2000) Effects of agricultural chemicals on the DNA sequence diversity of microbial communities. Microbiol Ecol 39:72–79
Yergeau E, Kang S, He Z, Zhou J, Kowalchuk GA (2007) Functional microarray analysis of nitrogen and carbon cycling genes across an Antarctic latitudinal transect. Int Soc Microb Ecol J 1:163–179
Yin B, Crowley D, Sparovek G, De Melo WJ, Borneman J (2000) Bacterial functional redundancy along a soil reclamation gradient. Appl Environ Microbiol 66:4361–4365
Zelles L (1999) Fatty acid patterns of phospholipids and lipopolysacharides in the characterization of microbial communities in soil: a review. Biol Fertil Soil 29:111–129
Zhou J (2003) Microarrays for bacterial detection and microbial community analysis. Curr Opin Microbiol 6:288–294
Zumstein E, Moletta R, Godon JJ (2000) Examination of two years of community dynamics in an anaerobic bioreactor using fluorescence polymerase chain reaction (PCR) single-strand conformation polymorphism analysis. Environ Microbiol 2:69–78
Zvyaginstsev DG, Kochkina GA, Kojhevin PA (1984) New approaches to the investigation microbial succession in soil. In: Mischustin EN (ed) Organisms as components of biogeocenosis soil. Nauka, Moscow, pp 81–103

Integrating Silvopastoralism and Biodiversity Conservation

A. Rigueiro-Rodríguez, M. Rois-Díaz, and M.R. Mosquera-Losada

Abstract Silvopastoral systems reflect some aspects of multifunctionality of European forests. Sylvopastoral systems combine timber production with pastoral activities and associated animal products while concurrently preserving different aspects of biodiversity that has been reduced in Europe in the last century. For the first time within EU policy, the Council Regulation to support rural development by the European Agricultural Fund for Rural Development (EAFRD) within the second pillar of the Common Agricultural Policy (CAP), supports the establishment of agroforestry systems by farmers. Silvopastoral system implementation could provide productive, social and environmental benefits. Keeping extensive livestock is an integrated and environmental-friendly land use system, as it requires minimum infrastructures and buildings that would not degrade the landscape and it mostly uses natural resources for the alimentation of the livestock. This article starts with a review of the history of silvopastoral systems, then focusing on their benefits, particularly concerning biodiversity. The rule of silvopastoral systems on the establishment of heterogeneous micro-environment, forest-agricultural land corridor creation and forest fire prevention are discussed, as well as, the importance of silvopastoralism in the conservation of indigenous domestic breeds, half of which are considered to be at risk of extinction. Finally, the article places silvopastoralism within a policy context and considers the potential impact of the management of these systems.

Keywords Biological diversity • Multifunctionality • Sustainable development

---

A. Rigueiro-Rodríguez and M.R. Mosquera-Losada (✉)
Departamento de Producción Vegetal, Escuela Politécnica Superior, Universidad de Santiago de Compostela, 27002 Lugo, Spain
e-mails: antonio.rigueiro@usc.es and mrosa.mosquera.losada@usc.es

M. Rois-Díaz
MEDFOREX Project Center, Passeig Lluis Companys 23, 08013 Barcelona, Spain

E. Lichtfouse (ed.), *Biodiversity, Biofuels, Agroforestry and Conservation Agriculture*, Sustainable Agriculture Reviews 5, DOI 10.1007/978-90-481-9513-8_12,

1 Introduction

Silvopastoralism is an ancient and traditional way of land management where trees, animals and pasture were integrated to meet human needs. Nowadays, silvopastoralism is promoted as a type of an agroforestry system due to economic, environmental and societal benefits derived from its multifunctionality (Rigueiro et al. 2008). Multifunctionality can be understood at both a temporal and spatial scale. It is feasible to combine the preservation of different aspects of biodiversity and fulfilling economic (higher profitability) and social requirements (rural population stabilisation) at the same time. The Agenda 21 global agreement of the UN Conference on Environment and Development in Rio de Janeiro proposed that agroforestry systems, including silvopastoralism are a form of sustainable land management that should be promoted (UN 1993). Moreover, a COUNCIL REGULATION on support for rural development by the European Agricultural Fund for Rural Development (EAFRD) was released (15 September 2005), which establishes that "measures targeting the sustainable use of forestry land through the first establishment of agroforestry systems on agricultural land" should be taken.

Silvopastoral systems can provide ecosystem services delivering biodiversity conservation and enhancement within an integrated and sustainable framework. This article starts with a review of the history of silvopastoral systems, then focusing on their benefits, particularly concerning biodiversity. Finally, it places silvopastoralism within a policy context and considers the potential impact of the management of these systems.

2 Brief History of Silvopastoral Systems

Silvopastoralism is one of the oldest agroforestry practices (Etienne 1996). In historical ecology several authors support the theory that natural western European forests may have been to a large degree grazed by wild large herbivores (Putman 1996; Vera 2000; Bradshaw et al. 2003). Such grazing had an influence on the forest structure and development. Later animal species were replaced by domestic livestock. Considering this hypothesis silvopastoralism can be understood as a natural practice, if well planned beforehand and if it mimics the grazing of wild animals.

As civilisation progressed towards stable patterns of agriculture, woodland grazing and silvopastoral systems were abandoned. However there was a continuous transfer of biomass fertility from forests to cultivated land via manure (Piussi 1994; Eichhorn et al. 2006). Branches of ash, elm, poplar and other species were collected and stored and oak acorns were used to provide fodder for livestock (Meiggs 1982; Ispikoudis et al. 2004; Eichhorn et al. 2006). Another reason for maintaining trees in the landscape was the production of fruit such as chestnuts for human consumption as an essential part of the diet (Herzog 1998).

Until the Middle Ages the maintenance of soil fertility was based upon a strict connectivity between agriculture, animal husbandry and forestry. Later on and especially in the twentieth century with the introduction of chemical fertilisers in most parts of Europe, soil fertility became less dependent upon the transfer of nutrients from woods (Eichhorn et al. 2006).

At that time silvopastoral systems were considered to have low productivity, hence forestry and agriculture developed quite independently. The industrialisation of the 1800s initiated the abandonment of traditional methods of land management favouring a more intense and homogeneous system (EEA 2006). The current 'biodiversity crisis' in western Europe results mainly from this drastic change in the long-term interaction between human activities and nature (Bradshaw and Emanuelsson 2004). It was not until the end of twentieth century when silvopastoral systems were again recognized as favourable as they ensured key ecosystem functions, such as soil improvement, nutrient recycling, biodiversity promotion as well as marketable products (SSM 2004; WFC 2004).

In European Mediterranean countries, silvopastoralism is a traditional and widespread land-use activity. For example, half of the so-called 'forested area' in Spain is covered by natural grasslands, shrub lands and open forests used traditionally for livestock. If productive forests with extensive livestock use are taken into account, this area increases up to 70%. Only a quarter of the Spanish forest area is exclusively used for timber production, where no livestock is allowed (MMA 2000). Keeping extensive livestock is an integrated and environmental-friendly land use system as it requires minimum infrastructures and buildings that would degrade the landscape and it mostly uses natural resources for the nutrition of the livestock.

3 Benefits Derived from Silvopastoral Systems

The benefits of silvopastoral systems have been widely reported and it is recognized that they coincide with the principles of multifunctionality and sustainability (Sibbald et al. 1994; Etienne 1996; Papanastasis et al. 1999; Mosquera-Losada et al. 2005; Rigueiro-Rodríguez et al. 2008).

Multiple use of forests may be defined in different ways. Two commonly accepted definitions are: (1) management of forests to obtain multiple products and benefits such as production, protection and conservation, (2) multiple use forestry takes an integrated approach towards the different categories of forests and encompasses the scientific, cultural, recreational, historical and amenity values of forest resources (Schuck et al. 2002). Thus, the concept of silvopastoralism falls clearly within the category of multifunctionality as it combines the timber production in an integrated system with livestock and/or pastures.

The derived benefits of silvopastoralism can be grouped according to their impact on the three sustainability pillars (Nair et al. 2008):

1. Economic pillar
   (a) Diversification of products (meat, wool, mushrooms, timber, cork …) ensures short- and long-term income
   (b) Enabling faster production of high-quality timber, e.g. *Quercus robur* L., *Juglans regia* L., *Fraxinus excelsior* L., *Prunus avium* L. etc.

   (c) Reducing the costs of silvicultural treatments as needs for mechanical clearing are reduced and pruning can provide extra fodder for animals
   (d) Livestock welfare and increased quality of animal products from diverse fodder resources and a spatially heterogeneous living environment
2. Social pillar
   (a) Helping to keep the population in rural areas and managing natural resources
   (b) Increasing landscape value for local residents and tourists
3. Environmental pillar
   (a) Contributing to biodiversity preservation
   (b) Recycling nutrients within the system, especially reducing nitrate and phosphorous leaching which is an important problem in the European Union
   (c) Preventing desertification and erosion
   (d) Reducing greenhouse gas emissions and acting as a carbon sink, when compared to intensively-managed agricultural systems
   (e) Decreasing risk of forest fires

The relative importance of each of these aspects may vary across different European climatic regions, e.g. Mediterranean regions, Atlantic with dry summers and Atlantic with humid summers (McAdam et al. 2005).

4 Biodiversity and Silvopastoral Systems

4.1 Preserving Biodiversity

Preserving biodiversity is maintaining its intrinsic value and the services that natural systems provide: food, fuel, fibre and medicines, regulation of water, air and climate, soil fertility, cycling of nutrients. Therefore the concern for biodiversity is integral to sustainable development and supports competitiveness, economic growth and employment and improved livelihoods (CEE 2006). According to the Millennium Ecosystem Assessment some two-thirds of ecosystem services worldwide are in decline due to over-use and the loss of species richness (Hassan et al. 2005), which is needed for their stability (EEA 2006).

Silvopastoral systems favour biodiversity from different aspects. In Europe there is variety of different combinations of tree and under-storey species and livestock, e.g. *Eucalyptus globulus* Labill, *Pinus radiata* D. Don, *P. nigra* JF Arnold or *P. pinaster* Aiton with shrubs of *Ulex* spp., *Cytisus* spp., *Erica* spp., and/or grasses grazed by horses, goats, sheep and cattle in northern Spain (Rigueiro-Rodriguez et al. 1997; McAdam et al. 1999; Casasús et al. 2007), Spanish and Portuguese dehesas of *Quercus suber* L. or *Q. ilex* L. with pigs or bulls (San Miguel 1994), pollarded *Fraxinus excelsior* with cattle in Spain and Sweden (Ispikoudis and Sioliou 2006), *Quercus coccifera* L. with goats in Greece (Schultz et al. 1987), wooded pastures in England (Rackham 2001), windbreaks in Denmark and Russia

(Nair 1993), reindeer husbandry in *Picea abies* (L.) H. Karst or *Pinus sylvestris* L. forests in Finland and Siberia (Kumpula 2001; Harrop 2007).

This variety of different combinations can be partly explained by the gradients of stand and site characteristics (e.g. light, moisture, fertility) that create different microclimatic niches and spatial heterogeneity (Mosquera-Losada et al. 2006). However the development of the under storey species and biomass is also driven by factors such as the edaphoclimatic characteristics, predominant tree species, the pressure exerted by different animal species and the management practices carried out by the farmer. Silvopastoral systems can host agricultural and forest species, in this way they act as biological corridors between the different ecosystems, and as such they contribute to species maintenance and biodiversity enhancement (Luoto et al. 2003). This is of special relevance in fragmented landscapes.

Until recently, when conserving biodiversity, the emphasis has been on preserving habitats in natural areas by excluding human communities, rather than on protecting practices that have resulted from relationships between humans and nature (Harrop 2007). This has been the response to the observed exceeding of the carrying capacity of the ecosystem in some cases. Nowadays it has been realized that management should mimic traditional agricultural practices to secure the persistence of species that depend upon the human-induced habitat (Harrop 2005a, 2007). The FAO project Globally Important Agricultural Heritage Systems (GIAHS) emphasises the protection of: (1) human practices which contribute to the creation and maintenance agricultural diversity, (2) landscapes that have resulted from the long lasting relationships between humans and the natural world, and (3) the human cultural heritage, that forms the foundations for the traditional practices. Examples of GIAHS sites include among other, silvopastoral systems like *reindeer herding* in Siberia or the *dehesas* in Southern Spain and Portugal (Harrop 2005b).

4.2 Birds and Silvopastoral Systems

Certain areas in Europe, between intensively-managed agricultural land and the abandoned farmland, contain a patchwork of semi-natural and natural habitats and varied farmland, hosting a high diversity of species. Those areas are known as 'high nature value farmlands (HNV)' and are considered as key areas for wildlife. Such farmlands occur in association with traditional cropping systems as well as with livestock grazing systems on semi-natural habitats (EEA 2006). Among those are, e.g. the above mentioned *dehesas*, which are open Iberian forests and are considered to be one of the most important habitats for biodiversity in Europe (Moreno and Pulido 2008). They contain many species listed under the Habitats Directive (EC 1992), especially birds and mammals (Bunce et al. 2004). Holm oak (*Quercus ilex*) dehesas are one example, being one of the last breeding refuges for endangered bird species like the imperial eagle (*Aquila adalberti* Brehm) and the black vulture (*Aegypius monachus* L.) (Díaz et al. 1997), and a favoured wintering site of some species like the common crane (*Grus grus* L.) (Avilés 2004). If those practices are abandoned some of the species (e.g. vultures) might decline in abundance, as has already happened with the bearded vultures (*Gypaetus barbatus* L.) (EC 2000; EEA 2006).

The diversity of certain bird species is known to be a good indicator of ecosystem biodiversity. While the number of forest-related bird species has barely changed, bird species commonly associated with farming have been continuously decreasing (EEA 2006). Today only 70% of the species found in Europe in 1980 remain. The decline can be largely ascribed to agricultural intensification. It is estimated that 40% of the endangered bird species in Europe are threatened because of agricultural intensification and 20% by abandonment (EEA/UNEP 2004). Farmland birds are widespread across Europe, but there are particularly large numbers associated with extensive systems in Southern Europe. With expected large-scale land abandonment of marginal agricultural land it is likely that these bird species will decline (Casals et al. 2008). While the most favourable conditions for the diversity of farmland species are considered to occur under extensive and/or traditional agricultural management (EEA 2006).

Also many different passerines and raptors are found in these extensive systems. McAdam et al. (1999) observed that pastures with wide spaced trees and grazed by sheep could enhance higher diversity of birds, beetles and spiders than pastures without trees.

Another important factor related to the diversity of birds is the number of beetle species as they represent one of their important food sources. Silvopasture has been shown to encourage greater numbers of beetles than comparable grasslands (McAdam et al. 1999). The invertebrates are important components of the ecosystems, enhancing the species variety and population size of farmland and forest birds.

Preserving diversity will contribute to the maintenance and resilience of silvopastoral systems, enabling the production of goods and services (cultural, food, medicines, carbon sequestration, water, etc.) and contributing to the European target of halting the loss of biodiversity by the year 2010.

4.3 Livestock Biodiversity

Both wild and domesticated animal species are suitable for silvopastoralism. Especially with regard to conservation of indigenous domestic breeds silvopastoralism can play an important role. Mosquera-Losada et al. (2005) stated that half of the European indigenous domestic breeds are threatened by extinction. Further, the EEA (2006) reports that, in western Europe, 91% of mountain breeds of sheep are threatened by crossbreeding for improved of meat and milk production qualities, and by abandonment of traditional husbandry systems. The switch to modern breeds has led to the abandonment of remote pastures in many areas and the loss of biodiversity that depends on grazing (EEA 2006). High nature value pastoral grazing systems depend on locally adapted old livestock breeds which are better adapted to harsh natural conditions and extensive practices.

4.4 Grazing Damage Versus Biodiversity Enhancement

The Temperate and Boreal Forest Resources Assessment (UNECE/FAO 2000) reports that close to 2 million hectares, has been damaged by wildlife and grazing, i.e. approximately 0.2% of total forest and other wooded land area. Although it is not specified what share of damage is due to domestic livestock and the species involved, nor is the degree of the damage or the age of damaged trees listed.

There are broad concerns that excessive grazing in forests can cause harmful changes through reduction in structural complexity and species richness (Anderson and Radford 1994; Summers et al. 1997; Milne et al. 1998; Kuiters and Slim 2002; Milner et al. 2002; Mysterud and Østbye 2004). The intensity of the browsing impact determines whether forests are capable of regeneration. Regenerating trees will develop into high canopy trees or they remain as scrub and bushes (Pollock et al. 2005). To avoid a high degree of browsing of trees, the stocking rates have to be related to site characteristics and the forage quality and quantity considered (Pakeman et al. 2003; Mosquera-Losada et al. 2006). Damage depends on the type and number of animals, the fodder available and the seedling species and abundance.

Grazing removes some dominant, strong competitors, leaving space for weak competitors thus the dominance pattern becomes more even (Virkajärvi et al. 1996). The patches of bare ground caused by grazing, trampling and dunging offer a niche for seed germination. As already mentioned overgrazing can cause problems with regeneration and reduce ecosystem quality. But also complete exclusion of grazing can diminish species diversity, as some ground-dwelling species can dominate and shade out the tree seedlings or other ground species. A light grazing regime can help to regulate the competition between species and the shade from shrubs, allowing establishment of grass. For example a stand could be grazed for 1–2 years and then left ungrazed for around 10 years to allow tree saplings to establish (McEvoy et al. 2005).

For stock and vegetation management it must be borne in mid that different animal species have different preferences regarding vegetation. Goats and horses mainly feed on shrubby vegetation and might create more damage to trees, although problems mainly emerge when they are stocked at high density. Cows and sheep in general feed on herbaceous vegetation or young woody vegetation. However sometimes even within the same animal species there are different behavioural traits exhibited by breeds, e.g. there is a cattle breed from Navarra, Spain, that never damages young poplar and a sheep breed from England that does not browse coniferous trees.

Generally, livestock damage trees, so in some countries grazing is not permitted within the forests, e.g. in Slovenia (Official Journal 1993). Damage usually occur during establishment of the forest or the regeneration period though adult trees may also suffer damage to some extent. Damage can be alleviated through appropriate choice of animal and tree species, forest management and adequate stocking rate and grazing pattern (e.g. continuous vs intermittent). Keeping damage levels to a bearable limit may not hinder the sustainability of the system, providing a balance between production and conservation of natural resources. Large herbivores coexisted in the forest before human activity, creating branch or stem wounds needed as food and habitat for certain insects and other decomposing organisms and opening gaps needed for regeneration of light-demanding species (Bradshaw et al. 2003). Sometimes the wounds created do not affect timber quality. In some other minor cases, debarking for instance could cause the death of some trees, usually associated with incorrect selection of animal species and overgrazing. Nowadays the role of the standing and/or forest deadwood is increasingly recognized as up to 30% of the European forest species depend on old trees and deadwood for their survival (Dudley and Vallauri 2004). Deadwood provides habitat, shelter and food for birds, bats and other mammals and is particularly important for the less visible majority of forest dwelling species: insects, especially beetles, fungi and lichens. Increasing the amounts of deadwood in managed forests and allowing natural dynamics would contribute to sustaining Europe's biodiversity (Dudley and Vallauri 2004). Therefore a certain level of damage to the trees by the animals should not always be considered as a problem, particularly if biodiversity is one of the objectives.

Grazing in woodlands or forests has been recognized as beneficial for biodiversity (Mitchell and Kirby 1990; Kirby et al. 1994; Mayle 1999; Bengtsson et al. 2000; Kampf 2000). Studies from Northern Europe show the positive effect of cattle grazing within the forests. In Finland forest pastures were a common practice until the 1960s when intensification of farming took place. Tuupanen et al. (1997) observed that cows grazing in the forest pastures could increase vegetation diversity compared to un-grazed forest when kept at a low stocking rate while also permitting economic levels of cattle production. Tuupanen et al. (1997) also noticed that forest areas encourage a more diverse animal behaviour (foraging, exploration, autogrooming) than in open meadows where the livestock spent more time standing inactive. Hence silvopastoral systems are beneficial for animal welfare, which is considered one of the main objectives within the Common Agricultural Policy of the European Union (CEC 2005). Hokkanen et al. (1998) studied the effect of grazing on Carabidae, which are species very sensitive to environmental changes, in meadows and forest pastures under open birch forest and dense forest with birch, pine and alder. Within forest pastures the number of both individuals and species of Carabidae increased in grazed compared to ungrazed areas. The Shannon diversity index was consistently higher in birch forest that the pine and alder and there was a larger total number of Carabidae species compared to the non-grazed boreal forests. Selective grazing seems then to increase the number and diversity of Carabidae species in the forest, even after a few years. This might be due to an opening of suitable niches, because many species favour open biotopes, while keeping the stocking rate low. McEvoy et al. (2005) have found that regeneration of oak seedlings under pine stands may be facilitated by previous grazing. It is in the first stages of the regeneration period or plantation establishment when most attention must be paid to livestock using individual protectors or temporal exclusion to allow the development of the saplings. In Australia light grazing in buloke (*Allocasuarina leuhmannii* Baker) forests encouraged the livestock to feed on the weed competing with the trees. These tend to be mainly exotic species and the reduction in herbicide, alternating with ungrazed periods for the regeneration of trees benefits the environment (Maron and Lill 2005). Moreover, Mediterranean wood pastures have traditionally maintained their open structure with livestock grazing. Well-planned grazing can have a positive impact on structure and species diversity in forests (Putman 1996; Hadjigeorgiou et al. 2005).

Livestock production has to be considered along with socio-economic and ecological objectives when designing sustainable grazing management. It is first necessary to determine the type of animals and grazing regimes to optimise the use of the available forage resources (Hadjigeorgiou et al. 2005; Casasús et al. 2007).

4.5 Silvopastoralism Preventing Forest Fires

Integrated systems such as these, where pastoralism is an active part of the forest management, help also to preserve biodiversity as they reduce forest fire risk. Forest fire policy has up to now been based on extinction, not on prevention. But due to the current trends of increasingly more damaging forest fires than previously climate change and desertification, a new policy attitude might address a more prevention-oriented management, e.g. clearing the forest either mechanically or, as it occurs already in some forests, using cattle or other livestock species to reduce understorey fuel. Some of the variables that influence occurrence of fires are weather, ignition source and characteristic and amount of fuel material. The latter can be controlled by livestock. Fires can cause loss of human life, biodiversity loss (species, habitats, landscape) and economic losses including ecotourism etc. and contribute to global warming. As consequence of the policy initiatives based on prevention, the amount of land burnt in France each year has been reduced by half between the 1980s and 1990s due to land use management practices with stronger fire prevention measures. It was considered that rural land use should aim at maintaining the traditional land mosaic of Mediterranean areas (forest, pastures and agricultural land), as perhaps the best option to prevent the propagation of large fires (EEA 2003). The option of using livestock instead of (or complementing) mechanical or chemical control to maintain firebreaks cleared of shrubs and trees is potentially important and is already subsidized in some regions, e.g. in eastern and southern Spain (Dopazo and Suárez 2004; Robles et al. 2008).

Some studies have shown that after several years of grazing in rangelands or forest pastures the herbage biomass is lower and had a smaller proportion of dead vegetation and therefore higher quality for grazing over a longer period than in the areas where grazing was excluded. It also prevents scrub encroachment, reducing the fire risk and maintaining the recreational value of forests (Hope et al. 1996; Adezábal, 2001; Bernués et al. 2005; Casasús et al. 2007). Grazing is a cost-effective instrument to manage abandoned land and prevent the appearance of different and often irreversible environmental hazards to which these areas are highly susceptible (Casasús et al. 2007).

5 Silvopastoralism Within a Policy Context

One of the priority objectives of the EU is optimising the use of available measures under the reformed Common Agrarian Policy (CAP), notably to prevent intensification or abandonment of High-Nature-Value farmland, woodland and forest and supporting their restoration (EEA 2006).

It has traditionally been considered that silvopastoralism and agroforestry are more important land-use systems in developing countries where subsistence farming is a priority. In the last decades the value of silvopastoralism and agroforestry are becoming more widely recognized in developed countries due to the benefits of less intensive land uses (Nair et al. 2008). Threats to biodiversity in Europe are a high degree of habitat fragmentation, intensive agriculture, land and biodiversity-friendly traditional practices abandonment, climate change, desertification and fires (EEA 1998, 2006). Silvopastoral systems may contribute to combat these as well as to preserving domestic livestock breeds. There is therefore a slight change towards recognizing the cultural, social, economic and environmental value of traditional systems, e.g. silvopastoralism, as it fulfils most of the policy goals of sustainable land management. At a global level, silvopastoral systems are recognised in the Agenda 21 (UN 1993), in the Orlando Declaration on agroforestry systems (WFC 2004) and in the Lugo Declaration specific on silvopastoralism (SSM 2004). All these documents are landmarks in the recognition of the multiple benefits and future needs of these systems. Neither society, forest owners nor policy-makers should overlook the opportunity silvopastoral activities create, beyond those of traditional forests. They contribute to keeping people in rural areas, combat forest fires from the prevention point of view and they contribute to the "2010 target" of halting the loss of biodiversity which the EU is committed to. The most important policy instrument for the implementation of silvopastoral systems is the *Council Regulation on support for rural development by the European Agricultural Fund for Rural Development (EAFRD)* (Council Regulation 1698/2005) and the *Proposal for a Council Decision on Community Strategic Guidelines for Rural Development (Programming period 2007–2013)* (COM(2005)304 final) from the European Union as for the first time there is an opportunity to fund farmers for the establishment of agroforestry systems (art. 41). This policy also allows finance to be directed the conservation local breeds in danger of being lost to farming extinction (art. 37). The decline in traditional livestock breeds has negative implications for the management of semi-natural habitats that have been shaped by agricultural practices. On average 18% of the Natura 2000 area belongs to habitat categories which depend on a continuation of extensive agricultural practices. Such practices can be supported via agri-environment schemes and other agricultural policy instruments (EEA 2006).

The Action Plan on the Protection and Welfare of Animals of European Community aims to promote animal welfare in the future. General minimum standards for the protection of farm animals have already been set in the Directive 98/58/EC. These rules reflect the "five freedoms": freedom from hunger and thirst, freedom from discomfort, freedom from pain, injury and disease, freedom to express normal behaviour, and freedom from fear and distress. In 2003, CAP reform introduced certain measures to promote the better handling and treatment of animals (EC 2006).

According to a Eurbarometer survey, EU consumers are willing to make an extra effort to buy animal welfare friendly products. Within the EU-27 and Turkey and Croatia, 62% of respondents would change their shopping habits in order to access more animal welfare friendly goods. The survey demonstrates general support for financially rewarding EU farmers who use better animal welfare practices. The well-being of animals during the production of food appears to be strongly associated with the healthiness and quality of products. Half of the people consider that high animal welfare standards produce healthier and better quality food (EC 2007).

Compensation in the form of rewards or payments to farmers who through their conservation efforts provide ecosystem services to wider society may also foster conservation (Jackson et al. 2007; Pascual and Perrings 2007).

6 Conclusion

While grazing in forests is still a controversial practice due to the potential for damage to the tree stock from the livestock, it is feasible under certain conditions to maintain a sustainable forest management where livestock and biodiversity conservation are part of an integrated system, which is multifunctional and which has several advantages over each of the individual components (forest and agriculture). Such multifunctional management will require more complex management input but will help promote sustainable development in rural areas, as a win-win alternative for Europe in the three elements of sustainability: economic, social and environmental. One of the key issues it to match the appropriate stocking rate for each site depending on the vegetation, animal species, productivity, environmental benefits including biodiversity conservation and enhancement and the multifunctionality of a suite of integrated outputs.

Acknowledgements The authors would like to thank the European Forest Institute and the European Environment Agency for the support received.

References

Adezábal A (2001) El sistema de pastoreo en el Parque Nacional de Ordesa y Monte Perdido. Interacción entre la vegetación supraforestal y los grandes herbívoros. Publicaciones del Consejo de Protección de la Naturaleza de Aragón, Serie Investigación 28, Zaragoza

Anderson P, Radford E (1994) Changes in vegetation following reduction in grazing pressure on the National Trust's Kinder Estate, Peak District, Derbyshire, England. Biol Conserv 69:55–63

Avilés JM (2004) Common cranes *Grus grus* and habitat management in holm oak dehesas of Spain. Biodivers Conserv 13:2015–2025
Bengtsson J, Nilsson SG, Franc A, Menozzi P (2000) Biodiversity, disturbances, ecosystem function and management of European forests. For Ecol Manage 132:39–50
Bernués A, Riedel JL, Asensio MA, Blanco M, Sanz A, Revilla R, Casasús I (2005) An integrated approach to study the role of grazing farming systems in the conservation of rangelands in a protected natural park (Sierra de Guara, Spain). Livest Prod Sci 96:75–85
Bradshaw R, Emanuelsson U (2004) History of Europe's biodiversity. Background note in support to a Report on 'Halting Biodiversity Loss'. European Environment Agency, Copenhagen (unpublished)
Bradshaw RHW, Hannon GE, Lister AM (2003) A long-term perspective on ungulate-vegetation interactions. For Ecol Manage 181:267–280
Bunce RGH, Pérez-Soba M, Jongman RHG, Gómez-Sal A, Herzog F, Austad I (2004) Transhumance and biodiversity in European mountains. Report of the EU-FP5 project TRANSHUMOUNT (EVK2-CT-2002-80017). IALE publication series nr 1, Alterra UR in collaboration with IALE, Wageningen
Casals P, Baiges T, Bota G, Chocarro C, de Bello F, Fanlo R, Sebastià MT, Taull M (2008) Silvopastoral systems in the northeastern Iberian peninsula: a multifunctional perspective. In: Rigueiro-Rodríguez A, Mcadam J, Mosquera-Losada MR (eds) Agroforestry in Europe: current status and future prospects. Advances in Agroforestry. Springer, Dordrecht, The Netherlands
Casasús I, Bernués A, Sanz A, Villalba D, Riedel JL, Revilla R (2007) Vegetation dynamics in Mediterranean forest pastures as affected by beed cattle grazing. Agr Ecosyst Environ 121:365–370
CEC (2005) Proposal for a council decision on community strategic guidelines for rural development (programming period 2007–2013) (COM(2005) 304 final). Commision of the European Communities, Brussels
CEE (2006) Halting the loss of biodiversity by 2010 and beyond. Sustaining ecosystem services for human well-being. COM(2006) 216final. Communication from the Commission, Brussels
Díaz M, Campos P, Pulido F (1997) The Spanish dehesas: a diversity in land-use and wildlife. In: Pain DJ, Pienkowski MW (eds) Farming and birds in Europe: the common agricultural policy and its implications for bird conservation. Academic, London
Dopazo C, Suárez J (2004) Fuel control management experiences with livestock grazing in fire-break areas in the region of Valencia (Spain). In: Mosquera-Losada MR, McAdam J, Rigueiro-Rodríguez A (eds) Silvopastoralismo y manejo sostenible. Libro de Resúmenes, Unicopia, Lugo
Dudley N, Vallauri D (2004) Deadwood – living forests: the importance of veteran trees and deadwood to biodiversity. WWF World Wide Fund for Nature, Gland
EC (1992) Council Directive 92/43/EEC of 21 May 1992 on the conservation of natural habitats and of wild fauna and flora. Official Journal L 206, 22/07/1992 P. 0007 – 0050
EC (2000) Pooling resources to save the bearded vulture. Natura2000 Newsletter 13:8–10
EC (2005) Council Regulation (EC) No. 1698/2005 of 20 September 2005 on support for rural development by the European Agricultural Fund for Rural Development (EAFRD). Official Journal of the European Union. L 277
EC (2006) Questions and Answers on the Action Plan on the Protection and Welfare of Animals. MEMO/06/21. Brussels, 23 Jan 2006
EC (2007) EU consumers willing to pay for better animal welfare. IP/07/398. Brussels, 22 March 2007
EEA (1998) Europe's environment: the second assessment, 1998. Dobris 3+. European Environment Agency, Copenhagen
EEA (2003) Mapping the impacts of recent natural disasters and technological accidents in Europe. Environmental issue report. No. 35. European Environment Agency, Copenhagen
EEA (2006) Progress towards halting the loss of biodiversity by 2010. EEA report No. 5/2006. European Environment Agency, Copenhagen
EEA/UNEP (2004) High nature value farmland: characteristics, trends and policy challenges. EEA report No. 1/2004. European Environment Agency, Copenhagen
Eichhorn MP, Paris P, Herzog F, Incoll LD, Liagre F, Mantzanas K, Mayus M, Moreno G, Papanastasis VP, Pilbeam DJ, Pisanelli A, Dupraz C (2006) Silvoarable systems in Europe – past, present and future prospects. Agroforest Syst 67:29–50

Etienne M (ed) (1996) Western European silvopastoral systems. INRA, Montpellier
Hadjigeorgiou I, Osoro K, Fragoso de Almeida JP, Molle G (2005) Southern European grazing lands: production, environmental and landscape management aspects. Livest Prod Sci 96:51–59
Harrop SR (2005a) The role and protection of traditional practices in conservation under the Convention on Biological Diversity – an enquiry into the UK's implementation of Article 8(j) Convention on Biological Diversity. Environ Law Manage 16(5):244–251
Harrop SR (2005b) Globally important ingenious agricultural heritage systems – an examination of their context in existing multilateral instruments. Report to United Nations Food and Agriculture Organisation. FAO, Rome
Harrop SR (2007) Traditional agricultural landscapes as protected areas in international law and policy. Agric Ecosyst Environ 121:296–307
Hassan R, Scholes R, Ash N (eds) (2005) Ecosystems and human well-being: current state and trends. Findings of the condition and trends working group. The millennium ecosystem assessment series, vol 1. Island Press, Washington DC
Herzog F (1998) Agroforestry in temperate Europe: history, present importance and future development. In: Mixed Farming Systems in Europe. Workshop Proceedings, A.P. Minderhoudhoeve-Series, Dronten, The Netherlands, pp 47–52, 25–28 May 1998
Hokkanen TJ, Hokkanen H, Tuupanen R, Virkajärvi P, Huhta H (1998) The effect of grazing on Carabidae in meadow and forest pastures. Grassland Sci Eur 3:413–417
Hope D, Picozzi N, Catt DC, Moss R (1996) Effects of reducing sheep grazing in the Scottish Highlands. J Range Manage 49:301–310
Ispikoudis I, Sioliou KM (2006) Cultural aspects of silvopastoral systems. In: Mosquera-Losada MR, McAdam J, Rigueiro-Rodríguez A (eds) Silvopastoralism and sustainable land management. CABI, Wageningen
Ispikoudis I, Soulis MK, Papanastasis VP (2004) Transhumance in Greece: past, present and future prospects. In: Bunce RGH, Perez-Soba M, Jongman RHG, Gomez Sal A, Herzog F, Austad I (eds) European mountains. IALE, Wageningen
Jackson LE, Pascual U, Hodgkin T (2007) Utilizing and conserving agrobiodiversity in agricultural landscapes. Agric Ecosyst Environ 121:196–210
Kampf H (2000) The role of large grazing animals in nature conservation – a Dutch perspective. Br Wildl 37:37–46
Kirby KJ, Mitchell FJ, Hester AJ (1994) A role for large herbivores (deer and domestic stock) in nature conservation management in British semi-natural woods. Arboricult J 18: 381–399
Kuiters AT, Slim PA (2002) Regeneration of mixed deciduous forest in a Dutch forest-heathland following a reduction of ungulate densities. Biol Conserv 105:65–74
Kumpula J (2001) Winter grazing of reindeer in woodland lichen pasture. Effect of lichen availability on the condition of reindeer. Small Rum Res 39:121–130
Luoto M, Pykälä J, Kuussaari M (2003) Decline of landscape scale habitat and species diversity after the end of cattle grazing. J Nat Conserv 11:171–178
Maron M, Lill A (2005) The influence of livestock grazing and weed invasion on habitat use by birds in grassy woodland remnants. Biol Conserv 124:439–450
Mayle B (1999) Domestic stock grazing to enhance woodland biodiversity, Rep. No. Information Note 28. Forestry Commission, Edinburgh
McAdam J, Hoppé GM, Toal L, Whiteside T (1999) The use of wide-spaced trees to enhance faunal diversity in managed grasslands. Grassland Sci Eur 4:293–296
McAdam J, Mosquera-Losada MR, Papanastasis V, Pardini A, Rigueiro-Rodríguez A (2005) Silvopastoral systems: analyses of an alternative to open swards. In: O'Mara FP, Wilkins RJ, Mannetje L, Lovett DK, Rogers PAM, Boland TM (eds) XX international grassland congress: offered papers. Academic, Wageningen, The Netherlands, pp 758–759
McEvoy PM, McAdam JH, Mosquera-Losada MR, Rigueiro-Rodríguez A (2005) Tree regeneration and sapling damage of pedunculate oak *Quercus robur* in a grazed forest in Galicia, NW Spain: a comparison of continuous and rotational grazing systems. Agroforest Syst 66:85–92
Meiggs R (1982) Trees and timber in the ancient Mediterranean world. Oxford University Press, Oxford

Milne JA, Birch CPD, Hester AJ, Armstrong HM, Robertson A (1998) The impact of vertebrate herbivores on the natural heritage of the Scottish uplands – a review Scottish natural heritage review no. 95. Scottish Natural Heritage, Edinburgh

Milner J, Alexander J, Griffinn C (2002) A highland deer herd and its habitat. Red Lion House, London

Mitchell FJG, Kirby KJ (1990) The impact of large herbivores on the conservation of semi-natural woods in the British uplands. Forestry 63:333–353

MMA (2000) Estrategia forestal española. Ministerio de Medio Ambiente. Secretaría General Técnica. Dirección General de Conservación de la Naturaleza, Madrid

Moreno G, Pulido FJ (2008) The function, management and persistence of dehesas. In: Rigueiro-Rodríguez A, Mcadam J, Mosquera-Losada MR (eds) Agroforestry in Europe: current status and future prospects. Advances in agroforestry. Springer, Dordrecht, The Netherlands

Mosquera-Losada MR, Rigueiro-Rodríguez A, Rois-Díaz M, Schuck A, Van Brusselen J (2005) Assessing biodiversity on silvopastoral systems across Europe. In: Lillak R, Viiralt R, Linke A, Geherman V (eds) Integrating efficient grassland farming and biodiversity. Grassland Sci Eur 10, EGS, Tartu, pp 44–47

Mysterud A, Østbye E (2004) Roe deer (*Capreolus capreolus*) browsing pressure affects yew (*Taxus baccata*) recruitment within nature reserves in Norway. Biol Conserv 120:545–548

Mosquera-Losada MR, Pinto-Tobalina M, Rigueiro-Rodríguez A (2006) The herbaceous component in temperate silvopastoral systems. In: Mosquera-Losada MR, McAdam J, Rigueiro-Rodríguez A (eds) Silvopastoralism and sustainable land management. CABI, Wallingford

Nair PKR (1993) An introduction to agroforestry. Kluwer, Dordrecht, The Netherlands

Official Journal (1993) Slovenian Forest Act. Official Journal of Republic of Slovenia, 30/93

Nair PKR, Gordon AM, Mosquera-Losada MR (2008) Agroforestry. In: Jorgensen SE, Brian DF (eds) Encyclopedia of ecology. Elsevier, Oxford

Pakeman RJ, Hulme PD, Torvell L, Fisher JM (2003) Rehabilitation of degraded dry heather *Calluna vulgaris* (L.) Hull] moorland by controlled sheep grazing. Biol Conserv 114:389–400

Papanastasis VP, Frame J, Nastis AS (eds) (1999) Grasslands and woody plants in Europe. British Grassland Society, Berskhire

Pascual U, Perrings C (2007) Developing incentives and economic mechanisms for in situ biodiversity conservation in agricultural landscapes. Agric Ecosyst Environ 121:256–268

Piussi P (1994) Selvicoltura generale. UTET, Torino, Italy

Pollock ML, Milner JM, Waterhouse A, Holland JP, Legg CJ (2005) Impacts of livestock in regenerating upland birch woodlands in Scotland. Biol Conserv 123:443–452

Putman RJ (1996) Ungulates in temperate forest ecosystems: perspectives and recommendations for future research. For Ecol Manage 88:205–214

Rackham O (2001) Wood-pasture and cultural savannas in Europe. In: UK Agroforestry Forum: Annual Meeting 2001. Trees, farms, rural development. University of Leeds School of Biology, Leeds

Rigueiro-Rodríguez A, Mcadam J, Mosquera-Losada MR (eds) (2008) Agroforestry in Europe: current status and future prospects, Advances in agroforestry series 6. Kluwer, Dordrecht, The Netherlands

Rigueiro-Rodriguez A, Silva-Pando FJ, Rodriguez-Soalleiro R, Castillon-Palomeque PA, Alvarez-Alvarez P, Mosquera-Losada MR, Romero-Franco R, Gonzalez-Fernandez MP (1997) Manual de sistemas silvopastorales. ADAPT, Santiago de Compostela, Spain

Robles AB, Ruiz-Mirazo J, Ramos ME, González-Rebollar JL (2008) Role of livestock grazing in sustainable use, fire prevention and naturalization of marginal ecosystems of southeastern Spain (Andalusia). In: Rigueiro-Rodríguez A, Mcadam J, Mosquera-Losada MR (eds) Agroforestry in Europe: current status and future prospects. Advances in agroforestry. Springer, Dordrecht, The Netherlands

San Miguel S (1994) La Dehesa Española. Origen, tipología, características y gestión. ETS Ingenieros de Montes de Madrid. Fundación Conde del Valle Salazar, Madrid

Schuck A, Päivinen R, Hytönen T, Pajari B (2002) Compilation of forestry terms and definitions. EFI Internal Report 6, 48 p. European Forest Institute, Joensuu

Schultz A, Papanastasis V, Katelman T, Tsiouvaras C, Kandrelis S, Nastis A (1987) Agroforestry in Greece. Working Document. Aristotle University of Thessaloniki, Greece
SSM (2004) Silvopastoralism declaration http://www.diswebline.com/congreso/declaration.doc. Accessed 30 Sept 2005
Sibbald AR, Griffiths JH, Elston DE (1994) Herbage yield in agroforestry systems as a function of easily measured attributes of the tree canopy. For Ecol Manage 45:71–77
Summers RW, Proctor R, Raistrick P, Taylor S (1997) The structure of Abernethy Forest, Strathspey, Scotland. Bot J Scot 49:39–55
Tuupanen R, Hokkanen TJ, Virkajärvi P, Huhta H (1997) Grazing suckler cows as managers of vegetation biomass and diversity on seminatural meadow and forest pasture. Management for grassland biodiversity. Grassland Sci Eur 2:165–170
UN (1993) Agenda 21: Earth Summit – The United Nations Programme of Action from Rio. http://www.un.org/esa/sustdev/documents/agenda21/index.htm. Accessed 10 Feb 2008
UNECE/FAO (2000) Forest Resources of Europe, CIS, North America, Australia, Japan and New Zealand (TBFRA 2000). Main report. UNECE/FAO contribution to the global forest resources assessment 2000. United Nations/New York/Geneva
Vera FWM (2000) Grazing ecology and forest history. CABI, Wallingford
Virkajärvi P, Hokkanen T, Koponen S, Uusi-Kämppä J, Mannerkorpi P, Castren H, Huhta H (1996) Forest pastures and semi-natural meadow for suckler cows in Finland. Grassland Sci Eur 1:665–669
WFC (2004) Orlando declaration http://conference.ifas.ufl.edu/WCA/orlando.pdf. Accessed 26 Sept. 2008

Index

MIX
Papier aus verantwortungsvollen Quellen
Paper from responsible sources
FSC® C105338

If you have any concerns about our products,
you can contact us on
ProductSafety@springernature.com

In case Publisher is established outside the EU,
the EU authorized representative is:
Springer Nature Customer Service Center GmbH
Europaplatz 3, 69115 Heidelberg, Germany

Printed by Libri Plureos GmbH
in Hamburg, Germany